Biomimetic Robotics

Biomimetic Robotics

Edited by **Rowland Wilson**

Willford Press
New York

Published by Willford Press,
118-35 Queens Blvd., Suite 400,
Forest Hills, NY 11375, USA
www.willfordpress.com

Biomimetic Robotics
Edited by Rowland Wilson

International Standard Book Number: 978-1-68285-181-4 (Hardback)

Printed in China.

Contents

Preface

This book aims to highlight the current researches and provides a platform to further the scope of innovations in this area. This book is a product of the combined efforts of many researchers and scientists from different parts of the world. The objective of this book is to provide the readers with the latest information in the field.

Robotics is a rapidly growing branch of engineering dealing with the designing, modelling and application of robots. Biomimetics is bringing forth new breakthroughs in the field of bio-engineering and bio-inspired robotics. Biomimetic robotics has diverse applications and is a growing field of robotic engineering. It involves the principles and techniques of nanotechnology, fibre optics and other engineering methods for developing bio-morphic systems, artificial implants and bionic implants. Researches and case-studies by internationally acclaimed experts are collated to provide a comprehensive insight into the techniques, modules and the new challenges emerging in the field. It is helpful for students, researchers and professionals associated with robotic engineering and biomimetics.

I would like to express my sincere thanks to the authors for their dedicated efforts in the completion of this book. I acknowledge the efforts of the publisher for providing constant support. Lastly, I would like to thank my family for their support in all academic endeavors.

Editor

1

Performance analysis of 3-D shape measurement algorithm with a short baseline projector-camera system

Jianyang Liu and Youfu Li*

Abstract

A number of works for 3-D shape measurement based on structured light have been well-studied in the last decades. A common way to model the system is to use the binocular stereovision-like model. In this model, the projector is treated as a camera, thus making a projector-camera-based system unified with a well-established traditional binocular stereovision system. After calibrating the projector and camera, a 3-D shape information is obtained by conventional triangulation. However, in such a stereovision-like system, the short baseline problem exists and limits the measurement accuracy. Hence, in this work, we present a new projecting-imaging model based on fringe projection profilometry (FPP). In this model, we first derive a rigorous mathematical relationship that exists between the height of an object's surface, the phase difference distribution map, and the parameters of the setup. Based on this model, we then study the problem of how the uncertainty of relevant parameters, particularly the baseline's length, affects the 3-D shape measurement accuracy using our proposed model. We provide an extensive uncertainty analysis on the proposed model through partial derivative analysis, relative error analysis, and sensitivity analysis. Moreover, the Monte Carlo simulation experiment is also conducted which shows that the measurement performance of the projector-camera system has a short baseline.

Keywords: 3-D shape measurement; Projector-camera; Short baseline; Binocular stereovision; FPP

Introduction

Noncontact optical measurement methodology has been widely used in many industrial applications, such as industry inspection and 3-D printing manufacturing. Among these mature optical 3-D measurement techniques, the structured light technique has been widely used in recent years due to its good characteristics of high precision, flexibility, and robustness to texture-less object surface reconstruction. Numbers of works have been presented in this issue [1-3]. According to a different model, the methods for obtaining 3-D shape information with a structured light system can be simply divided into two main categories: one common way is using conventional binocular stereo vision or named 'CSV' model and the other strategy is adopting fringe projection profilometry (FPP) technique. In the first category, the projector is always treated like a camera. In this model, projector and camera in the system are always required to be pre-calibrated before 3-D shape measurement task. Many camera calibration methods are available to be utilized directly [4,5]. However, for projector calibration, even with the latest accurate calibration methods using active target and phase-shifting technique [17,18], the accuracy of projector calibration can hardly reach as the same level as the camera. One of the intuitive reasons is that the parameters of a projector cannot be calibrated individually without the help of camera. Thus, the error propagation from camera calibration process is unavoidable and the overall system calibration accuracy is limited. The biased calibrated parameters of camera and projector will decrease the measurement accuracy. Particularly, in a short baseline arrangement system, the bias from feature point localization on the image will also be magnified with biased parameters. Therefore, how to accurately calibrate a short baseline arrangement system is more critical than a general configuration system which usually has a much larger baseline. In the second category, the projector is

* Correspondence: meyfli@cityu.edu.hk
Department of Mechanical and Biomedical Engineering, City University of Hong Kong, Kowloon, Hong Kong

commonly regarded as grating optical device. The height information is obtained from the phase-to-height mapping relationship between phase distribution and geometric parameters of the system. Hence, the projector is not needed to be pre-calibrated anymore. In some presented phase-to-height model, even the camera is also not required to be pre-calibrated [7,8].

In this paper, our method falls into the second category. We propose a generic FPP-based projecting-imaging model and explore the relationship between the phase distribution, height information, and geometric parameters of the system. Based on the proposed model, we then study the problem of how the uncertainty of relevant parameters, the length of baseline in particular, affects the 3-D shape measurement accuracy. In other words, we focus on the performance analysis of the 3-D shape measurement according to our proposed model particularly when the system has a short baseline.

Background and literature review

Conventional stereovision model

The schematic diagram of a projector-camera (Pro-Cam) structured light system is illustrated in Figure 1. The key problem in the 3-D shape measurement process is how to determine the corresponding relationship between the point on camera's image plane, projector's image plane, and the point on the object's surface. It is worthy noticing that these three points also represent the three vertices of a triangle, respectively. Generally, the structured light pattern projected from a projector plays a role of bridge. Once the relationship between the projector image plane and camera image plane is established, the short baseline arranged 3-D shape measurement system, which is illustrated in Figure 1, can be unified with a classic binocular stereovision system.

Given a point on the camera's image plane and its corresponding point on the projector's image plane, the coordinate of its corresponding point on the object's surface can be determined by conventional triangulation algorithm [16,19]. The relationship between nondistorted point $m_u(u_c, v_c)$ on the camera's image plane and its corresponding point $P_W = [X^W, Y^W, Z^W]^T$ on the object's surface can be described as follows:

$$s_c \begin{bmatrix} u_c \\ v_c \\ 1 \end{bmatrix} = \begin{bmatrix} f_{cx} & \lambda_c & u_{c0} \\ 0 & f_{cy} & v_{c0} \\ 0 & 0 & 1 \end{bmatrix} \begin{bmatrix} R^c_{11} & R^c_{12} & R^c_{13} & t^c_1 \\ R^c_{21} & R^c_{22} & R^c_{23} & t^c_2 \\ R^c_{31} & R^c_{32} & R^c_{33} & t^c_3 \end{bmatrix} \begin{bmatrix} X_W \\ Y_W \\ Z_W \\ 1 \end{bmatrix} \tag{1}$$

where (u_{c0}, v_{c0}) is the coordinate of the principle point. f_{cx} and f_{cy} are the focal length in pixels of the camera image plane along u and v axes, λ_c denotes the skewness of the two image axes on the camera's image plane. Then perspective transformation of camera imaging process can be simplified from Equation 1 and denoted as

$$\begin{bmatrix} s_c u^c_2 \\ s_c v^c_2 \\ s_c \end{bmatrix} = \begin{bmatrix} A_{11} & A_{12} & A_{13} & A_{14} \\ A_{21} & A_{22} & A_{23} & A_{24} \\ A_{31} & A_{32} & A_{33} & A_{34} \end{bmatrix} \begin{bmatrix} X_W \\ Y_W \\ Z_W \\ 1 \end{bmatrix}. \tag{2}$$

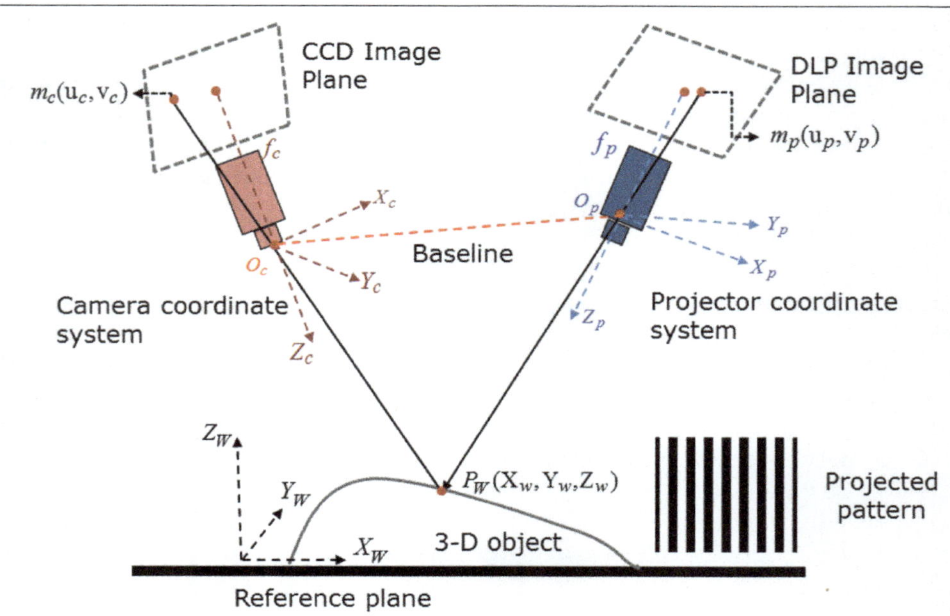

Figure 1 Schematic diagram of a Pro-Cam structured light system.

Similar for the projector, using phase-shifting technique [17], we can obtain a similar relationship:

$$\begin{bmatrix} s_p u_1^p \\ s_p v_1^p \\ s_p \end{bmatrix} = \begin{bmatrix} B_{11} & B_{12} & B_{13} & B_{14} \\ B_{21} & B_{22} & B_{23} & B_{24} \\ B_{31} & B_{32} & B_{33} & B_{34} \end{bmatrix} \begin{bmatrix} X_W \\ Y_W \\ Z_W \\ 1 \end{bmatrix}. \quad (3)$$

Arranging the variables, we can get the following relationship from Equations 2 and 3.

$$K \cdot P = Q, \quad (4)$$

where $K = \begin{bmatrix} A_{11}-A_{31}u_2^c & A_{12}-A_{32}u_2^c & A_{13}-A_{33}u_2^c \\ A_{21}-A_{31}v_2^c & A_{22}-A_{32}v_2^c & A_{23}-A_{33}v_2^c \\ B_{11}-B_{31}u_1^p & B_{12}-B_{32}u_1^p & B_{13}-B_{33}u_1^p \\ B_{21}-B_{31}v_1^p & B_{22}-B_{32}v_1^p & B_{23}-B_{33}v_1^p \end{bmatrix}$ is a 4×3 matrix, $Q = \begin{bmatrix} A_{34}u_2^c-A_{14} \\ A_{34}v_2^c-A_{24} \\ B_{34}u_1^p-B_{14} \\ B_{34}v_1^p-B_{24} \end{bmatrix}$ is a vector, and $P = [X^W, Y^W, Z^W]^T$. It is worth to note that if we are given the real distorted image point m_d, it has to be transformed to the nondistorted point m_u first. However, it is difficult to directly get the analytical inversion from distorted image point to nondistorted image point. In this work, the iterative method [12] is adopted and the iteration relationship is given as follows:

$$\begin{cases} m_u = m_d - (k_1 r^2 + k_2 r^4) m_u + \begin{bmatrix} 2k_3 x_u y_u + k_4 (r^2 + 2{x_u}^2) \\ k_3 (r^2 + 2{y_u}^2) + 2k_4 x_u y_u \end{bmatrix}. \\ m_d = m_u \end{cases} \quad (5)$$

Therefore, if we know the real pixel point pair $m_d^c(u_d^c, v_d^c)$ and $m_d^p(u_d^p, v_d^p)$, the remainder point in the triangle which is also the corresponding point in 3-D space $P_W(X_W, Y_W, Z_W)^T$, can be obtained by

$$P = (K^T K)^{-1} K^T Q. \quad (6)$$

It is well known that 3-D shape measurement accuracy can be improved through appropriately enlarging the baseline between two optical devices [6,15]. One intuitive reason is that by enlarging the baseline, the ambiguity of the correspondence problem which lies between the pixel on the right camera's image plane and left camera's image plane can be alleviated. A brief schematic of a short baseline stereovision system setup is shown in Figure 2. Because the pixel on the image plane has a certain physical size, the feature point A and the feature point B lie in the same uncertainty area (UA) which is denoted as blue rhombus. All the points lying in this area are corresponding to the same pixel point on the right camera and left camera's image plane, respectively. In other words, all the points lying in the same UA present the same depth information. Hence, the short baseline problem can be defined, that is, the depth measurement error, which can be denoted as UA shown in Figure 2, will be enlarged in a short baseline system.

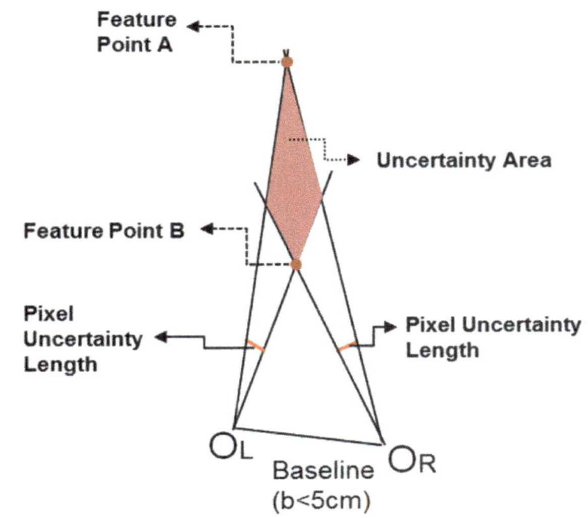

Figure 2 A brief structure of a short baseline stereovision system setup.

Equation 6 shows that an accurate measurement result is depending on the four factors: (1) accurate determination of the pixel point $m_{cu}(u_c, v_c)^T$ on camera's image plane, (2) unbiased camera's parameters, (3) unbiased projector's parameters, and (4) accurate determination of the pixel point which corresponds to the pixel point on the projector's image plane. The first two conditions are easy to be satisfied with available camera calibration algorithm [4,18] and well-developed image processing technique [19,20]. However, the latter two are much more difficult to be achieved. Furthermore, since the projector cannot 'capture' like a camera, how to determine the corresponding pixel point on the projector's image plane is a challenge. If the pixel point on the projector's image plane is biased, as Figure 2 shows, the measurement error will be enlarged in a short baseline arranged Pro-Cam system.

FPP-based phase-to-height mapping model

Fringe projection techniques have been used for 3-D object surface measurement for years because of its flexibility and good performance characteristic. In these techniques, the projector is commonly regarded as grating optical device, and series of fringe patterns (commonly sinusoidal fringe patterns) are projected onto an object's surface and then captured from other direction by a camera. The captured fringe patterns are deformed with respect to the geometry of the object's surface. Hence, the intensity distribution of the deformed pattern on the image plane can be retrieved through phase-measuring techniques. One of the classic techniques is Fourier-transform analysis [7]. The other widely used technique is phase-shifting algorithm [2]. Whichever technique is adopted, the critical final step is to create a mapping relationship between the pixel point on the image plane, its corresponding phase, and the height information. One of the basic geometry setup of the measuring system is shown in Figure 3. In the setup, the optical axis I_cO of the imaging system camera system is perpendicular to the reference plane. The optical axis I_pO of the projection system intersects with the optical axis I_cO at point O and makes an angle $\pi/2 - \theta$ with the reference plane. The line joining the two optical centers is the baseline b, and the projector and the camera have equal height $L_p = L_c = L$ with respect to the reference plane. In the work in [8], the phase and height relationship is simply derived using the triangulation method, which is

$$h(\mathrm{x}, \mathrm{y}) = \lambda|\Delta\varphi_{AD}|, \tag{7}$$

$$\lambda = Lp/2\pi b. \tag{8}$$

However, there are two hypotheses assumed in this work. Firstly, the distance between the camera's optical center and the reference plane is much larger than the height of the object, which is $L \gg h$ in general case. The other assumption is that the periodicity of projected fringe pattern, which is denoted as p on the reference plane, is evenly distributed along with axis X_w. However, in the practical case, the above hypothesis conditions are just two ideal conditions. A thorough analysis in [8] indicates that the periodicity of projected fringe pattern which distributes on the reference plane is in fact a function of the lateral coordinate x, the periodicity p_0 of the projecting pattern on the LCD image plane, and the angle θ between projector and camera. Hence, in the work in [9], based on the same geometry system setup shown in Figure 3, a more practical expression for phase and height relationship is given as follows:

$$h(\mathrm{x}, \mathrm{y}) = L/\left[\frac{2\pi L^2 b\cos\theta}{p_0|\Delta\varphi_{AD}|(L + x\cos\theta\sin\theta)^2} - \frac{b\cos\theta\sin\theta}{L + x\cos\theta\sin\theta} + 1\right]. \tag{9}$$

More accurate results were reported in this work. However, the requirement of parallelity and orthogonality in the work in [8,10] still limited the generality and flexibility in the actual measurement. Therefore, an improved structure of the measurement system is presented by the work in [3]. In this structure, which is shown in Figure 4, the line joining the optical centers of the projector system and camera system is not parallel to the reference plane but makes an angle with the reference plane. In addition, both optical axes are not required to be orthogonal to the reference plane. Comparing with the structure in the basic setup shown in Figure 3, it is more general and closer to practical situation.

After transforming the World Coordinate System to the charge-coupled device (CCD) imaging coordinate system, a final phase-to-height relationship in given as follows [3]:

$$h(\mathrm{x}, \mathrm{y}) = \frac{C_1|\Delta\varphi_{AD}| + C_2 u|\Delta\varphi_{AD}|}{1 + C_3 u + C_4|\Delta\varphi_D| + C_5|\Delta\varphi_{AD}| + C_6 u|\Delta\varphi_D| + C_7 u|\Delta\varphi_{AD}|} \tag{10}$$

where coefficients C_1 to C_7 are related to the geometric parameters of the measuring system and the intrinsic and extrinsic parameters of the imaging system. It is worthy noticing that there is another work presented by Du and Wang [1]. In this work, the two optical devices

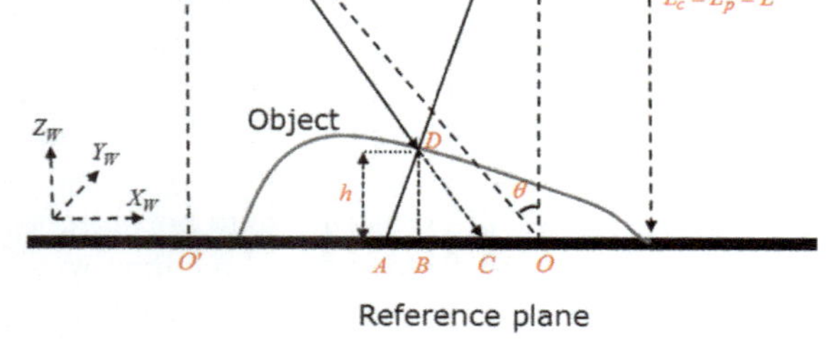

Figure 3 The basic geometry setup of a Pro-Cam system.

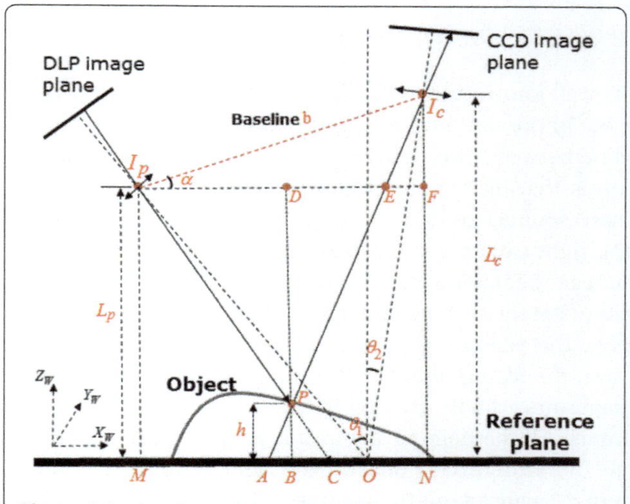

Figure 4 A generic geometry setup of a Pro-Cam system.

(camera and projector) are arbitrarily arranged; in other words, the geometry structure is also generic and has no special restriction. In other words, it implies that their model fits the case where the baseline between two optical devices could be as small as it could. The phase-to-height mapping relationship was given similarly like in Equation 10, which is described as follows:

$$h(x,y) = \frac{C_0 + C_1\varphi_D + (C_2 + C_3\varphi_D)I_D + (C_4 + C_5\varphi_D)J_D}{D_0 + D_1\varphi_D + (D_2 + D_3\varphi_D)I_D + (D_4 + D_5\varphi_D)J_D}, \tag{11}$$

where coefficients C_0 to C_5 and D_0 to D_5 are related to the geometric parameters of the measuring system and the intrinsic and extrinsic parameters of the imaging system. As we can see from Equations 10 and 11, in either of the works [1] and [3], the parameters which are physically meaningful (i.e., the length of the baseline) in the presented phase-to-height mapping model are hardly analyzed due to the difficulty of isolating these parameters from the calibrated coefficients. Moreover, they utilized a least-square method to calibrate the coefficients, yet not the related geometric parameters. Therefore, it is necessary to present a practical and analyzable model for convenient analyzing the geometry parameters. In particular, due to the specificity of our proposed short baseline arrangement system, the baseline's length influence is given the priority to be analyzed. Based on the generic setup in the work [3], we derive a new and different model for accurate phase-to-height mapping determination and parameters analysis.

Research design and methodology

In the following, a phase-to-height mapping model is presented for parameters' influence analysis. In this model, a rigorous mathematical relationship that exists between the height of an object's surface, the phase difference distribution map, and the parameters of the setup is firstly derived. Based on this model, we then study the problem of how the uncertainty of relevant parameters, particularly the baseline's length, affects the 3-D shape measurement accuracy. The uncertainty analysis on the proposed model including partial derivative analysis, relative error analysis, and sensitivity analysis are performed. Moreover, the Monte Carlo simulation experiment is also conducted.

Methods

Our proposed projecting and imaging model

The geometric optical geometry of our setup is shown in Figure 4. I_p and I_c are the exit pupil and entrance pupil of the projector and camera, respectively. The optical axes I_pO and I_cO cross the reference plane at point O and make angles θ_1 and θ_2 with Z_W axis (i.e., the normal direction of the reference plane), respectively. The baseline between these two optical centers is $I_pI_c = b$, which is not parallel to the reference plane. M is the perpendicular projection point of I_p on the reference plane, and the distance between them is $I_pM = L_p$. N is the perpendicular projection point of I_c on the reference plane, and the distance between them is $I_cN = L_c$. Point A on the reference plane and point P on the object surface correspond to the same image pixel location on the CCD plane. Point C on the reference plane and point P on the object surface are on the same pixel ray projecting from the projector. We add several dashed lines in the figure as guidelines for analysis. The dashed line I_pF is parallel with the reference plane and crosses the extension line of BP ($BP = h$) at point D, which intersects with lines I_cP and I_cN at points E and F respectively. In this work, we mainly focus on the measurement performance with respect to the influence of one parameter, which is the baseline b. Hence, similarly to the work in [10], we can assume that the fringe patterns formed by the projector are parallel to Y_W. From the geometry setup in Figure 4, we can get that the triangle APB is similar with the triangle EI_cF,

$$AB/EF = PB/I_cF = PB/b\sin\alpha. \tag{12}$$

Similarly, from the fact that triangle APB is similar with triangle ANI_c and triangle ACP is similar with triangle I_pPE, we can get

$$AB/AN = PB/I_cN = PB/L_c, \tag{13}$$

$$\begin{aligned} AC/PB &= I_pE/PD = \left(I_pF - EF\right)/PD \\ &= \left(L_p\tan\theta_1 + L_c\tan\theta_2 - EF\right)/\left(L_p - PB\right). \end{aligned} \tag{14}$$

Submitting Equation 12 into Equation 14, we can get

$$AC/PB = \left[L_p\tan\theta_1 + L_c\tan\theta_2 - (AB\cdot b\sin\alpha/PB)\right]/\left(L_p - PB\right). \tag{15}$$

Note that $AN = AC + OC + ON = AC + x + L_c\tan\theta_2$, submitting this relationship and Equation 16 into 18, we obtain

$$\begin{aligned} PB = L_pL_cAC/[&ACL_c + L_pL_c\tan\theta_1 + L_c^2\tan\theta_2 \\ &- (AC + x + L_c\tan\theta_2)b\sin\alpha] \end{aligned} \tag{16}$$

where p denotes the periodicity of the fringe patterns on the reference plane under divergent illumination. According to the work in [3], we can get

$$AC = p\cdot|\varphi_C - \varphi_A|/2\pi = p\cdot|\Delta\varphi_{PA}|/2\pi \tag{17}$$

Submitting Equation 17 into Equation 16, we can get the final relationship between the phase distribution $\varphi(x,y)$ and the height information $h(x,y)$, which is expressed as

$$h(\mathrm{x},\mathrm{y}) = \left(L_\mathrm{p}L_\mathrm{c}p|\Delta\varphi_{PA}(x,y)|\right)/[(p-pb\sin\alpha)|\Delta\varphi_{PA}(x,y)|-2\pi bx\sin\alpha + 2\pi L_\mathrm{p}L_\mathrm{c}\tan\theta_1 + 2\pi L_\mathrm{c}^2\tan\theta_2-2\pi L_\mathrm{c}\tan\theta_2 b\sin\alpha]. \tag{18}$$

It can also be written in a concise form as

$$h(x,y) = c_1|\Delta\varphi_{PA}(x,y)|/[c_2|\Delta\varphi_{PA}(x,y)| + c_3x + c_4], \tag{19}$$

where parameters c_1, c_2, c_3, c_4 are related with geometric parameters $L_\mathrm{p}, L_\mathrm{c}, p, b, \alpha, \theta_1, \theta_2$ and can be denoted as

$$\begin{cases} c_1 = L_\mathrm{p}L_\mathrm{c}p, c_2 = p-pb\sin\alpha \\ c_3 = -2\pi b\sin\alpha, c_4 = 2\pi L_\mathrm{p}L_\mathrm{c}\tan\theta_1 + 2\pi L_\mathrm{c}^2\tan\theta_2-2\pi L_\mathrm{c}\tan\theta_2 b\sin\alpha \end{cases}. \tag{20}$$

Performance analysis

Influence of the length of baseline b

The geometric parameters of the system setup include the angle between the optical axis of the projector and the camera, the distance between optical center of camera system and reference plane, the focal length of camera system and the periodicity of projected fringe patterns, etc. In this paper, we regard the baseline's length as the priority factor and focus on the length of baseline's influence on the final measurement result. From Equation 19, we can transform it into another form, which takes the baseline b as an input variable and is expressed as follows:

$$h(x,y) = K_1(x,y)/(K_2(x,y)-K_3(x,y)\cdot b) \tag{21}$$

$$\text{where} \begin{cases} K_1(x,y) = L_\mathrm{p}L_\mathrm{c}p|\Delta\varphi_{PA}(x,y)| \\ K_2(x,y) = p|\Delta\varphi_{PA}(x,y)| + 2\pi L_\mathrm{p}L_\mathrm{c}\tan\theta_1 + 2\pi L_\mathrm{c}^2\tan\theta_2 \\ K_3(x,y) = p\sin\alpha|\Delta\varphi_{PA}(x,y)| + 2\pi x\sin\alpha + 2\pi L_\mathrm{c}\tan\theta_2\sin\alpha. \end{cases}$$

From Equation 19, we get the relationship between the phase difference and the height information:

$$|\Delta\varphi_{PA}(x,y)| = \frac{h(\mathrm{x},\mathrm{y})(c_3x + c_4)}{hc_2-c_1}. \tag{22}$$

Similar to the derivative method in the work in [10], the partial derivative of Equation 23 with respect to the baseline b is calculated, and Equations 18 and 19 are submitted into the result. We get

$$\partial h(\mathrm{x},\mathrm{y})/\partial b = \left(Q_1h^2(x,y) - Q_2h^2(x,y)b - Q_3h(x,y)\right)/(Q_4 - Q_5b) \tag{23}$$

where

$$\begin{cases} Q_1 = 2\pi p\sin\alpha(L_\mathrm{p}L_\mathrm{c}\tan\theta_1 + L_\mathrm{c}^2\tan\theta_2 + x - xb\sin\alpha + L_\mathrm{c}\tan\theta_2 - L_\mathrm{c}\tan\theta_2 b\sin\alpha) \\ Q_2 = 2\pi p\sin^2\alpha(L_\mathrm{c}\tan\theta_2 + xb) \\ Q_3 = 2\pi pL_\mathrm{p}L_\mathrm{c}(x\sin\alpha + L_\mathrm{c}\tan\theta_2\sin\alpha) \\ Q_4 = 2\pi pL_\mathrm{p}{L_\mathrm{c}}^2(L_\mathrm{p}\tan\theta_1 + L_\mathrm{c}\tan\theta_2) \\ Q_5 = 2\pi pL_\mathrm{p}L_\mathrm{c}(L_\mathrm{c}\tan\theta_2\sin\alpha + x\sin\alpha). \end{cases}$$

Equation 23 shows that the height error $\partial h(x,y)/\partial b$ is a function of the parameters $Q_1, Q_2, Q_3, Q_4, Q_5, b, h$. The dependence of $\partial h(x,y)/\partial b$ on h is shown in Figure 5 with respect to the variation of other parameters Q_1, Q_2, Q_3, Q_4, Q_5. The red curve and blue curve in Figure 5 indicate the lengths of baseline b = 30 mm and b = 120 mm, respectively. A real experimental system setup consists of a pico-projector (Optoma PK301; Optoma USA, Fremont, CA, USA) and a mini-camera (Point Grey FL3-U3-13S2M-CS; Point Grey Research KK, Chiyoda-ku, Tokyo, Japan) with a 6-mm focal length. Hence, the parameters variation are in the following ranges: L_p from 390 to

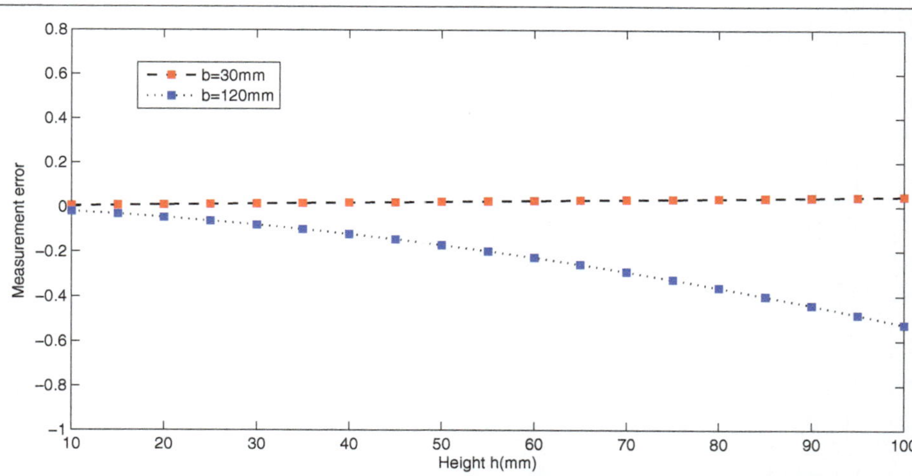

Figure 5 Plot of the measurement errors *∂h/∂b* versus the height with respect to the different length of baseline.

420 mm, L_c from 400 to 450 mm, p from 10 to 20 mm, θ_1 from 0° to 15°, θ_2 from 0° to 10°, α from 0° to 30°, x from −150 to 150 mm.

It is important to note that the parameters in the following analysis are also falling into this range. The results shown in Figure 5 indicate that in the two cases (the baseline $b = 30$ mm and the baseline $b = 120$ mm), the measurement error becomes larger as the height of target object increases. However, in the red curve which indicates the shorter baseline case ($b = 30$ mm), the measurement error is smaller (less than 0.1); meanwhile, the relationship between $\partial h(x,y)/\partial b$ and h is almost linear. On the other hand, when the baseline changes to a larger value ($b = 120$ mm), the relationship between the measurement error $\partial h(x,y)/\partial b$ and h is nonlinear and as the height of the object increases the measurement error increases faster than the shorter baseline ($b = 30$ mm). In particular, when the height of target object equals to 100 mm, the maximum value of the measurement error is larger than 0.5.

Relative measurement error analysis

A relative measurement error analysis is also conducted with respect to other parameters K_1, K_2, K_3 while the baseline is assumed fixed. Suppose there are small errors $\delta K_1, \delta K_2, \delta K_3$ existing in the parameters K_1, K_2, K_3, respectively. The relative measurement error of height $\delta h/h$ can be expressed by the following generic approximation:

$$\delta h/h = \sqrt{\left(\frac{\partial h}{\partial K_1}\delta K_1\right)^2 + \left(\frac{\partial h}{\partial K_2}\delta K_2\right)^2 + \left(\frac{\partial h}{\partial K_3}\delta K_3\right)^2} / h \tag{24}$$

Submitting Equation 21 into Equation 24, we can obtain

$$\delta h/h = \sqrt{\delta K_1{}^2 + h^2 \delta K_2{}^2 + b^2 h^2 \delta K_3{}^2} / K_1. \tag{25}$$

Equation 25 indicates that the relative measurement error $\delta h/h$ is a function of the length of baseline b, the parameters K_1, K_2, K_3, and their variation $\delta K_1, \delta K_2, \delta K_3$.

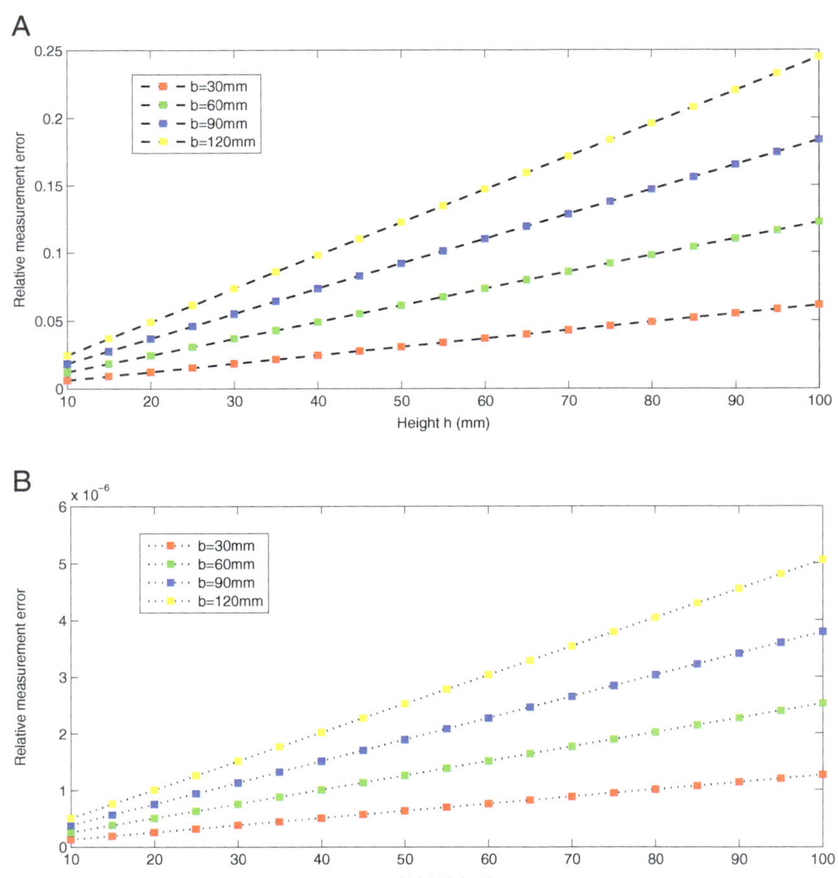

Figure 6 Plot of the relative measurement error $\delta h/h$ versus the baseline b. **(A)** For parameters K_1,K_2,K_3 go to the minimum value. **(B)** For parameters K_1,K_2,K_3 go to the maximum value.

The results are shown in Figure 6. Without loss of generality, we set the variation of parameters $\delta K_1, \delta K_2, \delta K_3 = 0.01$. The results tell us that when the height of the target object increases to 100 mm, the yellow curve which represents the largest baseline configuration system ($b = 120$ mm) yields the biggest relative measurement error, which is 25%, 0.0005% for minimizing and maximizing the parameters K_1, K_2, K_3 respectively. Meanwhile, the red curve ($b = 30$ mm) presents the smallest relative measurement error, which is 5%, 0.0001% for minimizing and maximizing the parameters K_1, K_2, K_3 respectively. Hence, we can get the same conclusion that if the length of the baseline increases, the relative error of the measured height becomes bigger, when the same errors $\delta K_1, \delta K_2, \delta K_3$ are introduced into the system.

Sensitivity analysis

Furthermore, we take a sensitivity analysis with respect to the baseline b. In the following part, b_e represents the estimates of b and $\Delta b/b = (b_e - b)/b$ indicates the relative discrepancy with respect to the nominal values. The error Δh can be expressed as the difference between the depth values calculated by substituting the two values b_e and b into Equation 21:

$$\Delta h = \frac{K_1}{K_2 - K_3 b_e} - \frac{K_1}{K_2 - K_3 b}. \tag{26}$$

When Equations 22 and 26 are combined, relative error $\Delta h/h$ results:

$$\Delta h/h = \frac{K_1}{K_2 - K_3 b_e} \Big/ \frac{K_1}{K_2 - K_3 b} - 1 = -\frac{\Delta b}{b} \frac{1}{1 + \frac{\Delta b}{b} - \frac{K_2/K_3}{b}} \tag{27}$$

Equation 27 expresses $\Delta h/h$ as a hyperbolic function of $\Delta b/b$, but for small values of $\Delta b/b$, the function is almost linear, as shown in Figure 7. The yellow curve which represents the largest baseline configuration system ($b = 120$ mm) yields the smallest relative variation of height with respect to the same relative discrepancy of

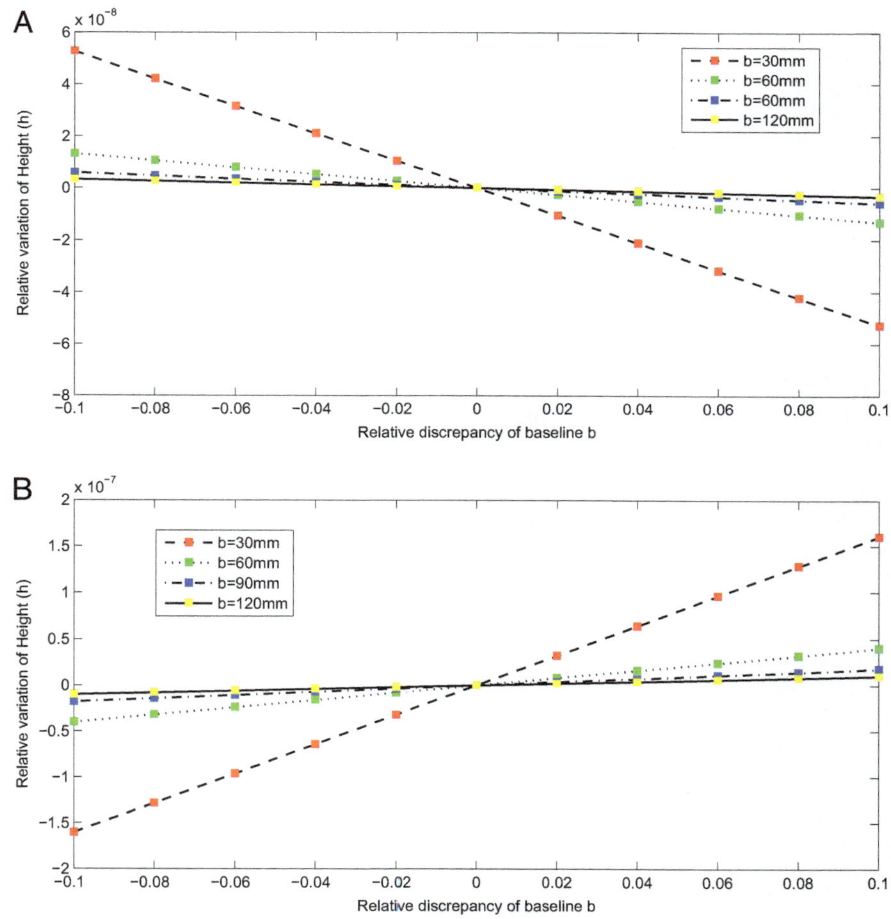

Figure 7 Plot of relative variation of height $\Delta h/h$ as a function of relative discrepancy of baseline $\Delta b/b$. (A) With respect to the minimum value of parameters K_2 K_3. **(B)** With respect to the maximum value of parameters K_2 K_3.

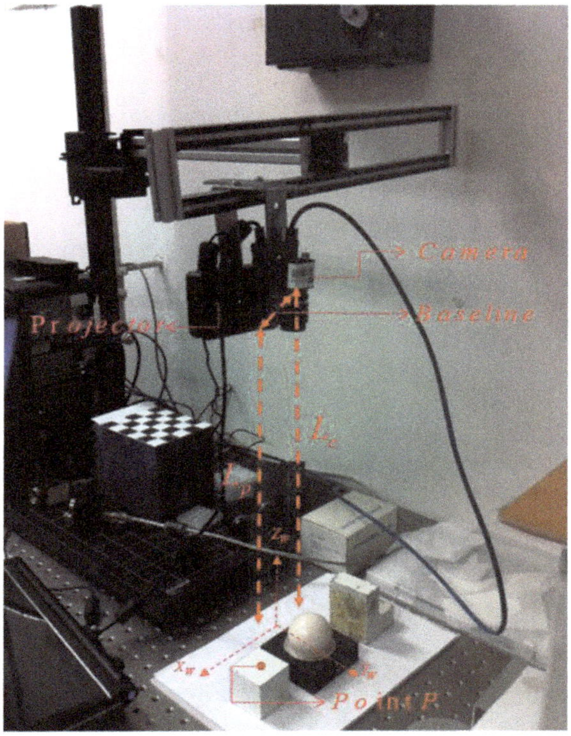

Figure 8 The experiment setup.

baseline, while the red curve which represents the shortest baseline arrangement system ($b = 30\ mm$) presents the biggest relative variation of height. This means that for a system with a shorter baseline, the proposed model is more sensitive to the small variation of other parameters. In other words, we can get the conclusion that the larger the baseline, the less sensitive is the system with respect to the same bias in the calibrated parameters. This conclusion is the same as presented in the discussion part of the work [11].

Results and discussion

A measurement error analysis with respect to the variation of baseline has been performed using partial derivative analysis, relative measurement error analysis, and sensitivity analysis. These analysis methods have been limited to parameters $L_p, L_c, \theta_1, \theta_2, \alpha, p, x$ and can be accurately calibrated. Parameters p, x can be easily evaluated with small uncertainty during the measurement by exploitation of the scale factor from pixels to millimeters in the reference frame. On the contrary, accurate determination of parameters $L_p, L_c, \theta_1, \theta_2, \alpha$ is hard due to the difficulty of precisely measuring the position of the pupils of the projector and of the camera. Meanwhile, it is also hard to determine the accurate relative orientation of the DMD image plane in a DLP-based projector. Hence, in order to eliminate the effect of other unknown uncertainty factors introduced into the analysis process, one way of doing the uncertainty propagation estimation for nonlinear systems is using Monte Carlo analysis [13,14]. The real experimental system setup (shown in Figure 8) consists of a pico-projector (Optoma PK301) and a mini-camera (Point Grey FL3-U3-13S2M-CS) with a 6-mm focal length. Hence, the variation of the parameters can be defined the same as in the previous part.

We present a global sensitivity analysis that permits the evaluation of the uncertainty distribution from other input parameters (i.e., $L_p, L_c, \theta_1, \theta_2, \alpha$) to the output (the height information) with respect to the different baseline lengths. In this method, we use four objects with different heights for the experiment. After that, in order to obtain the distribution map of height value, we change the length of baseline and its relevant parameters (i.e., $\theta_1, \theta_2, \alpha$), but keep the other parameters unchanged. It is worth noticing that the unchanged parameters in the evaluating process are randomly selected from the given range.

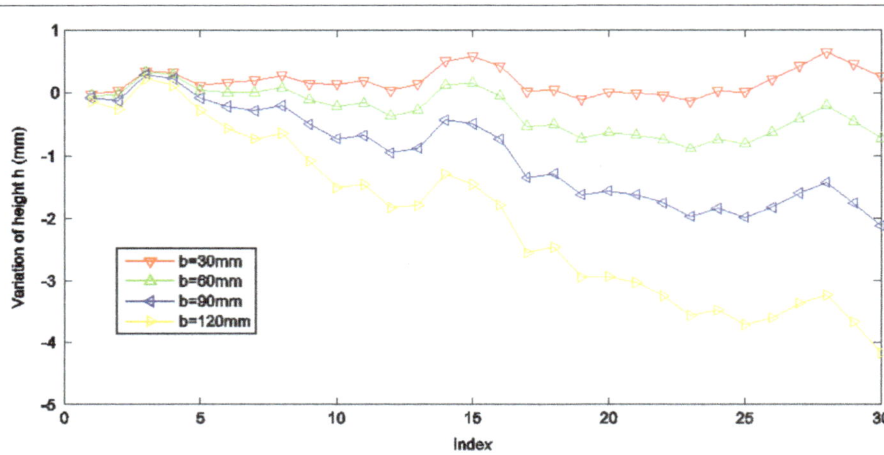

Figure 9 Plot the variation of measured height with respect to the different baseline lengths.

Figure 9 illustrates the measured height variation (comparing with selected ground truth value) which is corresponding to the random variation of other parameters when the baseline changes from 30 to 120 mm. The initial parameters are randomly selected in the pre-measured range for these four tests. We conduct each set of running for 30 times; in each set, we get the height information from the same sampled point on the object's surface (i.e., $x = 188.63$ mm). As we can see from the figure, the biggest variations of height are 0.6404, 0.8835, 2.1207, and 4.1729 mm for given baselines of 30, 60, 90, and 120 mm, respectively. Moreover, at each experiment set, when the baseline changes are bigger, the variation of height also increases.

Conclusions

In this paper, 3-D shape measurement error analysis is performed based on a short baseline Pro-Cam system. The first model is based on conventional stereovision technique. Through analysis, we obtain that as the baseline becomes shorter, the two main factors, which are the inherent biased parameters of the projector and unavoidable biased pixel point localization on the projector's image plane, have more uncertainty. Therefore, the measurement accuracy is further destroyed. In the second one, we propose a FPP technique-based projecting-imaging model. After deriving a new phase-to-height mapping relationship, measurement error which mainly refers to the height error is analyzed with respect to the length of baseline through partial derivative analysis, relative measurement error analysis, and sensitivity analysis. From the analysis result, we conclude that the smaller the baseline, the more sensitive the system is and the relative measurement error is smaller when if the same biases are introduced in the calibrated parameters. The Monte Carlo simulation experimental results also demonstrate the same measurement result using the proposed model under the short baseline configuration.

Competing interests

The authors declared that they have no competing interests.

Authors' contributions

JL carried out the numerical simulation, participated in the model derivation and drafted the manuscript. YFL proposed the short-baseline problem, participated in the model derivation and revised the draft. Both authors read and approved the final manuscript.

Authors' information

JL received B.S. and M.S. degrees from the Department of Applied Physics, Sichuan University, Chengdu, China, in 2008 and 2011, respectively. Currently, he is a Ph.D. student in the Department of Mechanical and Biomedical Engineering at City University of Hong Kong, Hong Kong. His research interests include Photoelectric Information processing, 3-D measurement, 3-D reconstruction, and robot vision. YFL received the Ph.D. degree in robotics from the Department of Engineering Science, University of Oxford, Oxford, U.K., in 1993. From 1993 to 1995, he was a Postdoctoral Research Associate in the Department of Computer Science, University of Wales, Aberystwyth, U.K. He joined City University of Hong Kong, Hong Kong, in 1995 where he is currently a Professor in the Department of Mechanical and Biomedical Engineering. His research interests include robot vision, sensing, and sensor-based control for robotics.

Acknowledgements

This work was supported by City University of Hong Kong (Project No. 7002829) and the National Natural Science Foundation of China (No. 61273286).

References

1. Du H, Wang Z (2007) Three-dimensional shape measurement with an arbitrarily arranged fringe projection profilometry system. Opt Lett 32(16):2438–2440
2. Zhang S (2009) Phase unwrapping error reduction framework for a multiple-wavelength phase-shifting algorithm. Opt Eng 48(10):105601–105608
3. Xiao Y, Cao Y, Wu Y (2012) Improved algorithm for phase-to-height mapping in phase measuring profilometry. Appl Optics 51(8):1149–1155
4. Zhang Z (2000) A flexible new technique for camera calibration. IEEE Trans Pattern Anal Mach Intell 22(11):1330–1334
5. Janne H (1992) Geometric camera calibration using circular control points. IEEE Trans Pattern Anal Mach Intell 14(10):965–980
6. Kytö M, Nuutinen M, Oittinen P (2011) Method for measuring stereo camera depth accuracy based on stereoscopic vision. IS&T/SPIE Electronic Imaging 7864:786401–786401
7. Mao X, Chen W, Su X (2007) Fourier transform profilometry based on a projecting-imaging model. JOSA A 24(12):3735–3740
8. Quan C, He XY, Wang CF (2001) Shape measurement of small objects using LCD fringe projection with phase shifting. Opt Commun 189(1):21–29
9. Spagnolo GS, Guattari G, Sapia C (2000) Contouring of artwork surface by fringe projection and FFT analysis. Opt Lasers Eng 33(2):141–156
10. Zhang Z, Zhang D, Peng X (2004) Performance analysis of a 3D full-field sensor based on fringe projection. Opt Lasers Eng 42(3):341–353
11. Zappa E (2011) Sensitivity analysis applied to an improved Fourier-transform profilometry. Opt Lasers Eng 49(2):210–221
12. Hammersley JM, Handscomb DC, Weiss G (1965) Monte carlo methods. Phys Today 18:55
13. Fishman GS (1996) Monte Carlo: concepts, algorithms, and applications. Springer, New York
14. Saltelli A, Ratto M, Andres T (2008) Global sensitivity analysis: the primer. Wiley.com
15. Delon J, Rougé B (2007) Small baseline stereovision. J Math Imaging Vis 28(3):209–223
16. Hong BJ, Park CO, Seo NS (2012) A Real-time Compact Structured-light based Range Sensing System. Semicond Sci Tech 12(2):193–202
17. Li Z, Shi Y, Wang C (2008) Accurate calibration method for a structured light system. Opt Eng 47(5):053604–053604-9
18. Huang L, Zhang Q, Asundi A (2013) Camera calibration with active phase target: improvement on feature detection and optimization. Opt Lett 38:1446–1448
19. Hartley RI, Sturm P (1997) Triangulation. Comput Vis Image Underst 68(2):146–157
20. Schalkoff RJ (1989) Digital image processing and computer vision. Wiley, New York

2

Towards developing a robot assistant for uterus positioning during hysterectomy: system design and experiments

Hiu Man Yip*†, Peng Li†, David Navarro-Alarcon† and Yun-hui Liu†

Abstract

Background: In traditional laparoscopic hysterectomy, an assistant is typically assigned to hold a uterus manipulator to facilitate the surgical procedures. The responsibility of the assistant is to position the uterus according to the primary surgeon's instructions. Throughout the surgery, which typically lasts more than 90 min, the assistant has to support the uterus manipulator with his/her hands, which easily causes fatigue to the assistant and eventually affects the manipulation performance and threatens the patient. Moreover, the manipulation done by the assistant may not always be satisfactory from the primary surgeon's point of view. Thus, having a robot assistant which can release the human assistant's hands from the manipulation task and allow the primary surgeon to directly position the patient's uterus can be a solution.

Methods: A four degrees of freedom (DOF) robot assistant is designed for positioning the patient's uterus during hysterectomy. The robot assistant is composed of three parts, a 3-DOF robotic positioning arm, a 1-DOF motorized uterus manipulator, and a supporting stand of the robot. To improve safety, the remote center of motion (RCM) mechanism is applied to the design of the robotic positioning arm. The positioning arm generates a partial spherical workspace and allows the manipulation of uterus to be done in a decoupled manner. To enhance the ability of the robot assistant to perform anteversion and retroversion motions, the motorized uterus manipulator is designed. The robot assistant is mounted to the operating table through the supporting stand.

Results: A prototype of the robot assistant is built, and experiments are conducted to verify the performance of the prototype. It is shown that the prototype is able to respond to the user's instructions accordingly and manipulates the uterus of a manikin to a desired position. Performances of robot manipulation and human manipulation are also compared. It is shown that the robot assistant can retain the manikin's uterus in a specified position in a more stable manner.

Conclusions: Our experimental study shows that the use of a robot is a feasible alternative to position the uterus. By using this type of system, costs can be reduced and safety can be improved.

Keywords: Robot-assisted surgery; Laparoscopic hysterectomy; Robot design; Remote-center-of-motion mechanism; Decoupled motion; Prototype development; Surgical tool manipulator

*Correspondence: hmyip@mae.cuhk.edu.hk
†Equal contributors
Department of Mechanical and Automation Engineering, The Chinese University of Hong Kong, Shatin, N.T., Hong Kong

Background

Excluding the pregnancy-related procedures, hysterectomy is the most commonly performed gynecological procedure [1]. Since the 1990s, there has been an increasing trend in the number of hysterectomy procedures performed in laparoscopic approach [2]. In the 2000s, with the introduction of the daVinci surgical system, surgeons started performing robot-assisted hysterectomy [1,3,4]. During a robot-assisted laparoscopic hysterectomy, the primary surgeon controls the robotic arms located at the patient's bedside from the surgeon's console, which is located away from the patient's bed; the assistant surgeon at the patient's bedside assists the primary surgeon in operating the uterus manipulator and using the accessory port of the daVinci system [3].

In [3], it is reported that the typical operative time for this kind of robot-assisted laparoscopic hysterectomy lasts no less than 190 min; in [1], it is reported that out of 16 cases, the median of the operative time is 242 min. However, for traditional laparoscopic hysterectomy conducted in the 2000s, it is reported that the median operative time is in general not more than 135 min [5]. By comparison, it is observed that the operative time required to perform traditional laparoscopic hysterectomy (i.e., performed by the primary surgeon directly at the patient's bedside) is in general shorter than those performed robotically with the daVinci system.

In terms of effectiveness, it is worth keeping the primary surgeon at the operating table to perform surgery. Besides, during this kind of robot-assisted surgery, operating the uterus manipulator is still the responsibility of the assistant surgeon. No matter in traditional or robot-assisted laparoscopic hysterectomy, asking an assistant surgeon to operate the uterus manipulator causes the problem that the manipulated position may not satisfy the primary surgeon's need. Furthermore, if no supporting device is used to assist the assistant surgeon in holding the uterus manipulator, the assistant surgeon would suffer from the problem of fatigue as he/she is required to hold the uterus manipulator throughout the surgery. Fatigue may decrease the performance of the assistant surgeon and may eventually threaten the patient. Thus, a robotic system which is capable of assisting in this tool manipulation while keeping the primary surgeon at the operating table side during hysterectomy is necessary; this is precisely our aim in this article.

Existing surgical robots

Robots have been introduced to the operating theater in recent decades; the daVinci surgical system is a well-known example. The daVinci system is a multipurpose surgical robotic system designed for laparoscopic surgeries. Four robotic arms are included in the system for manipulating a laparoscope and other surgical tools during surgery. Though tele-surgery is allowed with the daVinci system, the primary surgeon who operates the robot is away distant from the patient's bedside [6] due to its large footprint. Furthermore, as the robotic arms are designed for laparoscopic surgeries, the robotic arms are generally setup to operate above the patient's abdomen; thus, when performing laparoscopic hysterectomy, an assistant is still required to position the patient's uterus from the end of the operating table.

In general, surgical robots for minimally invasive surgery are mainly designed for endoscope manipulation or surgical tool manipulation during laparoscopic surgery in which several small incisions are opened on the patient's abdomen and the endoscope and the surgical tools are inserted into the patient's body through these incisions. Examples of robotic endoscope manipulators include AESOP [7], LARS [8], EndoAssist [9], and FreeHand [10].

EndoControl, the manufacturer of ViKY [11], a robot designed for endoscope or surgical tools manipulation, developed a system named ViKY UP [12] which aims at hysterectomy procedures. ViKY UP is a motorized uterus positioner which helps to manipulate the uterus manipulator during surgery. Though ViKY UP serves the aim of taking up the task of the assistant who is responsible for manipulating the uterus manipulator, the mechanism of the robot generates a cone-shaped workspace [13]. With this configuration, joint motions are coupled when the surgeon would like to manipulate the patient's uterus laterally or vertically.

Contributions

In this article, we present a robotic prototype for uterus positioning during laparoscopic hysterectomy. Different from the majority surgical robots, the remote center of motion of our prototype is designed to be placed at the cervix (i.e., inside the patient's body), which is more preferable in hysterectomy. With the use of the robot, safety can be improved as the robot can hold the uterus manipulator more stable than a human in general. Furthermore, a valuable manpower can be released with the use of the robot; thus, surgical costs can be reduced.

Methods

Design requirements

The robotic system presented in this article is aimed at providing assistance in positioning the uterus manipulator during laparoscopic hysterectomy. To achieve this, the proposed robotic system should have the following abilities:

1. The robot should be able to grasp and hold the uterus manipulator securely.

2. The robot should have the ability to manipulate the uterus manipulator in a natural manner as an assistant surgeon does.
3. The robot should be able to receive commands from the primary surgeon and response correspondingly.

In traditional laparoscopic hysterectomy, the assistant surgeon who is responsible for operating the uterus manipulator generally manipulates the patient's uterus according to the primary surgeon's order in three ways:

1. Anteversion/retroversion of uterus,
2. Laterally manipulation of uterus, and
3. Tensioning of uterus.

Design of the robot

The robotic system presented in this article is comprised of three parts, a robotic positioning arm, a motorized uterus manipulator, and a supporting stand of the robot.

Schematic of the robotic uterus manipulator positioning arm is shown in Figure 1. In the schematic diagram, the motorized uterus manipulator is used as the surgical tool manipulated by the positioning arm. The positioning arm has three degrees of freedom, yaw, pitch, and insertion. As in medical robotics, safety is the most important issue, a remote center of motion (RCM) mechanism [14] is used in the robotic positioning arm to improve safety. With the RCM mechanism, a point which is to be placed at or close to the patient's cervix will be mechanically forced to remain stationary no matter how the arm moves. To create the remote center of motion, the three actuated joints are designed to satisfy specific geometric arrangement. Here, denote the joint for yaw motion by joint 1, the joint for pitch motion by joint 2, and the joint for insertion by joint 3. The geometric arrangement of the joints is as follow:

- Joint 1 is a revolute joint.
- Joint 2 is a sliding joint which moves along a circular trajectory. The axis of rotation of joint 1 intersects with the center of the circular trajectory of joint 2. This intersection point gives the remote center of motion.
- Joint 3 is a translational joint. The axis of translation of joint 3 intersects with the axis of rotation of joint 1 and the center of the circular trajectory of joint 2. The surgical tool considered as a straight rod lies on the axis of translation of joint 3 and intersects with the RCM.

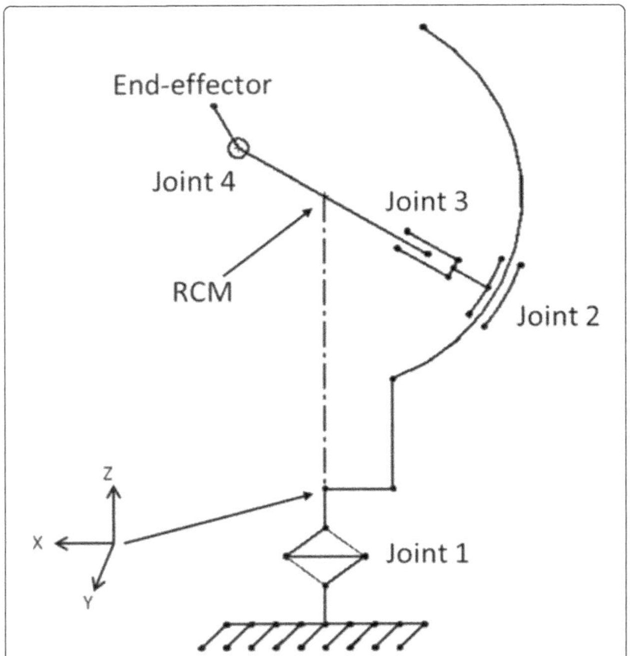

Figure 1 Schematic diagram of the design.

The motorized uterus manipulator is a device which enhances the ability of the robot assistant in performing anteversion and retroversion motions. It is a device with a tilting tip which has a moving direction similar to joint 2. Here, we denote this joint by joint 4.

Based on the schematic, a prototype of the robot assistant is designed and developed. Figure 2 shows the computer-aided design (CAD) model of the design and the CAD model of the robot assistant when it is used in the operating theater.

Actuation and mechanisms

To realize the motion of joint 2, pinion and rack mechanism is used. A block with a spur gear installed to the actuator is allowed to run along an arc-shaped guide rail with gear teeth on the rail as shown. A 20 W DC motor with gearbox is used as the actuator to ensure enough power can be delivered to drive the system. Using a motor with larger power than what it actually needs may also prevent the motor from damage due to overloading.

The structure of joint 2 is mounted on top of joint 1 directly, that is, joint 1 drives the entire structure including the arc-shaped guide rail of joint 2. Joint 1 is a revolute joint; the rotation motion from the actuator drives the structures mounted to it directly. A 20 W DC motor with gearbox is also used as the actuator for joint 1.

Joint 3 is mounted to the actuated block of joint 2. To realize the inserting motion, ball-screw mechanism is used. A 2 W DC motor with gearbox is used as the actuator as the ball-screw mechanism provides large reduction ratio, and the power required to drive the system is greatly reduced. An adaptor for grasping the surgical tool is attached to this joint. The adaptor is designed to hold uterus manipulators with a diameter of 8 or 12 mm.

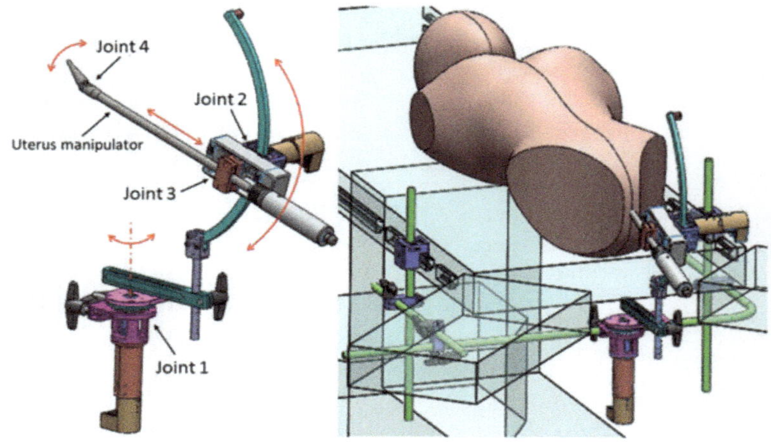

Figure 2 CAD model of the robot assistant (left) and when it is use (right).

Joint 4 is a joint of the motorized uterus manipulator which enables tilting motion. The motorized uterus manipulator is composed of two parts, a disposable unit and a reusable driving unit. The two units are connected together through a lock as shown in Figure 3. In this joint, a pair of bevel gears is used to convert the actuation direction of the tip to a configuration which can be driven by the actuator installed in the reusable driving unit. To drive the tip, a 6 W DC motor with gearbox is used as the actuator. All motors used for building this prototype are from Maxon.

Control system architecture

A schematic representation of the control system architecture is shown in Figure 4, which is composed of a user interface, a control system, and the robot actuators. The user interface consists of a standard joystick device, which allows the user to directly control the motion of the robot. The control system is the one in charge to regulate and drive the DC motors of the robot; it is composed of a Linux-based PC, an industrial motion controller, and the power amplifiers. We use a Galil DMC-4040 (DMC-4040, Galil Motion Control, Rocklin, CA, USA) industrial motion controller with an embedded four-axis 20 W current amplifier. The motion controller has implemented a speed tracking servo loop for each of the joints.

The prototype

A prototype of the robot assistant is developed according to the design shown in Figure 2; the metal parts were fabricated using standard aluminum alloy and stainless steel. The moving ranges of joint 1 to joint 4 are −65° to 65°, 30° to 138°, 0 to 50 mm, and 0° to 180°, respectively. Motions of individual joints of the robot assistant are shown in Figure 5. Figure 6 shows the manipulation experiments conducted with manikin.

Results and discussion

To verify the performance of the prototype, four experiments are conducted. Two experiments are conducted to verify the closed-loop response of the robot's motion, and two are conducted to compare the manipulation

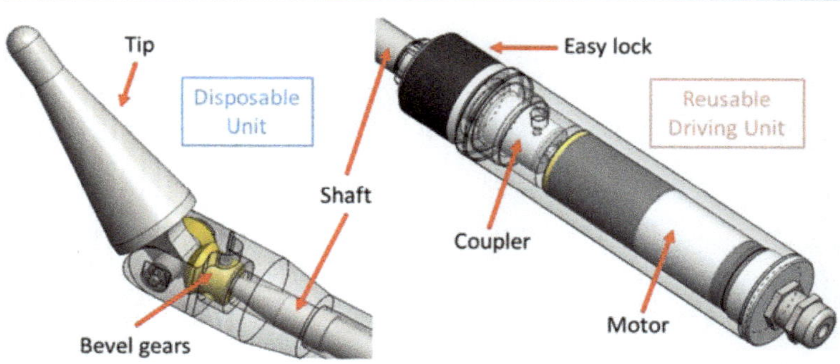

Figure 3 Structure of the motorized uterus manipulator.

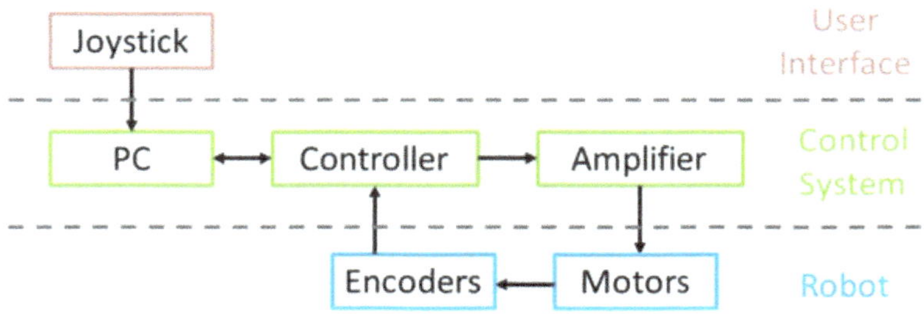

Figure 4 Control system architecture of the robot assistant.

performance of the robotic assistant with of a human assistant.

Velocity tracking experiment (auto-feed)

The objective of this experiment is to verify the performance of the robot in responding to an assigned velocity profile. In particular, we would like to verify the robot's performance when moving with non-constant speeds. In this experiment, the four degrees of freedom of the robot are tested separately. To each joint, a time-varying velocity profile (a sinusoidal function of time) is assigned, which the robot must follow as accurately

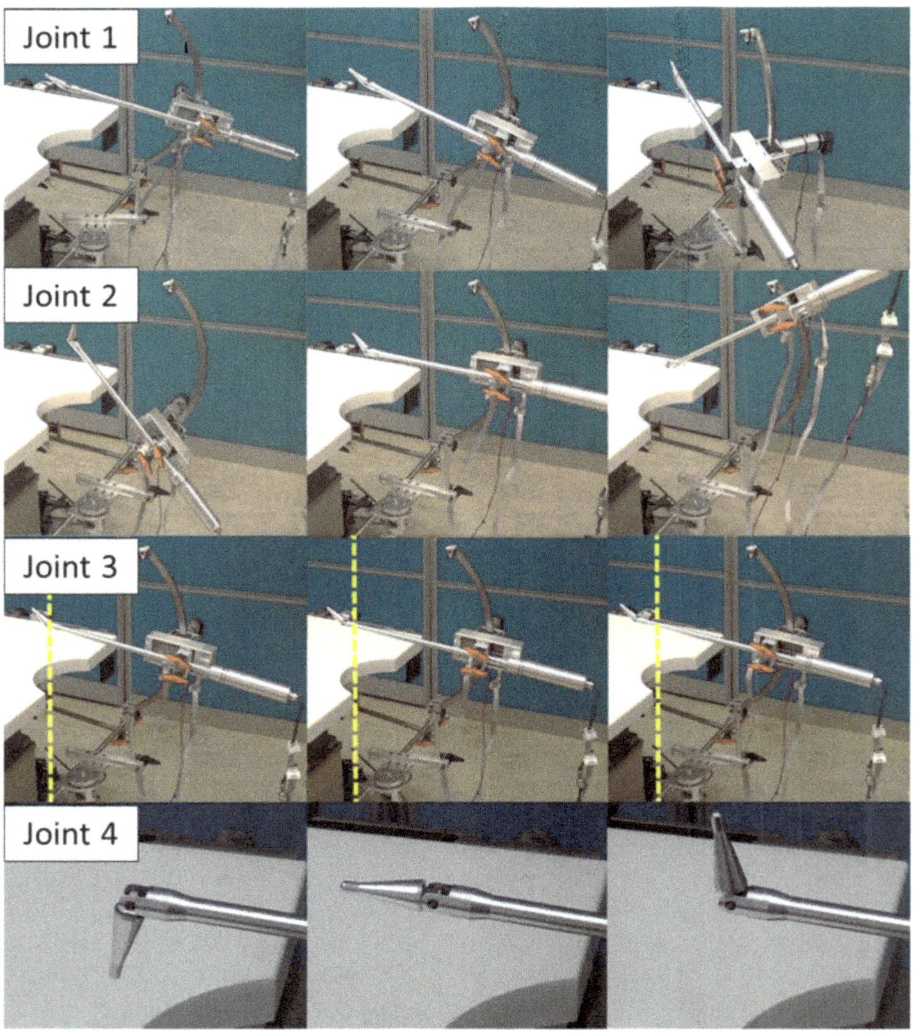

Figure 5 Joint motions of the robot assistant.

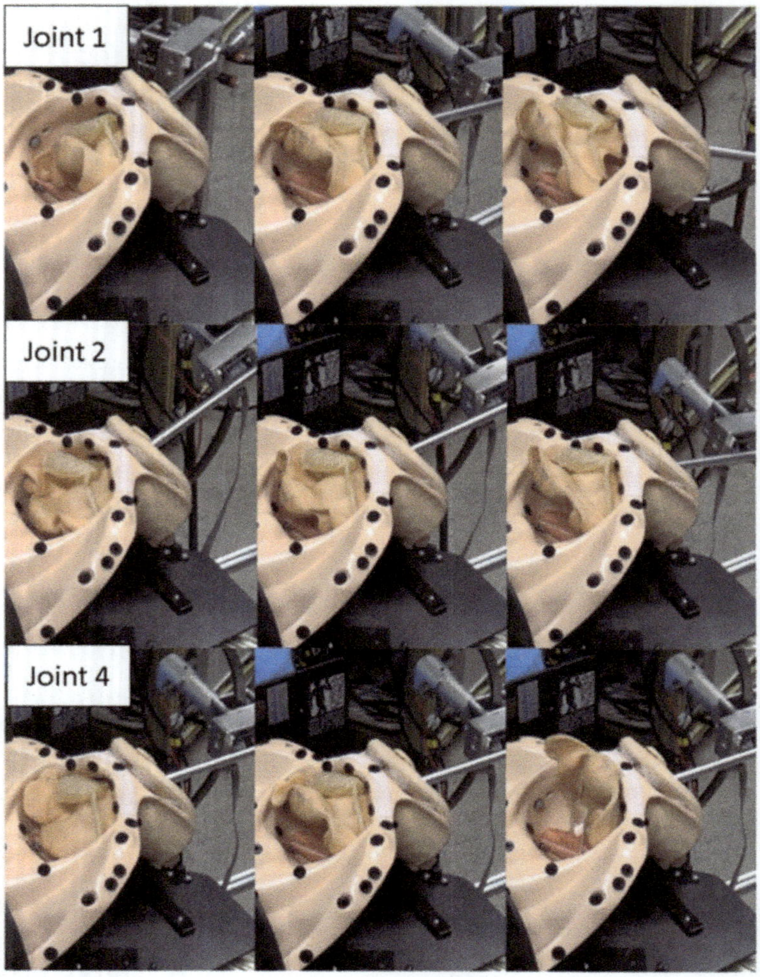

Figure 6 Manipulating the uterus of a manikin with the robot.

as possible. We denote the *i*th velocity input to the robot by

$$v_i = k_i \sin(t), \quad (1)$$

where v_i represents the joint velocity and k_i is a positive user-definable velocity gain.

The results of this velocity tracking experiment are shown in Figure 7. In the figure, the desired velocity profiles fed to the robot are represented by the black lines while the measured velocities of the joints are represented by the red lines. These experiments are obtained with the following velocity gains: $k_1 = 13$, $k_2 = 27$, $k_3 = 1.7$, and $k_4 = 50$. From these results, it is shown that the robot can closely track the time-varying velocity inputs.

Velocity tracking experiment (manual input from joystick)

As the robot is designed to be controlled and operated by the primary surgeon, this experiment is conducted in a manual-control mode. The objective of this experiment is similar to the auto-feed one, i.e., to verify the performance of the robot in responding to the user's input commands.

In this experiment, the robot is manually controlled by human through a joystick interface. For these experiments, we denote the *i*th velocity input to the robot by

$$v_i = k_i J_i, \quad (2)$$

where J_i represents a discrete (i.e. $-1, 1, 0$) input provided by the joystick interface. The user is asked to control the robot by pressing the buttons on the control joystick which correspond to the robot joints and their moving directions.

Similar to the auto-feed experiment, the joints of the robot are tested separately in this experiment. Figures 8 and 9 show the results of this experiment; Figure 8 shows the steady state response of the system while Figure 9 shows the transient responses. In the figures, the black

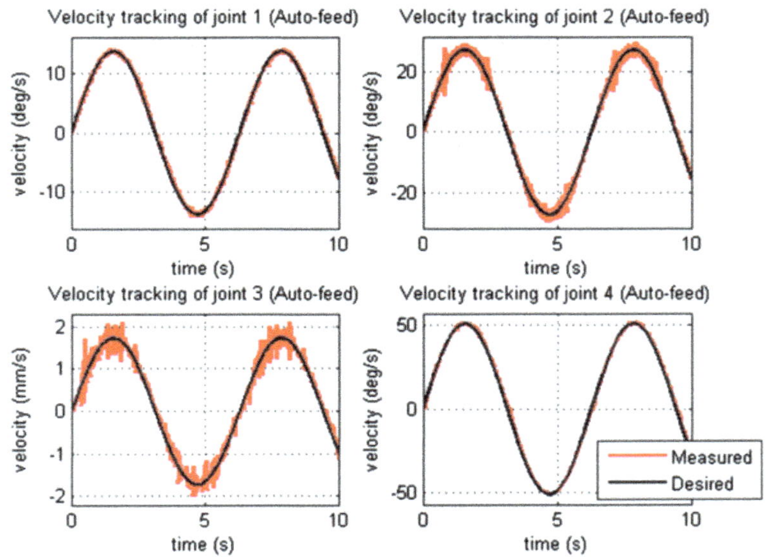

Figure 7 Results of the velocity tracking experiment (auto-feed).

lines indicate the input velocity profiles converted from the input commands from the joystick while the red lines indicate the measured joint velocities. These experiments are obtained with the following velocity gains: $k_1 = 20$, $k_2 = 50$, $k_3 = 3.5$, and $k_4 = 60$. From the results, it is shown that the robot can accurately regulate a constant speed profile provided by the user.

Manipulation experiment (one-dimensional distance)

In traditional laparoscopic hysterectomy, visual feedback is obtained from the laparoscopic camera inserted into the patient's body through a small incision opened on the patient's abdomen. In general, 2D visual feedback can be obtained and the surgery is performed with the use of this 2D visual information. This scenario is emulated in this experiment, and the experimental setup is shown in Figure 10. The objective of this experiment is to compare the manipulation performance of the robot with that of human.

In this experiment, a medical manikin is used as the simulated environment. To emulate the laparoscopic camera, a webcam is used to obtain the visual feedback from

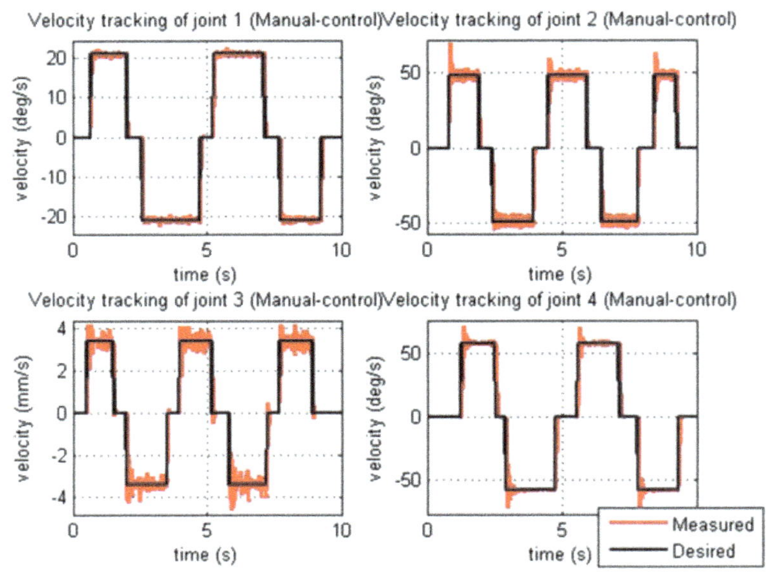

Figure 8 Results of the velocity tracking experiment (manual control).

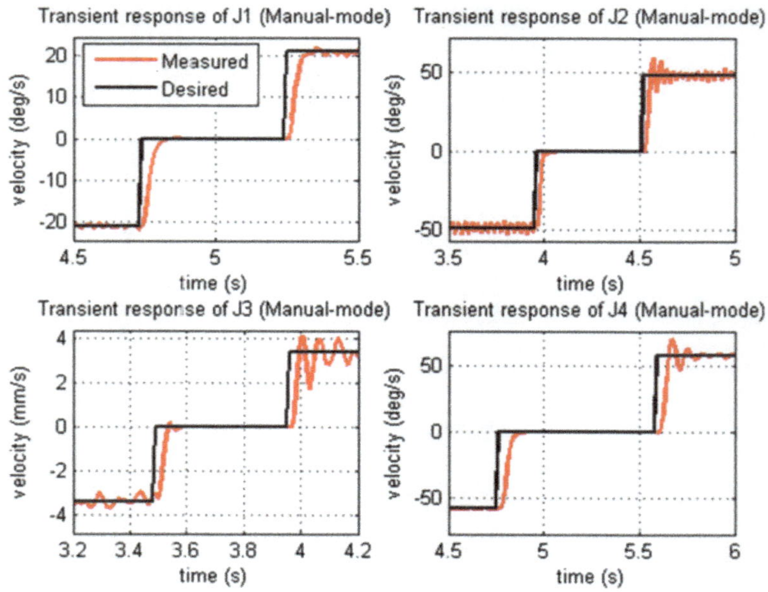

Figure 9 Transient responses of the joints in manual control mode.

the manikin. With the webcam, images with resolution of 640 × 480 pixel are obtained. An artificial marker is attached to the uterus as a visual feedback tracking point. In this experiment, a standard non-motorized uterus manipulator (produced by Apple Med [15]) is used as the surgical tool to position the uterus of the manikin.

On the robot side, the uterus manipulator is grasped by the robotic positioning arm. Human participants are asked to move the marker to a desired image configuration with the assistance of the robot and retain that position. The robot is controlled by the participants using a joystick.

To compare the performance of the robot with the performance of a human, the same experiment is conducted without the assistance of the robot. The human participants are asked to hold the uterus manipulator by themselves and keep a specified image position without the use of any supporting device. All participants involved in this experiment are without medical background. Short training for both manipulating with and without the robot is given to the participants before the experiment started.

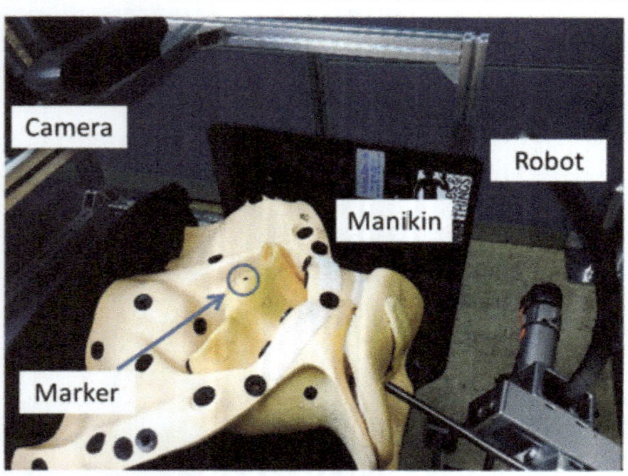

Figure 10 Experimental setup of the manipulation experiments.

In this experiment, the desired image position is represented by a one-dimensional line as shown in Figure 11. The participants are asked to align the marker attached to the uterus of the manikin to any point of the line specified on the screen both with and without the use of the robot. We choose this task as it is a common procedure to move the uterus aside during laparoscopic hysterectomy.

Pixel error between the marker and the line is measured. As the image position of the desired destination is known and the image position of the marker attached to the manikin can be obtained by feature tracking, the pixel error can be calculated by

$$e = p_d - p_m, \quad (3)$$

where e is the pixel error, p_d is the pixel position of the desired destination, and p_m is the pixel position of the marker. In this one-dimensional experiment, p_d and p_m can be understood as the x-coordinates of the desired destination and the marker on the image plane, respectively.

In both assisted and unassisted experiments, ten trials are conducted. The results are shown in Figure 12.

In Figure 12, the plot on the left shows the measured pixel error obtained the during manipulation with

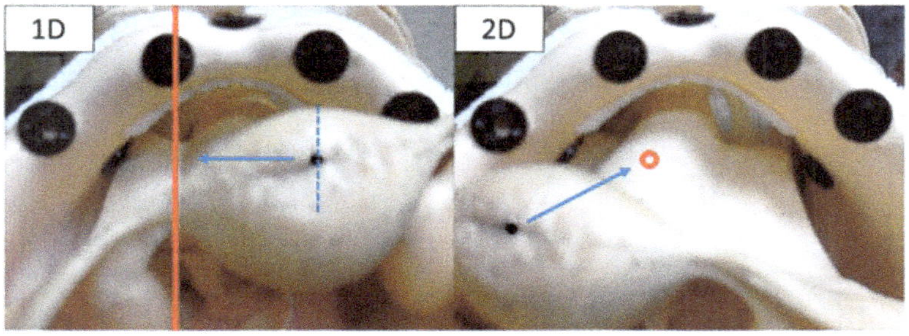

Figure 11 The desired position in the 1D (left) and 2D (right) image space.

the assistance of the robot; the plot on the right shows the measured pixel error obtained without the use of the robot assistant. From these results, it is shown that with the use of the robot, the participants are able to manipulate the uterus to a specified image position in a more stable way compared to the traditional hand manipulation.

Manipulation experiment (two-dimensional point control)

To further verify the performance of the robot, the manipulation experiment is extended to the two-dimensional image space. In this experiment, the desired position is specified by an image point instead of a line, as shown in Figure 11. The pixel errors in both x and y directions on the image plane are measured to evaluate the performances.

Figures 13 and 14 compare the positioning performance of the robot with that of solely human manipulation. These results show that while both human and robot are able to manipulate the uterus to a desired position, the robot can retain the desired configuration in a more stable manner, as expected.

Conclusions

In this article, a four degrees of freedom robot assistant for uterus positioning during laparoscopic hysterectomy is designed and developed. This system is composed of a 3-degrees of freedom (DOF) uterus manipulator positioning arm, a 1-DOF motorized uterus manipulator, and a supporting stand which mounts the robot to the operating table. The 3-DOF uterus manipulator positioning arm adopts the RCM mechanism, which is designed to improve safety. Instead of placing the RCM out of the patient's body as most surgical robots do, the RCM of the presented positioning arm is designed to be placed inside the patient's body. Note that when applying to hysterectomy procedures, the RCM must be placed at or close to the cervix. The positioning arm has a spherical workspace, and the three actuated joints are designed to satisfy specific geometric arrangement so that the manipulation of the uterus can be achieved in a decoupled manner. A 1-DOF motorized uterus manipulator with a tilting tip is also designed and developed to enhance the ability of the robot assistant to perform anteversion and retroversion motions.

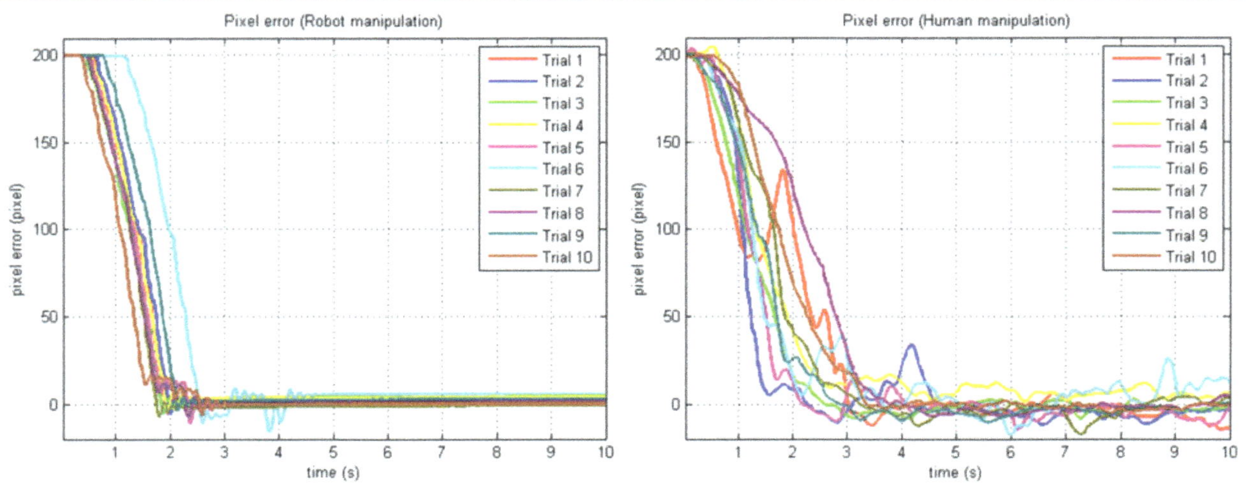

Figure 12 One-dimensional pixel error when manipulated with the robot (left) and without using the robot (right).

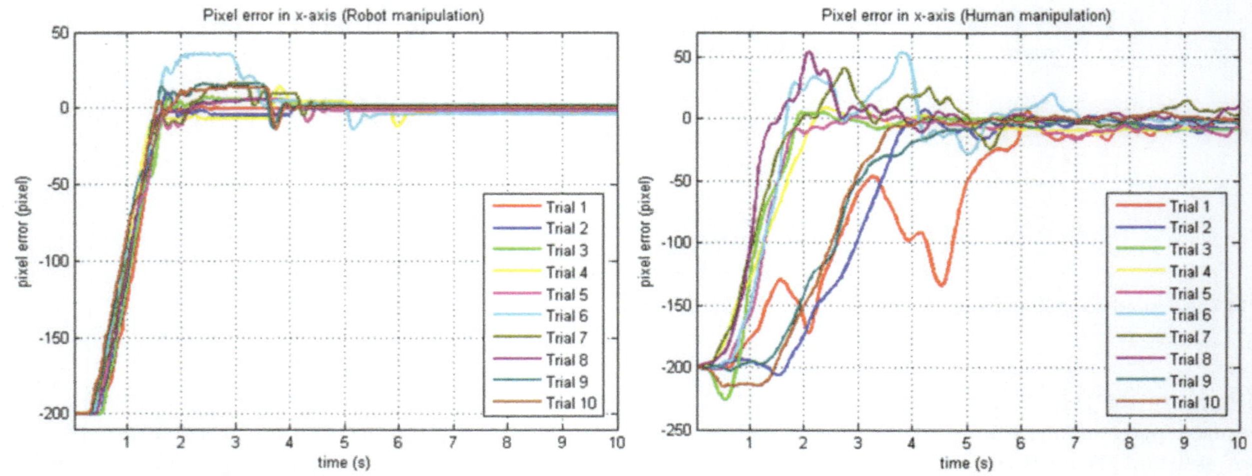

Figure 13 Pixel error comparison in the *x*-axis.

Experiments are conducted to verify the performance of the robot assistant, and the experimental results are presented. It is shown that the developed robot assistant can respond to the user's commands and assist the user in uterus manipulation. Furthermore, when compare with solely human manipulation, the robot assistant shows the advantage of being able to retain a desired position in a more stable manner, which is precisely the goal of this prototype. Our robotic system can release a human assistant from the task of uterus positioning. This helps saving the valuable human resources in the operating theater. Besides, more reliable and stable positioning of uterus can also be achieved.

However, there are several disadvantages in our system, for example, space under the operating table is required when installing the robot assistant to the operating table. When encountering with operating tables with this space unavailable, the RCM of the robotic positioning arm will be unable to align with the patient's cervix, which is undesirable. Also, due to the mechanical structure, the motion of joint 2 is comparatively not smoothly enough (see Figures 7 and 8) and the vibration frequency of the robot is high than that of human (see Figures 12, 13, and 14).

In the future, we would like to eliminate the necessity of under-bed space for installation so that the installation of robot can be more feasible when encountering with different operating tables. Also, we would like to improve the structure of the robotic system to increase its rigidity and reduce the system's vibration. Improvements on the controller are also expected for the same purpose. One step further, image-based automatic control can also be implemented to provide more autonomy to the system.

Last but not the least, in our experiments, a manikin is used. Though medical manikin is the closest simulated

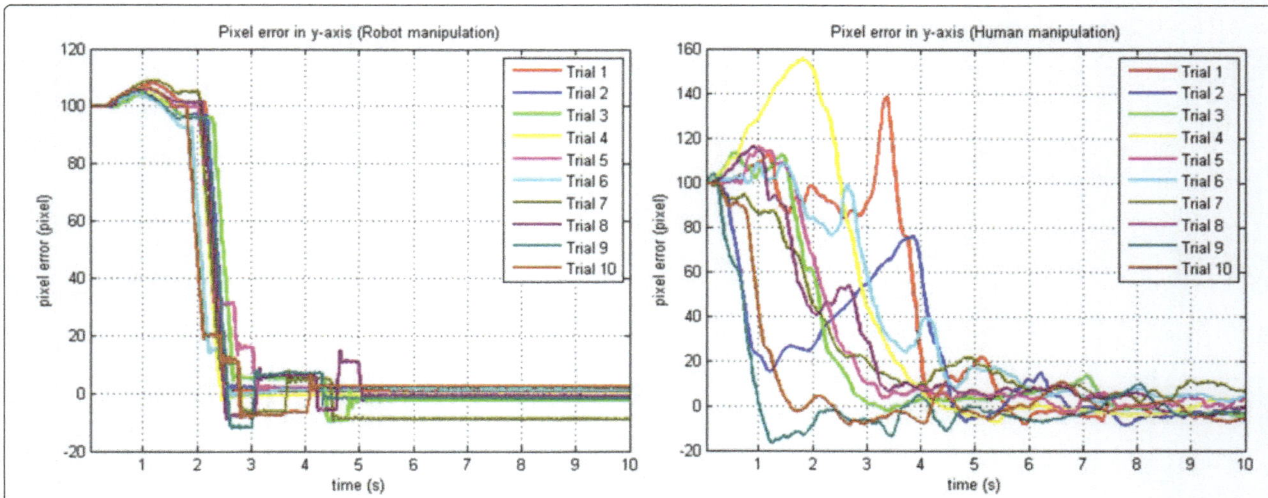

Figure 14 Pixel error comparison in the *y*-axis.

environment when comparing to human, differences exist. Indeed, differences also exist among different patients. These differences may eventually affect the workspace and payload required for the robot. Thus, to obtain more realistic and reliable data for designing future robots, ex vivo experiments with cadavers and clinical trials are expected.

Competing interests
The authors declare that they have no competing interests.

Authors' contributions
HMY designed the robot, conducted the experiments, and drafted the manuscript. PL built the robot and participated in the advisory of the design of the robot. DN-A constructed the control system and conducted the experiments. Y-hL conceived the study and supervised and coordinated the project. All authors read and approved the final manuscript.

References

1. Reynolds RK, Advincula AP (2006) Robot-assisted laparoscopic hysterectomy: technique and initial experience. Am J Surg 191:555–560
2. Farquhar CM, Steiner CA (2002) Hysterectomy rates in the United States 1990–1997. Obstet Gynecol 99:229–234
3. Beste TM, Nelson KH, Daucher JA (2005) Total laparoscopic hysterectomy utilizing a robotic surgical system. J Soc Laparoendosc Surg 9:13–15
4. Diaz-Arrastia C, Jurnalov C, Gomez G, Townsend CJr (2002) Laparoscopic hysterectomy using a computer-enhanced surgical robot. Surg Endoscopy 16:1271–1273
5. Shimizu DJ (ed) (2011) Hysterectomy: procedures, complications, and alternatives. Nova Science Publishers, Inc., New York, pp 102–103, 146–147, 176–177
6. Guthart GS, Salisbury, J Jr (2000) The intuitive telesurgery system: overview and application. In: Robotics and automation, 2000. Proceedings. ICRA '00. IEEE international conference on, vol. 1. pp 618–621
7. Sackier JM, Wang Y (1994) Robotically assisted laparoscopic surgery. Surg Endoscopy 8:63–66
8. Taylor RH, Funda J, Eldridge B, Gomory S, Gruben K, LaRose D, Talamini M, Kavoussi L, Anderson J (1995) A telerobotic assistant for laparoscopic surgery. IEEE Eng Med Biol Mag 14:279–288
9. Aiono S, Gilbert JM, Soin B, Finlay PA, Gordan A (2002) Controlled trial of the introduction of a robotic camera assistant (endoassist) for laparoscopic cholecystectomy. Surg Endoscopy 16:1267–1270
10. Sharma D, Brown C, Kouriefs C, Sood H, Grange P, Patel H (2009) Initial experience with the freehand robotic camera holder in laparoscopic urology. J Endourol 23(1):A249
11. Berkelman P, Ma J (2009) A compact modular teleoperated robotic system for laparoscopic surgery. Int J Robot Res 28:1198–1215
12. Swan K, Kim J, Advincula AP (2010) Advanced uterine manipulation technologies. Surg Tech Int 20:215–220
13. Long JA, Cinquin P, Troccaz J, Voros S, Berkelman P, Descotes JL, Letoublon C, Rambeaud JJ (2007) Development of miniaturized light endoscope-holder robot for laparoscopic surgery. J Endourol 21:911–914
14. Kuo C-H, Dai J (2009) Robotics for minimally invasive surgery: a historical review from the perspective of kinematics. In: Yan H-S, Ceccarelli M (eds). International symposium on history of machines and mechanisms. Springer, Netherlands. pp 337–354
15. Ng C, Chern B (2007) Total laparoscopic hysterectomy: a 5-year experience. Arch Gynecol Obstet 276:613–618

3

Tubular micromotors: from microjets to spermbots

Veronika Magdanz[1*], Maria Guix[1*] and Oliver G Schmidt[1,2,3*]

Abstract

In the last decade, it was demonstrated that tubular micromotors are of potential interest for environmental, sensing, and medical applications. Even though catalytic micromotors rely on toxic fuel for their propulsion, many proof-of-concept applications were shown. Currently, the research community of micromotors is searching for biocompatible fuel solutions in order to extend the field of applications to tasks in physiological conditions. This short review gives an overview over the advances in the field of tubular micromotors explaining the different fabrication methods, fuels, and applications. The article points out the utilization of catalytic microjets in sensing, medical, and environmental applications, as well as the journey towards biocompatible tubular motors driven by motile sperm cells.

Keywords: Microjets; Rolled-up; Microtubes; Micromotors; Spermbot; Nanotechnology; Microfabrication

Review

The development of tiny machines that are able to function on the microscale to perform various tasks are of special interest nowadays. This review focuses on the state of the art in the field of tubular micromotors spanning from catalytically driven microjets to hybrid micromotors propelled by motile sperm cells. The article stresses especially the fabrication methods and describes control mechanisms, fuels, applications, and further progress towards biocompatible alternatives that have ensued in the recent years. This article is only reviewing tubular micromotors and does not include micromotors of other shapes, such as spheres, rods, helices, or other designs, being out of scope of this short review.

There are two distinct ways to fabricate tubular micromotors: roll-up of nanomembranes and electrodeposition into the pores of membranes, being used as template. Although the fabrication procedures strongly differ, both methods offer a reliable mass production of catalytic microjets. The following paragraphs describe rolled-up nanotech and template-based deposition of microjets in detail and also present a comparative table (Table 1). In addition, the proof-of-concept applications that were demonstrated in the last decade for rolled-up and template-based microjets will be discussed.

* Correspondence: v.magdanz@ifw-dresden.de; m.guix.noguera@ifw-dresden.de; o.schmidt@ifw-dresden.de
[1]Institute for Integrative Nanosciences, Leibniz Institute for Solid State and Materials Research Dresden, 01069 Dresden, Germany
[2]Material Systems for Nanoelectronics, Chemnitz University of Technology, 09126 Chemnitz, Germany
Full list of author information is available at the end of the article

Rolled-up nanotech: 'a tool for many trades'

Rolled-up nanotech is a method of microfabrication that has opened a wide field of applications of microtubes [1,2]. Recent publications demonstrate the utility of rolled-up microtubes, e.g., as highly sensitive sensors [3,4], for microbatteries [5], for transistors [6], and as a tool to study cells in confinement [7,8]. Rolled-up nanotech was also utilized to fabricate the first tubular micromotors [2,9] with an inner platinum surface for bubble propulsion by catalytic decomposition of hydrogen peroxide.

Inorganic microjets from rolled-up nanomembranes

Rolled-up nanotech offers a versatile design of tubular micromotors. Geometrical parameters such as the radius, length, and shape of the micromotors can be varied, on the one hand by changing the pattern dimensions that are used for the deposition of nanomembranes on the substrate and on the other hand by the deposition angle, rate, and thickness of the material. This method can be used to fabricate conical or rather cylindrical microtubes of lengths ranging from the nanometer [10] to millimeter scale [11]. Almost any material can be integrated into rolled-up microtubes, and therefore, the technology can

Table 1 Comparative table of template-based electrodeposition and rolled-up nanotech as fabrication methods for tubular micromotors

	Template-based electrodeposition	Rolled-up nanotech
Required equipment	• Porous membranes as templates	• Photolithography
	• Electrodeposition setup	• Thin film deposition setup
Suitable materials	• Conducting materials as outer layer	• Materials compatible with vacuum deposition
Advantages	• Easily accessible method	• Wide range of materials
	• Compact fabrication setup	• Fabrication of tubular structures with wide range of length and diameter possible
		• Fine control of layer thickness down to nanometer precision
		• Dynamic structures (stimuli-responsive microtubes)
		• Fixed tube array possible for mechanistic studies (i.e., bubble nucleation)
Limitations	• Tube shape preset by template pores	• Often requires vacuum deposition machine
	• Requires conductive outer layer	
	• Release of tubes after dissolution of membrane	

This table compares the two distinct fabrication methods for tubular micromotors according to the current state of the art. However, this might change in the future as the processes are optimized and developed further.

implement a variety of new functionalities into the tubular micromotors.

Rolled-up nanotech is based on the strain engineering of inorganic nanomembranes that roll into microtubes when they are released from a sacrificial layer by etching with organic solvents [2]. First of all, a substrate, most commonly a glass slide or silicon wafer, is spin-coated with a photoresist. After exposure to ultraviolet light through a photomask, the photoresist is developed with an etching agent. Left on the substrate are certain photoresist patterns, e.g., squares or circles, which are the base for depositing nanomembranes of certain sizes and shapes via electron beam evaporation. The electron beam releases atoms from the target material under high-vacuum conditions. These atoms solidify on the surface of the substrate and form a thin film. During glancing angle deposition, a narrow gap remains open at the far end of the patterned photoresist where no material is deposited. This narrow gap, also called 'window,' is a point of attack for the etching solution that removes the photoresist. When the photoresist starts dissolving, nanomembranes roll up into tubes due to the internal strain as they release from the substrate (Figure 1). A certain thickness of each material, as well as a defined tube radius, can be achieved depending on the rate and time of deposition of each layer. An alternative to the etching of the sacrificial layer for achieving roll-up is the dry-releasing approach by thermal annealing using self-assembled metal nanodroplets [12].

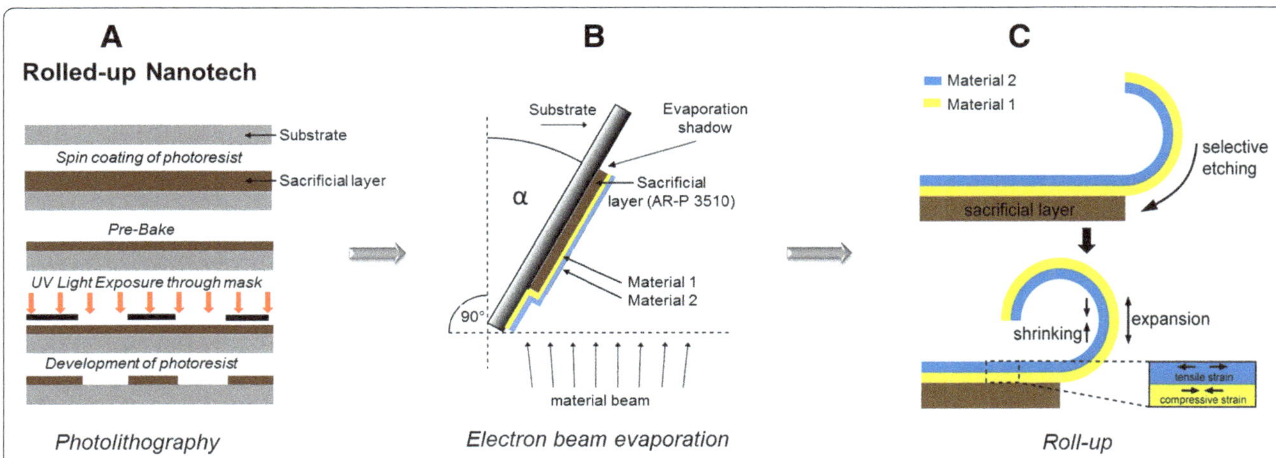

Figure 1 Fabrication process of microtubes by rolled-up nanotech. During the first step (photolithography) **(A)**, the photoresist is spin-coated onto a substrate, prebaked, and exposed to UV light through a mask before it is developed. During the electron beam evaporation **(B)**, the nanomembranes are deposited on an angle onto the sacrificial photoresist layer. The roll-up **(C)** is achieved by etching the sacrificial layer. **(B)** and **(C)** modified from [13,14].

For the creation of tubular bubble-propelled microjets, an inner platinum surface is essential; therefore, a deposition of nanomembranes such as titanium (adhesion layer), chromium (induces additional strain), iron (for magnetic remote control), and platinum (catalytic surface) is chosen. An inner or outer gold nanomembrane can be integrated in the microjet's design by means of sputtering or evaporation techniques for its further functionalization. Immobilizations of molecules on gold surfaces are accomplished by surface chemistry methods such as self-assembling monolayers (SAMs) of thiols [15] and/or other chemical linkers that selectively attach to the active groups of the target molecule. This method provides the ability to pick up and transport particles or biological entities, leading to their isolation in complex samples (e.g. serum, urine, saliva). The first example of pick-up and delivery of single cells by tubular micromotors was demonstrated by Sanchez et al. [16]. The Pt microjets with an outer gold layer can also serve as sensors: some clear examples are the functionalization of such SAMs with single-strand DNA probe [17], anti-carcinoembryonic antigen monoclonal antibody [18], or aptamers [19], permitting the selective capture of nucleic acids, cancer cells, or target proteins, respectively.

The incorporation of a magnetic nanomembrane provides the remote control via external magnetic fields [9]. When the tubular micromotors are magnetized by exposure to a permanent magnet, they act as compasses and sense the direction of the external magnetic field (Figure 2BIII) [20]. The magnetic layer also allows the utilization of the microtubes as micro-/nanodrillers [10,21]. In addition, the shape of the tubular micromotors can be altered to be slightly conical with a sharp tip when thin films are deposited on trapezoidal photoresist structures. This leads to the ability of the drillers to make incisions in cells and tissue [21]. It was demonstrated that microjets are powerful enough to swim in a controlled manner against flow in microfluidic channels [22,23]. Two- and three-dimensional (3D) closed-loop control [24,25] was achieved using a setup of electromagnetic coils that direct the magnetic field lines towards reference positions (Figure 2AII). This technique offers an accurate remote positioning and motion control of the microjets in 3D space. Light and temperature also serve as control parameters to change the speed of microjets during their operation [26]. Light diminishes locally the concentration of hydrogen peroxide, reducing the motility of the microjets in this area (Figure 2BI). This leads to a stop-and-go motion of the microjets when light is applied sequentially. An increase in temperature up to 37°C also improves the catalytic decomposition of peroxide to oxygen and water and leads to an acceleration of the microjets (Figure 2BIIa) [27,28].

Furthermore, rolled-up micromotors can be fabricated from 20- to 200-nm-thick iron nanomembranes, which actively decompose organic pollutants in an acidic environment [29]. It was also shown that micromotors can be trapped in microchips containing chevron and heart-shaped structures without any external

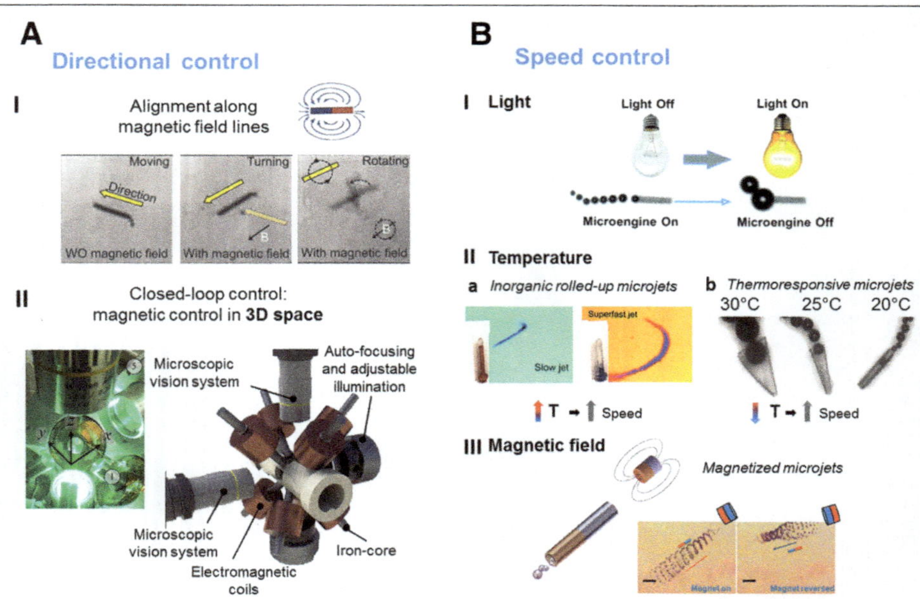

Figure 2 Control mechanisms for direction and speed of motion of microjets. (A) Directional control through alignment of microjets with external magnetic fields [9] (I) and closed-loop control of microjets in 3D by setup of eight electromagnetic coils [25] (II). **(B)** Speed control of microjets by changing light intensity [26] (I) or temperature [27,34] (II). A magnetic field can be used to attract and accelerate or repel and reverse the motion of magnetized microjets (III) [20].

mechanism due to their self-propulsion and can be used as an alternative system to concentrate components in the case of working with functionalized micromotors [30,31]. A particular treat of strain-engineered rolled-up micromotors is the integration of electronic functionality as envisioned in Mei et al. [32], rendering these promising engines for future fully autonomous and multi-functional microsystems.

Stimuli-responsive rolled-up microjets based on thermoresponsive polymers

Polymers are attractive materials because of their flexibility, transparency, and sensitivity to external stimuli. Stimuli-responsive polymer tubes were shown to open and close when a small temperature change is applied [33]. Polymeric bilayers were utilized to fabricate thermoresponsive microjets that fold and unfold upon temperature change [34]. Polycaprolactone as outer passive layer, poly(N-isopropylacrylamide) (PNIPAM) as middle thermo-active layer, and platinum as inner catalytic layer form the triple film of the microjet. PNIPAM swells below 28°C and shrinks above this temperature, so that a slight temperature difference leads to a change in the radius of the multilayer jet (Figure 2BIIb). By doing so, the microjet can be reversibly accelerated and stalled on demand, because the radius determines the bubble propulsion and, in turn, the speed of the microjet. Polymers are also, as mentioned above, easily functionalized with bioactive molecules. In addition, such flexible microjets have potential for studies of the mechanism of bubble generation inside tubular microjets and for biodegradable micromotors.

Quick and easy: template-based microjets by electrodeposition

Although the very first tubular micromotors fabricated by using electrochemical techniques were obtained by depositing platinum and gold onto a silver wire which was then subjected to further etching [35], breakthrough to much more attractive implementation came with the template-based tubular micromotors electrochemically deposited into porous membranes. The fabrication of template-based tubular microjets is based on sequentially depositing the materials of interest into a template (generally a tubularly or conically shaped porous membrane), followed by its dissolution and the release of the tubular micromotors. Such procedure is carried out by applying an electrical current that reduces cations from a desired material of the solution, leading to the deposition of such material onto the walls of the porous membrane.

The setup to perform such electrochemical depositions is composed by a counter electrode (i.e., platinum wire), a reference electrode (i.e., Ag/AgCl electrode), and a working electrode, which is based on a conductive layer that is in contact with one face of the porous membrane (see Figure 3A). Both aluminum oxide membranes (AAO membranes) [36] and polycarbonate membranes [37] are used for such purposes so as to obtain tubular or conical microtubes, respectively. In both cases, it is

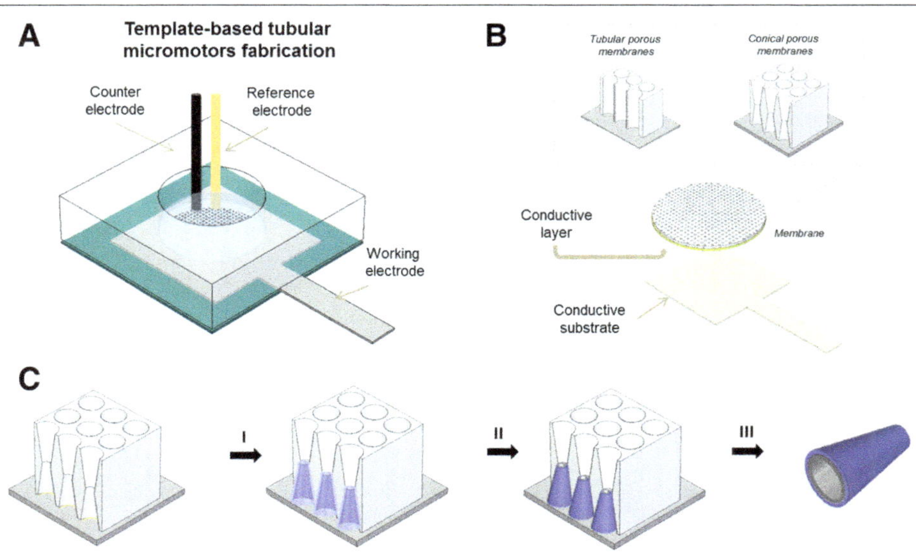

Figure 3 General schematic of an electrochemical plating cell and main steps to fabricate conical template-based tubular micromotors. **(A)** General schematic of an electrochemical plating cell and **(B)** zoom-in of the working electrode system, based on a conductive substrate in contact with the conductive face of the tubular or conical porous membrane. **(C)** Main steps followed to fabricate conical template-based tubular micromotors are deposition of the outer polymeric conductive layer (I), followed by the sequential deposition of the inner catalytic surface (II), and the final dissolution of the membrane (III) to release the micromotors.

especially important to sputter one side of the membrane with a conductive metal layer so as to obtain the proper contact between the conductive substrate and the template (Figure 3B).

The fastest and most efficient mass production procedure to obtain conical micromotors of about 8 μm in length and 1 μm in diameter is based on the previous deposition of a polyaniline (PANI) conductive polymeric layer onto polycarbonate membranes with conically shaped pores (Figure 2AI), which will be further used as the conductive layer template for the next sequential electrodeposition steps. Therefore, the sequential electrochemical deposition of solid metals (i.e., platinum) will be performed on the previous conductive layer, leading to the fabrication of perfectly conical microjets embedded into the polycarbonate membrane. After removing the previously sputtered conductive layer by hand polishing with alumina slurry, the polycarbonate membrane is dissolved and the micromotors will be released (Figure 3C) [38].

Among the different possibilities to deposit the first conductive template, some alternatives to PANI had been reported. Polypyrrole (PPy) or poly(3,4-ethylenedioxythiophene) (PEDOT) suits the requirements to obtain the outer layer (Figure 4AI) [37]. Zhao et al. had also reported a similar procedure, where the first deposited layer on the walls of the template (conductive layer template) is based on a copper layer (Figure 4AII) [39]. It is of special interest that the outer layer of the micromotors can be directly functionalized with certain functional groups inherently present in the polymer, as the carboxylic PEDOT used for on-chip immunoassays [40] or poly(3-aminophenylboronic acid) (PAPBA), which permits the recognition of yeast cells [41].

Two key parameters to properly manipulate the movement and the successful development of template-based microjets in environmental and sensing applications are their magnetic guidance and the functionalization of their outer surface. Such features are generally achieved by the integration of a magnetic layer (e.g. nickel (Ni) or iron (Fe) layer) and an external gold layer, which is obtained by sputtering a gold layer once the microjets are released from the template (covering only one side of the micromotor). In the case of polymeric outer layers, the Au layer is not required since they can be functionalized directly [40] or do not require any functionalization [41,46].

Additionally, the use of ultrasound field has recently been introduced as fast speed control method. Ultrasound induces the disruption of normal bubble evolution and ejection and thereby changes the movement of PEDOT/Ni/Pt micromotors in a sharp and reproducible manner [47].

One of the intriguing features of template-based micromotors is their size, which is in the same order of certain analytes of interest, such as bacteria or cancer cells, allowing a clear visualization of the pick-up and transport events by using optical microscopy. By convenient functionalization of the outer surface of the template-based tubular micromotors, the selective capture of *E. coli* bacteria [48] and a complete on-chip immunoassay were achieved [40]. However, some reported configurations required no functionalization to perform

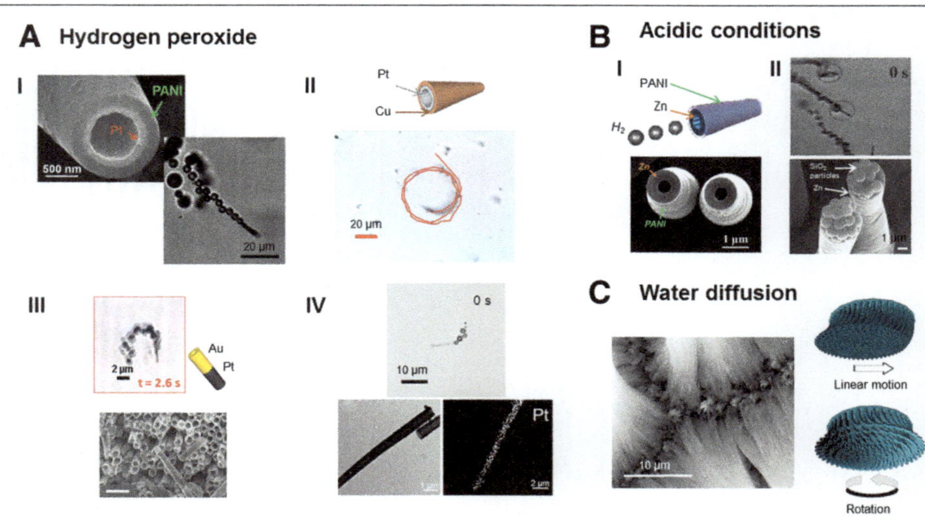

Figure 4 Template-based micromotors driven by chemical fuels. (A) Micromotors moving due to the decomposition of H_2O_2 at the platinum surface: PANI/Pt conical micromotors [38] (I), Cu/Pt conical micromotors [39] (II), tubular nanometric Au-Pt nanomotor [36] (III), and multilayer tubular nanomotors with a Pt inner surface [42] (IV). **(B)** Micromotors moving in the presence of acidic environment: PANI/Zn [43] (I) and fully loaded zinc micromotors [44] (II). **(C)** Linear and rotational movement of $(Pb_{0.25}Ba_{0.15}Sr_{0.6})TiO_3$ diffusiophoresis nanomotors in deionized water due to concentration difference energy [45].

sensing applications, for example, (PAPBA)/Ni/Pt tubular micromotors capture and release yeast cells [41], or molecularly imprinted polymer base catalytic tubular micromotors, which permitted the selective protein transport and isolation [46]. On the other hand, the enhanced fluid convection that is created around the tubular micromotors due to the bubble recoil mechanism has recently been studied and characterized by passive microsphere tracers [49], resulting in a higher mixing efficiency compared to catalytic nanowires or Janus particle motors. Such effect has been closely related to sensing applications, showing the potentiality of using PANI/Pt tubular micromotors for continuous and non-invasive mixing in microarray-based immunosensing applications [50].

The use of template-based micromotors towards environment issues had also been of special importance. Firstly, the viability of using micromotors presenting superhydrophobic surfaces to adsorb and capture oil was demonstrated, opening the door to new remediation approaches for oil-contaminated water [51]. In addition, the oxidative detoxification of organophosphate nerve agents was achieved by using catalytic micromotors in peroxide-activated contaminated samples, which leads to more efficient neutralization processes and shorter reaction times [42]. Although the most extensively used template-based micromotors are the ones which present an inner platinum surface, alternative novel configurations had been recently reported based on polyelectrolyte multilayer tubular nanomotors. Such nanoengines present an easy-to-functionalize outer surface based on biodegradable natural polysaccharides and platinum nanoparticles assembled by layer-by-layer (LbL) techniques into the template pores of their inner surface (see Figure 4AIV). Their controlled movement was achieved by integrating iron oxide (Fe_3O_4) nanoparticles, permitting their convenient motion towards HeLa cells and proving their capacity as a drug delivery system by using their outer layers for drug encapsulation [52].

Nanometric tubular micromotors had been recently reported by Zhao et al. [36] opening the doors to scale down the fabrication of such tubular microstructures to the nanoscale by means of electrodeposition techniques. In this particular case, the tubular nanomotors are fabricated using AAO membranes, obtaining tubular bimetallic nanomotors. Firstly, a sacrificial copper layer is deposited, followed by two more deposition steps: one of platinum and one of gold, resulting in a tubular nanometric motor with two defined sections (see Figure 4AIII) [36].

One of the main advantages that the template-based approach offers compared to strain engineering technology is that it is cheap and that it is a mass production method which leads to high-performance catalytic microtubular engines. However, we have to bear in mind that such micromotors will be restricted by the dimensions of the pores of the membrane, becoming specially complicated to tailor their shape and size for specific applications.

Jet propulsion: fuels and motion requirements

The motion of tubular microjets is due to the presence of certain fuels which promote catalytic reactions on the inner surface of the microjets, inducing the generation and expulsion of bubbles [2,9]. Such micromotors are moving by the bubble recoil mechanism, and their most extended configuration generally presents an inner platinum surface permitting their movement in the presence of hydrogen peroxide and certain surfactants. However, alternative fuels such as water or acids are particularly desirable nowadays, especially concerning their future potential implementation in the biomedical field for drug delivery.

Hydrogen peroxide

The first reported template-based micromotor moving in the presence of hydrogen peroxide was based on the sequential deposition of PANI, nickel, and platinum, resulting in a highly efficient tubular micromotor [38]. In spite of the toxic nature of the hydrogen peroxide fuel, at low concentrations, many feasible applications had been reported being basically focused on its use as a sensing platform for analytical purposes and its implementation towards environmental issues. Surface tension-reducing agents are crucial components of the fuel for the optimum swimming performance of microjets. Several ionic and non-ionic surfactants accelerate the velocity of microjets because they facilitate the bubble ejection from the microtube [53].

Acidic conditions

Tubular micromotors are of special interest both for industrial applications and their potential use in medicine, especially in extreme pH ranges like in the stomach environment. The first micromotor moving in acidic conditions was based on a PANI outer layer and a catalytic zinc inner layer and moved due to the thrust of hydrogen bubbles, showing a sensitive motion-based pH in extreme conditions (from −0.2 to 1.4) (see Figure 2BI) [43]. An advanced autonomous release system has been recently developed by slightly tuning the design of the PANI/Zn micromotor moving in acidic conditions. In this case, silica and gold nanoparticles were used as model cargo analogues, and they were tightly packed into the double-conical polycarbonate membrane before the zinc electrodeposition took place. After membrane dissolution, fully loaded zinc micromotors that can perform different cargo deliveries at once while they are moving in the presence of extreme acidic conditions were obtained (see Figure 4BII) [44].

Water

Although there is no tubular micromotor reported moving by the bubble recoil mechanism in pure water, alternative mechanisms have offered the possibility of moving such structures in water. On the one hand, diffusion induced the linear and rotational movement of $(Pb_{0.25}Ba_{0.15}Sr_{0.6})TiO_3$ nanotube growth in AAO membranes. While one end of the tubes was closed, the opened one was firstly filled with potassium hydroxide solution leading to its motion by means of a diffusion process once immersed in deionized water (see Figure 4C) [45]. On the other hand, tubular motors have been reported moving by the surface-induced Marangoni effect for decontaminating phenolic and azo dye compounds. Such motors were based on commercial pipette tips filled with enzyme solution and surfactant (sodium dodecyl sulfate (SDS)). The spontaneous release of the mixture once it is in the presence of the polluted water results in the movement of the motor, which at the same time releases the bioremediation agent and enhances its interaction with the pollutant due to the effective fluid convection related to its motion [54].

Recent developments have shown that Janus micromotors can be propelled in water, thanks to the water-reactive Al-Ga binary alloy [55] or magnesium [56], serving as inspiration for future alternative propulsion sources of tubular micromotors.

External stimuli

Recently, it was shown that photoactive, rolled-up titania microtubes can act as micromotors when they are exposed to ultraviolet light [57]. In comparison to the bubble-propelled microjets, the titania microtubes are thought to move by diffusiophoresis [58]. Another external energy source for tubular micromotors is ultrasound. Kagan et al. [59] demonstrated that acoustic droplet vaporization can lead to the displacement of perfluorocarbon-loaded microtubes for targeted tissue penetration and deformation.

Hybrid micromotors: towards biological power sources

The fact that powerful and durable catalytically driven microjets require toxic fuel for their actuation significantly obstructs their potential in medical and environmental applications and limits them to sensing and diagnostic procedures. There is a vast urge for finding a biocompatible fuel for the microjets which will promote the research field towards the dream of wireless microrobots that can be controlled and moved inside the human body. Using biomolecules and motile cells for the propulsion of microtubes is one first step towards biocompatible solutions for the actuation of devices on the microscale.

Hybrid microjets

The synergy between the synthetic tubular structures and biological entities gave rise to powerful micromotor configurations. When gold is integrated as an inner nanomembrane, it can likewise be functionalized with enzymes such as catalase that performs the catalytic reaction with hydrogen peroxide more efficiently than Pt. It requires lower concentrations of fuel for the locomotion of the microtubes. The first reported micromotor moving by means of H_2O_2 decomposition by catalase was based on rolled-up microtubes whose inner gold surface was functionalized by self-assembling monolayers, permitting the covalent immobilization of catalase. The micromotor moved at fairly low concentrations of H_2O_2 (1.5% *w/w*) by bubble recoil, showing a trajectory based on turns and circles due to variable viscous drag forces (Figure 5AI) [15]. The first application of catalase-based micromotors was reported by a similar configuration based on template-based micromotors (PEDOT)/Au-catalase, being used in toxicity assays for water assessment. The efficiency of the catalase enzyme was attenuated in the presence of certain toxins, inhibiting its activity and leading to lower speed and lifetime. Such behavior could be correlated to the concentration of certain pollutants of interest, leading to environmentally friendly assays which avoid the use of living aquatic organisms for water quality testing (Figure 5AII) [60].

Spermbots

Biological motors or motile cells are attractive sources for actuation of microstructures [61,64,65] and generally target the development of new tools for minimal invasive surgery, diagnostics, and drug delivery. Biological motors are attractive as driving sources because of their biocompatibility, autonomous motion, and high efficiency. Recently, a new generation of tubular micromotors was invented by harnessing motile spermatozoa as propulsion force for rolled-up microtubes (spermbots) [63]. This biocompatible approach circumvents the usage of toxic fuels and offers the motion of micromotors under physiological conditions. In the case of the spermbot, motile bovine spermatozoa are trapped inside 50-μm-long microtubes that have a diameter that is slightly larger than the head of the sperm cell. The flagellum serves as driving force that propels the microtube forward. Since the microtube consists of iron nanomembranes, it can be steered externally by magnetic fields (Figure 5BI) [63]. It was recently demonstrated that a setup of electromagnetic coils can be used in combination with an optical microscope to perform closed-loop control of the spermbots. Image recognition allows the selection of a sperm-driven tube and the targeted delivery to a selected reference point [62], being the first example of wirelessly controlling a non-magnetic cell by a magnetic field (Figure 5BII).

The spermbot has potential for biomedical applications and opens up a new field of use for micromotors: the spermbot can serve as an explorative tool in reproductive technologies to study infertility and develop

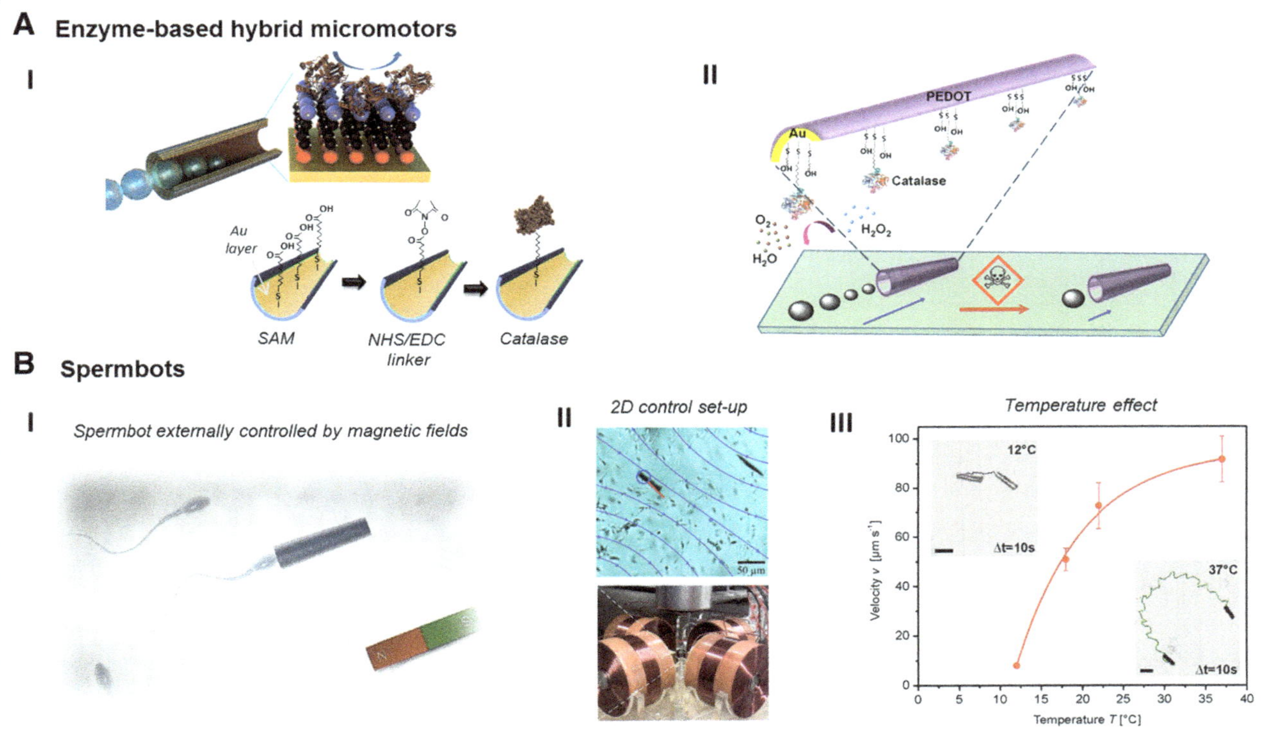

Figure 5 Enzyme-based hybrid micromotors and spermbots. (A) Enzyme-based hybrid micromotors consisting of a rolled-up microtube [15] (I) and template-based microtube [57] (II) with functionalized catalase. **(B)** The spermbots consist of a single sperm cell captured inside a rolled-up magnetic microtube [63] (I), wireless magnetic closed-loop control of spermbot [62] (II), and speed control by changing temperature (III). The insets show trajectories of a spermbot at 12°C and at 37°C over a time period of 10 s [63]. Scale bars, 50 µm.

new assisted *in vivo* reproduction techniques. However, certain requirements need to be accomplished for the application in this field: the sperm cell needs to maintain its motility and fertilization capability throughout the whole delivery process. In addition, a proper cell release mechanism needs to be applied to deliver the sperm cell to the fertilization site as well as suitable imaging and control techniques. There are several challenges that must be addressed in future studies when harnessing spermatozoa as transporters, some of which were discussed in a recent editorial [66]. One of the challenges that spermbots will be facing when used as drug delivery vehicles is the immune response of the host organism. The sperm cell as a foreign cell to the body might be engulfed by leukocytes. This process of phagocytosis will limit the lifetime of the spermbots. However, the microtube as an envelope for the sperm cell can be equipped in ways to prevent the leukocytosis in the same fashion that bacterial pathogens are able to overcome the phagocytotic engulfment and killing by appropriate blocking methods [67].

Conclusions

There have been significant advances during the last decade in the development of tubular micromotors. Both fabrication methods are well established and enable the fast and reliable mass production of microtubes. Methods to implement further functionalities in the tubes, such as the integration of enzymes or polymers, have been demonstrated. Hydrogen peroxide as fuel is currently the limiting factor for applications and further development of the catalytically driven jets. The use of alternative biocompatible fuels that are inherently present in the physiological environment is of special interest to promote their motion. Many proof-of-concept applications on the microscale demonstrate the potential of micromotors in the field of environmental science, sensing, and medical intervention. Hybrid micromotors, such as spermbots, offer an alternative approach to tubular micromotors and at the same time open up a new field of potential impact in reproduction research and in the fight against infertility by targeted delivery of sperm cells to the fertilization site.

Abbreviations

3D: three-dimensional; AAO: aluminum oxide; Fe_3O_4: iron oxide; H_2O_2: hydrogen peroxide; LbL: layer-by-layer; PANI: polyaniline; PEDOT: poly(3,4-ethylenedioxythiophene); PPy: polypyrrole; PAPBA: poly(3-aminophenylboronic acid); SAMs: self-assembling monolayers; SDS: sodium dodecyl sulfate.

Competing interests

The authors declare that they have no competing interests.

Authors' contributions

VM, MG, and OGS conceived and designed the manuscript. VM and MG wrote the draft. All authors read and approved the final manuscript.

Acknowledgements

The authors thank the Volkswagen Foundation (# 86 362) and the DFG Priority Programme (SPP 1726). VM thanks B. Koch for the fruitful discussion.

Author details

[1]Institute for Integrative Nanosciences, Leibniz Institute for Solid State and Materials Research Dresden, 01069 Dresden, Germany. [2]Material Systems for Nanoelectronics, Chemnitz University of Technology, 09126 Chemnitz, Germany. [3]Center for Advancing Electronics Dresden, Dresden University of Technology, 01187 Dresden, Germany.

References

1. Schmidt OG, Eberl K (2001) Nanotechnology: thin solid films roll up into nanotubes. Nature 410:168
2. Mei YF, Huang GS, Solovev AA, Bermúdez Ureña E, Moench I, Ding F, Reindl T, Fu RKY, Chu PK, Schmidt OG (2008) Versatile approach for integrative and functionalized tubes by strain engineering of nanomembranes on polymers. Adv Mater 20:4085–4090
3. Martinez-Cisneros CS, Sanchez S, Xi W, Schmidt OG (2014) Ultracompact three-dimensional tubular conductivity microsensors for ionic and biosensing applications. Nano Lett 14:2219–2224
4. Vervacke C, Bof Bufon CC, Thurmer DJ, Schmidt OG (2014) Three-dimensional chemical sensors based on rolled-up hybrid nanomembranes. RSC Adv 4:9723–9729
5. Sun X, Yan C, Chen Y, Si W, Deng J, Oswald S, Liu L, Schmidt OG (2014) Three-dimensionally "curved" NiO nanomembranes as ultrahigh rate capability anodes for Li-ion batteries with long cycle lifetimes. Adv Energy Mater doi:10.1002/aenm.201300912
6. Grimm D, Bof Bufon CC, Deneke C, Atkinson P, Thurmer DJ, Schäffel F, Gorantla S, Bachmatiuk A, Schmidt OG (2013) Rolled-up nanomembranes as compact 3D architectures for field effect transistors and fluidic sensing applications. Nano Lett 13:213–218
7. Koch B, Sanchez S, Schmidt CK, Swiersy A, Jackson SP, Schmidt OG (2014) Confinement and deformation of single cells and their nuclei inside size-adapted microtubes. Adv Healthcare Mater doi:10.1002/adhm.201300678
8. Xi W, Schmidt CK, Sanchez S, Gracias DH, Carazo-Salas RE, Jackson SP, Schmidt OG (2014) Rolled-up functionalized nanomembranes as three-dimensional cavities for single cell studies. Nano Lett 14:4197–4204
9. Solovev AA, Mei Y, Urena EB, Huang G, Schmidt OG (2009) Catalytic microtubular jet engines self-propelled by accumulated gas bubbles. Small 5:1688–1692
10. Solovev AA, Xi W, Gracias DH, Harazim SM, Deneke SM, Sanchez S, Schmidt OG (2012) Self-propelled nanotools. ACS Nano 6:1751–1756
11. Mönch I, Makarov D, Koseva R, Baraban L, Karnaushenko D, Kaiser C, Arndt KF, Schmidt OG (2011) Rolled-up magnetic sensor: nanomembrane architecture for in-flow detection of magnetic objects. ACS Nano 5:7436–7442
12. Li J, Zhang J, Gao W, Huang G, Di Z, Liu R, Wang J, Mei Y (2013) Dry-released nanotubes and nanoengines by particle-assisted rolling. Adv Mater 25:3715–3721
13. Harazim SM, Bolanos Quinones VA, Kiravittaya S, Sanchez S, Schmidt OG (2012) Lab-in-a-tube: on-chip integration of glass optofluidic ring resonators for label-free sensing applications. Lab Chip 12:2649–2655
14. Harazim SM (2012) Rolled-up microtubes as components for lab-on-a-chip devices. Technische Universität Chemnitz, Dissertation
15. Sanchez S, Solovev AA, Mei Y, Schmidt OG (2010) Dynamics of biocatalytic microengines mediated by variable friction control. J Am Chem Soc 132:13144–13145
16. Sanchez S, Solovev AA, Schulze S, Schmidt OG (2011) Controlled manipulation of multiple cells using catalytic microbots. Chem Commun 47:698–700
17. Kagan D, Campuzano S, Balasubramanian S, Kuralay F, Flechsig GU, Wang J (2011) Functionalized micromachines for selective and rapid isolation of nucleic acid targets from complex samples. Nano Lett 11:2083–2087
18. Balasubramanian S, Kagan D, Hu CMJ, Campuzano S, Lobo-Castañon MJ, Lim N, Kang DY, Zimmerman M, Zhang L, Wang J (2011) Micromachine-enabled capture and isolation of cancer cells in complex media. Angew Chem Int Ed 50:4161–4164
19. Orozco J, Campuzano S, Kagan D, Zhou M, Gao W, Wang J (2011) Dynamic isolation and unloading of target proteins by aptamer-modified microtransporters. Anal Chem 83:7962–7969
20. Zhao G, Sanchez S, Schmidt OG, Pumera M (2012) Micromotors with built-in compasses. Chem Commun 48:10090–10092
21. Xi W, Solovev AA, Ananth AN, Gracias DH, Sanchez S, Schmidt OG (2013) Rolled-up magnetic microdrillers: towards remotely controlled minimally invasive surgery. Nanoscale 5:1294–1297
22. Sanchez S, Solovev AA, Harazim SM, Schmidt OG (2011) Microbots swimming in the flowing streams of microfluidic channels. J Am Chem Soc 133:701–703
23. Khalil ISM, Magdanz V, Sanchez S, Schmidt OG, Misra S (2014) The control of self-propelled microjets inside a microchannel with time-varying flow rates. IEEE Trans Robot 30:49–58
24. Khalil ISM, Magdanz V, Sanchez S, Schmidt OG, Misra S (2014) Wireless magnetic-based closed-loop control of self-propelled microjets. PLoS One 9:e83053
25. Khalil ISM, Magdanz V, Sanchez S, Schmidt OG, Misra S (2013) Three-dimensional closed-loop control of self-propelled microjets. Appl Phys Lett 103:172404–4
26. Solovev AA, Smith EJ, Bof Bufon CC, Sanchez S, Schmidt OG (2011) Light-controlled propulsion of catalytic microengines. Angew Chem Int Edit 50:10875–10878
27. Sanchez S, Ananth AN, Fomin VM, Viehrig M, Schmidt OG (2011) Superfast motion of catalytic microjet engines at physiological temperature. J Am Chem Soc 133:14860–14863
28. Soler LC, Martínez-Cisneros CS, Swiersy A, Sánchez S, Schmidt OG (2013) Thermal activation of catalytic microjets in blood samples using microfluidic chips. Lab Chip 13:4299–4303
29. Soler L, Magdanz V, Fomin VM, Sanchez S, Schmidt OG (2013) Self-propelled micromotors for cleaning polluted water. ACS Nano 7:9611–9620
30. Restrepo-Pérez L, Soler L, Martínez-Cisneros CS, Sánchez S, Schmidt OG (2014) Trapping self-propelled micromotors with microfabricated chevron and heart-shaped chips. Lab Chip 14:1515–1518
31. Restrepo-Pérez L, Soler L, Martínez-Cisneros CS, Sánchez S, Schmidt OG (2014) Biofunctionalized self-propelled micromotors as an alternative on-chip concentrating system. Lab Chip 14:2914–2917
32. Mei YF, Solovev AA, Sanchez S, Schmidt OG (2011) Rolled-up nanotech on polymers: from basic perception to self-propelled catalytic microengines. Chem Soc Rev 40:2109–2119
33. Stoychev G, Zakharchenko S, Turcaud S, Dunlop JWC, Ionov L (2012) Shape-programmed folding of stimuli-responsive polymer bilayers. ACS Nano 6:3925–3934
34. Magdanz V, Stoychev G, Magdanz V, Stoychev G, Ionov L, Sanchez S, Schmidt OG (2014) Stimuli-responsive microjets with reconfigurable shape. Angew Chem Int Edit 126:2711–2715
35. Manesh KM, Cardona M, Yuan R, Clark M, Kagan D, Balasubramanian S, Wang J (2010) Template-assisted fabrication of salt-independent catalytic tubular microengines. ACS Nano 4:1799–1804
36. Zhao G, Ambrosi A, Pumera M (2013) Self-propelled nanojets via template electrodeposition. Nanoscale 5:1319–1324
37. Gao W, Sattayasamitsathit S, Uygun A, Pei A, Ponedal A, Wang J (2012) Polymer-based tubular microbots: role of composition and preparation. Nanoscale 4:2447–2453
38. Gao W, Sattayasamitsathit S, Orozco J, Wang J (2011) Highly efficient catalytic microengines: template electro-synthesis of polyaniline-platinum microtubes. J Am Chem Soc 133:11862–11864
39. Zhao G, Pumera M (2013) Concentric bimetallic microjets by electrodeposition. RSC Adv 3:3963–3966
40. Garcia M, Orozco J, Guix M, Gao W, Sattayasamitsathit S, Escarpa A, Merkoci A, Wang J (2013) Micromotor-based lab-on-chip immunoassays. Nanoscale 5:1325–1331
41. Kuralay F, Sattayasamitsathit S, Gao W, Uygun A, Katzenberg A, Wang J (2012) Self-propelled carbohydrate-sensitive microtransporters with 'built-in' boronic-acid recognition for isolating sugars and cells. J Am Chem Soc 134:15217–15220
42. Orozco J, Cheng G, Vilela D, Sattayasamitsathit S, Vazquez-Duhalt R, Valdés-Ramírez G, Pak OS, Escarpa A, Kan C, Wang J (2013) Micromotor-

based high-yielding fast oxidative detoxification of chemical threats. Angew Chem Int Ed 52:13276–13279
43. Gao W, Uygun A, Wang J (2012) Hydrogen-bubble propelled zinc-based microrockets in strongly acidic media. J Am Chem Soc 134:897–900
44. Sattayasamitsathit S, Kou H, Gao W, Thavarajah W, Kaufmann K, Zhang L, Wang J (2014) Fully-loaded micromotors for combinatorial delivery and autonomous release of cargoes. Small 10:2830–2833
45. Liu W, He R, Zhu H, Hu H, Li M, Zhao XZ (2010) Ultrafast nanotube based diffusiophoresis nanomotors. Appl Phys Lett 96:053114–053117
46. Orozco J, Cortés A, Cheng G, Sattayasamitsathit S, Gao W, Feng X, Shen Y, Wang J (2013) Molecularly imprinted polymer-based catalytic micromotors for selective protein transport. J Am Chem Soc 135:5336–5339
47. Xu T, Soto F, Gao W, Garcia-Gradilla V, Li J, Zhang X, Wang J (2014) Ultrasound-modulated bubble propulsion of chemically powered microengines. J Am Chem Soc 136:8552–8555
48. Campuzano S, Orozco J, Kagan D, Guix M, Gao W, Sattayasamitsathit S, Claussen JC, Merkoçi A, Wang J (2012) Bacterial isolation by lectin-modified microengines. Nano Lett 12:396–401
49. Orozco J, Jurado-Sánchez B, Wagner G, Gao W, Vazquez-Duhalt R, Sattayasamitsathit S, Galarnyk M, Cortés A, Saintillan D, Wang J (2014) Bubble-propelled micromotors for enhanced transport of passive tracers. Langmuir 30:5082–5087
50. Morales-Narváez E, Guix M, Medina-Sánchez M, Mayorga-Martinez CC, Merkoçi A (2014) Micromotor enhanced microarray technology for protein detection. Small 10:2542–2548
51. Guix M, Orozco J, Garcia M, Gao W, Sattayasamitsathit S, Merkoci A, Escarpa A, Wang J (2012) Superhydrophobic alkanethiol-coated microsubmarines for effective removal of oil. ACS Nano 6:4445–4451
52. Wu Z, Wu Y, He W, Lin X, Sun J, He Q (2013) Self-propelled polymer-based multilayer nanorockets for transportation and drug release. Angew Chem Int Ed 52:7000–7003
53. Simmchen J, Magdanz V, Sanchez S, Chokmaviroj S, Ruiz-Molina D, Baeza A, Schmidt OG (2014) Effect of surfactants on the performance of tubular and spherical micromotors - a comparative study. RSC Adv 4:20334–20340
54. Orozco J, Vilela D, Valdés-Ramírez G, Fedorak Y, Escarpa A, Vazquez-Duhalt R, Wang J (2014) Efficient biocatalytic degradation of pollutants by enzyme-releasing self-propelled motors. Chem Eur J 20:2866–2871
55. Gao W, Pei A, Wang J (2012) Water-driven micromotors. ACS Nano 6:8432–8438
56. Gao W, Feng X, Pei A, Gu Y, Li J, Wang J (2013) Seawater-driven magnesium based Janus micromotors for environmental remediation. Nanoscale 5:4696–4700
57. Giudicatti S, Marz SM, Soler L, Madani A, Jorgensen MR, Sanchez S, Schmidt OG (2014) Photoactive rolled-up TiO_2 microtubes: fabrication, characterization and applications. J Mater Chem C 2:5892–5901
58. Hong YY, Diaz M, Cordova-Figueroa UM, Sen A (2010) Light-driven titanium-dioxide-based reversible microfireworks and micromotor/micropump systems. Adv Funct Mater 20:1568–1576
59. Kagan D, Benchimol MJ, Claussen JC, Chuluun-Erdene E, Esener S, Wang J (2012) Acoustic droplet vaporization and propulsion of perfluorocarbon-loaded microbullets for targeted tissue penetration and deformation. Angew Chem Int Ed 51:7519–7522
60. Orozco J, García-Gradilla V, D'Agostino M, Gao W, Cortés A, Wang J (2013) Artificial enzyme-powered microfish for water-quality testing. ACS Nano 7:818–824
61. Martel S (2012) Bacterial microsystems and microrobots. Biomed Microdevices 14:1033–1045
62. Khalil ISM, Magdanz V, Sanchez S, Schmidt OG, Misra S (2014) Biocompatible, accurate, and fully autonomous: a sperm-driven micro-bio-robot. J Micro-Bio Robot 9:79–86
63. Magdanz V, Sanchez S, Schmidt OG (2013) Development of a sperm-flagella driven micro-bio-robot. Adv Mater 25:6581–6588
64. Behkam B, Sitti M (2007) Bacterial flagella-based propulsion and on/off motion control of microscale objects. Appl Phys Lett 90:23902–23904
65. Soong RK, Bachand GD, Neves HP, Olkhovets AG, Craighead HG, Montemagno CD (2000) Powering an inorganic nanodevice with a biomolecular motor. Science 290:1555–1558
66. Magdanz V, Schmidt OG (2014) Spermbots: potential impact for drug delivery and assisted reproductive technologies. Expert Opin Drug Deliv 11:1125–1129
67. Hornef MW, Wick MJ, Rhen M, Normack S (2002) Bacterial strategies for overcoming host innate and adaptive immune responses. Nat Immunol 3:1033–1040

4

Multirobot rendezvous with bearing-only or range-only measurements

Ronghao Zheng* and Dong Sun

Abstract

This paper studies distributed rendezvous strategies for multiple nonholonomic wheeled mobile robots with the aim of testing their practicality on real robots. We investigate control strategies which use just bearing-only or range-only measurements and do not need inter-robot radio communication to share the measurements. For the bearing-only case, two control laws proposed in our previous study are recalled and adapted. For the range-only case, rendezvous control laws for a two-robot system are proposed first and it is shown analytically a two-robot system achieves rendezvous globally under these control laws. Then the range-only-based control laws are extended to multirobot systems. Monte Carlo simulations indicate that a multirobot system achieves practical convergence under the range-only-based control laws. Experimental results illustrate the applicability and performance of the proposed control strategies for multiple wheeled-robot systems.

Keywords: Multirobot systems; Bearing-only measurement; Range-only measurement

Background

Recent theoretical and technological advances have spurred a broad interest to develop practical multirobot systems [1-5]. For mobile robots, navigation skill is one of their fundamental capabilities. Different navigation strategies are appropriate for different contexts. Different sensor types and sensing modules are used depending on the application scenarios. However, a problem of common interests and practical significance is how to perform tasks with less information and simpler sensors, such as using binary sensors, or using bearing-only or range-only sensors. This problem has attracted much attention in the multirobot community because an advantage of cooperation of teams of robots is that simple agents are able to perform complex tasks through mutual cooperation.

Bearing-only-based navigation is an approach to simplify the sensing system and is very useful in some cases. For example, for mobile robots equipped with single omnidirectional camera, or radar and sonar operating in passive listening mode, it is a practical requirement to design control strategies based only on measurements of bearings [6,7].

*Correspondence: rzheng3-c@my.cityu.edu.hk
Department of Mechanical and Biomedical Engineering, City University of Hong Kong, 83 Tat Chee Avenue, Hong Kong 999077, China

Range-only-based navigation is another approach and it is useful especially in those scenarios when the robots can only sense the intensity of the signals emitted by other robots, or, for example, underwater applications often use acoustic equipments to measure ranges by registering the time of flight of an echo request and reply. Generally speaking, the control design and analysis become more challenging when each robot can only measure distances to other robots without bearing information. This approach has a lot of applications, such as localization and mapping [8,9], formation control [10,11], and target enclosing [12,13]. If a robot has knowledge of its location in a coordinate frame, then under some persistent excitation condition, the robot may be able to get an estimation of the positions of other robots. This idea has been exploited in works such as [12,14]. However, in this paper we are interested in those scenarios where there is no global localization system so that absolute position information is not available.

Recently, a lot of researchers have studied the use of range-only-based technique to address the target tracking problem. This task becomes more challenging when wheeled mobile robots are used. Because of the nonholonomic constraints, wheeled mobile robots have restrictions in mobility and typically cannot be controlled by linear controllers. For example, recently, Matveev et al. [15] proposed a sliding mode control law to drive wheeled mobile robots towards a target and circumnavigate the target at a predefined distance. In [16], a two-phase switched logic-based control strategy was proposed.

Inspired by the research on bearing-only and range-only navigation, in this paper we consider another common task, i.e., the rendezvous problem in which multiple nonholonomic mobile robots are required to meet at a single point (see [17,18] and references therein). Rendezvous control is very useful in a variety of applications for multirobot systems. For example, a group of robots can be deployed to collect samples from a region, and after the task is finished, they are required to get together so that we can collect them and transport them to a new place. However, most existing rendezvous strategies require that every robot knows both bearing and range of its neighboring robots, which restricts many practical applications. In this paper, we study distributed rendezvous strategies using just either bearing-only or range-only measurements. The bearing-only-based control schemes are adapted from our previous work [19] and experimentally validated in this paper. Range-only-based control schemes are also developed. While target tracking often assumes that the pursuer is more maneuverable than the target, in the rendezvous application all robots employ a same control strategy and have same maneuver. This makes the control more difficult especially when only range information is available. The proposed control schemes are then experimentally implemented and validated on a group of wheeled mobile robots.

The work presented in this paper is an extension of our previous work reported in a conference version [20]. Although using a similar idea, the range-only controllers in this paper are redesigned. The first difference is, in [20], that to prove the convergence of the distance between the robot pair, the forward velocity assumes an infinite bound. However, in this paper, a more practical design is provided, which allows both forward and angular velocities to lie in bounded intervals. The second difference is that the convergence of the multirobot system under the range-only controllers is verified by using Monte Carlo simulations. The third difference is that the experimental platform is improved and a behavior-based collision avoidance algorithm is used, which allows us to use more robots in our experiments and allows us to test the performance of the proposed controllers in more realistic scenarios.

The rest of the paper is structured as follows. In 'Methods' section, 'Problem description' subsection introduces the problem studied in this paper; 'Bearing-only controllers' subsection presents two bearing-only-based control schemes adapted from our previous work; 'Range-only controllers' subsection proposes two range-only-based control schemes and proves their global stability for the two-robot case. These two control schemes are then naturally extended to the multirobot case. In 'Results and discussion' section, Monte Carlo simulations are then carried out to test the convergence of the range-only-based control schemes in multirobot systems. 'Results and discussion' section also discusses the experimental detail and results under the proposed control schemes. Concluding remarks are given in 'Conclusions' section.

Methods

Problem description

Assume N-wheeled mobile robots moving in a 2D obstacle-free workspace. In this paper, we model each robot as a kinematic unicycle. Then the posture of the ith robot is represented by a triple (x_i, y_i, θ_i) where (x_i, y_i) specifies the Cartesian coordinate of the center of the robot's body and θ_i gives its orientation. The motion of each robot $i \in \{1, 2, \ldots, N\}$ is governed by

$$\dot{x}_i = v_i \cos\theta_i, \tag{1a}$$

$$\dot{y}_i = v_i \sin\theta_i, \tag{1b}$$

$$\dot{\theta}_i = \omega_i. \tag{1c}$$

Here, control signals $v_i \in [-v_{\max}, v_{\max}]$ and $\omega_i \in [-\omega_{\max}, \omega_{\max}]$ are robot i's forward and angular speeds, respectively, and $v_{\max}$ and $\omega_{\max}$ (both positive) are bounds for forward and angular speeds, respectively. From (1), it can be seen that the mobile robot is subjected to the nonholonomic constraint $\dot{y}_i \cos\theta_i - \dot{x}_i \sin\theta_i = 0$.

Our goal is to design distributed control schemes (v_i, ω_i) to get all the N nonholonomic robots to congregate at a common location. With the distributed architecture, the controller of robot i only uses locally measurable information without a common reference frame, i.e., global position information is unavailable. In this paper, we consider two types of measurements. The first one is bearing-only measurement, i.e., each robot can only measure the bearing angles of the detectable robots in its local frame. The second is range-only measurement, i.e., each robot is able to measure only its distances from other robots that it can detect.

In this paper, we assume that the interaction between robots is bidirectional, i.e., if robot i can detect robot j then it can also be detected by robot j and we say that robots i and j are *neighbors*. This assumption is reasonable, for example, in the case where all the robots use omnidirectional sensors with identical parameters. We then represent the bidirectional interaction topology among robots with an undirected graph $\mathcal{G} = (V, E)$ where V is the node set with each node corresponding to each robot and E is the edge set such that $(i, j) \in E$ implies that robots i and j are neighbors. We denote the set of robot

i's neighbors as $\mathcal{N}_i$, i.e., $\mathcal{N}_i = \{j|(j,i) \in E\}$. A graph $\mathcal{G}$ is said to be connected if there is a path between every distinct nodes. If for any two distinct nodes i and j there is an edge connecting them, $\mathcal{G}$ is said to be completely connected.

Bearing-only controllers

In this subsection, we recall two bearing-only control schemes published in our previous work [19]. These two bearing-only control schemes are slightly modified here in order to take the speed bounds into account. The proofs of global convergence can be found in the original paper. We aim to examine their practical performances in this paper. We let $\rho_{ij} := \sqrt{(x_i - x_j)^2 + (y_i - y_j)^2}$ denote the distance between robots i and j. For each robot i, we define the anticlockwise angle difference between robot i's heading and the line-of-sight that would take it directly toward another robot j as the *bearing angle* of robot j in robot i's coordinate frame and denote it by α_{ij} (see Figure 1). To guarantee that α_{ij} is well-defined, it is assumed that for two distinct robots i and j, $\rho_{ij} > 0$ for all $t \geq 0$. If $\rho_{ij} \leq \varepsilon$, where ε is an arbitrary small positive number, then robots i and j achieve rendezvous and merge to a single robot, either i or j. Here, ε can be considered as a measure of the physical size of the robot.

We adapt the following control scheme from [19]. For each robot i,

Controller 1:

$$\begin{bmatrix} v_i \\ \omega_i \end{bmatrix} = \begin{cases} \frac{\sum_{j\in\mathcal{N}_i}}{|\mathcal{N}_i|} \begin{bmatrix} v_{\max}\cos\alpha_{ij} \\ \omega_{\max}\sin\alpha_{ij} \end{bmatrix}, & \text{if } \mathcal{N}_i \neq \emptyset, \\ [0,0]^T, & \text{otherwise.} \end{cases} \quad (2)$$

Here $|\mathcal{N}_i|$ denotes the cardinality of the neighbor set $\mathcal{N}_i$. When $\mathcal{G}$ is connected, we can see that $|\mathcal{N}_i| \geq 1$ for all i.

The convergence of a N-robot system under Controller 1 can be proved by using a Lyapunov-based method and is formally stated by the following result.

Theorem 1 ([19], Theorem 1). *A system of N mobile robots described by (1) rendezvous under Controller 1 provided that $\mathcal{G}$ is connected. Moreover, the energy function $V := \frac{1}{2}\sum_{i=1}^{N}\sum_{j\in\mathcal{N}_i}\rho_{ij}^2$ keeps decreasing until the robots achieve rendezvous.*

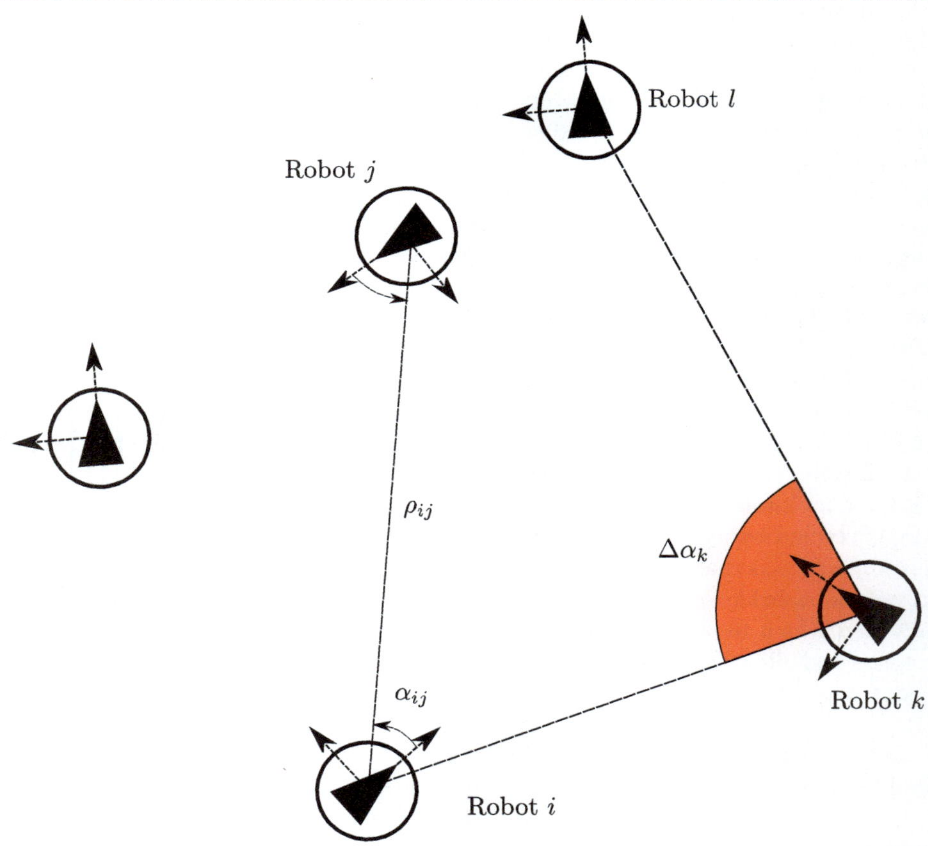

Figure 1 An illustration of ρ_{ij}, α_{ij} and the smallest circular sector of robot k. For robot k, α_{k+} and α_{k-} correspond to α_{ki} and α_{kl} in this figure.

When the interaction topology $\mathcal{G}$ between robots are completely connected, another bearing-only control scheme is proposed in [19] and adapted here. For each robot i,

Controller 2:

$$\begin{bmatrix} v_i \\ \omega_i \end{bmatrix} = \begin{cases} \frac{1}{2}\begin{bmatrix} v_{\max}(\cos\alpha_{i+} + \cos\alpha_{i-}) \\ \omega_{\max}(\sin\alpha_{i+} + \sin\alpha_{i-}) \end{bmatrix}, & \text{if } \Delta\alpha_i \in [0, \pi], \\ [0, 0]^T, & \text{otherwise.} \end{cases} \tag{3}$$

Here, $\Delta\alpha_i \in [0, 2\pi)$ is defined to be the central angle of the smallest circular sector of robot i which contains all the vectors $(\cos\alpha_{ij}, \sin\alpha_{ij})$ for $j \in \mathcal{N}_i$, and α_{i+} and α_{i-} are defined to the bearing angles which correspond to the two radii enclosing robot i's circular sector (see Figure 1).

The convergence of a N-robot system under Controller 2 is formally stated by the following theorem.

Theorem 2 ([19], Theorem 4). *A system of N mobile robots described by (1) rendezvous under Controller 2 provided that $\mathcal{G}$ is completely connected, and the perimeter of the convex hull defined by the positions of robots keeps decreasing until the robots achieve rendezvous.*

The idea behind Controller 2 can be explained as follows. When $\Delta\alpha_i \leq \pi$, robot i is located at a vertex or on an edge of the convex hull defined by the positions of all robots. In that case, a robot tries to shorten the distances to its neighbors which are also at the vertices or on the edge; otherwise, it just keeps stationary. In this way, the perimeter of the convex hull will shrink to a point and the robots can meet each other at that point.

Remark 1. Using the pseudo-linearization technique, it can be proven that under Controllers 1 and 2, the meeting point is located in a bounded region which is determined by the robots' initial postures [19].

Range-only controllers

In this subsection, we propose two range-only control schemes to drive wheeled mobile robots to rendezvous. We first investigate the two-robot case and prove its global convergence. The control schemes are then generalized to deal with the N-robot case.

To facilitate the analysis, we consider the relative coordinates ρ_{12}, α_{12}, and α_{21} between two robots 1 and 2. After some algebraic manipulation, we get the following equations:

$$\dot{\rho}_{12} = -v_1\cos\alpha_{12} - v_2\cos\alpha_{21}, \tag{4a}$$

$$\dot{\alpha}_{12} = -\omega_1 + \frac{v_1\sin\alpha_{12} + v_2\sin\alpha_{21}}{\rho_{12}}, \tag{4b}$$

$$\dot{\alpha}_{21} = -\omega_2 + \frac{v_1\sin\alpha_{12} + v_2\sin\alpha_{21}}{\rho_{12}}. \tag{4c}$$

Note that (4) is valid when $\rho_{12} \neq 0$.

Both robots are assumed to have access to only the current distance $\rho_{12}(t)$ and calculate its derivative $\dot{\rho}_{12}(t)$ by using a memory unit. The bearing information, i.e., α_{12} and α_{21}, is not available. In the following, we propose two control schemes based on $\rho_{12}(t)$ and $\dot{\rho}_{12}(t)$, which drive two robots to rendezvous, i.e., $\rho_{12}(t) \to 0$ as $t \to +\infty$.

The first range-only control scheme we propose is

Controller 3:

$$v_i = k_v\chi\{\rho(t), \rho_v\}, \tag{5a}$$

$$\omega_i = \begin{cases} k_\omega\chi\{\rho(t), \rho_\omega\}\omega_s/\rho(t), & \text{if } \dot{\rho}(t) < 0, \\ k_\omega\chi\{\rho(t), \rho_\omega\}\omega_f/\rho(t), & \text{if } \dot{\rho}(t) \geq 0, \end{cases} \tag{5b}$$

with $\omega_f > \omega_s > 0$. Here $\chi(x, \bar{x})$ is a saturation function

$$\chi(x, \bar{x}) := \begin{cases} x, & \text{if } x \leq \bar{x}, \\ \bar{x}, & \text{if } x > \bar{x}. \end{cases}$$

In our case, the parameter x of $f(x, \bar{x})$ is nonnegative and $\bar{x}$ is positive.

The rationale of Controller 3 can be explained by observing that when a robot finds itself approaching another robot, i.e., it is moving in the right direction, it rotates slowly (at the speed ω_s) to try to keep on that direction as long as possible; otherwise, it rotates fast (at the speed ω_f) to try to get to the right direction.

Under (5), $v_1 = v_2 := v$ and $\omega_1 = \omega_2 := \omega$. Define $\rho := \rho_{12}$, $\alpha := (\alpha_{12}+\alpha_{21})/2$ and $\beta := (\alpha_{12}-\alpha_{21})/2$. Then (4) can be rewritten as

$$\dot{\rho} = -2v\cos\alpha\cos\beta, \tag{6a}$$

$$\dot{\alpha} = -\omega + \frac{2v\sin\alpha\cos\beta}{\rho}, \tag{6b}$$

$$\dot{\beta} = 0. \tag{6c}$$

From (6c), we can see that β is a constant under Controller 3.

Theorem 3. *Design Controller 3 such that*

- $k_v\rho_v \leq v_{\max}$, $\rho_v = \rho_\omega$, *and*
- $k_\omega\omega_f \leq \omega_{\max}$, $\omega_f > \omega_s > 2k_v/k_\omega$.

A two-robot system described by (1) rendezvous under Controller 3 provided $\cos\beta \neq 0$.

Proof. It is clear that when $k_v\rho_v \leq v_{\max}$ and $k_\omega\omega_f \leq \omega_{\max}$, both v and ω are in their admissible intervals.

Since $\rho_v = \rho_\omega$ and $\omega_f > \omega_s > 2k_v/k_\omega$, then according to (5a) we have $\omega > 2v\sin\alpha\cos\beta/\rho$. Hence, $\dot{\alpha} < 0$ in (6b). Define t_k to be the instant when $\alpha(t_k) = -2k\pi + \pi/2$ for some integer k.

(i) Case I: $\cos\beta > 0$. According to (6),

$$\begin{aligned}\int_{\rho(t_k)}^{\rho(t_{k+1})}\frac{d\rho}{\rho} &= \int_{-2k\pi+\pi/2}^{-2k\pi-3\pi/2}\frac{2v\cos\alpha\cos\beta}{\rho\omega-2v\sin\alpha\cos\beta}d\alpha\\ &= -\int_{-2k\pi-3\pi/2}^{-2k\pi+\pi/2}\frac{2v\cos\alpha\cos\beta}{\rho\omega-2v\sin\alpha\cos\beta}d\alpha\\ &= -\int_{-2k\pi-3\pi/2}^{-2k\pi-\pi/2}\frac{\cos\alpha\cos\beta}{\frac{\rho\omega}{2v}-\sin\alpha\cos\beta}d\alpha\\ &\quad -\int_{-2k\pi-\pi/2}^{-2k\pi+\pi/2}\frac{\cos\alpha\cos\beta}{\frac{\rho\omega}{2v}-\sin\alpha\cos\beta}d\alpha.\end{aligned}$$

Because $v \geq 0$, when $\dot{\rho}(t) < 0$, we have $\cos\alpha > 0$, i.e., $\alpha \in (-2k\pi - \pi/2, -2k\pi + \pi/2)$. Similarly, when $\dot{\rho}(t) \geq 0$, we have $\cos\alpha \leq 0$, i.e., $\alpha \in [-2k\pi - 3\pi/2, -2k\pi - \pi/2]$. Hence,

$$\begin{aligned}\int_{\rho(t_k)}^{\rho(t_{k+1})}\frac{d\rho}{\rho} &= -\int_{-3\pi/2}^{-\pi/2}\frac{\cos\alpha\cos\beta d\alpha}{\frac{k_\omega\omega_f}{2k_v}-\sin\alpha\cos\beta}\\ &\quad -\int_{-\pi/2}^{\pi/2}\frac{\cos\alpha\cos\beta d\alpha}{\frac{k_\omega\omega_s}{2k_v}-\sin\alpha\cos\beta}\\ &= \ln\left(\frac{k_\omega\omega_f}{2k_v}-\sin\alpha\cos\beta\right)\Bigg|_{\alpha=-3\pi/2}^{-\pi/2}\\ &\quad + \ln\left(\frac{k_\omega\omega_s}{2k_v}-\sin\alpha\cos\beta\right)\Bigg|_{\alpha=-\pi/2}^{\pi/2}\\ &= \ln\frac{\left(\frac{k_\omega\omega_f}{2k_v}+\cos\beta\right)\left(\frac{k_\omega\omega_s}{2k_v}-\cos\beta\right)}{\left(\frac{k_\omega\omega_f}{2k_v}-\cos\beta\right)\left(\frac{k_\omega\omega_s}{2k_v}+\cos\beta\right)}.\end{aligned}$$

Because $0 < (\frac{k_\omega\omega_f}{2k_v} + \cos\beta)(\frac{k_\omega\omega_s}{2k_v} - \cos\beta) < (\frac{k_\omega\omega_f}{2k_v} - \cos\beta)(\frac{k_\omega\omega_s}{2k_v} + \cos\beta)$, we get $\ln\rho(t_{k+1}) - \ln\rho(t_k) < 0$, i.e., $\rho(t_{k+1}) < \rho(t_k)$; that is, $\rho(t)$ decreases in every period in which $\alpha(t)$ decreases by 2π. Because $\alpha(t)$ keeps decreasing, $\rho(t)$ will eventually approach 0.

(ii) Case II: $\cos\beta < 0$. The proof is similar to (i).

□

Using a similar idea as Controller 3, the second range-only controller is proposed. In Controller 4, instead of applying a switch controller on the angular velocity ω_i, we apply the switching control to the forward velocity v_i. Our second range-only controller is

Controller 4:

$$v_i = \begin{cases} k_s\chi\{\rho(t), \rho_v\}, & \text{if } \dot{\rho}(t) < 0,\\ k_f\chi\{\rho(t), \rho_v\}, & \text{if } \dot{\rho}(t) \geq 0,\end{cases} \tag{7a}$$

$$\omega_i = k_\omega\chi\{\rho(t), \rho_\omega\}\bar{\omega}/\rho(t), \tag{7b}$$

where $k_f > k_s > 0$.

The convergence of ρ under Controller 4 is stated in the following result.

Theorem 4. *Design Controller 4 such that*

- $k_f\rho_v \leq v_{\max}$, $\rho_v = \rho_\omega$, and
- $k_\omega\bar{\omega} \leq \omega_{\max}$, $k_\omega\bar{\omega} > 2k_f$.

A two-robot system described by (1) rendezvous under Controller 4 provided $\cos\beta \neq 0$.

Proof. The proof is similar to that of Theorem 3. Since $k_\omega\bar{\omega} > 2k_f$, thus $\dot{\alpha} < 0$.

(i) Case I: $\cos\beta > 0$. According to (6) and (7),

$$\begin{aligned}\int_{\rho(t_k)}^{\rho(t_{k+1})}\frac{d\rho}{\rho} &= -\int_{-3\pi/2}^{-\pi/2}\frac{\cos\alpha\cos\beta d\alpha}{\frac{k_\omega\bar{\omega}}{2k_s}-\sin\alpha\cos\beta}\\ &\quad -\int_{-\pi/2}^{\pi/2}\frac{\cos\alpha\cos\beta d\alpha}{\frac{k_\omega\bar{\omega}}{2k_f}-\sin\alpha\cos\beta}\\ &= \ln\left(\frac{k_\omega\bar{\omega}}{2k_s}-\sin\alpha\cos\beta\right)\Bigg|_{\alpha=-3\pi/2}^{-\pi/2}\\ &\quad + \ln\left(\frac{k_\omega\bar{\omega}}{2k_f}-\sin\alpha\cos\beta\right)\Bigg|_{\alpha=-\pi/2}^{\pi/2}\\ &= \ln\frac{\left(\frac{k_\omega\bar{\omega}}{2k_s}+\cos\beta\right)\left(\frac{k_\omega\bar{\omega}}{2k_f}-\cos\beta\right)}{\left(\frac{k_\omega\bar{\omega}}{2k_s}-\cos\beta\right)\left(\frac{k_\omega\bar{\omega}}{2k_f}+\cos\beta\right)} < 0,\end{aligned}$$

thus, $\rho(t_{k+1}) < \rho(t_k)$.

(ii) Case II: $\cos\beta < 0$. The proof is similar to (i).

□

To make Controllers 3 and 4 work, we need to avoid the case of $\cos\beta = 0$. One way to do this is to introduce some random behavior, e.g., stop moving for a while, into the controllers if the robots detect that ρ_{ij} keeps constant. The introduction of random behavior might also be helpful when the system suffers slow convergence, i.e., $|\cos\beta|$ is too small.

We generalize both Controllers 3 and 4 to the N-robot case. We call them Controllers 3e and 4e in this paper. To simplify the controller design of ω_i, here we set $\rho_\omega = +\infty$.

Controller 3e:

$$v_i = k_v\chi\left[\sum_{j\in\mathcal{N}_i}\rho_{ij}, \rho_v\right],$$

$$\omega_i = \begin{cases} k_\omega\omega_s, & \text{if } \sum_{j\in\mathcal{N}_i}(\rho_{ij}\dot{\rho}_{ij}) < 0,\\ k_\omega\omega_f, & \text{if } \sum_{j\in\mathcal{N}_i}(\rho_{ij}\dot{\rho}_{ij}) \geq 0,\end{cases}$$

where $\omega_f > \omega_s > 0$.

Controller 4e:

$$v_i = \begin{cases} k_s \chi \left[\sum_{j \in \mathcal{N}_i} \rho_{ij}, \rho_v \right], & \text{if } \sum_{j \in \mathcal{N}_i} (\rho_{ij} \dot{\rho}_{ij}) < 0, \\ k_f \chi \left[\sum_{j \in \mathcal{N}_i} \rho_{ij}, \rho_v \right], & \text{if } \sum_{j \in \mathcal{N}_i} (\rho_{ij} \dot{\rho}_{ij}) \geq 0, \end{cases}$$

$$\omega_i = k_\omega \bar{\omega},$$

where $k_f > k_s > 0$.

Work is still ongoing to discover the convergence results for the N-robot case. However, in 'Results and discussion' section, it is shown by means of both Monte Carlo simulations and experiments that these two range-only-based control schemes perform well provided that the interaction topology $\mathcal{G}$ among the robots is connected.

Results and discussion

Monte Carlo simulations

The widely used Monte Carlo simulations were utilized to infer the convergence of the N-robot system under both Controllers 3e and 4e with connected interaction topology. The readers are referred to [21] for more details of Monte Carlo methods. In the simulations, for team size $N = 3, 4, \ldots, 10$, each N is evaluated 100 times. All the initial configurations (positions and orientations) of the mobile robots and the control parameters for the controllers are generated randomly from uniform distributions over their domains respectively (see Table 1). Without loss of generality, we set $k_v = k_\omega = 1$ in the simulations. In each evaluation, the initial positions of the robots $(x_i(0), y_i(0))$ are chosen uniformly from a square with side W and the initial orientations $\theta_i(0)$ are chosen from $[0, 2\pi)$ uniformly.

The topology $\mathcal{G}$ is generated in the following way: If the distance between a pair of robots is less than d_c, these two robots are connected to each other. We then test the connectivity of the resulting $\mathcal{G}$, if $\mathcal{G}$ is connected, e.g., the corresponding Laplacian has a rank of $N - 1$, then we continue the evaluation; otherwise, this evaluation is abandoned and a new evaluation is carried out. To avoid using an over-connected $\mathcal{G}$, we chose $d_c = \sqrt{W^2/N}$ in the simulations. In each evaluation, we record the trajectories of the robots during 0 to 200 s. The sampling time T_s is set to be 0.1 s. It is tempting to chose a large $\omega_{\max}$, which causes no problem for a continuous-time system. However, a large $\omega_{\max}$ can cause a discrete-time realization of the system to be unstable. We found that keeping $\omega_{\max} < 10 \times 2\pi/T_s$ provides enough stability margin.

The outcome of each evaluation is the normalized average distance $d(t)$ to the centroid of the robot team, i.e., $d(t) = \frac{\sum_{i=1}^N \|(x_i(t)-\bar{x}(t), y_i(t)-\bar{y}(t))\|}{\sum_{i=1}^N \|(x_i(0)-\bar{x}(0), y_i(0)-\bar{y}(0))\|}$, where $\bar{x}(t) = \sum_{i=1}^N x_i(t)/N$ and $\bar{y}(t) = \sum_{i=1}^N y_i(t)/N$. We define the ratio of convergence as $\sigma = d(200\text{ s})$.

For the Controller 3e, we recorded that the average ratio of convergence is 3.0561% and $\max \sigma = 58.395\%$. For the Controller 4e, we recorded that the average ratio of convergence is 24.683% and $\max \sigma = 83.808\%$. Both results imply the Controllers 3e and 4e lead to practical convergence.

Table 1 Parameters used in the Monte Carlo simulations

Parameter	Value
N	$3, \ldots, 10$
k_v	1
k_ω	1
ρ_v	$v_{\max}$
W	$[10, 100]$
$x_i(0)$	$[0, W]$
$y_i(0)$	$[0, W]$
$\theta_i(0)$	$[0, 2\pi)$
d_c	$\sqrt{W^2/N}$
$v_{\max}$	$[0.9, 1.1]$
$\omega_{\max}$	$[5, 6]$
ω_s	$(2v_{\max}, 3]$
ω_f	$(\omega_s + 3, \omega_{\max}]$
$\bar{\omega}$	$(2v_{\max}, 3]$
k_s	$[0.4, 0.5]$
k_f	$[0.8, 1.0]$

Experimental platform

The proposed control schemes were implemented and tested for the case of five e-puck robots [22] shown in Figure 2. The experiments were conducted in a lab

Figure 2 Experimental platform.

environment. Because the sensing capability of the e-puck robot is very limited, in the current implementation, an overhead camera is used to track the absolute postures of the robots. The information from the tracking system is then used to produce the local bearing or range measurements, which are utilized in order to mimic local and distributed implementation of the proposed control schemes.

Collision avoidance

In the theoretical analysis, the robots are treated as moving points. However, real robot is never a point, and therefore collision avoidance cannot be negligible when several robots work together in a region. Collision avoidance itself, especially for nonholonomic mobile robots, is a challenging problem (see [23] and references therein). In our experiments, a behavior-based algorithm is adopted, which is straightforward and computationally effective.

It should be pointed out that using only bearing or range information is not enough to avoid collision. We need to assume that within a certain sensing zone $Q^{i} := \{j|\rho_{ij} \leq d_s, j \neq i\}$, a robot can detect both the bearing and range information, i.e., α_{ij} and ρ_{ij}, about other robots (see Figure 3). This can be done, for example, by using short-range infrared proximity sensors mounted around the body of the mobile robot. The forward speed v_i is multiplied by a variable m_i and become the new control signal

$$\tilde{v}_i = m_i v_i,$$

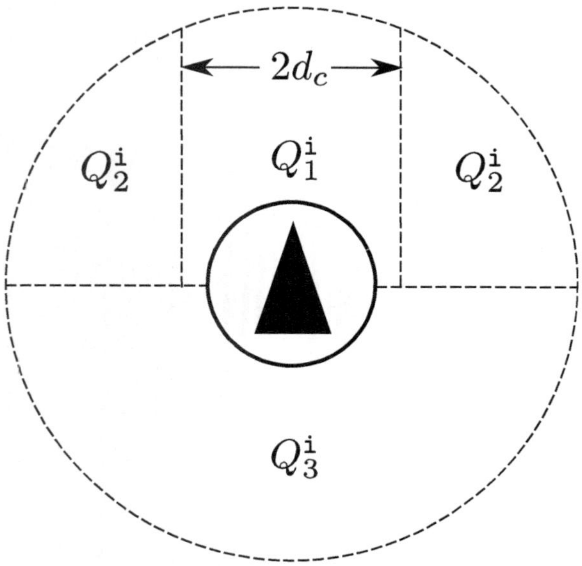

Figure 3 The partition of Q_1^i, Q_2^i, and Q_3^i when $v_i > 0$.

where

$$m_i = \begin{cases} 0, & \text{if } Q_1^i \neq \emptyset, \\ 0, & \text{if } Q_2^i \neq \emptyset \text{ and } |v_i| < |v_j|, \exists j \in Q_2^i, \\ 2, & \text{if } Q_1^i = \emptyset, Q_2^i \neq \emptyset \text{ and } |v_i| > |v_j|, \forall j \in Q_2^i, \\ 2, & \text{if } Q_1^i = \emptyset, Q_2^i = \emptyset, \text{ and } Q_3^i \neq \emptyset. \end{cases}$$

The partitions Q_1^i, Q_2^i, and Q_3^i are formally defined as

$$Q_1^i := \{j|\cos\alpha_{ij} \cdot v_i \geq 0 \wedge \rho_{ij}|\sin\alpha_{ij}| \leq d_c\} \cap Q^i,$$
$$Q_2^i := \{j|\cos\alpha_{ij} \cdot v_i \geq 0 \wedge \rho_{ij}|\sin\alpha_{ij}| > d_c\} \cap Q^i,$$
$$Q_3^i := \{j|\cos\alpha_{ij} \cdot v_i < 0\} \cap Q^i.$$

Notice that $\tilde{v}_i$ may not be achieved due to saturation of the actuator.

Since the working area is bounded, the robots should also avoid collision with the boundaries and keep inside the working area. To do that, once detecting the boundaries, a robot stops moving forward and rotates at the maximal angular speed for a short time span until its forward direction (the direction of v_i) points into the bounded area again.

Experimental results

A variety of experiments are conducted using discrete-time versions of the proposed controllers. The sampling time T_s is set to be 0.1 s. The parameters used in the experiments are given in Table 2. In the experiments, the robots are considered to achieve rendezvous if there is a path connecting all the robots and each edge of the path is smaller than $d_{\min}$. Despite the time delays caused by image processing and wireless communication and the inconsistent orientation values when the vision system may fail to identity the orientation of a robot

Table 2 Parameters used in the experiments

Parameter	Value
$v_{\max}$	6 cm/s
$\omega_{\max}$	2.43 rad/s
d_s	15 cm
d_c	9 cm
$d_{\min}$	12 cm
k_v	1
k_ω	1
ρ_v	6 cm/s
ω_s	0.63 rad/s
ω_f	1.94 rad/s
$\bar{\omega}$	1.46 rad/s
k_s	0.3
k_f	1.0

during a certain time period, the experimental outcome is very encouraging and indicates the robustness of the controllers.

Controller 1

Figure 4 shows the connected interaction topology $\mathcal{G}$ among the robots we used in our experiments.

Figure 5 shows the experimental result of the rendezvous under Controller 1 captured by the overhead camera. In Figure 5a, the initial postures of the robots are shown. In Figure 5b,c,d the actual trajectories of the robots, which are represented by sequences of colored dots, and their final postures at different time instants are also shown. Figure 6 confirms Theorem 1 that the energy function $V := \frac{1}{2}\sum_{i=1}^{N}\sum_{j\in\mathcal{N}_i}\rho_{ij}^2$ decays all the time until the robots rendezvous.

In our program, the routines run periodically based on a Windows timer. Sometimes the routines may run out of the time slot allocated then the positions of the robots cannot be recorded. This explains the missing parts of the trajectories. In analysis, for example, calculating the evolution of the energy function, we recover the missing parts of the trajectories through linear interpolation.

Controller 2

Figure 7 shows the robots rendezvous under Controller 2. The perimeter L of the convex polygon defined by the positions of robots is plotted as a function of time in Figure 8. Figure 8 validates Theorem 2 that the perimeter decays all the time until the robots meet.

Controller 3e

To apply Controllers 3e and 4e, $\dot{\rho}_{ij}(k)$ is approximated by $[\rho_{ij}(k) - \rho_{ij}(k-1)]/T_s$ where T_s is the sampling period. The measurement error of ρ_{ij} may introduce a rapid switching in the control signal. To prevent that, we use a simple filter $\rho_{ij}(k) = \sum_{l=0}^{n} a_l\hat{\rho}_{ij}(k-l)$ to reduce the measurement error effect. In our experiment, we choose $n = 2$ and $a_0 = 0.8, a_1 = a_2 = 0.1$.

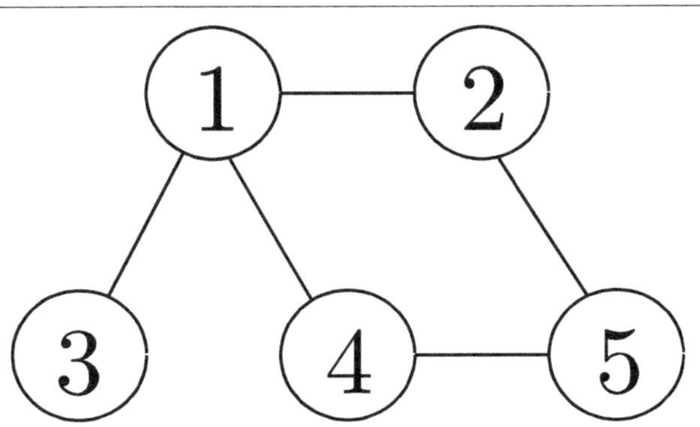

Figure 4 Interaction topology $\mathcal{G}$ among five robots.

We test the performance of Controller 3e under the connected topology of Figure 4. The result is presented in Figure 9. The evolution of the average distances to the group centroid is given in Figure 10.

Controller 4e

For the Controller 4e, we also use the interaction topology in Figure 4. The experimental result of rendezvous under Controller 4e is given in Figures 11 and 12.

It is not a surprising result that Controllers 3e and 4e take longer time to rendezvous than Controllers 1 and 2 do because Controllers 3e and 4e use only scalar information while Controllers 1 and 2 use vector information. By comparing the performances of Controllers 3e and 4e, it can be seen that a switching control of the angular speed (Controller 3e) performs better than the one of the forward speed (Controller 4e). By comparing the experimental results, those shown in Figures 10 and 12 for example, we find that Controller 3e reduces the number of oscillations of ρ_{ij} so it leads to faster convergence. This result also coincides with the observation from the Monte Carlo simulations.

Conclusions

In this paper, we study control schemes for driving a group of wheeled robots with nonholonomic constraints to a common location. The proposed control schemes use only the measurements of local bearing angles or only the measurements of distances among the robots. Our purpose is to examine whether the theoretical results obtained for bearing-only and range-only control schemes could be applied in practice to a real multirobot system. To this end, experiments are conducted on a team of e-puck robots. Given that there are unmodeled dynamic delays in the system due to sensing and information processing and the switching of controllers to deal with collision avoidance which are not accounted for in the theoretical analysis, the presented results are very positive.

However, these results are still preliminary. Further research will include developing a more sophisticated method of collision avoidance and implementing the proposed control schemes on more realistic scenarios, such as direct and dynamic interaction topology among mobile robots. Another research topic is to consider sensors with limited field-of-view and bounded range. We also plan to implement the proposed control schemes on mobile robots which have self-localization capability, such as those used in [3,5], and develop the local localization technique for multirobot systems [24,25]. There are still many open questions in regards to the convergence of a N-robot system. Monte Carlo simulations indicate that Controllers 3e and 4e achieve practical convergence, but a formal proof of this assertion remains an open problem.

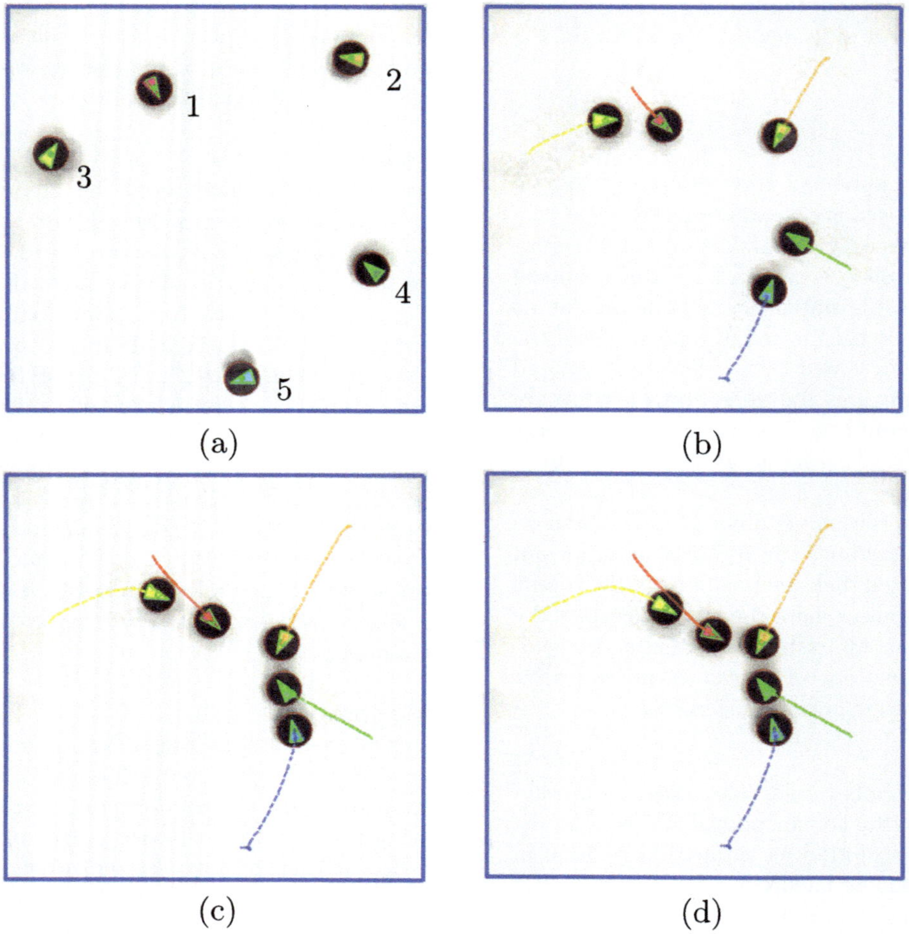

Figure 5 Rendezvous under Controller 1. **(a)** $t = 0$ s. **(b)** $t = 5$ s. **(c)** $t = 10$ s. **(d)** $t = 14$ s.

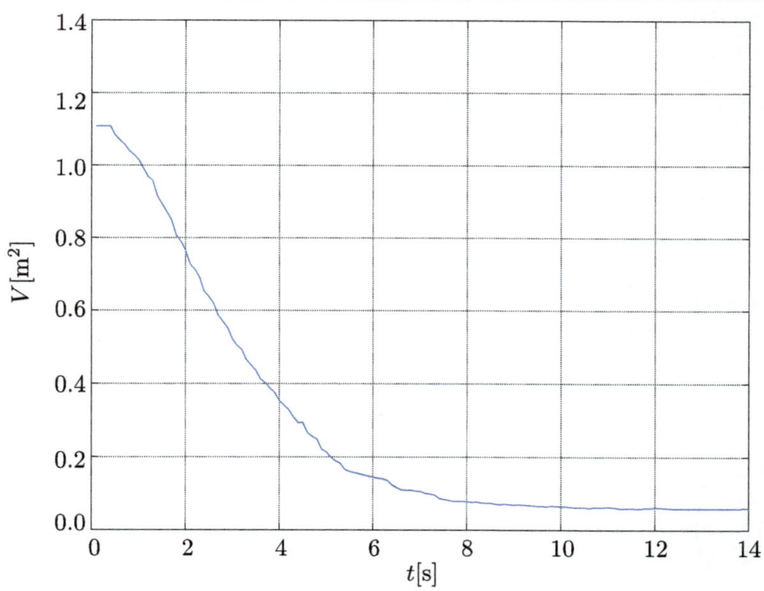

Figure 6 Evolution of the energy function $V(t)$. Here $V(t) := \frac{1}{2}\sum_{i=1}^{N}\sum_{j\in\mathcal{N}_i}\rho_{ij}^2(t)$.

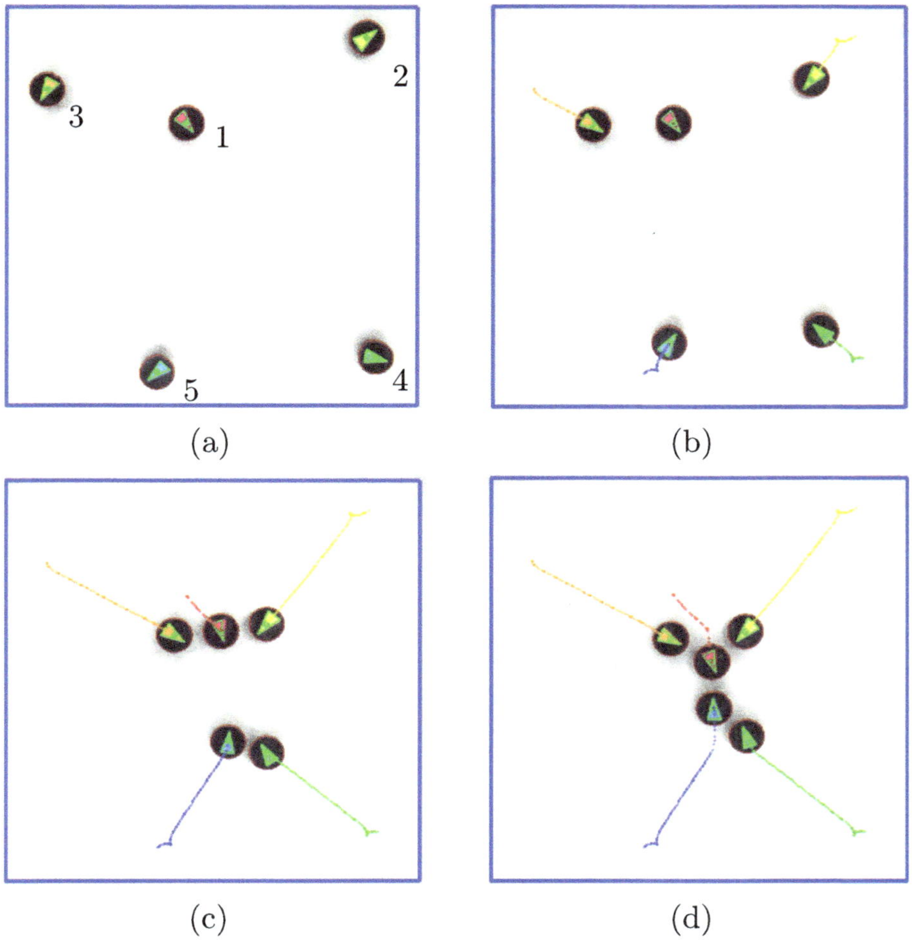

Figure 7 Rendezvous under Controller 2. (a) $t = 0$ s. **(b)** $t = 5$ s. **(c)** $t = 11$ s. **(d)** $t = 13$ s.

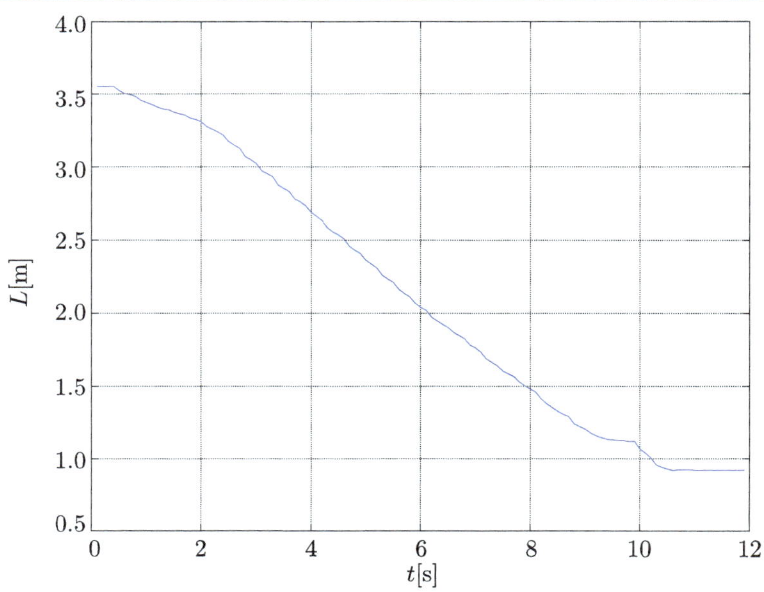

Figure 8 Evolution of the perimeter $L(t)$.

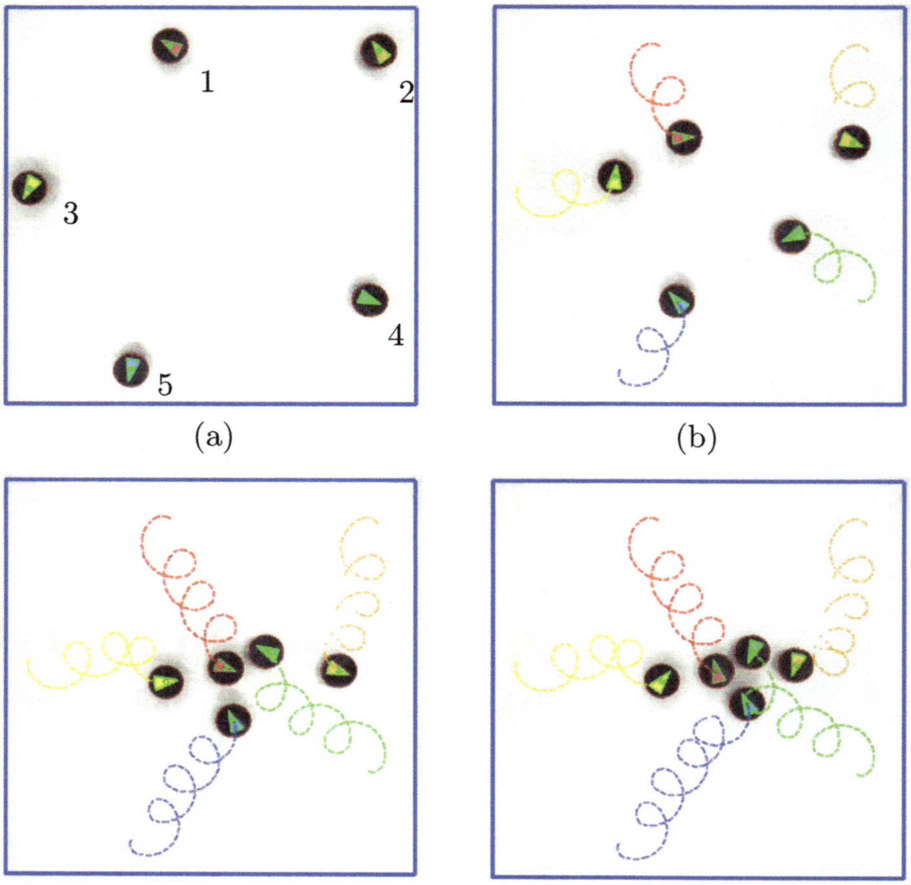

Figure 9 Rendezvous under Controller 3e. (a) $t = 0$ s. **(b)** $t = 5$ s. **(c)** $t = 11$ s. **(d)** $t = 13$ s.

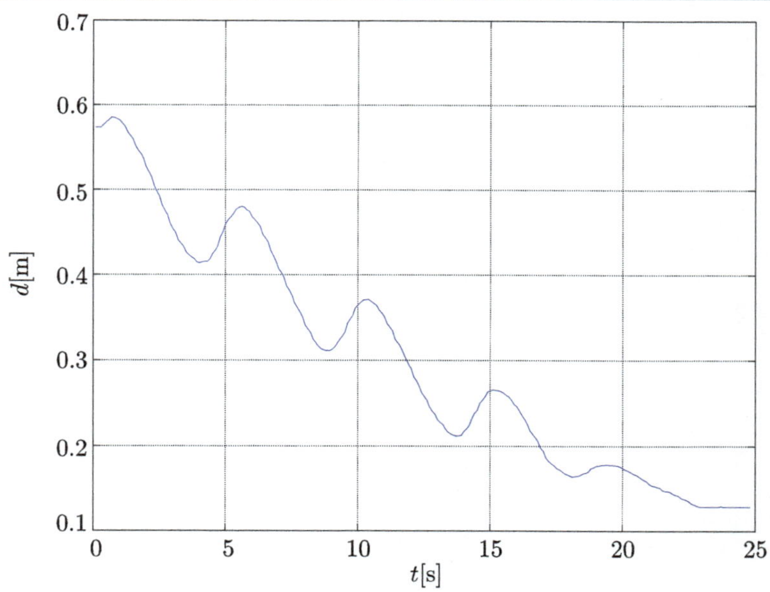

Figure 10 Evaluation of the average distance to group centroid $d(t)$.

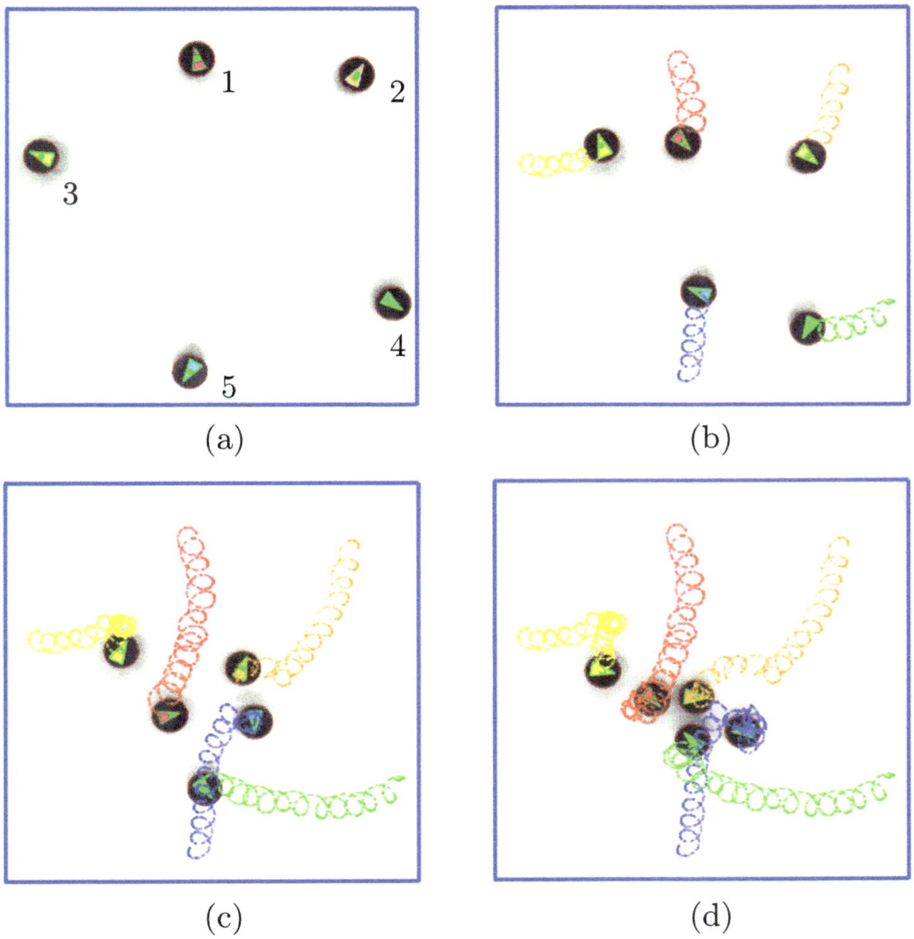

Figure 11 Rendezvous under Controller 4e. (a) $t = 0$ s. **(b)** $t = 20$ s. **(c)** $t = 50$ s. **(d)** $t = 73$ s.

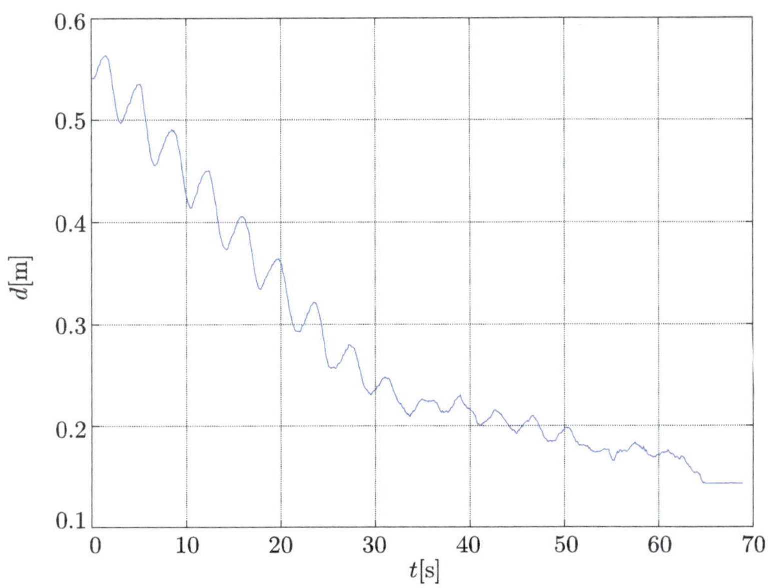

Figure 12 Evaluation of the average distance to group centroid $d(t)$.

Competing interests

The authors declare that they have no competing interests.

Authors' contributions

RZ and DS designed the study, developed the methodology, collected the data, performed the analysis, and wrote the manuscript. Both authors read and approved the final manuscript.

Acknowledgements

The work of RZ is supported through the Hong Kong PhD Fellowship Scheme. The work is also supported by a grant from the Research Grants Council, Hong Kong Special Administrative Region, China, under Grant CityU 119612 and a Collaborative Research Fund with Project no. CUHK6/CRF/13G.

References

1. Marshall JA, Fung T, Broucke ME, D'Eleuterio GMT, Francis BA (2006) Experiments in multirobot coordination. Robot Autonomous Syst 54(3):265–275
2. Ren W, Chao H, Bourgeous W, Sorensen N, Chen YQ (2008) Experimental validation of consensus algorithms for multivehicle cooperative control. IEEE Trans Control Syst Technol 16(4):745–752
3. Sun D, Wang C, Shang W, Feng G (2009) A synchronization approach to trajectory tracking of multiple mobile robots while maintaining time-varying formations. IEEE Trans Robot 25(5):1074–1086
4. Yan X, Chen J, Sun D (2012) Multilevel-based topology design and shape control of robot swarms. Automatica 48(12):3122–3127
5. Li X, Sun D, Yang J (2013) A bounded controller for multirobot navigation while maintaining network connectivity in the presence of obstacles. Automatica 49(1):285–292
6. Farina A (1999) Target tracking with bearings–only measurements. Signal Process 78(1):61–78
7. Guo J, Yan G, Lin Z (2011) Balanced circular formation control based on gossip communication and angle of arrival information In: Proceedings of the 2011 Chinese Control Conference. Yantai, 22–24 July 2011, pp 6036–6041
8. Cahoon L, Hinich M (1976) A method for locating targets using range only. IEEE Trans Inf Theory 22(2):217–225
9. Kantor G, Singh S (2002) Preliminary results in range-only localization and mapping In: Proceedings of 2002 IEEE International Conference on Robotics and Automation, Washington D.C., 11–15 May 2002, pp 1818–1823
10. Cao M, Yu C, Anderson B (2011) Formation control using range-only measurements. Automatica 47(4):776–781
11. Soares J, Aguiar A, Pascoal A, Martinoli A (2013) Joint ASV/AUV range-based formation control: theory and experimental results In: IEEE International Conference on Robotics and Automation, Karlsruhe, 6-10 May 2013, pp 5579–5585
12. Shames I, Dasgupta S, Fidan B, Anderson B (2012) Circumnavigation using distance measurements under slow drift. IEEE Trans Automatic Control 57(4):889–903
13. Cao Y, Muse J, Casbeer D, Kingston D (2013) Circumnavigation of an unknown target using UAVs with range and range rate measurements In: Proceedings of IEEE Conference on Decision and Control, Florence, 10-13 December 2013, pp 3617-3622
14. Bopardikar S, Bullo F, Hespanha J (2008) A pursuit game with range-only measurements In: Proceedings of IEEE Conference on Decision and Control, Cancun, 9–11 December 2008, pp 4233–4238
15. Matveev AS, Teimoori H, Savkin AV (2011) Range-only measurements based target following for wheeled mobile robots. Automatica 47(1):177–184
16. Namaki-Shoushtari O, Pedro Aguiar A, Khaki-Sedigh A (2012) Target tracking of autonomous robotic vehicles using range-only measurements: a switched logic-based control strategy. Int J Robust Nonlinear Control 22(17):1983–1998
17. Lin Z, Francis B, Maggiore M (2005) Necessary and sufficient graphical conditions for formation control of unicycles. IEEE Trans Automatic Control 50(1):121–127
18. Dimarogonas D, Kyriakopoulos K (2007) On the rendezvous problem for multiple nonholonomic agents. IEEE Trans Automatic Control 52(5):916–922
19. Zheng R, Sun D (2013) Rendezvous of unicycles: a bearings-only and perimeter shortening approach. Syst Control Lett 62(5):401–407
20. Zheng R, Sun D (2013) Rendezvous of wheeled mobile robots using bearings-only or range-only measurements In: Proceedings of IEEE International Conference on Robotics and Biomimetics, Shenzhen, 12–14 December 2013
21. Kalos M, Whitlock P (2008) Monte Carlo methods. Wiley-VCH, Weinheim
22. Mondada F (2014) e-puck education robot. http://www.e-puck.org/. Accessed 27 February 2014
23. Mastellone S, Stipanović D, Graunke C, Intlekofer KA, Spong M (2008) Formation control and collision avoidance for multi-agent non-holonomic systems: theory and experiments. Int J Robot Res 27(1):107–126
24. Chen H, Sun D, Yang J, Chen J (2010) Localization for multirobot formations in indoor environment. IEEE/ASME Trans Mechatronics 15(4):561–574
25. Chen H, Sun D, Yang J (2009) Global localization of multirobot formations using ceiling vision slam strategy. Mechatronics 19(5):617–628

A two-axis cable-driven ankle-foot mechanism

Evandro M Ficanha[1], Mohammad Rastgaar[1*] and Kenton R Kaufman[2]

Abstract

This paper describes a novel cable-driven ankle-foot mechanism with two controllable degrees of freedom (DOF) in dorsiflexion-plantarflexion (DP) and inversion-eversion (IE). The presented mechanism is a proof of concept to demonstrate feasibility. Ankle kinematic measurements demonstrate that ankle IE rotations during a step turn are significantly different from walking on a straight path. This suggests that the ankle-foot mechanisms used in prostheses, exoskeletons, and bipedal robots can be improved by controlling a second degree of freedom in the frontal plane. The proposed prototype mechanism is described in detail, and its design considerations and parameters are presented. The mechanism is capable of producing trajectories similar to the human ankle during a step turn. The device shows passive mechanical impedance close to the human ankle mechanical impedance, allowing its mechanical impedance to be controlled using an impedance controller. The presented mechanism is capable of providing key mechanical characteristics similar to the human ankle, including power, range of motion, and weight, suggesting the feasibility of this design concept.

Keywords: Ankle mechanism; Ankle impedance; Two-DOF ankle-foot; Turning; Ankle rotations; Cable-driven prosthesis; Human ankle kinematics; Multi-axis ankle-foot prosthesis; Powered lower extremity prosthesis

Background

Cable-driven mechanisms have been used on upper extremity exoskeletons and assistive robots such as prosthetic hands. In such mechanisms, pulling the cables results in motion, the same way that the muscles pull tendons in the musculoskeletal system. Exploiting a cable-driven system allows the DC motors to be placed at locations that are not necessarily close to the mechanisms' joints, hence, assisting in maintaining a low profile and reducing the size and weight of the device. One example of such design is lower extremity powered exoskeleton (LOPES) with eight actuated degrees of freedom at the hip, knee, and pelvis [1].

A similar cable-driven design concept can be followed in the lower extremity prostheses design to extend the controllable degrees of freedom (DOFs) beyond state of the art. Currently available ankle-foot prostheses are capable of controlling the ankle joint in only a single DOF in the sagittal plane. These prostheses improve the quality of life of amputees and are designed to increase mobility. However, turning plays a major role in activities of daily living and account for an average of 25% of steps, ranging from 8% to 50% of all steps depending on the activity [2]. Turning requires modulation of ankle impedance in the sagittal and frontal planes to control lateral and propulsive ground reaction forces in order to accelerate the body center of mass along the gait path, resulting in increased lateral and propulsive impulses when compared to a straight step [3]. As a result, turning strategies for amputees and non-amputees are different. Non-amputees rely on hip movement in the frontal plane and moment generation at the ankle joint, while amputees using passive prostheses rely on hip extension in the sagittal plane to compensate for the lack of forward propulsion from their passive prosthesis [4-7]. This data suggests that the next generation of ankle-foot prosthesis can benefit from designs that are capable of functioning in both the frontal and sagittal planes. A cable-driven mechanism seems to be a suitable solution to overcome the weight, size, and profile constraints while introducing a second actuation component. An ankle-foot mechanism capable of impedance modulation and power generation in both the sagittal and fontal planes allows the user to turn in arbitrary directions and walk on slopes that will increase their agility.

* Correspondence: rastgaar@mtu.edu
[1]Department of Mechanical Engineering-Engineering Mechanics, Michigan Technological University, Houghton, Michigan 49931, USA
Full list of author information is available at the end of the article

When physical systems interact with each other, they behave either as an impedance (i.e., accepts external motion inputs and generates force outputs) or an admittance (i.e., accepts external force inputs and generates motion outputs) [8]. The human ankle functions by adjusting its mechanical impedance and capability of producing net positive work during push-off. These capabilities have inspired the design of several new ankle-foot prostheses. One example is a transfemoral prosthesis with powered ankle and knee joints in the sagittal plane developed by Sup et al. where the controller is capable of regulating the impedance of each joint as needed [9-12]. SPARKy 1 and 2 have been developed by Hitt et al. as a tendon-actuated ankle-foot prostheses capable of producing the necessary push-off moments during walking and running, respectively [13,14]. Additionally, Au et al. developed the BiOM, an ankle-foot prosthesis with a finite state machine to determine the gait phase and generate the appropriate kinetic output [15-19]. All three of the aforementioned devices are capable of increasing mobility in the sagittal plane by producing net positive work in the sagittal plane. Finally, Bellman et al. conceived SPARKy 3, an ankle robot with two DOFs; however, a prototype design has not been developed [20]. This leads to the fact that a design to address the weight and size limitations is necessary to develop the next generation of ankle-foot prostheses capable of properly functioning in both the sagittal and frontal planes. These additional capabilities will allow a more natural gait, especially during turning and traversing inclined planes.

The paper is structured as follows: first, the mechanism design is described in details. Next, a trajectory control and instrumentation necessary for a torque feedback is described. Finally, the evaluation experiments are described showing the mechanism capability for reproducing the human ankle kinematics during a step turn. Additionally, a comparison of the passive mechanical impedance of the mechanism's ankle with the human ankle is provided.

Methods

A prototype cable-driven ankle mechanism with two controllable DOFs in DP and IE directions was designed and fabricated as a proof of concept prototype (Figure 1). The design goal was to demonstrate the feasibility of achieving angular displacements, angular velocities, and torques similar to the human ankle during straight walking and turning in both the sagittal and frontal planes.

Ankle kinematics and design considerations

A set of gait experiments were performed to compare the ankle kinematics during a step turn and a straight walking step. These experiments were performed to develop preliminary design parameters for the ankle angular displacements and speeds and eventually for evaluation of the prototype mechanism capabilities. A motion capture camera system was used to record the human subjects' ankle kinematics during the stance phase of the gait. The ankle motions during a step turn and straight walking were measured using five male subjects with no reported musculoskeletal disorders. The step turn consisted of pivoting around the leading leg and turning into a new direction approximately perpendicular to the initial direction similar to turning around a corner (Figure 2). Tests were performed at two different speeds to evaluate the effects of the gait speed. A slow speed was set as the preferred speed of the participant for the gait (average of 96 steps per minute); a fast speed was set as the fastest speed the subject felt comfortable to perform the step turn without running (average of 114 steps per minute). The participants provided written informed consent, as approved by the Michigan Technological University Institutional Review Board, prior to their participation in the study.

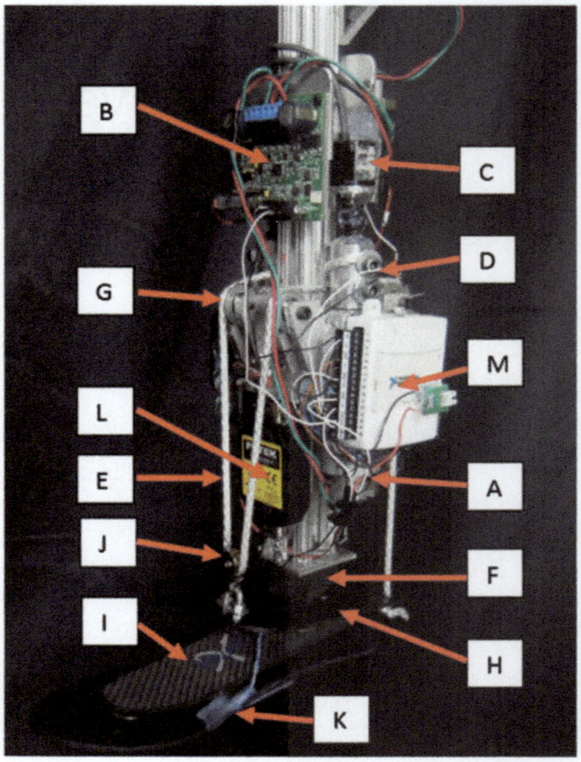

Figure 1 Two-DOF ankle-foot mechanism prototype.

Ankle kinematics in the frontal plane for the right ankle during the straight steps (Figure 2A) in both slow and fast speeds did not differ; however, during a step turn (Figure 2B), there was a significant increase in the ankle inversion. At straight step and low speed, the ankle

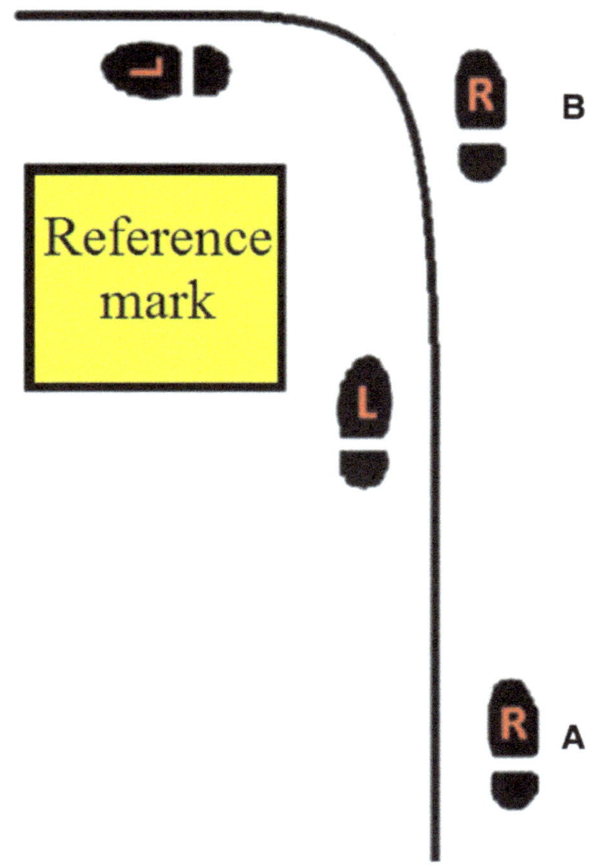

Figure 2 Foot positions during the step turn test. (A) Straight step. **(B)** Step turn.

was in 2.8° eversion during terminal stance, while during the step turn, it changed to 10.5° inversion. By increasing the speed, 2.5° eversion and 14.3° inversion were observed for straight step and step turn, respectively. The changes were equivalent to 248% and 303% of the straight step average range of motion (ROM) in the frontal plane for the slow and fast speed tests (5.37° ±0.6° and 5.52° ±0.8°, respectively). Additionally, the average maximum angular velocities among the participants during the fast speed tests were 120°/s in the sagittal plane and 101°/s in the frontal plane. The maximum angular displacements were 22° in plantarflexion and 20° in inversion. These values are important since an anthropomorphic ankle-foot mechanism should be able to reproduce them. There was also an increased rotation of the ankle in IE during turning. This demonstrated the significance of controllable frontal plane rotations in an ankle-foot mechanism for improved maneuverability.

In the transverse plane, the changes in ankle rotations were not as significant as in the frontal plane. The average ankle rotations during the sidestep cut (Figure 2A) decreased during turn at the terminal stance by −32% with respect to the average ROM during the straight walk in the transverse plane. However, hip rotations in the transverse plane also need to be accounted in the relative motion of the body with respect to the foot. Further investigation is required to quantify the contributions of the hip and ankle to the transverse plane moments and angles during locomotion before implementing it in the design of ankle-foot mechanisms.

Prototype design

The prototype design consists of two DC motors and planetary gear heads (A) powered by two motor controllers (B) connected to two quadrature encoders (C) (Figure 1). Two cable drums (D) transfer the required torque to the ankle through the shock-absorbing nylon rope (E). The rope is securely attached to the cable drums to avoid slippage. A universal joint (F) connects the pylon to the foot and is supported by an elastomer to provide passive stiffness and damping to the ankle. The cable is directed to the foot with pulleys (G). The cable is attached to a carbon fiber plate (H), which is connected to a commercially available prosthetic foot (Otto Bock Axtion®) (I). In the rear side of the carbon fiber plate, the cable is mounted to both sides of the longitudinal axis of the foot. At the front side of the carbon fiber plate, the cable passes through a pulley (J). Torque feedback, which will be explored later, is provided by six strain gauges in the foot (K) using two strain gauge amplifiers (L). An analog to digital converter (M) is connected to a remote computer and is used to acquire the sensors data and provide the motor controllers' inputs.

The mechanism is capable of moving in the sagittal plane (DP) when the motors rotate in opposite directions and in the frontal plane (IE) when the motors rotate in the same direction. Also, any combination of DP and IE can be obtained by combining different amounts of rotations in each motor. The proposed mechanical design with two DOFs uses the fundamental principle that three points are sufficient to define a plane in space. If the plane is constrained from translations and has a fixed point of rotation, by defining the position of those three points, all the rotations of the plane can be achieved. In the proposed design (Figure 3), three points (A, B, and C) can be moved in the Z direction, and thus generate the rotations about the *X*- and *Y*-axes that are equivalent to DP and IE, respectively, while the universal joint constrains the ankle from rotating about the *Z*-axis. If the two motors move in opposite direction, for example, the left motor rotates about the negative *Z*-axis and the right motor rotates about the positive *Z*-axis with equal magnitude, the resultant driving forces result in both the cable tensions $\overrightarrow{T_{RF}}$ and $\overrightarrow{T_{LF}}$ to be in the positive Z axis. This causes point C to move upwards, while points A and B move downwards, resulting in dorsiflexion.

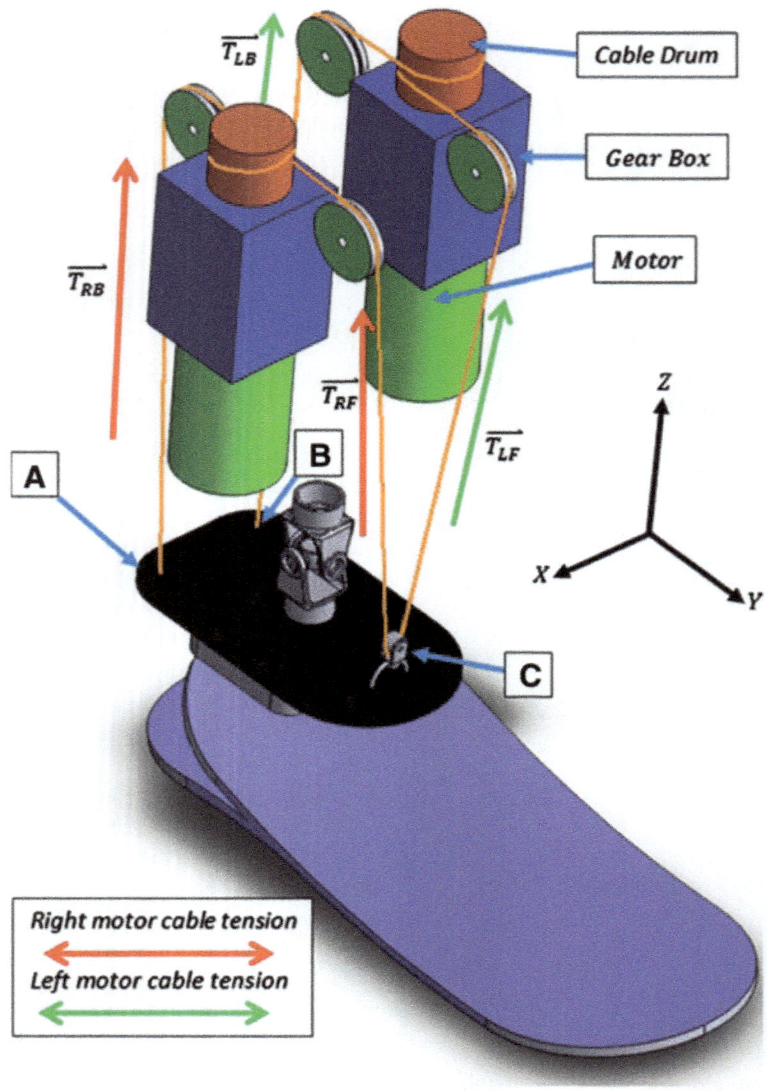

Figure 3 Simplified drawing of the cable-driven mechanism. Points **A**, **B**, and **C** are the interaction points between the cable (orange) and the carbon fiber plate. The cable tensions produced by the right motor $\overrightarrow{T_{RF}}$ and $\overrightarrow{T_{RB}}$ for the front and back of the foot respectively, and the cable tensions produced by the left motor $\overrightarrow{T_{LF}}$ and $\overrightarrow{T_{LB}}$ for the front and back of the foot, respectively, are shown. The shin and other components are not shown for simplicity.

If the motors move in the same direction, for example, about the negative Z axis, the cable tension $\overrightarrow{T_{RB}}$ increases and moves point A upwards; while the cable tension $\overrightarrow{T_{LB}}$ decreases allowing point B to move downwards. This generates a net eversion torque in the foot. In the front of the foot, since point C is located on the axis of symmetry of the carbon fiber plate along the X direction, the amount of cable released by one of cable drum is the same as the cable pulled by the other cable drum. Therefore, the total length of the cable in the front side of the foot does not change and point C does not move up or down. Additionally, since the cable goes through a pulley at point C, it neither moves in any direction nor constrains the cable from motion.

The proposed design benefits from using two motors simultaneously. The motor torques can be combined to produce DP torques and subtracted to produce IE torques. Considering the fact that more power is required in the sagittal plane than the frontal plane during walking, this approach is more appropriate than employing a larger motor for DP and one smaller motor for IE. Additionally, it does not require any extra hardware other than the universal joint and the pulley at the ankle.

The cable-driven design, besides the ability to control the ankle in two DOFs, can provide significant flexibility for managing the ankle-foot inertia. The motors and gear boxes can be located remotely and still provide the necessary characteristics and user preferences. The actuation system has the potential to move up or down on the pylon to accommodate the users' residual limb or to optimize the weight distribution and inertia characteristics.

Selection and design of the active components

For an average able-bodied human weighing 80 kg, the energy consumption required at each step is 36 J (250 watts peak power) [14] and torque as high as 140 Nm [21] during walking. These amounts are 35% higher for an individual with transtibial prosthesis [14,22], and the mechanism is estimated to have 40% losses, resulting in an anticipated peak power consumption of 470 watts, an energy consumption of 68 J, and a peak torque of 264 Nm in the sagittal plane. Also, the device needs to generate 120 deg/s for the sagittal plane and 101°/s for IE, as estimated from the gait experiment described previously.

Considering these requirements, the prototype ankle-foot mechanism uses two brushed DC motors and motor controllers that are capable of a continuous torque output of 0.275 Nm at 10,000 RPM (producing 240 watts each for a total of 480 watts). They are connected to gear boxes with a 104:1 reduction gear ratio. Two 11.1 V LiPo batteries connected in series with an energy density of 572 kJ/kg are estimated to provide energy for 5,800 steps. Even though the current prototype is designed as a proof of concept, it weighs 3 kg without the battery and socket which is the equivalent weight of the ankle-foot of a 52-kg person [21]. Considering the 104:1 gear ratio of the gear box and the 0.0127 m radius of the cable drum, the pulling force at the cables, F_{pull}, can be found as:

$$F_{pull} = \frac{T}{0.0127} \times 104 \tag{1}$$

where T is the instantaneous torque generated by the motor. The cable speed $\dot{L}$ is defined as:

$$\dot{L} = \frac{\omega}{60 \times 104} \times 2\pi \times 0.0127 \tag{2}$$

where ω is the instantaneous speed of the motor in RPM. At 0.275 Nm of torque and 10,000 RPM, the generated pulling force is 2,252 N at 0.127 m/s. The moment arm in the sagittal plane, the distance in the Y-axis direction from the center of the universal joint to the cables in the back of the carbon fiber plate, is 57 mm. The moment arm in the frontal plane, the distance in the X-axis from the center of the universal joint to the cables in the back of plate, is 28.4 mm. DP and IE torques can be found as follows:

$$\tau_{DP} = \left(F_{pull\text{-}left} + F_{pull\text{-}right}\right) \times 0.057 \times \cos(\theta_{DP}) \tag{3}$$

$$\tau_{IE} = \left(F_{pull\text{-}left} - F_{pull\text{-}right}\right) \times 0.0284 \times \cos(\theta_{IE}) \tag{4}$$

where τ_{DP} and τ_{IE} are torques in DP and the IE, respectively, and $F_{pull-left}$ and $F_{pull-right}$ are the pulling forces from the left and right motors.

Due to the design geometry, when the foot deviates from the neutral position, the cables do not connect to the carbon fiber plate at a 90° angle, resulting in losses of the generated torque. These losses are taken into the account by $\cos(\theta_{DP})$ and $\cos(\theta_{IE})$ terms in Equations 3 and 4, respectively. At the maximum angular displacements observed in the gait experiment, the prototype shows a 7% loss in DP torque at −21.6° plantarflexion (maximum DP angle) and a 6% loss in IE at 19.7° inversion (maximum IE angle). These nonlinear losses are small but should be considered during the control of the mechanism.

The angular velocities $\dot{\theta}_{DP}$ and $\dot{\theta}_{IE}$ are as follows:

$$\dot{\theta}_{DP} = \frac{\dot{L}_{Right} + \dot{L}_{Left}}{2 \times 0.057} \times \frac{180}{\pi} \tag{5}$$

and:

$$\dot{\theta}_{IE} = \frac{\dot{L}_{Right} - \dot{L}_{Left}}{2 \times 0.0284} \times \frac{180}{\pi} \tag{6}$$

where $\dot{L}_{Left}$ and $\dot{L}_{Right}$ are the cable velocities pulling from the left and right motors, respectively. The motors (at 0.275 Nm torque and 10,000 RPM) are capable of generating 256 Nm at 128°/s in DP and 64 Nm at 256°/s in IE. These values are close to the estimated 264 Nm of required torque in the sagittal plane based on the maximum torque required for walking [21] and the expected losses in the prosthesis and close to the angular velocities of 120°/s for DP and 101°/s for IE measured in the gait experiment described previously.

Selection and design of the passive component

The shock-absorbing nylon rope stretches to provide shock absorption while being soft to the touch. Nylon ropes are strong and flexible which makes it ideal for the design of this prototype. The carbon fiber plate is 3.175 mm thick and is a fundamental component of the design. It has been reported that the average plantarflexion stiffness at 50% MVC is 143 Nm/° and is near constant from 30% to 80% MVC [23]; thus, the carbon fiber plate was designed to provide a stiffness comparable to these amounts in the sagittal plane. The carbon fiber plate acts as a spring connected in series between the cable and foot. The cable needs to be always in tension to assure the proper control over the foot, causing the carbon fiber plate to be always under a bending moment.

Maximum force, torque, and range of motion

A Kistler® Type 5233A force plate was used to measure the maximum amount of force produced by the ankle-foot mechanism with a step input in plantarflexion. The mechanism's pylon was constrained in the *z* direction with the intention to measure the torque at the ankle while DC motors were applying the stall torques. The device was capable of generating a lift force of 712 N, which corresponds to 103.2 Nm torque. The maxim IE torque can be calculated directly from the DP torque and the mechanism's geometry to an estimated 25.8 Nm.

In the current configuration, the maximum range of rotation recorded from the position encoders was 49° in DP and 98° in IE, due to the range of rotation of the cable drums. Both DP and IE ranges are within the set goals. Testing the ROM in IE, it was found that the IE motion becomes unstable at angles above 62° (if forced by an external torque to this angle) similar to rolling of the ankle, which is an occurring injury among sport players and is in inversion [24]. In the prototype mechanism, ankle roll happens when either points A or B (Figure 3) crosses the *Y*-axis. At this configuration, the tension in the cables, which are applying a torque against the disturbance force, change the direction of the torque and makes the ankle unstable. The maximum expected rotation observed in IE during the gait tests using the camera system was 19.7° and thus instability should not be a problem during reproduction of a straight walk or turn.

Instrumentation and control

Ankle torque and angle feedback

To actively regulate the ankle impedance, torque and angle feedbacks are required. The prototype is equipped with strain gauges mounted to the foot, which can be correlated to the external torques and will be used in our future work for impedance modulation of the prototype. Strain gauges are commonly used in load cells to measure the strains in structures due to external forces. The strain gauges change the resistance as they stretch or contract, and their change in resistance is correlated to the strain of the object they are attached to and consequentially the force or torque applied to the object. Strain gauges are typically wired in a Wheatstone bridge configuration using four strain gauges (some of the strain gauges can be replaced by resistors). The increase or decrease in resistance of two of the strain gauges (placed in opposite sides of the bridge) causes the output voltage of the bridge to increase or decrease, respectively. The opposite is also true for the other two strain gauges where the increase or decrease in resistance decreases or increases the output voltage of the bridge, respectively.

For estimating the torque in DP, four strain gauges were attached to the sole of the foot (Figure 4A). Two strain gauges $S_{dp}2$ were placed behind the center of rotation of the ankle in the sagittal plane and were wired into opposite sides of the Wheatstone bridge (Figure 5A). Any ground reaction force at the hill caused a decrease in the voltage of the Wheatstone bridge, which could be correlated to the torque in the sagittal plane of the foot when the heel was interacting with the ground (e.g., heel strike). Two strain gauges $S_{dp}1$ were located in front of the center of rotation of the ankle in the sagittal plane and wired in the opposite sides of the Wheatstone

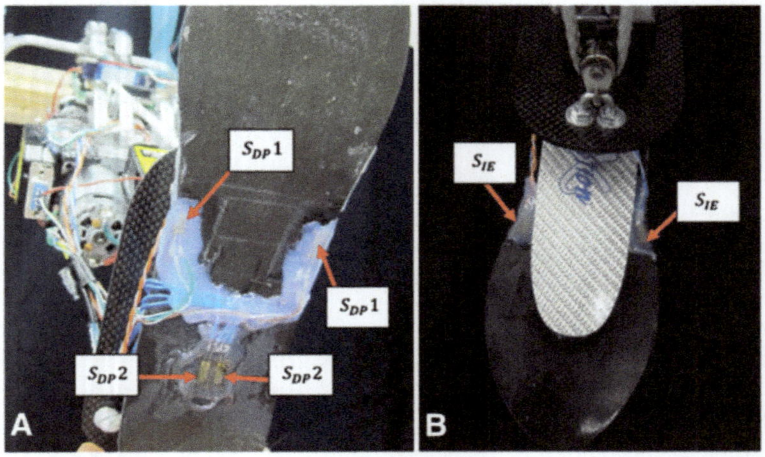

Figure 4 Strain gauge placement in the foot. (A) For the sagittal plane torque, two strain gauges are used for push-off torque estimation ($S_{DP}1$) and two strain gauges are used for heel-strike torque estimation ($S_{DP}2$). **(B)** For the frontal plane torque, two strain gauges are used for push-off torque estimation ($S_{IE}1$).

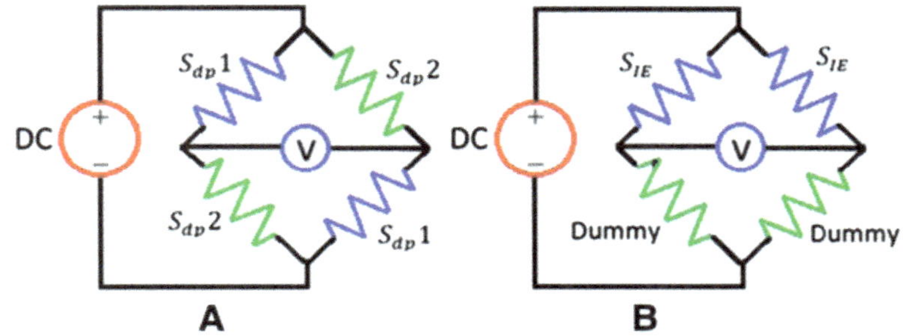

Figure 5 Wheatstone bridges. (A) Strain gauge placement in the Wheatstone bridge for the sagittal plane torque estimation. Two strain gauges are used for push-off torque estimation in the sagittal plane ($S_{DP}1$), and two strain gauges are used for heel-strike torque estimation in the sagittal plane ($S_{DP}2$). **(B)** Strain gauge placement in the Wheatstone bridge for the frontal plane torque estimation. Two strain gauges are used for push-off the frontal plane torque estimation (S_{IE}).

bridge (Figure 5A). Any ground reaction force at the front of the foot caused an increase in the output voltage of the Wheatstone bridge, which could be correlated to the torque in the sagittal plane of the foot when it was contacting the ground (e.g., push-off). Note that when the foot was flat on the ground, the output from the strain gauges $S_{dp}1$ canceled the output from the strain gauges $S_{dp}2$; therefore, the resultant voltage could always be correlated to the net sagittal plane torque in the ankle.

For estimating the torque in IE, two strain gauges were attached to the top surface of the foot as shown in Figure 4B. The other two strain gauges were attached to an inert piece of carbon fiber but could also be replaced by two resistors with resistance identical to the strain gauges. The strain gauges were placed on the outside edges of the foot and were on the same side of the Wheatstone bridge (Figure 5B); hence, the difference in strains caused an increase or decrease in voltage at the bridge. The voltage could be correlated to the IE torque in the foot when the front of the foot was contacting the ground (e.g., push-off). This configuration made the bridge insensitive to the torque in the sagittal plane, since the strain gauges were in the opposite ends of the bridge. Therefore, if they both contracted or stretched by the same amount as it would happen in the presence of a DP torque, the output would not be affected. This feature was important since it was necessary to have the frontal plane torque estimation to be decoupled from the sagittal plane torque.

To correlate the strain gauge readings to the actual disturbance torques, a Kistler® Type 5233A force plate was used to measure the external force applied during quasi-static loading tests. The tests consisted of loading the foot in different configurations and recording the applied force and the corresponding strain measurement. The tests were performed in 1 - plantarflexion by applying a load when the heel was in contact with the ground, 2 - dorsiflexion by applying a load when the forefoot was in contact with the ground, 3 - eversion by applying a load when the right edge of the forefoot was in contact with the ground, and 4 - inversion by applying a load when the left edge of the forefoot was in contact with the ground. From the external forces, the geometry of the foot, and the strain measurements, the applied torques were calculated. It is important to note that in the sagittal plane, the proportional factor between the external force and the strains measured at heel loading and forefoot loading were not the same, since the strain gauges were attached to two different areas of the foot. The proportional factors for the strain gauges at heel loading and forefoot loading were estimated at 1.41 and 19.52 Nm/volt, respectively. As a result, different proportional factors needed to be used depending if the strain measurement was positive or negative. In the frontal plane, the proportional factors for inversion and eversion torques were closer (4.43 and 3.55 Nm/volt, respectively), which was expected since the foot is near symmetrical about its sagittal plane.

Position controller

Presently, two optical quadrature encoders (200 pulses per revolution) provide position feedback to a proportional plus derivative (PD) controller to control the relative position of the foot with respect to the pylon. The overall control architecture of the ankle-foot mechanism can be seen in Figure 6. The block diagram of the position controller is shown in Figure 7.

The overall controller consists of two identical PD controllers to estimate the appropriate motor voltages. The input to the left motor controller is the sum of DP and IE angles, while the input to the right motor controller is the difference between the DP and IE angles. The desired angular displacements were the ankle angles measured during the gait experiment using the camera system that were stored in a lookup table to be reproduced by the ankle-foot mechanism for evaluation purposes as discussed in the following sections.

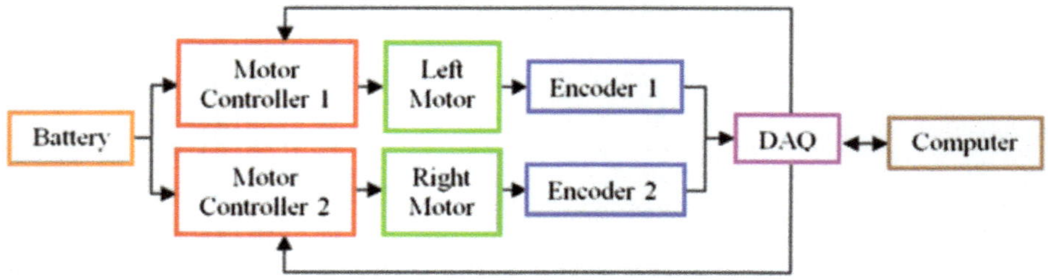

Figure 6 Overall control architecture of the ankle-foot mechanism.

Evaluation of the ankle-foot mechanism

Tracking performance

The aim of this evaluation study is to show that the mechanism is capable of reproducing the human ankle rotations during straight walk and turning. The input of the controller is the time history of a representative subject's ankle angles in DP and IE recorded with a camera system and stored in a lookup table. The output is the ankle angles calculated using the quadrature encoders.

The angular displacement of the ankle in DP and IE (θ_{DP} and θ_{IE}, respectively) can be calculated from the left and right quadrature encoders' feedback (θ_{Left} and θ_{Right}, respectively), where K_{DP} and K_{IE} are constant gains to define the ankle rotations as a function of the quadrature encoders' outputs and are based on the mechanism's geometry.

$$\theta_{DP} = K_{DP}\left(\frac{\theta_{Left} + \theta_{Right}}{2}\right) \tag{7}$$

$$\theta_{IE} = K_{IE}\left(\theta_{Right} - \theta_{Left}\right) \tag{8}$$

The input and output angles of the position controller can be seen in Figure 8 when the reference angle of the controller is updated at a rate of 125 Hz. For ease of comparison, the output plots have a time shift to remove the 24 ms delay of the output. Also, all signals are filtered with a low-pass filter with a cut-off frequency of 5 Hz to remove frequency components from the input that are beyond the bandwidth of the motors currently installed, as well as sensor noises from the output signal.

Due to the physical characteristics of the mechanism, small angular differences between the left and right motors caused larger changes in the foot rotations, making the system more sensitive to disturbances and noise in the frontal plane compared to the sagittal plane. Note that the current prototype was developed as a proof of concept to validate the design kinematics and uses low-cost brushed DC motors and gear boxes with limited bandwidth, which are the cause of the observed time delay. Therefore, faster motors and sensors with lower noise levels will be used in future designs. Nevertheless, both plots indicate that the mechanism is capable of reproducing the ankle rotations of a human subject during step turn with desirable accuracy.

Impedance test

Mechanical impedance of a system is the relationship between the displacement input and its force outputs. Control strategies have been developed for impedance control of active prostheses [25,10], where higher level control identifies the gait cycle and the lower level control regulates the actuators impedance characteristics. Ankle-foot robots can be designed to show passive and active mechanical impedance similar to the human ankle. In a stationary position, the passive mechanical impedance is dependent only to the passive components of the device. If the passive mechanical impedance of the mechanism is similar to the human ankle mechanical impedance, it may allow the impedance controller to easily regulate the active mechanical impedance of the mechanism to the target impedance for each specific gait cycle.

An impedance estimation test was conducted to evaluate the passive impedance of the ankle-foot mechanism

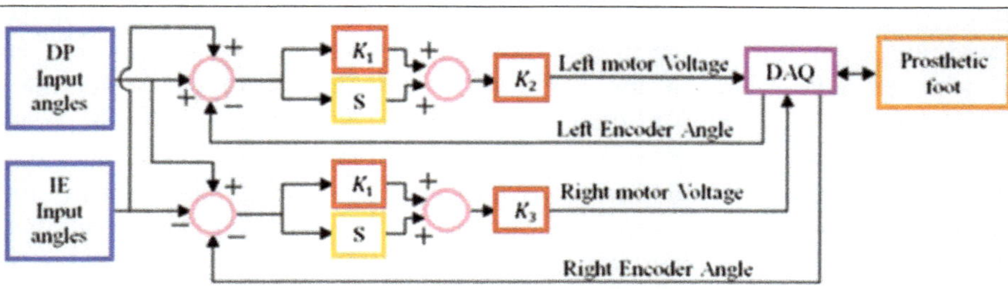

Figure 7 Block diagram of the ankle-foot mechanism position controller.

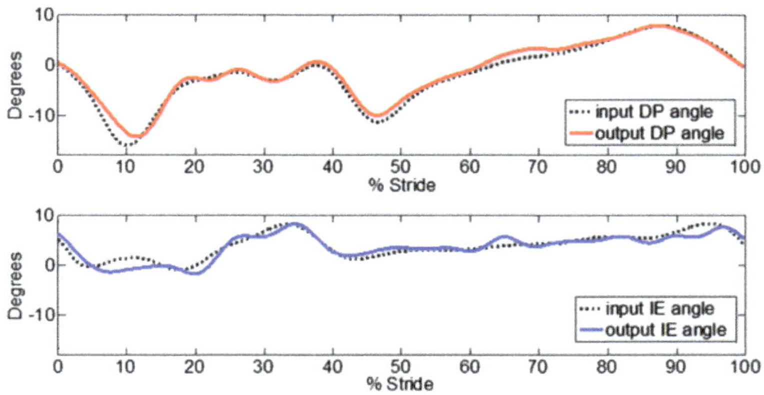

Figure 8 Input and compensated output for time delay (24 ms) of the ankle-foot mechanism. The input is the recorded right ankle rotations of a representative subject during swing and stance periods of the sidestep cutting at fast speed.

to compare it to the passive and active impedance of a human subject. The experiment was similar to the tests performed in [26,27] where the Anklebot, a multi-axis lower extremity therapeutic robot, was employed to apply pseudo-random torque perturbations to the ankle in DP and IE directions. The prototype mechanism attached to the Anklebot is shown in Figure 9, where the Anklebot and the ankle-foot mechanism were attached mechanically to each other. The passive mechanical impedance of the mechanism was measured with all the controllers of the mechanism turned off. For this test, the rubber foot shell was inserted in the same type of shoe used for human tests to ensure consistency in the experiments.

Anklebot provided the input torques to the ankle joint of the prototype mechanism (τ_{DP} and τ_{IE}) and also measured the resultant ankle displacements (θ_{DP} and θ_{IE}). A stochastic system identification method was used to estimate the mechanical impedance of the mechanism's ankle in DP and IE directions, $Z_{DP}(f)$ and $Z_{IE}(f)$, respectively.

$$\tau_{DP} = Z_{DP}(f)\theta_{DP} \tag{9}$$

$$\tau_{IE} = Z_{IE}(f)\theta_{IE} \tag{10}$$

Sixty seconds of data were sampled at 200 Hz yielding 12,000 samples for both Anklebot and for the combined Anklebot and ankle-foot mechanism. Welch's averaged, modified periodogram method of spectral estimation, as implemented in MATLAB, was used to estimate one-sided auto and cross-power spectral densities of the torque and angle sequences in the DP and IE directions. A periodic Hamming window with a length of 512 samples was used with an FFT length of 1,024 samples, yielding a spectral resolution of 0.195 Hz. Power spectral density functions were estimated by averaging their values calculated from 45 data windows with 50% overlap (256 samples).

The foot and Anklebot's shoe share the same motion; therefore, the mechanical impedance of the prototype mechanism's ankle was obtained by subtracting the Anklebot mechanical impedance from the total estimated mechanical impedance as follows.

$$Z_{ankle} = Z_{ankle+Anklebot} - Z_{Anklebot} \tag{11}$$

To compare the mechanical impedance of the ankle-foot mechanism with the mechanical impedance of human ankle, experiments were performed on a human subject with relaxed muscles and with 10% of maximum voluntary contraction (MVC) of the lower leg muscles, which was maintained by monitoring the surface electromyography (EMG) of the tibialis anterior muscle of the subject, following the procedures described in [26]. Figure 10 shows the Bode plots of the mechanical impedances in the sagittal plane of the prototype mechanism, the human subject's ankle with relaxed muscles, and the human subject's ankle with 10% MVC. The quasi-static stiffness magnitude of the prototype mechanism was 39.5 dB (94 Nm/radian) in the sagittal plane at below 1 Hz with relatively linear impedance and phase characteristics up to 5 Hz (the frequency range of interest [15]), with an average coherence value of 0.92. The human subject showed similar stiffness in the sagittal plane for the co-contraction test and a lower quasi-static stiffness magnitude in the passive test when compared to the ankle mechanism. The frontal plane impedance magnitude (Figure 11) of the mechanism was between the active and passive stiffness of the human subject's ankle impedance with a value of 24.24 dB (16 Nm/radian) at below 1 Hz. Considering that the magnitude of the ankle impedance is different among individuals, the results showed that it is feasible to design a mechanism with impedance

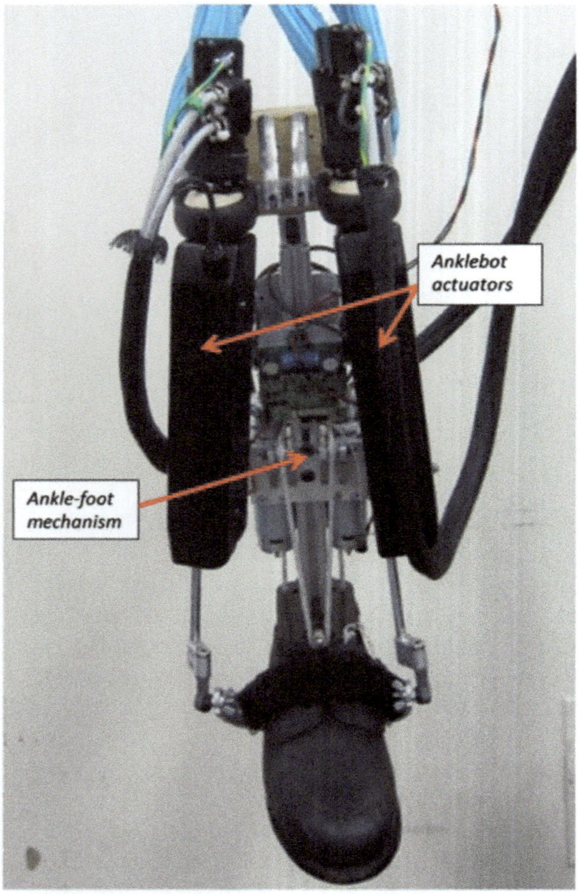

Figure 9 Anklebot attached to the ankle mechanism for estimation of the mechanical impedance of the ankle-foot mechanism.

characteristics close to the human ankle impedance, allowing for fine tuning of these values by an impedance controller that will be designed in the future.

Results and discussion

Ankle rotation

Ankle rotations during straight walk and step turn were measured using a camera system to study the ankle function in the frontal plane during the turn. The ankle rotations in the frontal plane during straight walk in both slow and fast speeds were close; however, there was a significant increase in the ankle's rotation in the IE direction during the step turn. The ankle angles in IE during the push-off increased from 2.8° eversion during straight walk at low speed to 10.5° inversion during the step turn at slow speed. These values were 2.5° eversion and 14.3° inversions during the step turn for straight walk and step turn, respectively. These changes were equivalent to 248% and 312% of the straight step average ROM in the frontal plane for the slow and fast speed tests, respectively. The changes observed in the angular displacement of the ankle in the frontal plane during step turn, when compared to straight walk, showed the importance of the IE rotation during turning.

Mechanical characteristics of the ankle mechanism

The prototype ankle-foot mechanism was designed to control the foot in two DOFs. The mechanism uses two identical motors to generate torques in the sagittal plane and in the frontal plane. When both motors rotate in the same direction, they generate DP motion at the foot. IE motion is also generated when the motors move in opposite directions. Sagittal plane torque requirements are larger than the frontal plane torque requirements, and the design benefits from its combined power from both motors to generate DP torques and difference in torques to generate IE torques, instead of using one large motor for DP and one smaller motor for IE. The cable design also allows for placement of the motors at different places on the pylon. This allows the design to be easily adjusted for amputees with different residual limb length or to optimize weight distribution. Additionally, the design may be significantly improved using Bowden cables, allowing the motors and gear boxes to be mechanically decoupled from the pylon and placed elsewhere. This will allow for great flexibility on the design and customization of the prostheses based on individual user needs, especially for amputees with long residual limbs. Also, having the flexibility of placing the motors and gearboxes, which carry the most of the weight of the mechanism, away from the distal end of tibia would allow for a lightweight prosthesis that in theory can result in more symmetric gait patterns.

The prototype ankle mechanism was designed to match the mechanical characteristics of the human ankle. This included power, range of motion, and weight. The maximum power consumption, based on the energy expenditure of humans and the expected losses in the mechanism, is 470 watts. The motors chosen for the ankle mechanism have a combined power of 480 watts. The motors are rated at 0.275 Nm torque at 10,000 RPM, which results in 256 Nm at 128°/s in DP and 64 Nm at 256°/s in IE. These values are similar to the estimated 264 Nm of required torque in plantarflexion, and the angular velocities are close to 120 and 101°/s required for the sagittal plane and the frontal plane motion, respectively. Using a force plate, the maximum lifting force of the ankle mechanism was found to be 712 N in the sagittal plane which is the equivalent of the weight of a 73-kg person.

Impedance controllers have been used in active prosthesis to control the position and torque generated by the device. An impedance controller requires both position feedback and force feedback of the interactive forces between the device and the environment. The

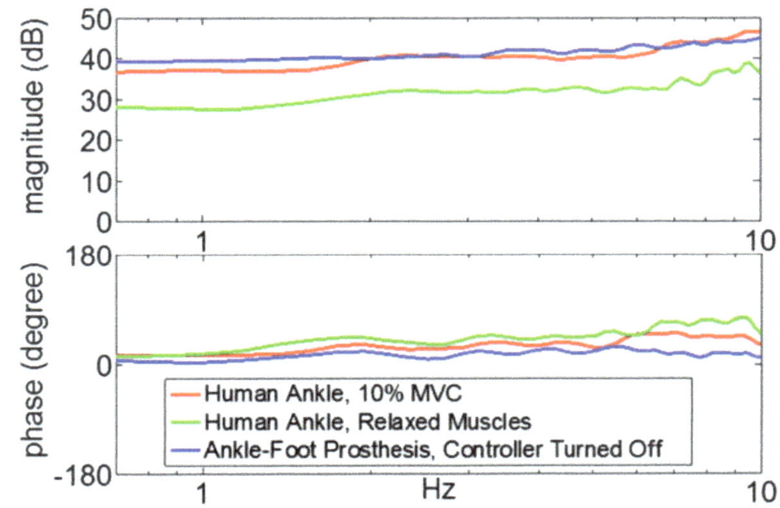

Figure 10 Bode plot of the ankle impedance in the sagittal plane. Impedance of the ankle-foot mechanism and a human subject with both relaxed muscles and 10% MVC. The top plot is the magnitude and the bottom plot is the phase. The quasi-static impedance of the mechanism was above the human ankle impedance magnitude with 10% MVC at low frequencies (around 1 Hz).

ankle mechanism in its current form uses quadrature encoders at the cable drums for position feedback. This provides an estimate of the ankle angles but does not account for the nonlinearities of the device caused by the rotation of the ankle, or the deflection of the carbon fiber plate, and stretch of the nylon rope. This can be remedied by using the encoders to measure the rotation of the foot directly, instead of the cable drums' rotations. The torques generated from the mechanism can be measured using strain gauges at the foot, as described in this paper. However, the nonlinearities caused by the rotation of the foot also affect the generated torque, which need to be accounted in the controller design in the future. This will allow the use of ground reaction torques feedback and the angular displacements of the foot in the implementation of impedance controllers for the two-axis ankle mechanism in both the sagittal plane and the frontal plane.

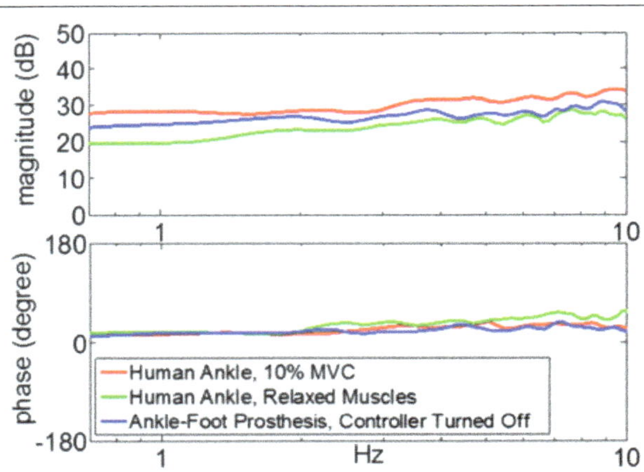

Figure 11 Bode plot of the magnitude and phase of the ankle impedance in the frontal plane. Impedance of the ankle-foot mechanism and a human subject with relaxed muscles and 10% MVC. The top plot is the magnitude, and the bottom plot is the phase. The quasi-static impedance of the mechanism was in between of the human ankle impedance magnitude with relaxed muscles and with 10% MVC at low frequencies (around 1 Hz).

Design evaluation

Using a simple position controller, the recorded ankle rotations of a subject were reproduced by the prototype ankle mechanism. It was observed that the device was capable of tracking the human ankle motion in both the sagittal plane and the frontal plane. Since the mechanism was designed to use the difference in motor torques and angles to generate frontal plane torques and angular displacements, respectively, the design was found to be more sensitive to disturbances and noise in IE in comparison to DP.

The impedance estimation test showed that the quasi-static stiffness (below 1 Hz) of the prototype mechanism was 39.5 dB (94 Nm/radian) in the sagittal plane with relatively linear impedance and phase characteristics up to 5 Hz, which is the frequency of interest. The linearity of the system, shown as the average coherence of the estimated impedance, was 0.92. These values were similar to the ankle stiffness of a human subject in the sagittal plane during the muscle co-contraction test and higher than the subjects' quasi-static stiffness with relaxed muscles. In the frontal plane, the mechanism impedance was between the active and passive stiffness of the human subject's ankle, with a stiffness of 24.24 dB (16 Nm/radian). Considering the variation in ankle stiffness among individuals, these experiments show that the mechanism can be designed with a passive impedance similar to the human ankle. This will facilitate controlling the impedance of the mechanism through an impedance controller to the necessary values based on the gait requirements.

The current prototype was developed as a proof of concept to validate the design kinematics. It uses low-cost brushed DC motors and compatible gear boxes, resulting in low bandwidth of the motors and gear box, limited efficiency and power to weight ratio of the mechanism. Even so, the mechanism was capable of reproducing the ankle rotations of a human subject during step turn, adequate torques, and desirable mechanical impedance similar to a human ankle, while weighing 3 kg. Therefore, using faster, brushless motors and sensors with lower noise levels in future designs will increase the efficiency, power, and bandwidth, while decreasing the device's weight.

Conclusions

In this paper, we described a novel two-axis, cable-driven ankle-foot mechanism. First, we described an experiment using a motion capturing camera system to study the human ankle during straight walk and turning, which showed that the inversion and eversion of the human ankle are significant during turning. This suggested that the design of ankle-foot mechanisms could be improved by controlling the degree of freedom of ankle mechanisms in the frontal plane. The proposed mechanism was described in detail, and its design parameters and considerations were presented. Moreover, a method for estimating the external torques applied to the ankle using strain gauges was presented, which is necessary in the development of impedance controllers. A simple position controller was implemented and used to demonstrate the capability of the device in tracking the same motion as the human ankle during a turning maneuver. It was shown that the prototype was capable of generating enough torque to lift a 73 kg person. An experiment was performed to explore the mechanism's passive mechanical impedance. The device showed passive mechanical impedance close to the impedance of a human subject, which prompts the feasibility of regulating the ankle-foot mechanism mechanical impedance using an impedance control. The presented mechanism was capable of providing key mechanical characteristics similar to the human ankle, including power, range of motion, and weight, while allowing instrumentation for the development of an impedance controller in both dorsiflexion-plantarflexion and inversion-eversion.

Abbreviations

DOF: Degrees of freedom; DP: Dorsiflexion-plantarflexion; EMG: Electromyography; IE: Inversion-eversion; LOPES: Lower extremity powered exoskeleton; MVC: Maximum voluntary contraction; ROM: Range of motion.

Competing interests

The authors declare that they have no competing interests.

Authors' contributions

All authors were fully involved in the study and preparation of the manuscript. All authors read and approved the final manuscript.

Acknowledgement

This material is based upon work supported by the National Science Foundation under CAREER Grant No. 1350154.

Author details

[1]Department of Mechanical Engineering-Engineering Mechanics, Michigan Technological University, Houghton, Michigan 49931, USA. [2]Motion Analysis Laboratory, Department of Orthopedic Surgery, Mayo Clinic and Mayo Foundation, 201 W. Center St., Rochester, MN 55902, USA.

References

1. Veneman JF, Kruidhof R, Hekman EEG, Ekkelenkamp R, Asseldonk EHFV, Hvd K (2007) Design and evaluation of the LOPES exoskeleton robot for interactive gait rehabilitation. IEEE Trans Neural Syst Rehabil Eng 15(3):379–386
2. Glaister BC, Bernatz GC, Klute GK, Orendurff MS (2007) Video task analysis of turning during activities of daily living. Gait Posture 25(2):289–294
3. Glaister BC, Orendurff MS, Schoen JA, Bernatz GC, Klute GK (2008) Ground reaction forces and impulses during a transient turning maneuver. J Biomechanics 41(4):3090–3093
4. Ventura JD, Segal AD, Klute GK, Neptune RR (2011) Compensatory mechanisms of transtibial amputees during circular turning. Gait Posture 34:307–312
5. Orendurff MS, Segal AD, Berge JS, Flick KC, Spanier D, Klute GK (2006) The kinematics and kinetics of turning: limb asymmetries associated with walking a circular path. Gait Posture 23(1):106–111
6. Hase K, Stein RB (1999) Turning strategies during human walking. J Neurophysiol 81(6):2914–2922
7. Segal AD, Orendurff MS, Czerniecki JM, Schoen J, Klute GK (2011) Comparison of transtibial amputee and non-amputee biomechanics during a common turning task. Gait Posture 33(1):41–47. doi:10.1016/j.gaitpost.2010.09.021
8. Hogan N (1985) Impedance control: an approach to manipulation: part I—theory. J Dyn Sys Meas Control 107(1):1–7
9. Goldfarb M (2010) Powered robotic legs – leaping toward the future. Bioengineering, National Institute of Biomediacl Imaging and
10. Sup F, Bohara A, Goldfarb M (2008) Design and control of a powered transfemoral prosthesis. Int J Rob Res 27:263–273
11. Sup F, Varol HA, Mitchell J, Withrow TJ, Goldfarb M (2009) Preliminary evaluations of a self-contained anthropomorphic transfemoral prosthesis. IEEE ASME Transaction of Mechatronics 14(6):667–676
12. Sup F (2009) A powered self-contained knee and ankle prosthesis for near normal gait in transfemoral amputees. Vanderbilt University, Nashville, Tennessee
13. Hitt J, Merlo J, Johnston J, Holgate M, Boehler A, Hollander K, Sugar T (2010) Bionic running for unilateral transtibial military amputees. In: 27th Army Science Conference (ASC). Orlando, Florida
14. Hitt JK, Sugar TG, Holgate M, Bellman R (2010) An Active foot-ankle prosthesis with biomechanical energy regeneration. J Med Devices 4(1):011003
15. Au SK (2007) Powered ankle-foot prosthesis for the improvement of amputee walking economy. Massachusetts Institute of Technology, Cambridge, MA
16. Au S, Herr H (2008) Powered ankle-foot prosthesis. Robot Automation Mag 15(3):52–59
17. Au S, Weber J, Herr H (2009) Powered ankle-foot prosthesis improves walking metabolic economy. IEEE Trans Robot 25(1):51–66
18. Au SK, Herr H, Weber J, Martinez-Villalpando EC (2007) Powered ankle-foot prosthesis for the improvement of amputee ambulation. International Conference of the IEEE, Engineering in Medicine and Biology Society, Lyon.
19. Herr HM, Grabowski AM (2010) Powered ankle-foot prosthesis improves metabolic demand of unilateral transtibial amputees during walking. Amer. Society of Biomechanics, Long Beach, CA.

20. Bellman RD, Holgate MA, Sugar TG, Bellman RD, Holgate MA (2008) Sugar TG SPARKy 3: design of an active robotic ankle prosthesis with two actuated degrees of freedom using regenerative kinetics. In: Proceedings of the 2nd Biennial IEEE/RAS-EMBS International Conference on Biomedical Robotics and Biomechatronics. Scottsdale, AZ, USA, pp 511–516
21. Samuel K, Au JW, Hugh H (2007) Biomechanical design of a powered ankle-foot prosthesis. In: Paper presented at the International Conference on Rehabilitation Robotics, Noordwijk. The Netherlands
22. Rao SS, Boyd LA, Mulroy SJ, Bontrager EL, Gronley JK, Perry J (1998) Segment velocities in normal and transtibial amputees: prosthetic design implications. IEEE Trans Rehabil Eng 6(2):219–226
23. Sinkjaer T, Toft E, Andreassen S, Hornemann BC (1998) Muscle stiffness in human ankle dorsiflexors: intrinsic and reflex components. J Neurophysiol 60(3):1110–1121
24. Saripalli A, Wilson S (2005) Dynamic ankle stability and ankle orientation. In: Paper presented at the 7th Symp. Footwear Biomech. Conf, Cleveland, OH
25. Eilenberg MF, Geyer H, Herr H (2010) Control of a powered ankle-foot prosthesis based on a neuromuscular model. IEEE Trans Neural Syst Rehabil Eng 18(2):164–173
26. Rastgaar M, Ho P, Lee H, Krebs HI, Hogan N (2010) Stochastic estimation of the multi-variable mechanical impedance of the human ankle with active muscles. In: ASME Dynamic Systems and Control Conference. Boston, MA
27. Rastgaar M, Ho P, Lee H, Krebs HI, Hogan N (2009) Stochastic estimation of multi-variable human ankle mechanical impedance. In: ASME Dynamic Systems and Control Conference. Hollywood, CA

6

Kinematic study and implementation of a bio-inspired robotic fish underwater vehicle in a Lighthill mathematical framework

Abhra Roy Chowdhury[1*], Vinoth Kumar[1], Bhuneshwar Prasad[1], Rajesh Kumar[2] and Sanjib Kumar Panda[1]

Abstract

This paper has focused on the formulation of the biological fish propulsion mechanism given by Sir J. Lighthill mathematical slender body theory for a bio-inspired robotic fish. A 2-joint, 3-link multibody vehicle model biologically inspired by a body caudal fin (BCF) carangiform fish propulsion is designed. The objective is to investigate and show that a machine mimicking real fish behavior can navigate efficiently over a given distance with a good balance of speed and maneuverability. The robotic fish model (kinematics and dynamics) is integrated with the Lighthill (LH) mathematical model framework. Different mathematical propulsive waveforms are combined with an inverse kinematics-based approach for generating fish body motion. Comparative studies are undertaken among a non-LH model, a LH model, and the proposed propulsive wave models based on a distance-based performance index. Proposed LH cubic and NURB quadratic functions are found to be 16.32% and 17.94% efficient than a non-LH function, respectively. With the help of the simulation results, closed-loop experiments are done and an operating region is established for critical kinematic parameters tail-beat frequency and propulsive wavelength. The simulation and experimental plots are compared and found to be similar to the kinematic behavior study of the biological yellowfin tuna.

Keywords: Biology inspired robotics; Body caudal fin; Lagrange-Euler equations; Lighthill equation; Operating region

Background

Biomimetics [1] reflects the features and capabilities of the biological evolution [2] of a system that could be efficiently replicated or mimicked in a human engineered system to the design of new technologies and the improvement of conventional ones. This approach has been proposed to be the answer to the improved performance and reliability for large scale complex systems by faster adaptation with dynamic environment. One of the focused technologies has been the development of autonomous underwater vehicles [3] as a greater part to the increasing interest in unmanned underwater surveillance and monitoring. Of particular interests are regions of the underwater environment which are unexplored and dynamic as well as underwater detection, pollution source tracking, underwater archeology, search and rescue, and so forth. The study of underwater evolution of life and its plethora of locomotion modes has long been a subject of interest to the biological community. Majority of conventional design of autonomous underwater vehicles used propellers as their principal mode of propulsion. The propeller-based locomotion [3], although rendered the initial answers to underwater locomotion, set issues on high-maneuverability, efficiency, and low power consumption. The scientific community and researchers also found that propeller-strikes produce greater amount of marine debris, marine creature's mortality and shallow water ecosystem disturbances. Biomimicked or fish-like robots are expected to be quieter, more maneuverable (lesser accidents), and possibly, more energy efficient (longer missions). Undulating-finned robot can preserve undisturbed condition of its surroundings for data acquisition and exploration (stealth). The movement of fish through water without creating ripples and eddies were more reasons to choose a bio-inspired design for underwater locomotion.

* Correspondence: abhra.roychowdhury@nus.edu.sg
[1]Department of Electrical and Computer Engineering, National University of Singapore, Singapore 117576, Singapore
Full list of author information is available at the end of the article

Considering the propulsive features [4] of existing fish modes, a novel propulsive mechanism that integrates fish-like swimming with modular links and fin movements has been proposed [3] where the modeling, simulation, and development studies of a body caudal fin (BCF) [5] carangiform-based prototype is built and tested. The robot will be able to implement speedy and efficient fish-like swimming. The present research work specifically identifies the usefulness of present model for purposes of both speed and maneuverability. The kinematics-based approach allows producing a dynamic body motion that can reproduce the fluid flow pressure field generating the undulatory motion of the fish. Further, the kinematics and dynamics study helped to frame the mathematical formulation of the fish body motion describing its dynamic behavior. From a robotics perspective, defining and enhancing the swimming efficiency is still a kernel issue in the study of robotic fishes. However, the fish robot swimming like a real fish does not guarantee that it would achieve the same high efficiency [3]. The undulatory (oscillatory) nature of the fish motion has been prominently mentioned in several works [4]. Another solution adopted by researchers [5-8] is to conduct large number of experiments and find empirical expressions to refine the body's motion functions and fin propulsion. Due to the complex nature of the mechanical system, the paper focuses on developing a linear system model using robot dynamics derivation. The simulation environment is in MATLAB©, Simulink©, and SolidWorks©. The contributions in the present work are enumerated as following:

- Mathematical input waveforms are proposed in LH framework to generate different types of (undulatory/oscillatory) body propulsive waves. Comparative study of each model with the fundamental LH quadratic wave model is done for a performance parameter based on the total trajectory length.
- Each of the bio-inspired wave function combines with an inverse kinematics-led trajectory planning resulting in a bio-inspired algorithm to produce laterally compressed waveforms of the caudal region.
- Integrate the present robotic fish mathematical model (kinematics and dynamics) with the proposed bio-inspired algorithms resulting from different mathematical inputs.
- Finding the operating region (ORE) for the identified kinematic parameters to facilitate closed-loop control based on the characterization of the biological fish swimming model.

The underlying subject of this research is to explore the remarkable ability of mathematical (applied mechanics) and physical concepts leading to fundamental insights behind fish biology. It has been found that sophisticated arithmetic calculations can bring practical benefits in revealing the fundamental biological systems. Darwinian-inspired evolution in modeling of biological systems is a strong partner of experimental work in physical sciences based on the principles in Newtonian and quantum frames. Present research investigates into this idea. The limitation of the paper is the linearized model of the robotic fish to undertake the study. Another assumption being made is stationary fluid, and therefore, the forces acting on a single link are due to the motion of that link. This may introduce inaccuracies in the manipulator model, while the discretized algorithm is implemented in a digital computer. The paper is organized as follows. In the 'System model' section, the design characteristics and features of the prototype are explained with the kinematics and dynamics modeling studies of a three link robotic fish. The 'Lighthill mathematical framework design' section discusses the Lighthill's existing mathematical framework and its integration with the present prototype. This section also investigates the proposal and study of the different mathematical input waveforms generating the fish undulatory and oscillatory motion. The 'Results and discussion' section presents the robotic fish body undulatory propulsion mechanism for a given trajectory, simulation, and experimental results with/without the Lighthill wave function as well as with different mathematical input in LH framework. Comparisons are based on a distance-based performance index. Closed-loop experiments [9] were made on the basis of comprehensive kinematic analysis governing the fish undulatory motion. An operating region is found for the dominant kinematic parameter TBF and propulsive wavelength each to facilitate future experiments for such bio-inspired systems. It is also verified with the reported literature for the real fish motion. In the 'Conclusions' the conclusions and directions for future work are discussed.

Methods

System model

The BCF mode carangiform style [5] of locomotion shown in Figure 1 is approximated using a 3-link (including the pectorals attached to the head) mechanism with two actuated joints as illustrated in Figure 2. The first link as the 'head' and second link as the body are roughly two-thirds of the length of the entire robot. The tail of the robot is formed by the third link connected to the caudal fin. The configuration of the robotic fish and non-fixed head is illustrated. Our 3-link mechanism ($n = 2$) is a reasonable approximation to the BCF carangiform locomotion, and therefore, small modifications of this model should have been used in the analysis of carangiform swimming. Moreover, the caudal peduncle and fin have been rendered in to a thunniform (shark) semi-crescent structure for an

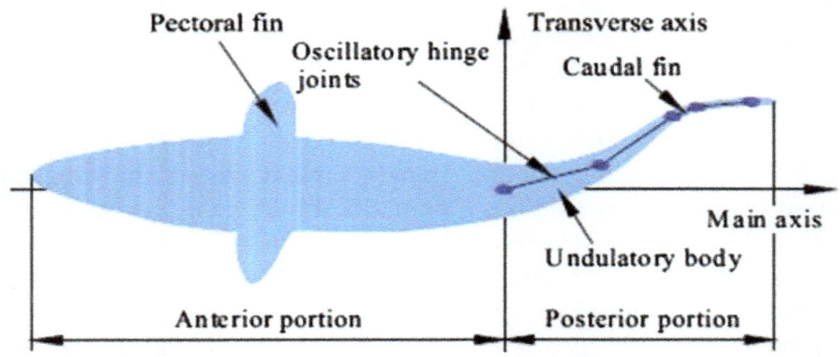

Figure 1 BCF mode of carangiform swimming showing undulatory motion in one-third part of the posterior body.

efficient thrust and therefore allowing high cruising speed for longer period of time [5]. While the kinematics study explains the geometry of the motion of robotic fish w.r.t, a fixed reference coordinate frame as a function of time, the dynamics of any rigid body [10-12] can be completely described by the translation of the centroid and the rotation of the body about its centroid. This leads to the ability to derive the actuator torques necessary to produce the tail motion that is desired. A linearized kinematic and dynamic model of the robotic fish system is developed. The present research work investigates into the system modeling as an n-joint manipulator-based mobile vehicle; the earth-fixed frame has been defined w.r.t. the fixed body reference

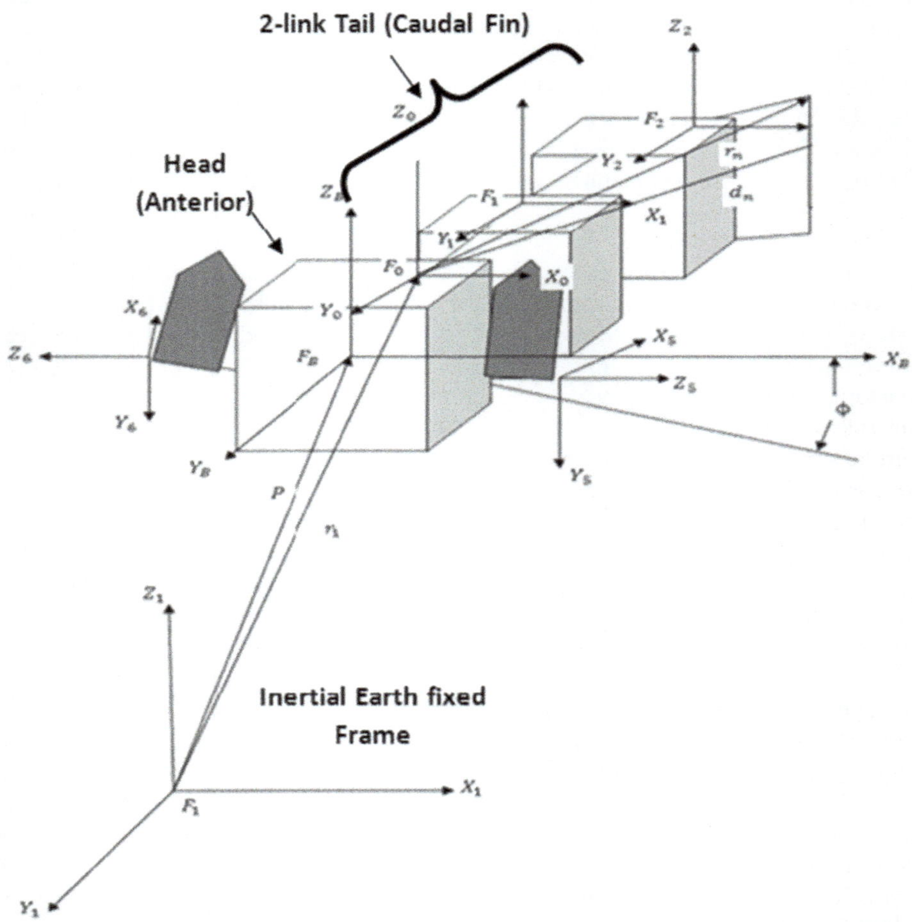

Figure 2 Relative orientations and locations of local coordinate frames at CM of the head and inertial earth-fixed reference frame.

frame. Relevantly, the two major sections of the present paper robotic fish system model are the following:

- Denavit-Hartenberg (DH) kinematics model [11]: The kinematics due to translation and rotation along the joints on the robotic fish system in fluid (water) environment.
- Lagrangian dynamics model [10,11]: The dynamics due to kinetic and potential energies generate along each link and inertia of the free-flow water entrained in the robotic fish system.

When describing the kinematics and dynamics of the model shown in Figure 2, the interlink actuator shaft constitutes the inertial frame of reference. A local coordinate frame is assigned to each degree of freedom (DOF). The coordinate frames are assigned according to the standard Denavit-Hartenberg notation [11] mentioned in Appendix A.

Generalized equations of motion

The dynamic equations of the robotic fish are obtained using the Lagrange-Euler formulation [10,11] given as:

$$\frac{d}{dt}\left(\frac{\partial L}{\partial \dot{q}_i}\right) - \frac{\partial L}{\partial q_i} = \tau_i \quad i = 1, 2, \ldots, n \tag{1}$$

The Lagrange (L) function is defined as the difference between the kinematic and potential energies expressed as:

$$L = K - P \tag{2}$$

where K is the total kinetic energy of the robot; P is the total potential energy of the robot; q_i is the joint variable of i_{th} coordinates; $\dot{q}_i$ is the first time derivative of the i_{th} joint variable, and τ_i is the corresponding generalized force (external torque) at i_{th} joint acting on the head. The manipulator dynamic equations have been developed in three dimensions for an n-link manipulator on a 6-DOF base, assuming that there is gravity acting on the system. The equations of motion for the present 2-link robotic fish based on the n-link serial manipulator can be set as:

$$\sum_{j=1}^{n} D_{ij}(q)\ddot{q}_j + \sum_{k=1}^{n}\sum_{m=1}^{n} h_{ikm}\dot{q}_k\dot{q}_m + c_i = \tau_i \qquad i = 1, 2 .., n \tag{3}$$

where $D\ (q)$ is the symmetric inertia matrix; $h(q, \dot{q})$ is the velocity coupling vector or Coriolis and centrifugal force vector; $c\ (q)$ is the gravitational vector; and τ is the generalized force of the Lagrange equations. As the above equation has been successfully used for investigating the dynamics of underwater vehicles [10,12] as well as robotic manipulators [11], this research therefore aims to develop an n-joint manipulator-based mobile vehicle. Following the analysis, the mechanical design in SolidWorks is implemented. SolidWorks as the mainstream software in virtual prototype field combines multibody dynamics modeling with large displacements as well as multifunctioning tools. It has a more powerful geometric modeling function. By utilizing SolidWorks, a kinematics model of a robotic fish, with coordinated motion of multiple propulsive mechanisms, is established as shown in Figure 3a,b,c and the real hardware prototype used for closed-loop experiments [9] is shown in Figure 3d. Details of the dynamic modeling of present bio-inspired robot can be found in [12].

Lighthill mathematical framework design

The initial work to understand the fish kinematics appeared prominently in the article by James Gray [13]. It mentions about the fish swimming by generating a traveling wave down their bodies from the anterior (leading surface) toward the posterior caudal tail (trailing surface). This propulsive wave travels faster than the fish body (center of mass) velocity. The wave amplitude also increases from the head to the tail, maximum at the narrow peduncle region. Lighthill in his analysis of Gray's work [13] extended and formulated the slender body theory, which proposed the fact that the overall fluid flow around the body is a compound of the steady flow around the straight body and the flow due to the lateral displacement $h(x, t)$. The second component (flow) $V(x, t)$ for a given cross sectional area S_x with fluid velocity U given by:

$$V(x, t) = \left(\frac{\partial h}{\partial t}\right) + U\left(\frac{\partial h}{\partial x}\right) \tag{4}$$

Lighthill's theory discusses the swimming efficiency in detail and discusses on the role of flow produced by the lateral displacement on the efficiency improvement during the undulatory propulsion. The overall swimming efficiency as a function of overall fluid flow $V(x, t)$ is termed as Froude's efficiency [13] η_f given by:

$$\eta_f = 1 - \frac{1}{2}\left[\frac{V(x, \bar{t})^2_{x=l}}{\left(\frac{\partial h}{\partial t}\right) V(x, t)_{x=l}}\right] \tag{5}$$

The theory uses a set of partial differential equations to calculate the thrust and efficiency of swimming with definite rhythmic and symmetric body motions. Supporting a laterally compressed wave and maintaining a high Froude number efficiency (as might be done by the biological fishes in real), the motion of a traveling wave displacement vector $h(x, t)$ was introduced by Lighthill as an empirical expression given by:

$$h(x, t) = f(x) \times g(t - x/c) \tag{6}$$

where the $f(x)$ term that indicates the amplitude, $g(x, t)$ is an oscillatory frequency dependent wave function, and c is the fish body velocity. Out of the many alternative

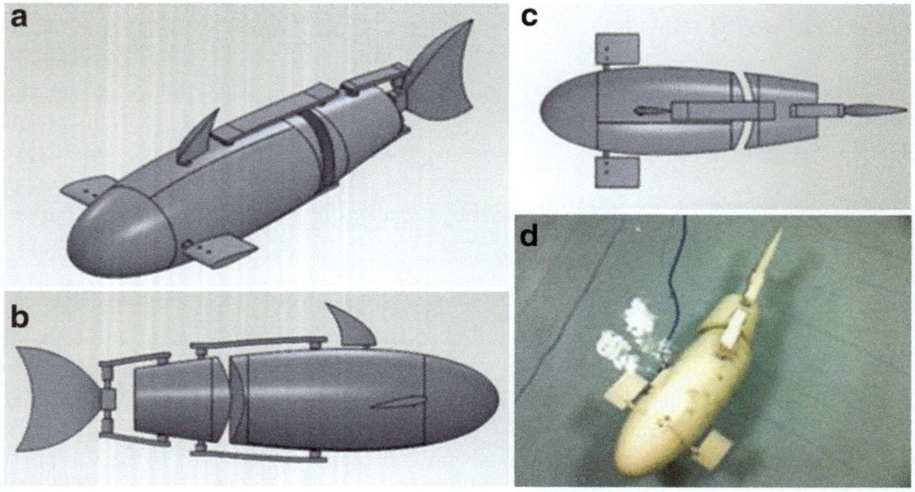

Figure 3 SolidWorks model and hardware prototype of the robotic fish. (a-c) Kinematics model of robotic fish with coordinated motion of multiple propulsive mechanisms. **(d)** Real hardware prototype used for closed-loop experiments.

forms that can be represented, the equation structure resembles primarily that of a time-dependent harmonic oscillation wave. It is found that propulsion patterns generated by fishes are dependent on purpose like food-search, hunting, migrating, mating, etc. [14]. The body dimensions (size) evolved over the ages as well as a specific environment also plays major role in these patterns. Based on the detailed study of the biological attributes of fish undulatory propulsion as well as Lighthill postulation on the oscillatory motion, a novel approach is made to extend and evaluate different mathematical functions that can fit in the frame of Lighthill. The integration of Lighthill mathematical model with robotic fish inverse kinematics and dynamics model as shown in Figure 4. The platform to validate the present research is the robotic fish designed and fabricated in our laboratory. Also, it is significant from the point of view that an evolutionary (biological) trait can be studied through the bio-inspired algorithms evolving from these functions, like different need based actions, for example, a minimum energy body motion to travel a fixed distance etc. Therefore, a variety of actions and environments in real world can be understood through this research on a robotic fish (machine world). We can term it as understanding organic evolution of fish swimming (a biological hypothesis) through bio-inspired machines (designed on principles of mathematics and physics) [15,16], in an inverse learning map. Another viewpoint presented through our 3 DOF robotic fish model is that, on varying this DOF, a proportionate variation can be seen in the resulting motion. For

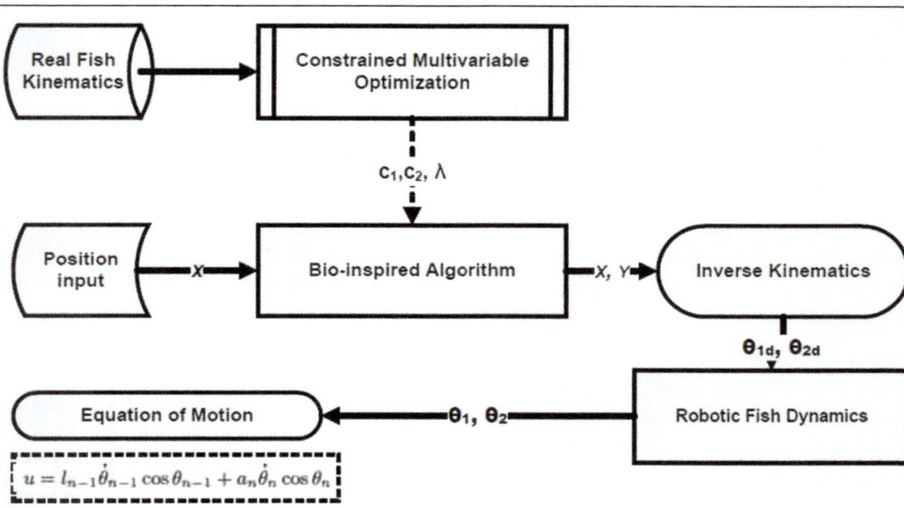

Figure 4 Block diagram showing integration of LH model in the robotic fish kinematics and dynamics model.

example, selecting one DOF and a sine oscillating function would generate a tadpole motion in a 2D plane. Similarly, selecting one DOF and a suitable spline wave equation can produce a tadpole helical motion in 3D plane. On the other hand, if we introduce a 2 DOF and the Lighthill quadratic wave function, it results into a planar BCF carangiform (undulatory) motion. Further, adding a DOF and running the median pectoral fin (MPF) anguilliform waveform can lead to a more maneuvering motion resembling between a tuna and an eel. Therefore, combination of the various DOFs and Lighthill frame wave function can together attain as well as sustain different undulatory actions to undertake various tasks in different environments. The different mathematical wave functions are given below.

Undulatory Lighthill quadratic amplitude body wave

Lighthill was the one of the pioneers in applying the methods of slender body theory [13] to an undulating body swimming in an inviscid fluid medium. Dewar's [17] practical studies on kinematics and energetics on carangiform fishes by direct observations of biological fish are imminent to support Lighthill's theory as they lie in a digitized record. It was later used in one of the pioneering works of biologically inspired robot called MIT Robotuna [6] to validate a carangiform robotic swimming mode and other prominent research findings [7,15]. In this work, the mechanical robotic fish body generates a planar progressive wave function as a function of lateral curvature in spine and musculature, showing the undulatory propulsive behavior.

$$y_{body}(x,t) = \left(c_1 x + c_2 x^2\right)\left[\sin(kx + \omega t)\right] \tag{7}$$

where y_{body} is the transverse displacement of body, x is the displacement along main axis, k is the body wave number ($k = 2\pi/\lambda$), λ is the body propulsive wave length, c_1 is the linear wave amplitude envelope, c_2 is the quadratic wave amplitude envelope, and ω is the body wave frequency. Taking this body wave, the present research does kinematic analysis to determine the proper body wave parameters (c_1, c_2, k, ω, etc.) for a desired and efficient undulatory swimming motion. We have further compared other mathematical functions in this frame (given below) to find their suitability to replace for a better undulatory wave function for robotic fish rectilinear swimming in a two-dimensional plane.

Undulatory Lighthill cubic amplitude body wave

In this paper, we propose the cubic body wave equation as an extension to Barrett's formulation by replacing the quadratic amplitude expression by a cubic spline polynomial. The motive behind this was to observe the efficacy of cubic polynomials in designing a continuous and smooth (non-jerky) motion through each rotational joint and therefore, passing more closely through the lateral displacement data points. It also accommodates more intermediate or via points in the plane. In the field of trajectory generation for robotics, cubic polynomials have been reported in majority due to its ability to track complicated trajectories efficiently [18].

$$y_{body}(x,t) = \left[\left(c_1 x + c_2 x^2 + c_3 x^3\right)\right]\left[\sin(kx + 2\pi f \times t)\right] \tag{8}$$

As the curvature of the wave primarily depends on the second derivative, it is found to be continuous here, to assist in velocity control. A major advantage of adding local intermediate points is not only to fit a smooth curvature for the harmonic motion but also flexibility to the body in physical terms. Throughout a given time period when the speed, orientation, and velocity changes gradually, these factors render the resulting trajectory smoothness. In physical terms, this smoothness means that there are lesser abrupt changes in power output for the robot's drive system, thus reducing navigational errors and helping to moderate the robot's energy consumption. Therefore, the function allows specifying undulatory motion that conserves both battery power while reducing travel time and minimizing navigational errors.

Non-uniform rational B-spline (NURB) quadratic and cubic body wave (tadpole-like motion)

In order to adapt the Lighthill frame harmonic waveform from control design perspective (with the presence of control points), a Bezier spline traveling waveform is proposed in this paper. Wu [19] has modified the methods of thin airfoil theory to analyze the motion of a waving 2D plate. The essential 2D and 3D methods developed by Lighthill and Wu, respectively, have formed the basis from which the turning models were developed. Singh and Pedley [20] claim that Lighthill's classical elongated body theory for fish swimming forms the fundamental basis for the 3D flow model modified from the 2D small amplitude model reported earlier. Literature reports of combining efficient swimming propulsive force (speed) and maneuverability in the tadpole-like swimming mode can also be found [21]. The survey further mentions 3D propulsive motion of a free swimming larvae generating and controlling helical trajectory (rotation while swimming) by Long et al. [22]. Morphological and kinematic asymmetries of tail motion generate a rolling and pitching action [23] in addition to the yaw motion. Similar work, by Krishnamurthy et al. [24] find a bio-robotic implementation (robotic electric ray) in planar motion by modulating the propeller speed as the control variable, but the implementation was done in 2D (as a sub-case of 3D) to avoid dynamic complexity issues. Instantaneous trajectories are computed by extending Lighthill's two-dimensional model in *xyz*-plane by Singh and Pedley [20] and the tadpole's cycloptic helical motion [22]. Inspired by the aforesaid work, a framework was incubated and finally modeled for fitting the present robotic fish kinematic model, generating the 3D turning maneuvers. An attempt is made in this research also to define a mathematical expression in Lighthill frame for the 3D undulatory motion. It is to be noted that the robotic fish model is actuated by rotational joints with 1 DOF which can lead to various singularities in response to the mathematical solutions generated by the NURB trajectory. This motion would aim to satisfy the combined requirements of propulsive forces (speed) and turning angles (maneuverability) from the perspective of an energy efficient trajectory. This section, therefore, proposes and formulates a NURB mathematical equation as the new parametric curve to represent the propulsive wave behavior as lateral curvature in spine and musculature in 3D. In addition to the control points, other advantages of using this model is that it can offer a common mathematical form for standard analytical shapes and also provide the flexibility to design a large variety of parametric shapes. Secondly, the geometrical evaluation can be reasonably fast by numerically stable and accurate algorithms. They are invariant under affine as well as perspective transformations. They are generalizations of non-rational B-splines and non-rational/rational Bezier surface curves. NURB shapes are not only defined by control points but also by their weights associated with each control point. This waveform assumes each reference point to be a control point as the traveling wave passes through only the first and the last control points therefore reducing the path length. This also adds to the trajectory planning in a way as the full control over the wave shape is achieved by tweaking a few parameters. They are also necessary to make it an optimal choice for the propulsive wave model. The weight calculation like parameter in other mathematical expressions is done based on the real kinematic data of carangiform fish. A NURB curve $C(u)$, which is also a vector-valued piecewise rational polynomial function representation, can be expressed as:

$$C(u,v) = \frac{\sum_{i=1}^{n}\sum_{i=1}^{n}\{w_i \times P_i \times N_{i,k}(u,v)\}}{\sum_{i=1}^{n}\sum_{i=1}^{n}\{w_i \times N_{i,k}(u,v)\}} \quad (9)$$

where w_i is the weights, P_i is the control points (vector), and $N_{i,k}$ is the normalized B-spline basis functions of degree k. The proposed mathematical model is in a normalized form. Therefore it reduces the scope of redundancy and dependency on one hand whereas increasing the consistency in results on the other. The proposed NURB mathematical model in Lighthill frame is verified for quadratic spline equations. To implement a curved trajectory, an arbitrary orientation for the robotic fish vehicle at its final position cannot be specified. To overcome this shortcoming, a somewhat more complicated curve in the form of a three-degree cubic polynomial is proposed, which uses an additional amplitude coefficient (weight) c_3 form. Like in the case of other two coefficients c_1 and c_2, its value is also dynamically calculated for the present robotic fish on the basis of the real fish study report [17]. Undulatory motions defined by the function in two-degree and three-degree NURB equations are written as:

$$y_{body}(x,t) = \left\{\frac{c_1x + c_2x^2}{\sqrt{c_1{}^2 + c_2{}^2}}\right\} \sin(kx + 2\pi f \times t) \quad (10)$$

$$y_{body}(x,t) = \left\{\frac{c_1x + c_2x^2 + c_3x^3}{\sqrt{c_1{}^2 + c_2{}^2 + c_3{}^2}}\right\} \sin(kx + 2\pi f \times t) \quad (11)$$

The smooth transition it provides helps preserve momentum, reduce navigation error due to loss in the robot's drive mechanisms, and allow the robot to maintain its speed throughout its maneuvering. The proposed form (and its derivative) shown in the above equation has the advantage of being both continuous and differentiable at all points in the interval across all segments. The second derivative is also continuous throughout and differentiable across each interval except at 'via' points between segments. This form can permit considerable freedom in selecting the robot's orientation (turning) at via points. It is now computationally easier to calculate velocity, rotational velocity, and acceleration at via points throughout the trajectory, therefore, allowing us to specify maneuvers that can reduce travel time and minimizing navigational errors while conserving the battery power. The mathematical modification is implemented in the form of a transition from the existing uniform no-rational spline equation (Lighthill quadratic and cubic spline wave equations) to the NURB. As discussed earlier, another major implementation issue challenged by this improved model is to test the dynamic undulatory motion of the fish in a three-dimensional plane which may not be well-defined under the existing two-dimensional plane Lighthill linear/quadratic amplitude equations. Its efficacy in the control performance remains to be verified in the future work. But based on its properties [25] and results obtained in the present research, it is validated that the equation causes sudden rise around the initial conditions (*jerk* at start point) as shown in Figure 5b, if control points do not have a specific arrangement. Except for the two (quadratic and cubic) Bezier splines, other waveforms reveal smooth curvature indicating an undisturbed oscillatory/undulatory. This effect causes unwanted offset distance of desired control points from the actual trajectory points.

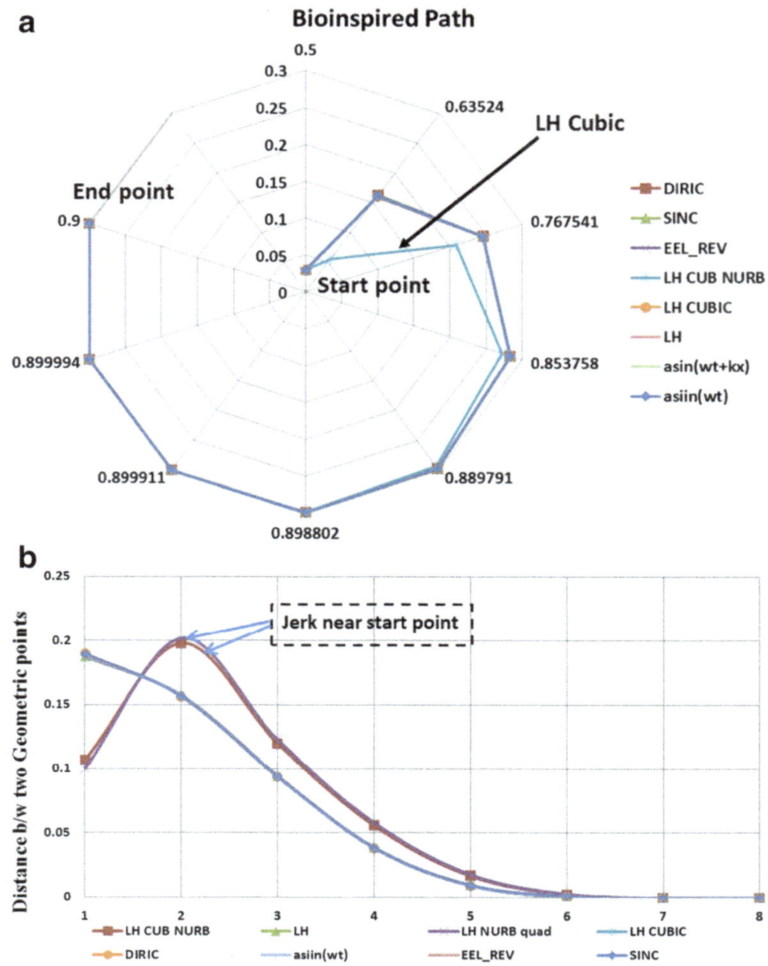

Figure 5 Time trajectory for different oscillatory/undulatory wave-functions. (a) Trajectory comparison graph for all waveforms. **(b)** Sudden rise around the initial conditions if control points do not have a specific arrangement.

Undulatory SINC and DIRIC body waves

The mathematical function called cardinal sine or sinc oscillates between positive and negative values with equal periodic repetition represents a sine wave that decays in amplitude as $1/x$. This oscillation represents a decrease in amplitude with an increasing frequency. This also depicts that the mean value at a point $x \rightarrow 0$ will have a higher magnitude when compared to neighborhood of $\partial x \rightarrow 0+$, $\partial x \rightarrow 0-$ as well as for other oscillating peaks. The function is continuous at all real values with a removable singularity at $x = 0$ (where first derivative is equal to 1). Figure 6a shows two unnormalized (blue) and normalized (red) mathematical expressions. The normalization causes the definite integral of the function over the real numbers equal to 1. It is a building block for a large function class in Fourier analysis (a major technique in the solution of differential equations). It is also seen to be a solution to the wave equations for transmission in communication theory. Physical meaning can be interpreted as a traveling wave signal propagated through body leading edge to the trailing edge for a sharp or gradual turning action. If we look into the traveling wave envelope proposed by Lighthill, sinc function suits into it such that it moves down the fish body with velocity c, and whose amplitude $c_1x + c_2x^2$ may vary with position along the fish body. The proposed undulating motion in Lighthill frame is given as:

$$y_{body}(x,t) = \left[(c_1x + c_2x^2)\right]\left\{\frac{\sin(kx + 2\pi f \times t)}{2\pi f \times t}\right\} \quad (12)$$

An extension to the sinc function, in the form of a Dirichlet function is defined in the Lighthill frame. This function is found to be a periodic sinc or aliased sinc function.

$$y_{body}(x,t) = f(x) \times h\left(t - \frac{x}{c}\right) \quad (13)$$

Where

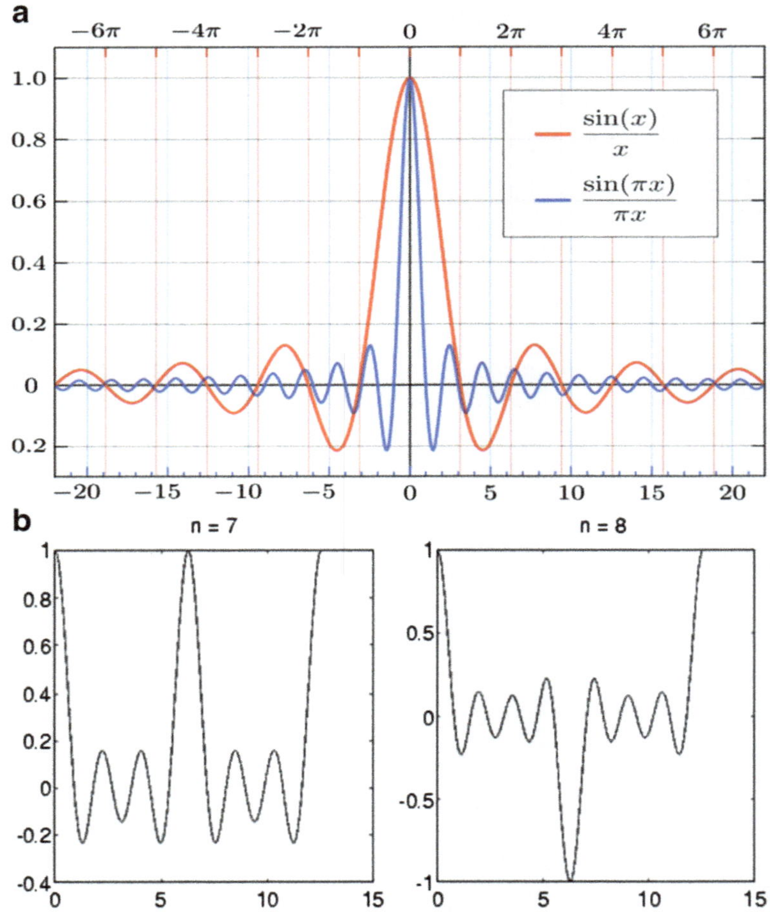

Figure 6 Undulatory cardinal sine traveling wave function. (a) Two unnormalized (blue) and normalized (red) mathematical expressions. **(b)** The tail beat of the caudal region as the traveling wave (midline) switches the amplitude mean value depending on the value of *N*.

$$h(x,t) = \left\{\frac{\sin\ N(kx + 2\pi f \times t)}{N\sin(2\pi f \times t)}\right\}. \tag{14}$$

For N as odd, the function has a period of 2; for N as even, its period is 4. It can be used to add orientation to the tail beat of the caudal region as the traveling wave (midline) switches the amplitude mean value depending on the value of N. It is verified in its discrete Fourier transform for an N-point rectangular window as shown in Figure 6b. The centerline mathematical equation of motion in Lighthill frame is given by:

$$y_{body}(x,t) = \frac{1}{(c_1x + c_2x^2)}\left\{\frac{\sin(kx + 2\pi f \times t)}{\sin(2\pi f \times t)}\right\} \tag{15}$$

It can be noted that the quadratic wave amplitude is in the denominator and is used to dampen the oscillating wave function unlike the sinc function where it is allowed to grow as the numerator.

Undulatory anguilliform body wave (EEL like/maneuvering model)

Replicating the kinematics of a silver lamprey, a MPF anguilliform motion was proposed by Tytell [26]. From the maneuvering perspective, this equation adds another dimension to the overall fish propulsion. The motion of the fish centerline described by an exponentially growing traveling wave in Lighthill frame is given by a function as follows:

$$y_{body}(x,t) = \left[\left(c_1x + c_2x^2\right)\exp\left(2.18\left({}^{x}/_{L}-1\right)\right]\right. \tag{16}$$

where y is the lateral position of the midline, x is the coordinate following the midline, α is the amplitude growth rate parameter [26], L is the body length, and λ is the propulsive wave length. Modifications were done in this equation as compared to the original equation due to the fact that the present robotic fish model is carangiform so the DOF is restricted to the body posterior (caudal) unlike anguilliform where the entire centerline participates in undulation.

The idea of testing a modified anguilliform equation in a carangiform frame is whether maneuverability can be improved in slow undulatory motion. Two changes have been accommodated in order to undertake this test. Firstly, by replacing the constant amplitude by a quadratic amplitude wave and secondly, by provisioning that α can act as both an amplitude growth and dampening factor. This mathematical expression shows that a main wave function (sine or cosine) is decomposed in two Fourier functions. The primary function is again an oscillating sine wave which is multiplied by an exponentially growing function (can be represented in Fourier series) to satisfy the boundary conditions. The continuous everywhere exponential function is used for damping the sine wave. This function also resembles the Laplace transform with open unity integral.

Results and discussion

The implementation in robotic fish prototype for the Lighthill model verification and kinematic analysis is done in MATLAB© and Simulink© environment. Simulation results including time histories of forward velocity as well as the corresponding trajectories are discussed. Tests were done to show the effectiveness of Lighthill's propulsion model. Two conditions were compared, where the fish is allowed to travel a trajectory (using inverse kinematics algorithm) from an initial point to a destination point, without the Lighthill undulatory action as compared to the natural mode of its body motion, i.e., with Lighthill mathematical model implementation, fixed at three identified major parameters, i.e., oscillating (tail-beat) frequency (ω), propulsive wavelength (λ) and caudal amplitude (c_1 and c_2). For more clarity, Figure 7 shows the relative trajectory traversed with respect to center (origin) comparing the fish undulatory motion trajectory, with and without the Lighthill model implementation, clearly depicting the shorter route maintained by the Lighthill model over the non-Lighthill model. It therefore validates the classical theory postulated by Lighthill for the present robotic fish model. The above simulation shows the results and graphs; the performance factor of the robotic fish is calculated for a given (fixed) distance and battery power (onboard energy). It is found that with Lighthill implementation, the motion performance improves by 14.68%. The motion performance (PI) is defined as follows:

$$PI = \left[\frac{distance_{w/o\ LH} - distance_{with\ LH}}{total\ \ distance}\right] \tag{17}$$

The distance between two geometric points on a plane is calculated by the straight line equation algorithm:

$$distance_{b/wpp} = \sqrt{(x_{i+1}-x_i)^2 + (y_{i+1}-y_i)^2} \tag{18}$$

The initial point is $x = 0.5$, $y = 0.0297$ and the final point is $x = 0.9$, $y = 0.3$. The coordinate points are normalized to a scale of 1: 100 (for distances measured in meters). The distance between these two points is found to be 0.482765046 units which is also the shortest distance (normal between the two planes). In the Lighthill quadratic wave equation, the amplitude constants c_1 and c_2 have been assumed to have a value [27] with little information on their choice. Our present research looks into the real kinematics of carangiform tuna fish in [17] as well as in the kinematic formulation of the Robotuna [17]. For two different conditions of caudal amplitude, the value of (c_1, c_2) is calculated to be 0.002, 0.00835 and 0.2, 0.01736. For undulatory cubic spline, the values of c_3 are determined to be 0.01 and 0.346 under the two constraints. Tables 1 and 2 give the detailed information on the geometric points covered during the path traveled by the robotic fish. The paths are also plotted as trajectories as shown in Figure 8a,b,c,d,e,f,g,h. In Table 1, the trajectory details have been tabulated for wave function of oscillating sine, undulatory Lighthill quadratic, Lighthill cubic and NURB quadratic waves in the form of geometric points on the traveling plane of the robotic fish (undulating with these mathematical input wave functions). Their trajectories are plotted in Figure 8a,b,c,d. Comparing these trajectories with respect to the shortest distance, it is found that the percentage difference between the shortest path and user-defined wave generated path was least for undulatory Lighthill cubic polynomial with 1.2%, followed by undulatory Lighthill quadratic polynomial with 1.22%, and standing oscillating sine wave with 1.25%; but a major deviation from the minimum value is observed in undulatory NURB quadratic wave with 4.19%. In Table 2, the trajectory details in the form of geometric points have been tabulated for mathematical input function of undulatory NURB cubic, SINC, DIRIC, and anguilliform waveforms. The trajectories drawn for these functions are shown in Figure 8e,f,g,h. It is noted that maximum percentage deviation from the shortest path continues with NURB cubic wave with 3.44% followed by SINC with 1.27%, DIRIC with 1.28%, and the lowest value (closest to the shortest distance) was with the anguilliform with 1.24%. The user-defined wave functions have been analyzed, and some of them have been used to generate undulating traveling wave through the robotic fish body during closed-loop feedback operations. From Tables 1 and 2, it can be deduced that out of all the user-defined wave functions, the undulatory Lighthill cubic polynomial emerges out to be the most effective wave input to generate the trajectory with an improvement of 17.94% over a non-Lighthill path and an improvement of 1.1% over the Lighthill quadratic wave trajectory. It is closely followed by undulatory Lighthill quadratic polynomial

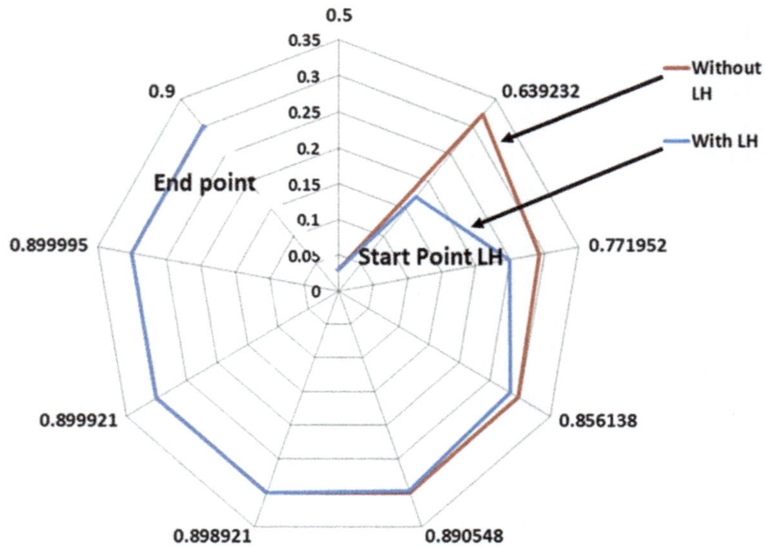

Figure 7 Relative path comparison relative to a center point (meters).

with 16.32% and anguilliform input with 14.7% mentioned in [26]. Plotting a trajectory comparison graph for all in Figure 5a, it is clear by looking into the blue line denoting the Lighthill cubic waveform that it generates as the path closest to the shortest distance between the two points of travel. During formulation, it has been strictly adhered that the function does not require computational complexity which will pose problem in real-time implementation. Another comparative study is undertaken relative to the primary Lighthill quadratic wave function. In Table 1, which mentions oscillating sine, Lighthill cubic, and NURB quadratic waveforms, (apart from Lighthill quadratic) it appears that only Lighthill cubic wave shows improvement, whereas, the two other wave functions namely oscillating fin and NURB quadratic wave stay close but record more than the Lighthill quadratic distance by 1.02% and 28.7%, respectively. A similar comparison is undertaken for Table 2. It brings forth the fact that out of all the input functions mentioned in Table 2, NURB cubic wave shows maximum deviation from the Lighthill quadratic distance with 22.2% followed by DIRIC wave to be 6.4%, SINC wave to be 5.8%, and the closest to the ideal value is anguilliform wave with 2.2%. Although all of them have shown a higher magnitude, w.r.t. Lighthill quadratic wave generated trajectory but some of them have managed to stay close to it. Therefore, two conclusions can be drawn from the complete analysis. Firstly, undulatory LH cubic waveform can prove to be an efficient algorithm for trajectory generation of a traveling propulsive body wave for robotic fish rectilinear motion. Secondly, other input functions proposed can also be used but mainly based on the required objective/action for, e.g., sinc or diric for turning or based on body dimensions for, e.g., NURB quad or cubic used for a tadpole-like motion. It can also be chosen on a purpose for, e.g., anguilliform/eel-like for maneuvering and carangiform for speed. These

Table 1 Trajectory (geometric) points for mathematical oscillatory/undulatory propulsive waveforms

Node number	Oscillatory sin		LH quad		LH cubic		NURB quad	
	x pos	*y* pos	*x* pos	*y* pos	*x* pos	*y* pos	*x* pos	*y* pos
1 (start point)	0.500000	0.029700	0.500000	0.029700	0.500000	0.029700	0.500000	0.029700
2	0.635240	0.160918	0.634913	0.159944	0.634781	0.159549	0.598402	0.046079
3	0.767541	0.245599	0.767169	0.245283	0.767018	0.245155	0.723631	0.204411
4	0.853758	0.283287	0.853553	0.283192	0.853471	0.283154	0.827144	0.270115
5	0.889791	0.296484	0.889726	0.296460	0.889699	0.296450	0.880214	0.292851
6	0.898802	0.299593	0.898791	0.299589	0.898787	0.299587	0.896969	0.298948
7	0.899911	0.299970	0.899910	0.299969	0.899910	0.299969	0.899745	0.299912
8	0.899994	0.299998	0.899994	0.299998	0.899994	0.299998	0.899982	0.299994
9 (end point)	0.900000	0.300000	0.900000	0.300000	0.900000	0.300000	0.900000	0.300000

Table 2 Trajectory (geometric) points for mathematical oscillatory/undulatory propulsive waveforms

	NURB cubic		SINC		DIRIC		Anguilliform	
	x pos	*y* pos	*x* pos	*y* pos	*x* pos	*y* pos	*x* pos	*y* pos
1 (start point)	0.500000	0.029700	0.500000	0.029700	0.500000	0.029700	0.500000	0.029700
2	0.603409	0.055233	0.635629	0.162066	0.635703	0.162282	0.635182	0.160746
3	0.728721	0.208521	0.767983	0.245970	0.768066	0.246040	0.767476	0.245543
4	0.830356	0.271654	0.853999	0.283398	0.854045	0.283419	0.853722	0.283270
5	0.881458	0.293319	0.889869	0.296513	0.889884	0.296518	0.889780	0.296480
6	0.897237	0.299041	0.898814	0.299597	0.898817	0.299598	0.898800	0.299592
7	0.899771	0.299921	0.899912	0.299970	0.899912	0.299970	0.899911	0.299970
8	0.899984	0.299995	0.899994	0.299998	0.899994	0.299998	0.899994	0.299998
9 (end point)	0.900000	0.300000	0.900000	0.300000	0.900000	0.300000	0.900000	0.300000

mathematical inputs are responsible to build different bio-inspired algorithms which later combine with the real-time kinematics and dynamics of the robotic fish. Henceforth, the robotic fish driven by these algorithms support adaptation policy shown by a biological fish in different environments.

Operating region

Using the system dynamic model formulated as Equation (3), MATLAB simulation result are generated for the robot closed-loop motion between two points in a three-dimensional plane. This mechanism uses a bio-inspired algorithm under Lighthill cubic wave function as it is found to be the best result for trajectory generation. The simulation results [28,29] are useful during the closed-loop experimental verification [9] and operation of the robotic fish prototype as we know the variation of forward velocity with major kinematic parameters like tail-beat frequency (TBF), caudal amplitude (CA), propulsive wavelength (PW), and yaw angle. To enhance the system repeatability therefore reliability, each parameter is plotted for various wavelengths and operating tail-beat frequency values, resulting in the operating region. Based on the simulation results discussed above, an ORE [30] that summarizes these major parameters in consideration is formed in the form of a plot as shown in Figure 9a,b. The vertical axis in usual notation mentions swimming (forward) velocity in cm/s while the horizontal axis is denoted by two parameters, TBF and PW. The range of TBF is from 0 to 2.8 units (Hz) while for *PW* it is from 1.2 to 2.2 units (m). Variation of swimming speed with TBF and PW is shown as oscillating bell-shaped curves with circle and square markers, respectively. The two peaks marked by a star (with close dashed lines) symbol and a triangle (with wide dashed lines) symbol show the peak velocities achieved for a given value of TBF (1.1 Hz) and PW (1.45 cm), respectively. It is to be observed that for the peak velocity of 17.35 cm/s, the choice of TBF is unique, i.e., 1.1 Hz; but there is no wavelength available in the permissible range. Contrary to this, if the peak velocity of 17.3 cm/s (triangle symbol) is chosen, for a unique value of propulsive wavelength at 1.45 cm, there exist two TBF values at 1 and 1.15 Hz. Therefore, for any choice of speed (based on the purpose like fast or moderate swimming), the combinational choice of the two parameters can be found for robotic fish. Interestingly, it can be observed that most of the times, for a given velocity, there exists two or more values of TBFs or PWs. This is due to the non-linear nature of the curve or many one function mapping. The final choice by the biological fish must be done based on the metabolic rates (related to Froude efficiency) or any other hidden parameter, which is our future course of study. The value of caudal amplitude is kept constant at $c_1 = 0.002$ and $c_2 = 0.00835$. Another plot, Figure 9c from Dewar's experiment on the biological equivalent (yellowfin tuna), shows the results of swimming velocity (cm/s) versus time. This plot also shows metabolic rate (triangular marker) trend with time as the fish moves. The swimming velocity pattern and magnitude are in agreement with the present robotic fish kinematic reports. This suggests that the kinematic design of present robotic fish has somewhat successfully emulated a real fish. Clearly, the kinematic studies relation of the four parameters (TBF, CA, PW, and yaw angle) for the present robotic fish underwater vehicle model is validated at par with Dewar's findings [17]. Looking from another dimension [16,24], this can be seen as an important and useful research to understand the biology of fish swimming, with applied mechanics replacing fish dynamics and arithmetic formulation substituting for the undulatory input motion, respectively. Robotic fish can therefore be used to study the fish biology since many parametric results were found to be in agreement with test data of tropical yellowfin tuna. The present ORE patterns therefore can also allow us to make predictions closer to the real fish kinematics. Another objective of using the ORE is to implement the closed-loop [9] control of the robotic fish with the help of comprehensive kinematic study undertaken the w.r.t. parameters. This would also make it

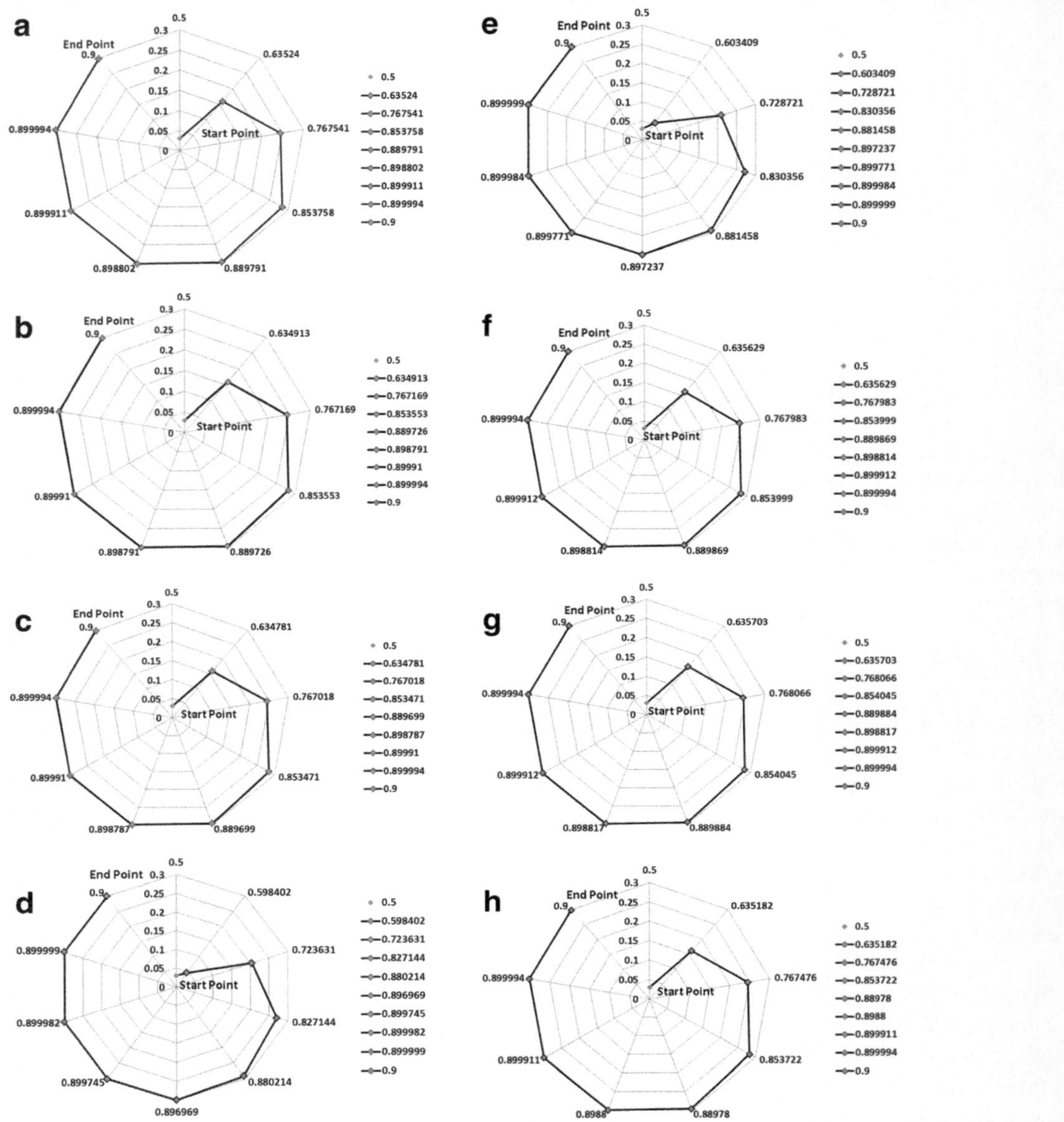

Figure 8 Trajectory traversal between two fixed points using different oscillatory/undulatory wave functions. (a-d) Plotted trajectories for wave function of oscillating sine, undulatory Lighthill quadratic, Lighthill cubic, and NURB quadratic wave in the form of geometric points on the traveling plane of the robotic fish. **(e-h)** Trajectory details in the form of geometric points tabulated for mathematical input function of undulatory NURB cubic, SINC, DIRIC, and anguilliform waveforms.

easier to operate the robot and less time consuming as the kinematic characteristic of robotic fish is presented in the form of a simple operating region chart [30]. This study stresses on the importance of major kinematic parameters while neglecting the minor parameters to avoid computational complexity. Present kinematic study does a value addition to the closed-loop control [9,31] technique as it discusses intriguing yet solvable arithmetic formulations (of input waveforms) in Lighthill frame to run a sophisticated biological process (fishlike undulatory/oscillatory motion). The simulation results also denote that amplitude and wavelength are found to be the less chosen parameters compared to the TBF if swimming speed is considered. Also a compromise (optimization) of these three important parameters is required for the most efficient undulatory fish body motion. Clearly, the relation of distance-based performance factor and forward velocity for the present robotic fish underwater vehicle model is validated at par with Dewar's findings [17]. The yaw axis servomotor is a HiTec HS-5646WP (11.3 Kg-cm/6 V) used in all the joints. Arduino-Uno with ATMEGA2560 serves as the central processing unit for the robotic fish. The mechanical

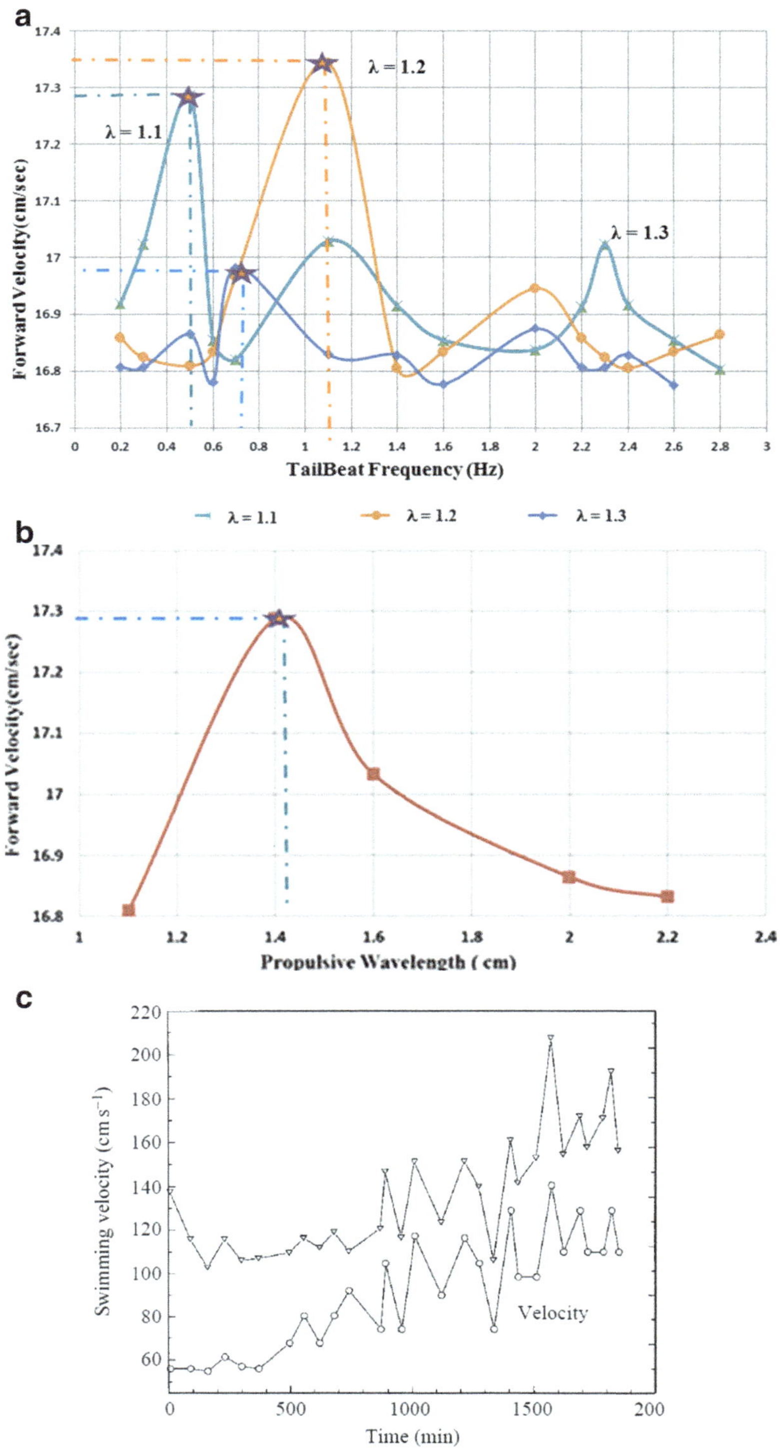

Figure 9 Operating region for tail beat frequency (TBF) and propulsive wavelength.

computer-aided (CAD) design and animation has been implemented with the help of the Solidworks (mechanical model) and MATLAB SimMechanics VRML (3D motion). The motivation of this paper is described by the remarkable swimming abilities of BCF mode carangiform style of swimming built as a robotic fish prototype. The NACA airfoil aerodynamic structure has been designed to boost the swimming efficiency by reducing the drag. Using the present dynamic model and derived steady kinematic simulation results [28,29], the closed-loop experiments [9] were done for different body motion configurations to emulate the undulation of the robotic fish in fluid environment shown in Figure 10.

Conclusions

The present research has focused on the relevance of Lighthill (LH) based biomimetic robotic propulsion of a proposed 2-joint, 3-link multibody vehicle model, biologically inspired by a BCF carangiform fish. The objective of this paper is to translate the BCF mode carangiform swimming behavior of a biological fish to a robotic fish to allow its energy efficient navigation over a given distance using a good balance speed and agility characteristics. The robotic fish model (kinematics and dynamics) is integrated with the LH mathematical model framework. Mathematical input waveforms are investigated in LH framework to generate posterior body undulatory movements. These functions are combined with robot inverse kinematics to generate various bio-inspired trajectories for the posterior robotic fish vehicle motion. Distance-based performance criteria for a given trajectory are proposed to do a comparative analysis for the input undulatory waveforms. Comparisons are done between a non-LH and a path defined in LH frame. Based on present kinematic model simulation of identified kinematic parameters, closed-loop experiments are done to establish operating region for two critical kinematic parameters TBF and PW. This finding also aims to facilitate future experiments for a robotic fish model. Interestingly, the robotic behavior in simulation and experiments (closed-loop) is showing swimming behavior similar to the biological fish mentioned by Lighthill slender body theory and Dewar's kinematic experiments on yellow fin tuna. The future work primarily focuses on the development of an improved inverse kinematics algorithm within the Lighthill framework. A behavior-based control strategy and its implementation for energy efficient undulatory fish motion would further strengthen the vision of the machines mimicking biology principles in a significant way.

Appendix A

(a) Link and joint parameters (shown in Figure 2)

- Joint angle (θ_i): the angle of rotation from the X_{i-1} axis to the X_i axis about the Z_{i-1} axis. It is the join variable if joint i is rotary.
- Joint distance (d_i): the distance from the origin of the *(i-1)* coordinate system to the intersection of the Z_{i-1} axis and the X_i axis along the Z_{i-1} axis.
- Link length (a_i): the distance from the intersection of the Z_{i-1} axis and the X_i axis to the origin of the i_{th} coordinate system along the X_i axis.
- Link twist angle (α_i): the angle of rotation from the Z_{i-1} axis to the Z_i axis about the X_i axis.

(b) Reference Frames

- F_I : inertial frame of manipulator-based system.
- F_B : base frame located at the center of mass of the base.
- F_i : coordinate frame of the i_{th} link of the system.

(c) Vectors

- r_B : position of frame F_O relative to and projected onto frame F_B.
- r_I : position of frame F_O relative to and projected onto frame F_O.
- d_i : position of frame F_i relative to and projected onto frame F_O.
- r_i : position of point on link i relative to frame F_O.
- i : position of point on link i relative frame F_i.
- b : position of point on the base relative frame F_B.

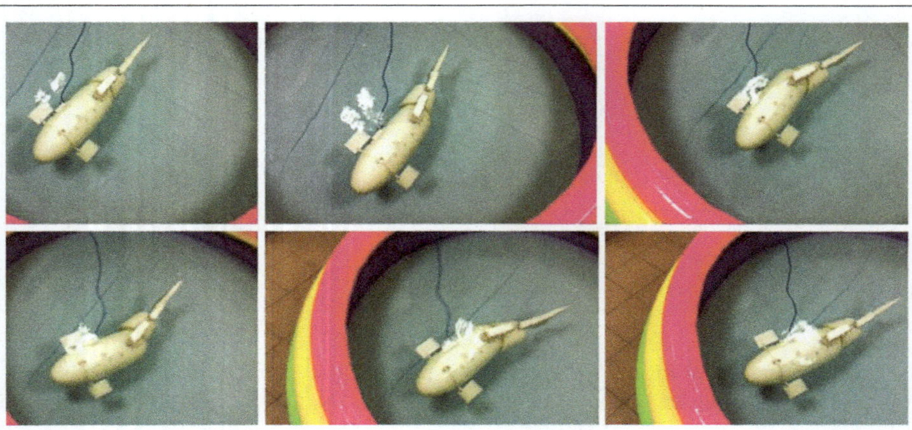

Figure 10 Closed-loop hardware prototype motion in different frames.

- b_B : position of point on the base relative to frame F_I.
- v_i : velocity of point on link i relative to frame F_I.
- v_B : velocity of point on the base relative to frame F_I.

Competing interests
The authors declare that they have no competing interests.

Authors' contributions
ARC has conceptualized and written the manuscript with contributions from VK, BP, and RK SKP has rendered critical comments and revised the manuscript. All authors have contributed to the final manuscript.

Acknowledgements
We would like to thank Mr. Alok Agrawal of Purdue University for his key contributions for the mechanical CAD design of the robotic fish. We would also like to acknowledge the useful suggestions and feedback given by Mr. Shailabh Suman of Acoustic Research Lab NUS, Dr. Mandar Chitre of Acoustic Research Lab NUS, Dr. Pablo Alvaro Valdivia of Singapore-MIT Alliance for Research and Technology (SMART), Professor Marcelo H. Ang Jr., and Professor Xu Jianxin of Department of Electrical and Computer Engineering. We thank the Office of Defense Science and Technology Agency (DSTA) for their support of the present research.

Funding
This work is presently supported by the Defense Science and Technology Agency (DSTA), under the Ministry of Defense (Singapore), Government of Singapore under Grant R-263-000-621-232-MINDEF-DSTA for Underwater Vehicle Technology Project STARFISH.

Author details
[1]Department of Electrical and Computer Engineering, National University of Singapore, Singapore 117576, Singapore. [2]Electrical Engineering Department, Malviya National Institute of Technology, Jaipur, India.

References
1. Triantafyllou MS, Techet AH, Hover FS (2004) Review of experimental work in biomimetic foils. IEEE J Ocean Eng 29:585–594. doi:10.1109/JOE.2004.833216
2. Bullinaria JA (2003) From biological models to the evolution of robot control systems. Phil Trans R Soc A 361:2145–2164. doi:10.1098/rsta.2003.1249
3. Bandyopadhyay PR (2005) Trends in biorobotic autonomous undersea vehicles. IEEE J Ocean Eng 30:109–139. doi:10.1109/JOE.2005.843748
4. Lighthill MJ (1970) Aquatic animal propulsion of high hydro-mechanical efficiency. J Fluid Mech 44:256–301. http://dx.doi.org/10.1017/S0022112070001830
5. Sfakiotakis M, Lane DM, Davies JBC (1999) Review of fish swimming modes for aquatic locomotion. IEEE J Ocean Eng 24:237–252. doi:10.1109/48.757275
6. Barrett D, Grosenbaugh M, Triantafyllou MS (1996) Optimal control of a flexible hull robotic undersea vehicle propelled by an oscillating foil. IEEE Symposium on Autonomous Underwater Technology, Monterey, CA, pp 1–9, doi:10.1109/AUV.1996.532833
7. Junzhi Y, Min T, Shuo W, Erkui C (2004) Development of a biomimetic robotic fish and its control algorithm. IEEE Trans Syst Man Cybern B Cybern 34:No. 4
8. Motomu N, Norifumi O, Kyosuke O (2003) A study on the propulsive mechanism of a double jointed fish robot utilizing self-excitation control. JSME Int J Series C 46:982–990. doi:10.1299/jsmec.46.982
9. Roy Chowdhury A, Prasad B, Kumar V, Kumar R, Panda SK (2014) Inverse dynamics kinematic control of a bio-inspired robotic-fish underwater vehicle propulsion based on Lighthill's slender body model. IEEE OES MTS OCEANS, Taipei, Taiwan
10. Fossen TI (2011) Handbook of marine craft hydrodynamics and motion control. John Wiley & Sons. doi:10.1002/9781119994138
11. Fu KS, Gonzalez RC, Lee CS (1987) Robotics control, sensing, vision and intelligence. McGraw-Hill New York
12. Roy Chowdhury A, Prasad B, Kumar V, Kumar R, Panda SK (2014) Bio-harmonized dynamics model for a biology inspired carangiform robotic fish underwater vehicle. In: Proc. 19th IFAC World Congress, Vol. 19. Cape town
13. Lighthill MJ (1960) Note on the swimming of slender fish. J Fluid Mech 9:305–317. http://dx.doi.org/10.1017/S0022112060001110
14. Videler JJ (1993) Fish Swimming. Chapman and Hall, London. doi:10.1007/978-94-011-1580-3
15. Liu J, Huosheng H (2010) Biological inspiration: from carangiform fish to multi-joint robotic fish. J Bionic Eng 7(1):35–48
16. Long JH Jr (2011) Biomimetics—robotics based on fish swimming. Chapter 244. In: Encyclopedia of fish physiology: from genome to environment. Elsevier Academic Press, San Diego, CA, ebook isbn: 9780080923239
17. Dewar H, Graham J (1994) Studies of tropical tuna swimming performance in a large water tunnel—kinematics. J Exp Biol 192:45–59
18. Thompson S, Koto Kagami S (2005) Continuous curvature trajectory generation with obstacle avoidance for car-like robots. IEEE International Conference on Computational Intelligence for Modeling, Control and Automation:863–870, doi:10.1109/CIMCA.2005.1631373
19. Wu T (1961) Swimming of a waving plate. J Fluid Mech 10:321–344, http://dx.doi.org/10.1017/S0022112061000949
20. Singh K, Pedley TJ (2008) The hydrodynamics of flexible-body maneuvers in swimming fish. Physica D: Nonlinear Phenom 237(1417):2234–2239, http://dx.doi.org/10.1016/j.physd.2008.02.002
21. Liu H, Kawachi K (1999) A numerical study of undulatory swimming. J Comput Phys 155:223–247, doi:http://dx.doi.org/10.1006/jcph.1999.6341
22. Long JH, Jr, Lammert AC, Pell CA, Kemp M, Strother J, Crenshaw HC, McHenry MJ (2004) A navigational primitive: biorobotic implementation of cycloptic helical klinotaxis in planar motion. IEEE J Ocean Eng 29:795–806. doi:10.1109/JOE.2004.833233
23. Chen B, Jiang SR, Liu YD (2010) Research on the kinematic properties of a sperm like swimming micro robot. J Bionic Eng 7:123–129, http://dx.doi.org/10.1016/S1672-6529(09)60225-0
24. Krishnamurthy P, Khorrami F, de Leeuw J, Porter M, Livingston K, Long JH Jr (2010) An electric ray inspired biomimetic autonomous underwater vehicle. American Control Conference 2010:5224–5229, INSPEC Accession Number: 11509303
25. Bock RK, Krischer W (1998) Data analysis briefbook. Springer, Berlin. doi:10.1007/978-3-662-03725-6
26. Tytell ED (2004) The hydrodynamics and eel swimming II: effects of swimming speed. J Exp Biol 207:3265–3279. doi:10.1242/jeb.01139
27. Zhou C, Low KH, Chong CW (2009) An analytical approach for better swimming efficiency of slender fish robots based on Lighthill's model. IEEE International Conference on Robotics and Biomimetics ROBIO:1651–1656, doi:10.1109/ROBIO.2009.5420418)
28. Roy Chowdhury A, Prasad B, Kumar V, Kumar R, Panda SK (2012) Implementation of a BCF mode bio-mimetic robotic fish underwater vehicle based on Lighthill mathematical model. IEEE ICROS International Conference on Control, Automation and Systems Jeju, Korea:437–442, (INSPEC Accession Number: 13225528). (Best Student Paper Award)
29. Roy Chowdhury A, Prasad B, Kumar V, Kumar R, Panda SK (2012) Kinematics study and implementation of a biomimetic robotic-fish underwater vehicle based on Lighthill slender body model. IEEE OES Autonomous Underwater Vehicles (AUV), National Oceanography Centre, Southampton UK:1–6, doi:10.1109/AUV.2012.6380721
30. Roy Chowdhury A, Prasad B, Kumar V, Kumar R, Panda SK (2013) Finding an operating region for a bio-inspired robotic fish underwater vehicle in the Lighthill framework. IEEE International Conference on Robotics and Biomimetics ROBIO Shenzhen China, 854 – 860, doi:10.1109/ROBIO.2013.6739569
31. Roy Chowdhury A, Panda SK (2014) Kinematic parameter based behavior modelling and control of a bio-inspired robotic fish. Proceedings of SICE Annual Conference 2014, Sapporo, Japan

7

Interactive actuation of multiple opto-thermocapillary flow-addressed bubble microrobots

Wenqi Hu[1], Qihui Fan[2] and Aaron T Ohta[1*]

Abstract

Opto-thermocapillary flow-addressed bubble (OFB) microrobots are a potential tool for the efficient transportation of micro-objects. This microrobot system uses light patterns to generate thermal gradients within a liquid medium, creating thermocapillary forces that actuate the bubble microrobots. An interactive control system that includes scanning mirrors and a touchscreen interface was developed to address up to ten OFB microrobots. Using this system, the parallel and cooperative transportation of 20-μm-diameter polystyrene beads was demonstrated.

Keywords: Microrobot; Bubble; Thermocapillary flow; Parallel actuation; Interactive control

Background

Microrobots, nontethered microstructures that can physically manipulate objects, are flexible tools for micro-transportation. Various types of microrobots have successfully transported objects such as microbeads [1-4], microgels [5], and single cells [4,5].

Microrobotic transportation has two advantages when compared to tools such as optical tweezers, optically induced dielectrophoresis (ODEP), and micromanipulators. First, microrobotic transportation does not rely on the optical properties or chemistry of the target objects and the surrounding medium. In contrast, optical tweezers require a refractive index difference between the objects and medium [6]. The medium properties can also play a role in ODEP manipulation, as the electrical conductivity of the medium can affect the electric field gradients that create the dielectrophoretic force [7]. Secondly, although the most widely used micromanipulators can achieve high velocities during manipulation, their throughput is still limited [8], as this is a serial form of manipulation. Parallel operation of multiple micromanipulators is limited by the working space, due to the macroscale components of the micromanipulators. In contrast, the parallel transportation of microrobots has been demonstrated [1,9-15].

One method for achieving parallel microrobot actuation used frequency-selective microrobots that have mechanical responses at different resonant frequencies [1]. Time-division-multiplexed control signals were sent to two magnetic microrobots to achieve simultaneous, independent locomotion. Other frequency-based addressing methods were applied to three magnetic stick–slip microrobots, independently addressed in parallel [10], and to helical microrobots to enable different swimming velocities [11]. Other methods of controlling multiple magnetic microrobots include having differing magnetization strengths [12] or temporarily disabling the magnetization of the microrobots [13]. Electrostatic actuation was another method used, where slight variations in microrobot dimensions enabled independent addressing [14,15].

However, to simultaneously actuate more microrobots independently by magnetic force or electrostatic force, more complicated control and fabrication methods will be necessary. It is difficult to generate localized magnetic fields at the microscale, so each microrobot has to be fabricated slightly differently in order to be addressed independently using the methods mentioned above.

It is straightforward to create multiple localized optical patterns by using optical elements such as micromirrors [16], spatial light modulators [17], or scanning mirrors [18]. Optical tweezers and OET take advantage of this capability to manipulate multiple targets at the same time [16,17]. This feature of optically addressed systems

* Correspondence: aohta@hawaii.edu
[1]Department of Electrical Engineering, University of Hawaii at Manoa, 2540 Dole Street, Holmes Hall 483, Honolulu, HI 96822, USA
Full list of author information is available at the end of the article

is also inherent in the opto-thermocapillary flow-addressed bubble (OFB) microrobot system [19,20]. However, unlike ODEP and optical tweezers, the OFB microrobots are gas bubbles in a liquid medium. The actuation of the OFB microrobots is less dependent on the material property and does not require direct laser or electrical field penetration through target objects, limiting the potential for damage to the objects under manipulation.

The actuation of the OFB microrobot has been discussed thoroughly in previous publications [19,21]. In summary, a heated region is generated by the selective illumination of the absorbing substrate with a light source. The heated region creates a temperature gradient on the bubble surface, driving a toroidal thermocapillary flow that can be divided into two principal components of interest (Figure 1). The lateral component of the thermocapillary flow is parallel to the substrate and moves the bubble towards the center of the light pattern. The desired effect from this flow component is that the bubble will follow a moving light pattern and will be trapped by a stationary light pattern. Another thermocapillary flow component, perpendicular to the substrate around the bubble perimeter, arises due to the temperature gradient along the vertical direction (z-direction).

OFB microrobots have transported microbeads [21], cell-laden hydrogels [19,20], and single cells [21,22]. Although possible, the parallel and independent addressing of more than five OFB microrobots has not been demonstrated. In this paper, multiple OFB microrobots were addressed by an interactive control system with a scanning mirror. Up to ten OFB microrobots are addressed and configured. In addition, four OFB microrobots were used to simultaneously pattern microbeads in hydrogel prepolymer. Microbeads were also handled by four OFB microrobots to demonstrate its cooperative working ability.

Methods

The optical setup of the OFB microrobot is depicted in Figure 2. A 980-nm laser (Laserlands 980MD-0.4 W-BL, Laserlands, Wuhan, China) was focused to a spot width of 4.4 μm by a 0.42-N.A. ×20 objective lens (Mitutoyo, Takatsu-ku, Kawasaki, Japan) onto the absorbing substrate. Unlike previous setups for single OFB microrobots [19,21], a dual-axis scanning mirror system (Newport GVSM002, Newport Corporation, Irvine, CA, USA) is integrated into the light path to direct the laser to different locations on the substrate.

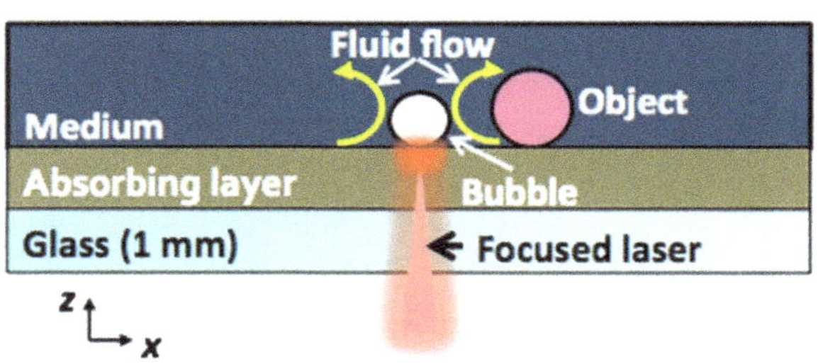

Figure 1 Side view of substrate and liquid medium containing OFB microrobots, showing the opto-thermal actuation mechanism. The bubble is stably trapped above the localized hot spot by the temperature gradient created by the laser, and a toroidal thermocapillary flow is formed.

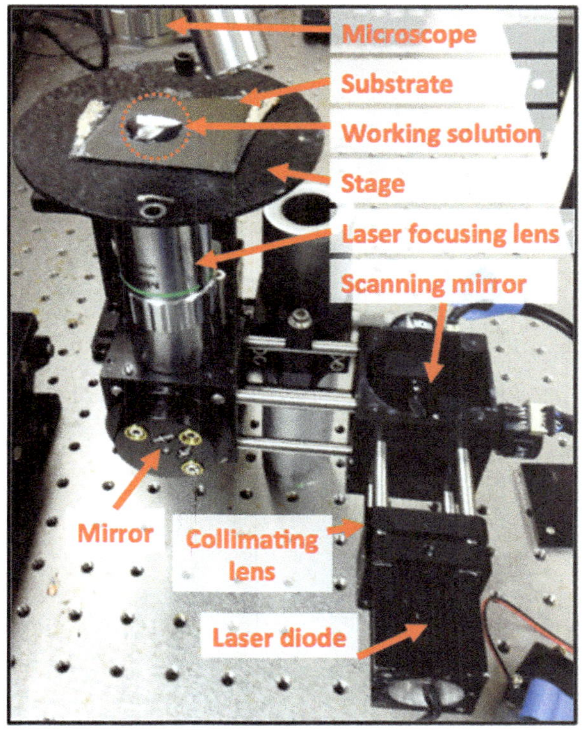

Figure 2 The optical setup of the interactive OFB microrobot system. The cross section of the substrate and working solution (marked by dotted circle) is shown in Figure 1.

The OFB microrobot actuation takes place on an absorbing substrate. The absorbing layer on the substrate is made of a layer of indium tin oxide (ITO) that is 100 nm in thickness, topped with a layer of 1-μm-thick amorphous silicon (α-Si). This absorbing layer is able to convert light energy into heat. The ITO layer also serves as an adhesion layer for the α-Si.

The control system of the OFB microrobot is made of two subsystems: pattern generation and scanning mirror control (Figure 3a). The pattern generation subsystem uses a touchscreen computer (Samsung Series 7, Samsung Electronics, Suwon, South Korea) as the interface to the system operator. An interactive control application was written using the simple multi-touch (SMT) library of the Processing programming tool [23]. The application created for the OFB microrobot control system has the ability to create, reposition, and remove light patterns within a

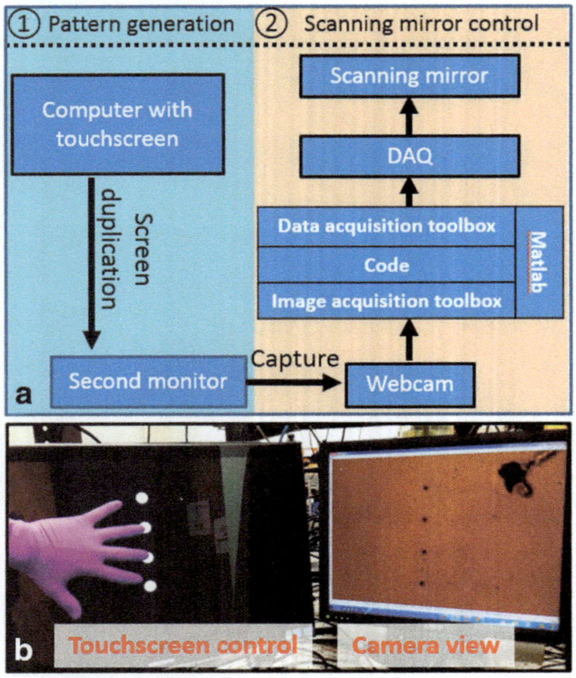

Figure 3 Interactive control system and interface. (a) The interactive control system of OFB microrobot, showing the two subsystems: pattern generation and scanning mirror control. **(b)** The interactive control interface. The positions of the four OFB microrobots visible in the camera view of the working space were controlled by the four white spots on the touchscreen.

set workspace, according to the touch input of the operator. These light patterns control the movement of the microrobots and appear on the touchscreen as white circles on a black background (Figure 3b). The touchscreen display was aligned with a live-view camera feed from the microscope, enabling the operator to simultaneously control the light sources while receiving instant visual feedback. The display on the touchscreen was duplicated on a second monitor, which is captured by a webcam (Logitech C170, Logitech International S.A., Lausanne, Switzerland) attached to the scanning mirror control subsystem.

The centerpiece of the scanning mirror control system is a custom script that uses the MATLAB Image Acquisition Toolbox to determine the positions of the light patterns on the second computer monitor. The x and y positions of the light patterns are recorded using units of pixels, which are converted to analog voltages by the MATLAB script. Subsequently, the script uses the MATLAB Data Acquisition Toolbox to queue the analog voltages into a data acquisition unit (DAQ; National Instruments USB-3653, National Instruments Corporation, Austin, TX, USA). The DAQ outputs the voltages to the control circuits of the scanning mirror system, thereby controlling the position of the mirrors. Thus, the scanning mirrors are adjusted to the correct angles to redirect the laser to the corresponding position on the substrate.

The scanning mirror finishes scanning through all the points of the light patterns in 0.05 s per cycle of the control script, independent of how many points there are in the cycle. Bubbles can be generated immediately in the liquid medium by the laser illumination, and longer illumination durations lead to larger bubble sizes [21]. To control the bubble size, the heat generation at each location can be further controlled by pulsing the laser at various frequencies and duty cycles. This is controlled by a TTL signal provided by a function generator (Agilent 33220A, Agilent Technologies, Santa Clara, CA, USA). In the experiments presented here, the laser pulse frequency was 400 Hz, and the laser pulse width was 100 μs. This resulted in bubbles with diameters ranging from 7 to 10 μm, with an average diameter of 8 μm. The bubbles do not collapse during the experiments, as the laser pulse rate is rapid enough to maintain the bubble. However, if the laser is switched off, the bubble will collapse within 2 s due to the Laplace pressure. Under these actuation conditions, the OFB microrobots can move up to 500 μm/s.

The current control system requires two computers, one running the Processing program for the pattern generation subsystem and one running the MATLAB script for the scanning mirror control subsystem, as well as an extra monitor. It is currently nontrivial to create MATLAB scripts that will support touchscreen input, so this hardware workaround was implemented to take advantage of the image processing of MATLAB and the multi-touch library of Processing. This control system can be easily replicated by any research group in the microrobotic or micromanipulation research areas.

Results and discussion

Control of multiple OFB microrobots

Using the interactive microrobot control system described in the previous section, ten bubble microrobots were configured into different patterns, including the letters 'UH' and arrays that may be useful in cell patterning [24] (Figure 4). This experiment and all of the following use an aqueous solution that contains 1% agarose prepolymer (type IX, ultra-low gelling temperature). Agarose is a naturally extracted biomatrix that can support cell growth [25], which aligns with the ultimate goal of using this microrobot system to assemble cells.

Transportation by multiple OFB microrobots

Multiple OFB microrobots can increase the throughput of the micro-transportation. This was demonstrated by assembling a 4 × 4 matrix of 20-μm-diameter polystyrene beads using a column of four OFB microrobots (Figure 5). To assemble each column of beads, the four microrobots

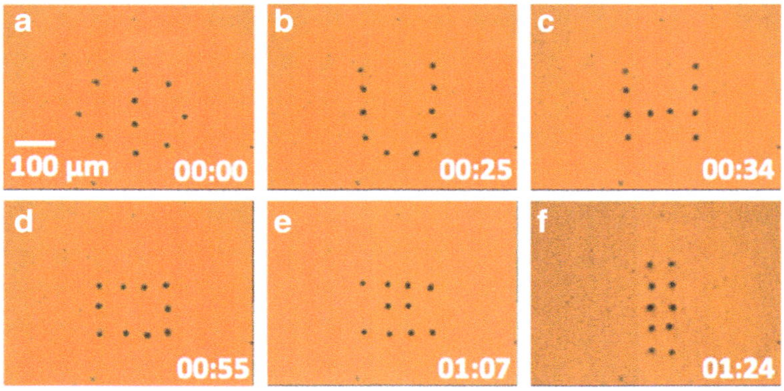

Figure 4 Ten OFB microrobots were sequentially configured into different patterns (a-f) using the interactive control system. The time stamp format is minutes: seconds.

first pulled four beads close to each microrobot, so that each microrobot was transporting a single bead. Then, the four microrobots moved to the patterning sites with the beads following behind them (Figure 5a). Finally, the laser was turned off, terminating the thermocapillary flow and releasing the beads from the microrobots. The process was then repeated three more times, completing the 4 × 4 matrix. The entire assembly process was completed in only 2.5 min, less than half the time needed by a single OFB to build the 3 × 3 cell matrix in the same medium (approximately 6 min) [21].

Moreover, cooperative transportation of micro-objects was also performed. This demonstrates that parallel control of individual microrobots can enable the cooperative use of microrobots to accomplish tasks. Four OFB microrobots were used to perform a cooperative 'hand-off' routine, where microbeads were passed between the microrobots (Figure 6). Two 20-µm beads were first respectively exchanged between the microrobots labeled A and C and the microrobots labeled B and D (Figure 6a,b,c). The two beads were then brought to the center of the screen (Figure 6d) and exchanged between microrobots B and

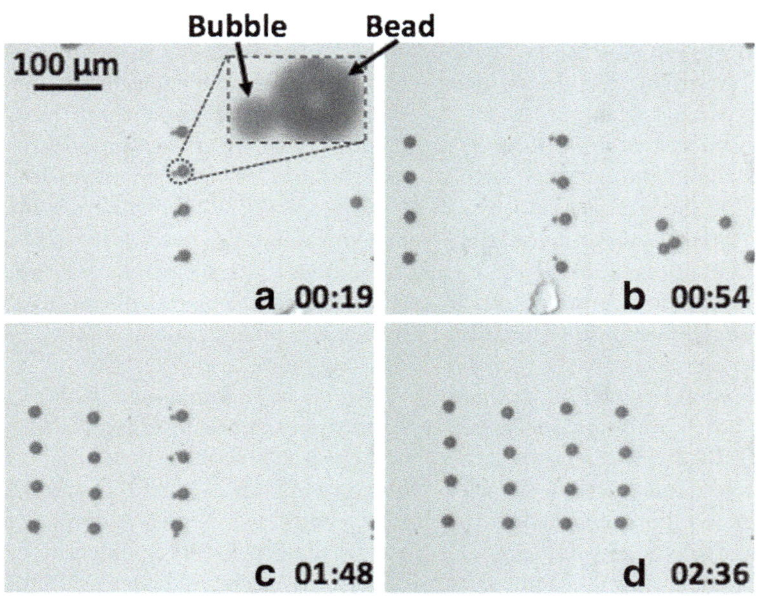

Figure 5 Matrix assembly. A 4 × 4 matrix of 20-µm-diameter polystyrene beads in 1% agarose solution was assembled by column of four OFB microrobots. The time stamp format is minutes: seconds. **(a)** Start of assembly. A magnified photo of an OFB microrobot with a bead is shown in the inset. **(b)** The leftmost column of the matrix has been assembled, and the four OFB microrobots are transporting the second column of microbeads. **(c)** Two columns of the matrix have been assembled, and the third column is undergoing transportation. **(d)** The assembled 4 × 4 matrix of beads.

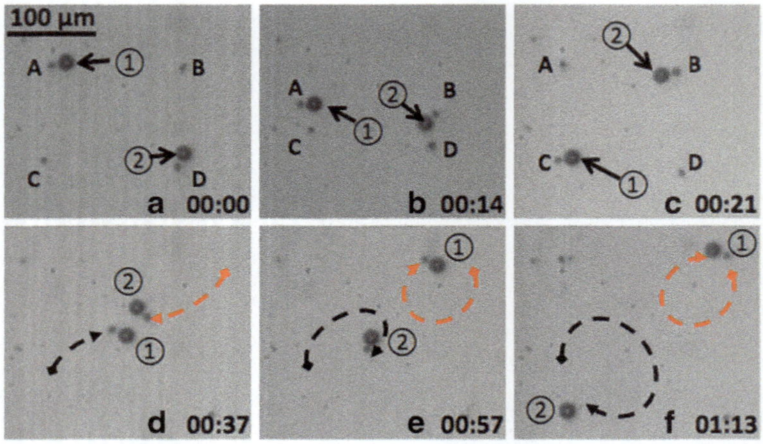

Figure 6 Two 20-μm-diameter beads were passed between four independently controlled OFB microrobots. The four OFB microrobots are marked as A, B, C, and D. The two 20-μm beads are marked by 1 and 2. The time stamp format is minutes: seconds. **(a-c)** OFB microrobot A transferred the no. 1 bead to microrobot C; at the same time, microrobot D transferred the no. 2 bead to microrobot B. **(d)** The beads are transported to the center of the field of view. **(e)** The beads were exchanged between microrobots B and C. **(f)** The OFB microrobots then transported the exchanged beads to the side opposite their starting locations in **(d)**.

C (Figure 6e). At last, as shown in Figure 6f, the beads were brought by the OFB microrobots to the position diagonally opposite to their locations shown in Figure 6c. This experiment demonstrates that multiple microrobots adjacent to a micro-object do not interfere with the manipulation of the object.

Conclusions

An interactive control system for the OFB microrobot was demonstrated by performing parallel and cooperative assembly of microbeads. This system has the potential to be scaled up even further, increasing the throughput and utility of the OFB microrobotic system.

Although the interactive control system performed well in the experiments described here, the system can be further simplified. The need for the webcam and extra monitor can be eliminated by directly outputting a signal with the light pattern location information from the pattern generation computer to the scanning mirror control computer. A single computer can also accomplish the functionality of the two computers in the current setup, if the MATLAB script is configured to accept input from the touchscreen of its host computer. Moreover, the live camera view of the OFB workspace can also be displayed directly on the touchscreen to make it easier for the operator to co-locate the target objects and the microrobots, similar to the control systems in [26,27].

It can be approximated that increasing the number of microrobots will linearly reduce the completion time of microassembly tasks until a saturation point is reached. Beyond this saturation point, the microrobot density will limit the amount of parallel micromanipulation that can occur. The maximum number of 10-μm-diameter OFB microrobots that theoretically can be generated in the field of view used here (600 μm by 450 μm) is 2,700 microrobots. However, the current system does not approach that limit, so the linear relationship can be used to estimate the impact on task completion time.

The maximum number of microrobots is limited in the current control system by the ability of the human operator to track multiple objects. In these experiments, a maximum of ten microrobots can be controlled by a single operator. In order to scale this system to control a much larger number of microrobots in parallel, automated controls have to be developed. An automated control system can build upon algorithms developed for path planning and macroscale robotic swarm research [28]. In addition to increasing the throughput of microassembly tasks, multiple microrobots can enable certain manipulation tasks, such as studying temporal dynamics of micro-object interactions. In this case, it may be necessary to have several objects moving at various velocities and trajectories relative to one another, which can be achieved by using multiple microrobots.

The micro-object used in the experiments above is a 20-μm polystyrene bead, which is near in size to many biological micro-objects, such as cells. Thus, the results here are expected to be transferrable to these biological micro-objects. As one example, patterning of single mammalian cells into different patterns can benefit research in tissue engineering [29]. When manipulating biological materials, temperature is a concern, as objects such as cells can die at temperatures above the physiological temperature of 37°C. In this system, the maximum temperature is localized to the very center of the laser spot and quickly decreases to a temperature that is safe for cells: the temperature is less than 30.5°C at a distance of 3.5 μm from the center of the laser spot [21]. As the bubble radii in these experiments are approximately 5 μm, this ensures that cells are manipulated at a safe temperature.

Competing interests
The authors declare that they have no competing interests.

Authors' contributions
WH programmed the microrobot control system, assembled the microrobot system, carried out the experiments, and drafted the manuscript. QF assisted with the experiments and the writing of the manuscript. AO conceived the microrobot system, participated in the coordination of the experiments, and helped to write the manuscript. All authors read and approved the final manuscript.

Acknowledgements
This project was supported in part by grant number 1R01EB016458 from the National Institute of Biomedical Imaging and Bioengineering of the National Institutes of Health (NIH) as part of the NSF/NASA/NIH/USDA National Robotics Initiative. These contents are solely the responsibility of the authors and do not necessarily represent the official views of the NIH.

Author details
[1]Department of Electrical Engineering, University of Hawaii at Manoa, 2540 Dole Street, Holmes Hall 483, Honolulu, HI 96822, USA. [2]Department of Mechanical Engineering, University of Hawaii at Manoa, 2540 Dole Street, Holmes Hall 300, Honolulu, HI 96822, USA.

References
1. Frutiger DR, Vollmers K, Kratochvil BE, Nelson BJ (2009) Small, fast, and under control: wireless resonant magnetic micro-agents. Int J Rob Res 29:613–636
2. Tottori S, Zhang L, Qiu F, Krawczyk KK, Franco-Obregón A, Nelson BJ (2012) Magnetic helical micromachines: fabrication, controlled swimming, and cargo transport. Adv Mater 24:811–816
3. Floyd S, Pawashe C, Sitti M (2009) Two-dimensional contact and noncontact micromanipulation in liquid using an untethered mobile magnetic microrobot. IEEE Trans Robot 25:1332–1342
4. Steager EB, Sakar MS, Magee C, Kennedy M, Cowley A, Kumar V (2013) Automated biomanipulation of single cells using magnetic microrobots. Int J Rob Res 32:346–359
5. Sakar MS, Steager EB, Kim DH, Kim MJ, Pappas GJ, Kumar V (2010) Single cell manipulation using ferromagnetic composite microtransporters. Appl Phys Lett 96:043705
6. Ashkin A, Dziedzic JM, Bjorkholm JE, Chu S (1986) Observation of a single-beam gradient force optical trap for dielectric particles. Opt Lett 11:288–290
7. Valley JK, Jamshidi A, Ohta AT, Hsu HY, Wu MC (2008) Operational regimes and physics present in optoelectronic tweezers. J Microelectromechanical Syst 17:342–350
8. Savia M, Koivo HN (2009) Contact micromanipulation—survey of strategies. IEEE/ASME Trans Mechatronics 14:504–514
9. Diller E, Floyd S, Pawashe C, Sitti M (2011) Control of multiple heterogeneous magnetic micro-robots on non-specialized surfaces. In: IEEE international conference on robotics and automation. pp 115–120
10. Diller E, Floyd S, Pawashe C, Sitti M (2012) Control of multiple heterogeneous magnetic microrobots in two dimensions on nonspecialized surfaces. IEEE Trans Robot 28:172–182
11. Mahoney AW, Nelson ND, Peyer KE, Nelson BJ, Abbott JJ (2014) Behavior of rotating magnetic microrobots above the step-out frequency with application to control of multi-microrobot systems. Appl Phys Lett 104:144101
12. Diller E, Giltinan J, Sitti M (2013) Independent control of multiple magnetic microrobots in three dimensions. Int J Rob Res 32:614–631
13. Diller E, Zhang N, Sitti M (2013) Modular micro-robotic assembly through magnetic actuation and thermal bonding. J Micro-Bio Robot 8:121–131
14. Donald BR, Levey CG, Paprotny I (2008) Planar microassembly by parallel actuation of MEMS microrobots. J Microelectromechanical Syst 17:789–808
15. Paprotny I, Levey C, Donald BR (2012) Turning-rate selective control: a new method for independent control of stress-engineered MEMS microrobots. In: Proceedings of robotics: science and systems conference
16. Chiou PY, Ohta AT, Wu MC (2005) Massively parallel manipulation of single cells and microparticles using optical images. Nature 436:370–372
17. Grier DG (2003) A revolution in optical manipulation. Nature 424:810–816
18. Mio C, Gong T, Terray A, Marr DWM (2000) Design of a scanning laser optical trap for multiparticle manipulation. Rev Sci Instrum 71:2196
19. Hu W, Ishii KS, Ohta AT (2011) Micro-assembly using optically controlled bubble microrobots. Appl Phys Lett 99:094103
20. Hu W, Ishii KS, Ohta AT (2012) Micro-assembly using optically controlled bubble microrobots in saline solution. In: IEEE international conference on robotics and automation. pp 733–738
21. Hu W, Fan Q, Ohta AT (2013) An opto-thermocapillary cell micromanipulator. Lab Chip 13:2285–2291
22. Hu W, Ishii KS, Fan Q, Ohta AT (2012) Hydrogel microrobots actuated by optically generated vapour bubbles. Lab Chip 12:3821–3826
23. Processing.org http://processing.org/. Accessed: 16 Sep 2011
24. Albrecht DR, Underhill GH, Mendelson A, Bhatia SN (2007) Multiphase electropatterning of cells and biomaterials. Lab Chip 7:702–709
25. Zorlutuna P, Annabi N, Camci-Unal G, Nikkhah M, Cha JM, Nichol JW, Manbachi A, Bae H, Chen S, Khademhosseini A (2012) Microfabricated biomaterials for engineering 3D tissues. Adv Mater 24:1782–1804
26. Onda K, Arai F (2012) Parallel teleoperation of holographic optical tweezers using multi-touch user interface. In: IEEE international conference on robotics and automation. pp 1069–1074
27. Bowman RW, Gibson G, Carberry D, Picco L, Miles M, Padgett MJ (2011) iTweezers: optical micromanipulation controlled by an Apple iPad. J Opt 13:044002
28. Pimenta LCA, Michael N, Mesquita RC, Pereira GAS, Kumar V (2008) Control of swarms based on hydrodynamic models. In: IEEE international conference on robotics and automation. pp 1948–1953
29. Albrecht DR, Underhill GH, Wassermann TB, Sah RL, Bhatia SN (2006) Probing the role of multicellular organization in three-dimensional microenvironments. Nat Methods 3:369–375

8

Learning by imitation with the STIFF-FLOP surgical robot: a biomimetic approach inspired by octopus movements

Milad S Malekzadeh[1*], Sylvain Calinon[1,2], Danilo Bruno[1] and Darwin G Caldwell[1]

Abstract

Transferring skills from a biological organism to a hyper-redundant system is a challenging task, especially when the two agents have very different structure/embodiment and evolve in different environments. In this article, we propose to address this problem by designing motion primitives in the form of probabilistic dynamical systems. We take inspiration from invertebrate systems in nature to seek for versatile representations of motion/behavior primitives in continuum robots. We take the perspective that the incredibly varied skills achieved by the octopus can guide roboticists toward the design of robust motor skill encoding schemes and present our ongoing work that aims at combining statistical machine learning, dynamical systems, and stochastic optimization to study the problem of transferring movement patterns from an octopus arm to a flexible surgical robot (STIFF-FLOP) composed of two modules with constant curvatures. The approach is tested in simulation by imitation and self-refinement of an octopus reaching motion.

Keywords: Learning from demonstration; Robot-assisted surgery; Continuum robots

Background

Time allowed biological organisms to refine and find smart ways to behave and survive in their environment. Hence, imitation of the models, systems, and elements of nature for the purpose of solving complex human problems (biomimetics) has given rise to new technologies in robotics, both at the level of hardware and software.

In minimally invasive surgery (MIS), widely used in the abdominal operations in the last 30 years [1], tools go through narrow openings and manipulate soft organs to perform surgical tasks. However, due to the limited flexibility and maneuverability of the available tools in MIS, up to four or five trocar accesses are required during operation. To cope with this issue, novel robotic surgical instruments have been developed to give higher flexibility and dexterity to the surgeons [2]. Such systems are actuated by motors moving rods [3], gears [4], or a combination of cables and rods [5]. Yet, stiffness control of such rigid mechanical structures remains a largely unexplored area. The actuation techniques presented in [3-5] are not able to have different level of stiffness in different parts of the robot. In [3], the robot consists of two snake-like robots, one covering the other. They use cable-driven mechanism to stiff the inner robot while the outer longer robot can be compliant at the tip. By using this technique, it is not possible to have a robot compliant at the proximal part and stiff at the distal part. A motor-driven robot has been designed in [4] with two arms and the motors embedded on the arms. This robot provides only stiff configuration at the tip suitable to perform surgical tasks in single-port laparoscopy (SPL).

The STIFF-FLOP project aims at developing biologically inspired manipulators and soft robotic arms to manipulate objects while controlling the stiffness of selected body part [6]. They are inspired by *biological systems* such as an octopus arm or elephant trunk. There are several examples of this kind of robots with high degrees of freedom (DOF) such as the OCTARM continuum robot [7] or the Clemson University elephant trunk [8]. In spite of being inherently compliant when interacting with objects, their rigid nature imposes limitations in their usability in the

*Correspondence: milad.malekzadeh@iit.it
[1] Department of Advanced Robotics, Istituto Italiano di Tecnologia (IIT), Via Morego 30, 16163 Genoa, Italy
Full list of author information is available at the end of the article

medical field. Hence, some other manipulators with no rigid structures were developed, e.g., the Air-Octor [9], which combines pneumatic tendon actuation or the arm developed within the OCTOPUS project [10].

None of the mentioned systems answer to the MIS requirements since flexibility and softness come generally with limited force; while high forces are difficult to be achieved without any rigid supports [11]. A different approach has been followed in the STIFF-FLOP project [12] based on variable and controllable stiffness techniques using composite granular jamming, membrane coupling, and pneumatic actuation [13].

There is a wide range of bioimitation mechanisms, ranging from the blind copy of actions (mimicry) to the understanding of the intent underlying an observed action (goal-level imitation), associated with self-training capability required to match the extracted intent. Several approaches in machine learning exist to cover this span of imitative behaviors.

The core idea of acquiring skills by imitation has been labeled over the years with various names such as *teaching by showing* [14], *robot coaching* [15], *programming by demonstration* [16], or *learning from demonstration* (LfD) [17]. A survey of LfD techniques addressing robotics control problems has been presented in [17]. In [16], Billard et al. present an overview of different machine learning approaches in robot programming by demonstration.

In programming by demonstration, providing *good demonstration* plays a key role to speed up the learning process. This is not always easy, specially when the contexts for the demonstration and the reproduction are not sufficiently similar. This may happen, for example, when the structure of the demonstrator agent and the robot are very different. The generalization capability is also often limited to a certain range of situations, which shows the importance of designing representations of skills and tasks that can be trained by various learning strategies. In particular, one key challenge is to allow the imitation system to continuously adapt the learned model when new demonstrations are available, by also exploiting in parallel exploration, auto-calibration, and adaptation strategies. In some situations, the design of efficient *self-refinement* algorithms is crucial for the robot to learn how to perform a task in new situations [16]. A robust and flexible representation of skills and movements is one of the keys to enable robots to jointly exploit several learning methodologies. In this article, we study this problem in the case of biological or robotic systems characterized by a hyper-redundant kinematic chain.

In [18,19], we introduced a high-level skill transfer approach by learning context-dependent reward functions in the same vein as in inverse reinforcement learning (IRL) [20] problems. We first used a statistical approach to extract the underlying intents of a demonstrated action (by kinesthetic teaching with a 7-DOF Barrett WAM), in the form of context-dependent objective functions, and then transferred these high-level goals to the STIFF-FLOP robot by using a stochastic reward-weighted self-refinement strategy so that the robot could find a policy matching the extracted objectives.

In this article, the skill transfer problem from an invertebrate system in nature (octopus) to a STIFF-FLOP robot with two modules is studied, see Figure 1. In particular, we introduce a novel representation of dynamical systems dedicated to continuum structures and study the problem in the case of octopus reaching movements. The proposed approach presents a new form of dynamic motion primitives for continuous bodies that can cope with the hyper-redundancy of the system and the perturbations in the environment.

We use a representation of the octopus arm movement which enables us to transfer motion skills to a STIFF-FLOP robot with two modules. The representation, based on Euler angles and offset values following the standard Denavit-Hartenberg convention in robotics, is invariant to rotation and translation in Cartesian coordinate system.

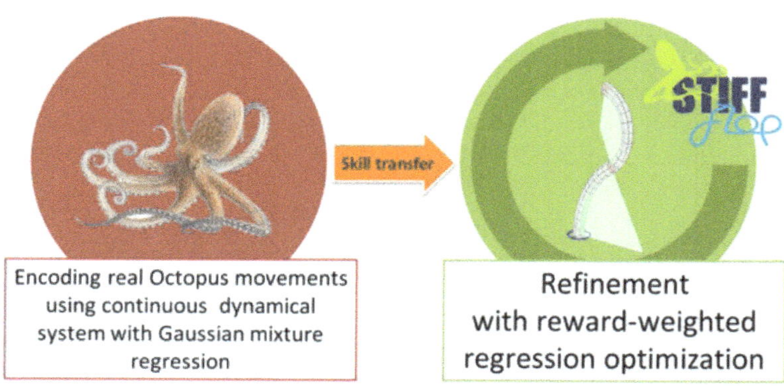

Figure 1 The STIFF-FLOP robot mimics the octopus reaching motion after skill transfer by using a probabilistic representation based on dynamical systems.

The use of rotation-translation invariant representations is indeed highly preferable if we want to exploit biological motion data in skill transfer applications. Otherwise, the extracted features can hardly be generalized to other situations.

Our approach then exploits the interaction between statistical models and continuous dynamical systems. In particular, the use of statistics allows the system to discover relevant invariants and correlations among the control variables in the form of joint probability distributions, information that is later exploited in the reproduction phase to reproduce appropriate movements when confronted with perturbations.

In this article, we present a first set of experiments to validate that the proposed approach can effectively enable the STIFF-FLOP robot to mimic a motion recorded from an octopus arm.

Octopus reaching motion

Octopuses are considered to be among the most developed and intelligent animals in the invertebrate kingdom, with impressive skills that can be attributed to the high maneuverability of their arms and the capacity of the peripheral nervous system to process sensory information and control arm movements [21]. The octopus can use its arms for various tasks such as locomotion, food gathering, hunting, and sophisticated object manipulation.

Kinematic analyses of octopus reaching and fetching movements have already revealed remarkable control principles underlying movement generation [21-24].

Motion primitives can be regarded as a minimal set of movements, which can be combined in different ways to give rise to rich movement repertoires and facilitating the acquisition of skills [25]. In [21], Zelman et al. demonstrated that the kinematics of octopus arm movements can be described by a reduced set of motion primitives, which required the analysis of different types of arm movements. They showed that a weighted combination of 2D Gaussian functions, in the form of a Gaussian mixture model (GMM), could be used to span arm behavior in 3D space.

The encoding approach that we propose takes inspiration from this work but provides two important novelties. It first extends the approach to a dynamical system to handle perturbations in the environment. It then exploits the GMM in a different way. In [21], each Gaussian is used as a parametric function to fit the local shape of a surface (as bell-shaped components), while it is used in our application to fit the joint distribution of the data. Combined with a regression approach, this allows us to exploit the statistical properties of the model if multiple demonstrations are available. Our work also differs in the sense that the analysis and encoding are not the final targets. Namely, we do not aim at classifying octopus movements or studying movements from a physiological perspective. Instead, we study the problem of designing technical approaches that could be used to transfer movement skills from biological systems to the STIFF-FLOP robot. This goal is reflected by the proposed encoding approach that has closer links to robotics systems, with a reversible representation with three Euler angles and a scalar offset, depending on two input variables (time and arm-index). The representation facilitates retargeting to the robot by representing the two agents as kinematic chains with Denavit-Hartenberg parameters.

In [26], a motion capture system for 3D tracking of octopus arm movements was developed to create a database containing several types of arm movements from octopuses of different ages and sizes. The problem is very challenging. Indeed, the soft property of the octopus arm and its flexible ability to change shape does not allow the use of conventional shape recognition algorithms. In contrast to human motion analysis, it is also not possible to attach markers on the octopus arms as it immediately rejects them [26].

In this paper, we exploited one of the most representative reaching movement from this database, gratefully provided by Prof. Binyamin Hochner, Shlomi Hanassy, and Alexander Botvinnik from the Department of Neurobiology, Hebrew University, Jerusalem, Israel, and Prof. Tamar Flash from the Department of Computer Science and Applied Mathematics, Weizmann Institute of Science, Rehovot, Israel.

Recording and storing such data require discrete sampling in both time and space. The raw data consists of discrete Cartesian position data with (on average) 100 points along the octopus arm and (on average) 100 time steps. To reduce the computational time, we re-sampled the data in both arm-index and time with 50 instances (surface of 2,500 points).

Methods

Spatiotemporal representation of octopus movements

In order to generate a mathematical model of the octopus movements, the arm was approximated as a robot with a high number of links. The continuous arm is approximated by a kinematic chain with a high number of revolute joints, describing the local curvature and torsion, alternated with prismatic joints describing the local elongation. The forward kinematics of such system can thus be described by a set of three Euler angles and a scalar offset. The *ZYX* decomposition of the Euler angles resulted in the most appropriate selection for the motion range in our dataset. This representation is reversible; i.e., by having the angles and offset, the points in Cartesian space can be retrieved and vice-versa.

We will define as continuous *arm-index* $s \in [0, 1]$ the position of any point along an arm, with $s = 0$ and $s = 1$

representing respectively the base and the tip. We will rescale the duration of a movement so that it can be represented by a continuous *time index* $t \in [0, 1]$. We thus have three Euler angles θ_x, θ_y, and θ_z and an offset ΔL for each t and s that can be represented as a set of surfaces as shown in Figure 2.

The original raw data (noisy Cartesian positions) are first resampled and smoothed through a two-dimensional polynomial fitting (surface fitting) with a 7-degree polynomial. The degree for the polynomial is set experimentally by testing different orders. The arm movement is translated to keep the base of the octopus at the origin. The data is then globally rescaled to match the size of the robot.

Dynamical system with evolution over time and arm-index

Dynamical movement primitives (DMP) is a popular tool in robotics to learn and reproduce movements while being robust to perturbations [27]. It consists of a spring-damper system modulated by a non-linear force profile encoded as a series of predefined basis functions associated with local force behaviors (or alternatively, by the path of a virtual spring-damper system producing a non-linear force profile). We introduced in [28] a probabilistic formulation of DMP by jointly learning the basis functions and the local force behaviors. The trajectory of the attractor point is then encoded by means of a statistical model (as explained later), whence the trajectory can be later extracted by regression.

The novelty in the current work is the use of surface attractors instead of trajectory attractors (generalization of the generic spring-damper system to a spatiotemporal dynamical system). We use both arm-index s and time t (rather than only time) as input variables, which enables the approach to encode the movement of invertebrate arm movements with a compact model (namely, by encoding the movement of all points along the arm).

The motion is reconstructed from the dynamical systems with

$$\ddot{\boldsymbol{x}} = \kappa^{\mathcal{P}}\left[\boldsymbol{y} - \boldsymbol{x}\right] - \kappa^{\mathcal{V}}\dot{\boldsymbol{x}} \quad \Rightarrow \quad \boldsymbol{y} = \frac{1}{\kappa^{\mathcal{P}}}\ddot{\boldsymbol{x}} + \frac{\kappa^{\mathcal{V}}}{\kappa^{\mathcal{P}}}\dot{\boldsymbol{x}} + \boldsymbol{x}, \quad (1)$$

where $\kappa^{\mathcal{P}}$ and $\kappa^{\mathcal{V}}$ are respectively the stiffness of the spring and the damping ratio of the damper. $\dot{\boldsymbol{x}}$ and $\ddot{\boldsymbol{x}}$ are the velocity and acceleration calculated by differentiating $\boldsymbol{x}$ for each s with respect to t. In other words, the observations $\boldsymbol{x}$ of the motion are converted into the trajectory of virtual attractors $\boldsymbol{y}$ for each arm-index. Similarly as $\boldsymbol{x} = \left[\theta_z, \theta_y, \theta_x, \Delta L\right]$ in Figure 2, $\boldsymbol{y}$ can thus be represented as a set of attractor surfaces.

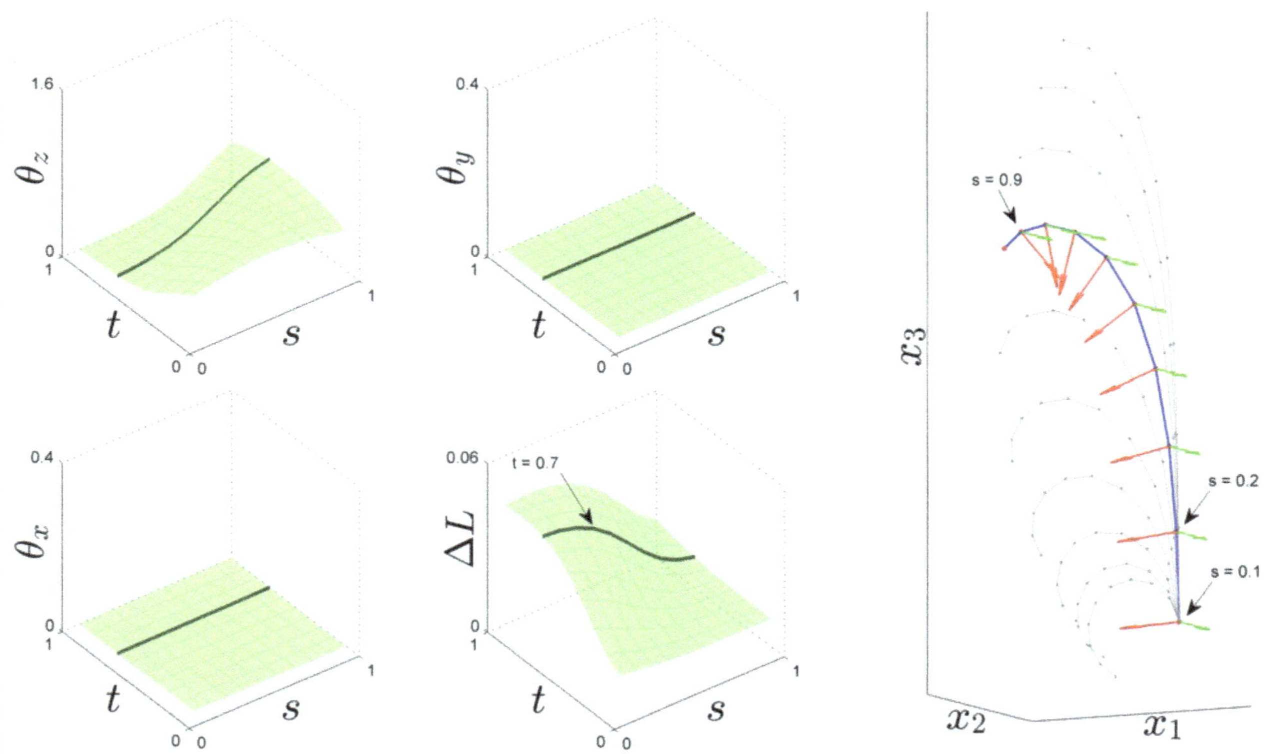

Figure 2 Spatiotemporal representation. *Left:* The black lines represent slices in the Euler angle surfaces $\boldsymbol{\theta}$ and offset surface $\boldsymbol{\Delta L}$, corresponding to a static pose described by all the links ($0 \leq s \leq 1$) at $t = 0.7$. *Right:* For the same time frame, some of the corresponding Frenet frames along the arm are depicted in a 3D Cartesian space.

An illustration of a continuum arm is presented in Figure 3 by using the proposed dynamical system with evolution over time and arm-index[a].

A probabilistic model of continuous movements

An *expectation-maximization* (EM) algorithm [29] is used to fit a Gaussian mixture model (GMM) to the augmented attractor dataset $\boldsymbol{\xi} = [t, s, \boldsymbol{y}]$, which is described as

$$\mathcal{P}(\boldsymbol{\xi}) = \sum_{k=1}^{K} \pi_k \mathcal{N}(\boldsymbol{\xi}|\boldsymbol{\mu}_k, \boldsymbol{\Sigma}_k),$$

$$\text{with} \quad \mathcal{N}(\boldsymbol{\xi}|\boldsymbol{\mu}_k, \boldsymbol{\Sigma}_k) = \frac{1}{(2\pi)^{\frac{D}{2}} |\boldsymbol{\Sigma}_k|^{\frac{1}{2}}} \times \exp\left(-\frac{1}{2}(\boldsymbol{\xi} - \boldsymbol{\mu}_k)^\top \boldsymbol{\Sigma}_k^{-1} (\boldsymbol{\xi} - \boldsymbol{\mu}_k)\right),$$

where D is the dimension of variables and K is the number of Gaussian components.

The parameters of the GMM including the priors π_k, centers $\boldsymbol{\mu}_k$, and covariance matrices $\boldsymbol{\Sigma}_k$ are learned by iteratively performing the following steps until convergence

$$\begin{aligned}
\textit{E-step:} \quad h_i(\boldsymbol{\xi}_j) &= \frac{\pi_i \mathcal{N}(\boldsymbol{\xi}_j | \boldsymbol{\mu}_i, \boldsymbol{\Sigma}_i)}{\sum_{k=1}^{K} \pi_k \mathcal{N}(\boldsymbol{\xi}_j | \boldsymbol{\mu}_k, \boldsymbol{\Sigma}_k)}, \\
\textit{M-step:} \quad \pi_i &\leftarrow \frac{\sum_{j=1}^{N} h_i(\boldsymbol{\xi}_j)}{\sum_{k=1}^{K}\sum_{j=1}^{N} h_k(\boldsymbol{\xi}_j)}, \\
\boldsymbol{\mu}_i &\leftarrow \frac{\sum_{j=1}^{N} h_i(\boldsymbol{\xi}_j)\, \boldsymbol{\xi}_j}{\sum_{j=1}^{N} h_i(\boldsymbol{\xi}_j)}, \\
\boldsymbol{\Sigma}_i &\leftarrow \frac{\sum_{j=1}^{N} h_i(\boldsymbol{\xi}_j)\, (\boldsymbol{\xi}_j - \boldsymbol{\mu}_i)(\boldsymbol{\xi}_j - \boldsymbol{\mu}_i)^\top}{\sum_{j=1}^{N} h_i(\boldsymbol{\xi}_j)},
\end{aligned} \tag{2}$$

where N is the number of datapoints. In the *E-step*, $h_i(\boldsymbol{\xi}_j)$ is the posterior probability $\mathcal{P}(k|\boldsymbol{\xi}_j)$, computed by the

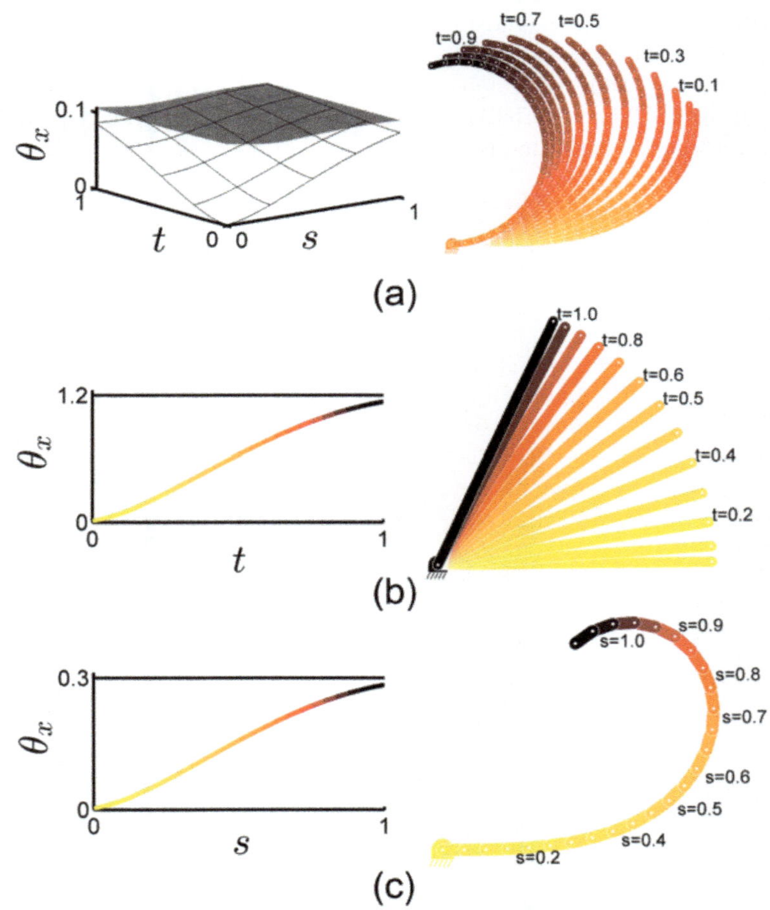

Figure 3 A dynamical system with surface attractor (evolution over time and arm-index). (a) The grey surface on the left represents the attractor surface corresponding to the observed motion of the continuum arm, while the white surface is the reproduced motion of the arm. The right figure shows the arm configurations in 2D Cartesian space at different time steps. **(b)** The left figure shows the evolution in time of a link with a given arm-index *s* along the kinematic chain and the corresponding configurations (on the right). **(c)** The figures show the pose of the continuum arm for a given time step *t*.

Bayes theorem. This is the probability that the Gaussian component k is responsible for $\boldsymbol{\xi}_j$. EM guarantees the improvement of the likelihood at each iteration, converging to a local optimum [30]. The EM algorithm is initialized with *k-means* clustering[b] [31] to start the iterative procedure from a good initial estimate. Here, K is determined empirically.

Gaussian mixture regression (GMR) is then used to estimate the conditional probability $\mathcal{P}(\boldsymbol{y}|s,t)$ with a Gaussian distribution $\mathcal{N}\left(\hat{\boldsymbol{\mu}}, \hat{\boldsymbol{\Sigma}}\right)$, with a fast computation independent from the number of datapoints used to train the model [32]. In our case,

$$\hat{\boldsymbol{\mu}} = \sum_{i=1}^{K} h_i\left(\boldsymbol{\xi}^{\mathcal{I}}\right)\left[\boldsymbol{\mu}_i^{\mathcal{O}} + \boldsymbol{\Sigma}_i^{\mathcal{OI}}\boldsymbol{\Sigma}_i^{\mathcal{I}^{-1}}\left(\boldsymbol{\xi}^{\mathcal{I}} - \boldsymbol{\mu}_i^{\mathcal{I}}\right)\right],$$
$$\text{and} \quad \hat{\boldsymbol{\Sigma}} = \sum_{i=1}^{K} h_i^2\left(\boldsymbol{\xi}^{\mathcal{I}}\right)\left[\boldsymbol{\Sigma}_i^{\mathcal{O}} - \boldsymbol{\Sigma}_i^{\mathcal{OI}}\boldsymbol{\Sigma}_i^{\mathcal{I}^{-1}}\boldsymbol{\Sigma}_i^{\mathcal{IO}}\right], \quad (3)$$

$$\text{where} \quad h_i\left(\boldsymbol{\xi}^{\mathcal{I}}\right) = \frac{\pi_i\mathcal{N}\left(\boldsymbol{\xi}^{\mathcal{I}}|\,\boldsymbol{\mu}_i^{\mathcal{I}}, \boldsymbol{\Sigma}_i^{\mathcal{I}}\right)}{\sum_{k=1}^{K}\pi_k\mathcal{N}\left(\boldsymbol{\xi}^{\mathcal{I}}|\,\boldsymbol{\mu}_k^{\mathcal{I}}, \boldsymbol{\Sigma}_k^{\mathcal{I}}\right)},$$
$$\boldsymbol{\mu}_i = \begin{bmatrix}\boldsymbol{\mu}_i^{\mathcal{I}}\\ \boldsymbol{\mu}_i^{\mathcal{O}}\end{bmatrix}, \quad \boldsymbol{\Sigma}_i = \begin{bmatrix}\boldsymbol{\Sigma}_i^{\mathcal{I}}\ \boldsymbol{\Sigma}_i^{\mathcal{IO}}\\ \boldsymbol{\Sigma}_i^{\mathcal{OI}}\ \boldsymbol{\Sigma}_i^{\mathcal{O}}\end{bmatrix},$$
$$\boldsymbol{\xi} = \begin{bmatrix}\boldsymbol{\xi}^{\mathcal{I}}\\ \boldsymbol{\xi}^{\mathcal{O}}\end{bmatrix}, \quad \boldsymbol{\xi}^{\mathcal{I}} = \begin{bmatrix}s\\ t\end{bmatrix}, \quad \boldsymbol{\xi}^{\mathcal{O}} = \boldsymbol{y},$$

where the superscripts $^{\mathcal{I}}$ and $^{\mathcal{O}}$ represent input and output variables.

In contrast to other regression methods, GMR does not model the regression function directly, but models the joint probability of the data. It then derives the regression function from the joint probability model. Density estimation can thus be learned in an off-line phase (with the GMM encoding approach presented above) and extracted in real time at reproduction, since GMR has a low computation time independent from the number of datapoints used to train the model.

Figure 4 illustrates the encoding and retrieval process. For regression, any subset of multivariate input and output dimensions can be selected, which can change in the course of the reproduction. Expectations on the remaining dimensions can be computed in real time, corresponding to a convex sum of linear approximations (with weights varying non-linearly). GMR can thus handle different sources of missing information, since the system is able to consider any combination of input/output mappings during reproduction.

Given the attractor surfaces $\boldsymbol{y}$, the whole motion can then be directly reconstructed by a double integration over t as

$$\dot{\boldsymbol{x}} = \int_t \left(\kappa^{\mathcal{P}}\left[\boldsymbol{y} - \boldsymbol{x}\right] - \kappa^{\mathcal{V}}\dot{\boldsymbol{x}}\right)dt \quad , \quad \boldsymbol{x} = \int_t \dot{\boldsymbol{x}}\,dt\,. \quad (4)$$

We will use the term DS-GMR to refer to this combination of dynamical system and Gaussian mixture regression.

Self-refinement with reward-weighted regression

The model parameters above can then be optimized through a self-refinement algorithm with an objective or reward function defined according to the different objectives of the task. The self-refinement process consists of searching for new solutions in the policy parameter space, associate a reward to each trial, and then sample from the exploration space by a weighted combination of the most successful trials obtained so far.

Dayan and Hinton originally suggested that a self-refinement problem can be tackled by EM to avoid gradient computation [33]. They introduced the idea of

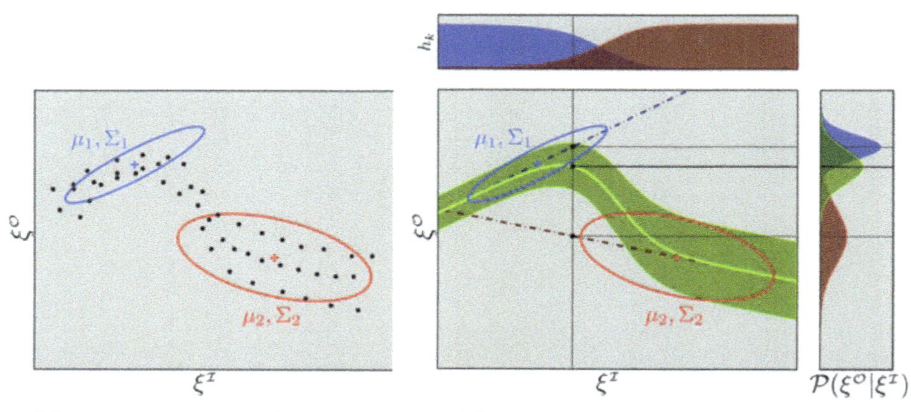

Figure 4 Illustration of the mapping representation as a mixture model and retrieval process. *Left:* Encoding of the input-output observations in a GMM. *Right:* Probabilistic estimation of output data from given input data with GMR.

treating immediate rewards as probabilities of a fictitious event, in which case probabilistic inference techniques can be used for optimization. They maximize the reward by solving a sequence of probability matching problems, where the task parameters are chosen at each step to match a fictitious distribution determined by the average rewards experienced on the previous steps. Although there can be large changes in the task parameters from one step to the next, there is a guarantee that the average reward is monotonically increasing. From this simple idea, various reward-weighted policy learning approaches emerged [34-38]. Indeed, several research fields converged to similar algorithmic solutions, with approaches such as the *cross-entropy method* (CEM) [39] or the *covariance matrix adaptation evolution strategy* (CMA-ES) [40].

One option for implementing such self-refinement is to use an EM-based stochastic optimization algorithm to refine the GMM parameters encoding the movement, similarly as in [18]. The procedure corresponds to an EM algorithm in which the reward signal is treated as a pseudo-likelihood, which can easily be extended to multi-optima policy search [41].

Another line of research is to explore methodologies that could move toward more structured techniques of exploration. In [42], it was proposed to speed up the search process by redefining it as an iterative reward-weighted regression problem. This is particularly relevant for the subclass of problems in which we have access to the goal or to the highest value that a reward can take (e.g., reaching the center of a target, be as close as possible to a reference trajectory, etc.).

In this case, the reward (or objective) is expressed in the form of a vector, and the aim of the optimization is treated as an iterative reward-weighted regression problem in the augmented space formed by the policy parameters and the achieved goals. At each iteration, a Gaussian distribution showing the joint probability is fit to the augmented data containing the best policies and the corresponding goals obtained so far. The input of the regression problem is represented by the desired goal $\boldsymbol{\zeta}^{I*}$. The output is a candidate optimal policy that is tested on the system and that will be associated with the corresponding outcome.

An illustrative example of the reward-weighted regression algorithm with 1D policy and goal is shown in Figure 5.

The augmented data is defined as $\boldsymbol{\zeta} = \left[\boldsymbol{\zeta}^{I\top}, \boldsymbol{\zeta}^{O\top}\right]^{\top}$. At each iteration, the ordered set of datapoints $\{\boldsymbol{\zeta}_m\}_{m=1}^{M}$ with $r(\boldsymbol{\zeta}_1) \geqslant r(\boldsymbol{\zeta}_2) \geqslant \ldots \geqslant r(\boldsymbol{\zeta}_M)$ is used as a form of importance sampling [37] to estimate a Gaussian distribution with parameters

$$\begin{aligned} \tilde{\boldsymbol{\mu}} &\leftarrow \frac{\sum_{m=1}^{M} r(\boldsymbol{\zeta}_m)\ \boldsymbol{\zeta}_m}{\sum_{m=1}^{M} r(\boldsymbol{\zeta}_m)}, \\ \tilde{\boldsymbol{\Sigma}} &\leftarrow \frac{\sum_{m=1}^{M} r(\boldsymbol{\zeta}_m)\ (\boldsymbol{\zeta}_m - \tilde{\boldsymbol{\mu}})(\boldsymbol{\zeta}_m - \tilde{\boldsymbol{\mu}})^{\top}}{\sum_{m=1}^{M} r(\boldsymbol{\zeta}_m)} + \boldsymbol{\Sigma}_0. \end{aligned} \tag{5}$$

where $\boldsymbol{\Sigma}_0$ is a predefined minimal exploration noise. With the joint probability $\mathcal{P}(\boldsymbol{\zeta}) \sim \mathcal{N}\left(\tilde{\boldsymbol{\mu}}, \tilde{\boldsymbol{\Sigma}}\right)$ in Eq. (5), the conditional probability $\mathcal{P}(\boldsymbol{\zeta}^{O}|\boldsymbol{\zeta}^{I*})$ can then be computed to obtain a subsequent policy (see Eq. (3)), by stochastic sampling from the retrieved Gaussian distribution in the policy parameter space.

Thus, at each iteration of the algorithm, a new policy $\boldsymbol{\zeta}^{O}$ is evaluated and associated with the achieved goal $\boldsymbol{\zeta}^{I}$ and the corresponding scalar reward r (calculated through a reward function of exponential form). This iterative process continues until convergence or a maximum number of iterations is attained.

Similarly to EM-based stochastic optimization, the process can be extended to a GMM/GMR representation for multi-policy search [41].

In the proposed experiment, the process starts from a set of randomly generated policies, where the initial number has been determined empirically. The objective function is also explicitly defined.

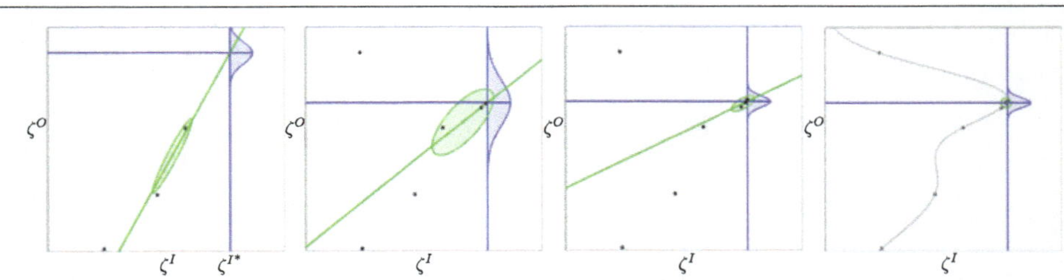

Figure 5 An example of reward-weighted regression. *(From left to right)* Subset of iterations during the refinement algorithm. At each iteration, a new Gaussian distribution, depicted by the green ellipse, is fit to the most promising augmented dataset (policy parameters and goal). A regression-based exploration strategy is then used in the augmented-space $\boldsymbol{\zeta}$ to iteratively find better policies to achieve the goal. For the regression, we assume that we know the desired goal (blue vertical bar $\boldsymbol{\zeta}^{I*}$) but we do not know how to achieve it (namely, the input of the regression is the desired goal).

STIFF-FLOP robot kinematics

The first prototype of the STIFF-FLOP robot, currently under development, will be composed of two cylindrical modules. Each section will consist of a soft cylinder with three chambers disposed concentrically around the axis, where air is inflated to bend the link in the desired orientation. A central chamber filled with hard grain-shaped particles is used to stiffen the link at a desired orientation by air suction.

In this section, we briefly summarize the forward and inverse kinematics of the STIFF-FLOP robot (see [18] for more details). We will use a constant curvature model to derive the forward kinematics along the whole kinematic chain (see Figure 6). The inverse kinematics is developed in a modular way, so that any number of links can be used within the model. We use a two-link robot for the experiment but the approach can be scaled to a higher number of modules.

In the local frame of each module, the rest position (no chamber is inflated) corresponds to the module aligned along the vertical axis $\boldsymbol{e}_3$, with a rest length L_0, see Figure 6-*left*.

The position $\boldsymbol{Q}_i$ of the tip of the i-th module can be written as a function of the angle α_i, the arc length β_i, and the curvature radius ρ_i (see Figure 6-*center*) as

$$\boldsymbol{Q}_i = [\rho_i\,(1-\cos(\beta_i))\cos(\alpha_i)\,, \quad \rho_i\,(1-\cos(\beta_i))\sin(\alpha_i)\,, \quad \rho_i\sin(\beta_i)]\,.$$

Both variables $\boldsymbol{Q}_i$ or $\{\rho_i, \alpha_i, \beta_i\}$ can be used to describe the kinematics of the module. The constant curvature coordinates of the single module can be obtained from the position $\boldsymbol{Q}_i$ of the tip by using the inverse relations

$$\rho_i = \frac{Q_{i,1}^2 + Q_{i,2}^2 + Q_{i,3}^2}{2\sqrt{Q_{i,1}^2 + Q_{i,2}^2}}, \qquad \alpha_i = \arctan\frac{Q_{i,2}}{Q_{i,1}},$$

$$\beta_i = \arccos\left(1 - \frac{\sqrt{Q_{i,1}^2 + Q_{i,2}^2}}{\rho_i}\right).$$

The constant curvature coordinates allow us to obtain the position of any point along the single module. Given the position of the tip, the constant curvature coordinates are obtained by the equation above. The Cartesian coordinates of a point positioned at a fractional position $\gamma \in [0, 1]$ of the module are then given by

$$\boldsymbol{F}_{\rho_i,\alpha_i,\beta_i}(\gamma) = [\rho_i\,(1-\cos(\beta_i\gamma))\cos(\alpha_i)\,, \quad \rho_i\,(1-\cos(\beta_i\gamma))\sin(\alpha_i)\,, \quad \rho_i\sin(\beta_i\gamma)]\,. \tag{6}$$

Here, γ corresponds to all possible points from the base of the module to the tip ($\gamma = 0$ corresponds to the base).

We will use the Cartesian position $\boldsymbol{Q}_i$ of the tip in the rest frame of the base as internal variables. They will replace the role of joint angle commonly used as internal variables in kinematic models of standard manipulators.

The constant curvature model also allows us to evaluate the orientation of the tip that only depends on the position of the tip, evaluated by rotating the base frame to make $\boldsymbol{e}_3$ tangent to the module at the tip, keeping the other axes rigidly displaced along the manipulator. The tip orientation of the i-th module in the $(i-1)$-th tip frame is defined by

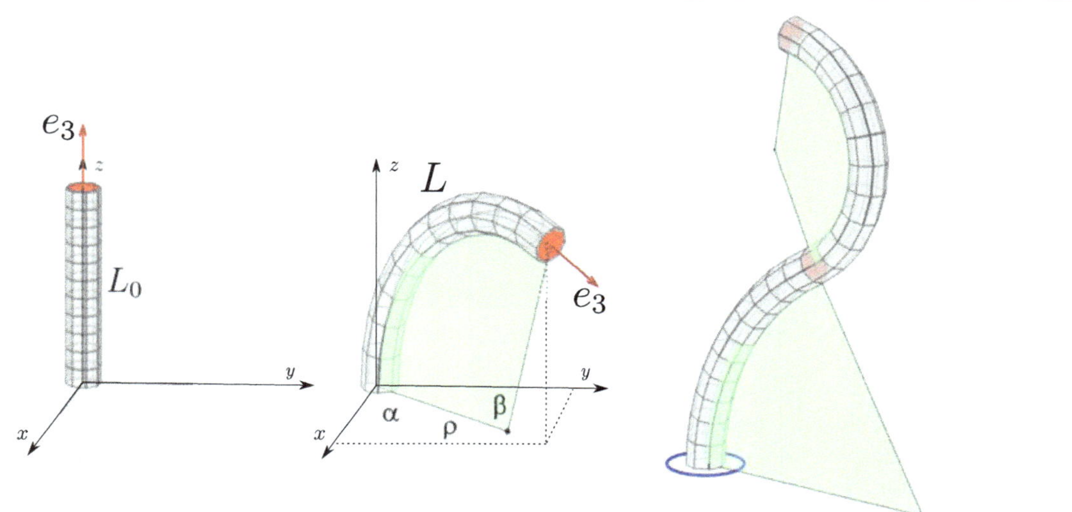

Figure 6 Model of the STIFF-FLOP robot. *(Left)* Single module of a constant curvature model at rest position with the rest length L_0, *(center)* with the pose of the tip described as a function of α, β, and L. *(Right)* STIFF-FLOP robot with two modules.

$$R_{(i-1)i} = \frac{1}{Q_i^\top Q_i} \begin{bmatrix} -Q_{i,1}^2 + Q_{i,2}^2 + Q_{i,3}^2 & -2Q_{i,2}Q_{i,1} & 2Q_{i,1}Q_{i,3} \\ -2Q_{i,2}Q_{i,1} & Q_{i,1}^2 + Q_{i,3}^2 - Q_{i,2}^2 & 2Q_{i,2}Q_{i,3} \\ -2Q_{i,1}Q_{i,3} & -2Q_{i,2}Q_{i,3} & Q_{i,3}^2 - Q_{i,1}^2 - Q_{i,2}^2 \end{bmatrix}.$$

The use of the constant curvature coordinates allows us to evaluate the possible limits that the hardware possesses. The only limitation on the possible configurations are placed in the fact that the inflation mechanism only allows a limited range of elongation, depending on the bending of the module and its orientation in space. The length of the robot can be obtained as a function of the constant curvature coordinates as $L_i = \rho_i \beta_i$.

Once the robot is bent in a given direction, a range of possible elongations can be obtained by fixing the curvature radius ρ_i and varying the arclength β_i, which is achieved by inflating the chambers. The geometry of the system suggests that this elongation also depends on the bending direction (i.e., which chamber is inflated to get that curvature). As a result, we will have joint limitations such as $\beta_{\min}(\alpha_i, \rho_i) < \beta_i < \beta_{\max}(\alpha_i, \rho_i)$, which will need to be obtained experimentally (no workspace analysis and joint limit measurements have been performed on the prototype so far). As a starting hypothesis, we will consider limitations corresponding to the nominal elongation when all the chambers are inflated (80%), thus limiting the possible lengths of the robot to $L_0 < L_i < L_0 + 0.8L_0$.

This setup allows an easy integration of multiple robot links, since any additional module can be thought as a constant curvature model applied on the previous. The position and orientation of the tip of the robot can be recursively evaluated, for any possible number of links K, by computing

$$y = \sum_{i=1}^{K} R_{0(i-1)} Q_i, \quad R_{0K} = \prod_{i=1}^{K} R_{(i-1)i},$$

with $R_{00} = I$. For two links, this results into $y = Q_1 + R_{01} Q_2$ and $R_{03} = R_{01} R_{12}$.

Results and discussion

The model is first tested to emulate two typical octopus reaching motion primitives by manually setting the model parameters. It is then tested if these parameters could be autonomously learned from real octopus motions and transferred to the STIFF-FLOP robot.

Emulation of octopus motion primitives

The model is first employed to emulate typical movements used by octopus to reach for food. These movements involve bend propagation and elongation of the proximal part of the arm. Note here that the reaching motion differs from the *fetching* of the grasped food to the mouth, where the octopus arm is configured into a quasi-articulated structure with three segments and three rotary joints [22,24,43]. In contrast, *reaching* movements are usually generated by a combination of two motor primitives: the propagation of a bend along the arm and an elongation of the arm [25,44]. These two motor primitives may be combined with different weights to create a broader spectrum of reaching movements.

The advantage of the DS-GMR representation to encode such kind of patterns is that it provides not only information about individual bending/elongation patterns but also information about their local correlations throughout the movement, in the form of a full covariance matrix.

The planar bend propagation movement from the base to the tip can easily be simulated by setting non-zero off-diagonal elements in a full covariance matrices (tilted elongated ellipsoids) acting as attractors in a curvature-elongation space. For example, for planar movements, one of the Euler angles and offset (e.g. θ_x and ΔL) can represent the curvature-elongation space, see Figure 7a-c. An elongation primitive can easily be obtained by translating one of the Gaussians in the elongation space, see Figure 7d-f.

Imitation of octopus motion by the STIFF-FLOP robot

In this section, we test if the proposed encoding technique can be used to autonomously transfer skills from real octopus movements to the STIFF-FLOP robot.

Figure 8 shows the recording of the representative octopus reaching movement that was selected from the database (see '*Octopus reaching motion*' section for details). We can observe that the angle θ_x has a wider range than the others, suggesting that the octopus reaching movement is mostly achieved in a plane.

The DS-GMR model is exploited to transfer the movement to a two-link STIFF-FLOP robot. The constant curvature constraint of each link results in a segmentation of the continuous surfaces into a given number of subsurfaces, corresponding to the number of modules (here, two).

In order to map the movement to the STIFF-FLOP robot, we consider the Frenet frames at the base of the robot (R_1), in the middle point between the two modules $(R_{n/2})$, and at the tip $(R_{n/2})$, where n is the number of segments used for the discretization of the continuous kinematic chain.

These frames can be easily calculated from the Euler angle representation at each time step t. The elongation of each module can be calculated by summing the elongation of the $n/2$ segments composing each module. The frames

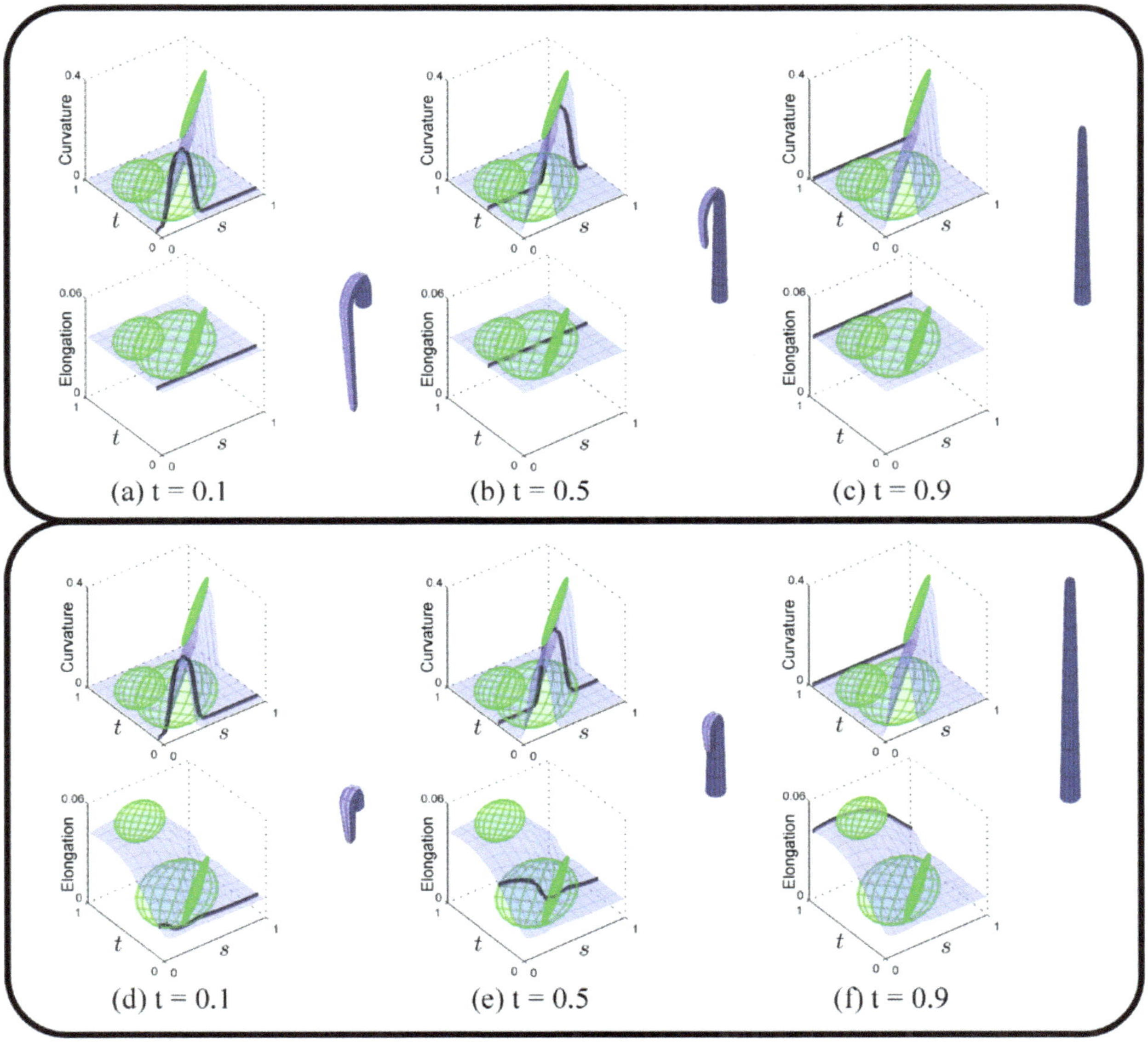

Figure 7 Emulation of reaching movements. Emulation of reaching movement with bend propagation **(a-c)** and with bend elongation **(d-f)**.

$\boldsymbol{R}_1$, $\boldsymbol{R}_{n/2}$, and $\boldsymbol{R}_n$ are then used to approximate the corresponding STIFF-FLOP module variables $\left[L_j, \alpha_j, \beta_j\right]_{j=1}^{2}$, see Figure 6. This is achieved by considering the rotation matrix $\boldsymbol{R}$, corresponding to the frame at the tip of each module, defined as

$$\text{For the first module:} \quad \mathcal{R} = \boldsymbol{R}_1^{\top} \boldsymbol{R}_{n/2}\,,$$
$$\text{For the second module:} \quad \mathcal{R} = \boldsymbol{R}_{n/2}^{\top} \boldsymbol{R}_n\,,$$

and calculating the constant curvature variables for each module as

$$\alpha = \arctan\left(\mathcal{R}_{23}/\mathcal{R}_{13}\right), \quad \beta = \arccos\left(\mathcal{R}_{33}\right) \quad \text{and} \quad L = \sum_{i=1}^{n/2} \Delta L_i\,,$$

where, for example, $\mathcal{R}_{23}$ is the scalar in the second row and third column of matrix $\mathcal{R}$, and ΔL is the offset.

The result of this procedure is shown in Figure 9, where Euler angles and offsets corresponding to the octopus (light gray) and STIFF-FLOP (green and red for the first and second modules, respectively) are presented.

This approach for the transfer of skills from a continuous arm to an arm with piecewise constant curvatures can provide a good initial estimate but does not guarantee the best fit. Indeed, even though the movement is quite similar, the different structure of the STIFF-FLOP robot can cause some dissimilarity in the learned movement.

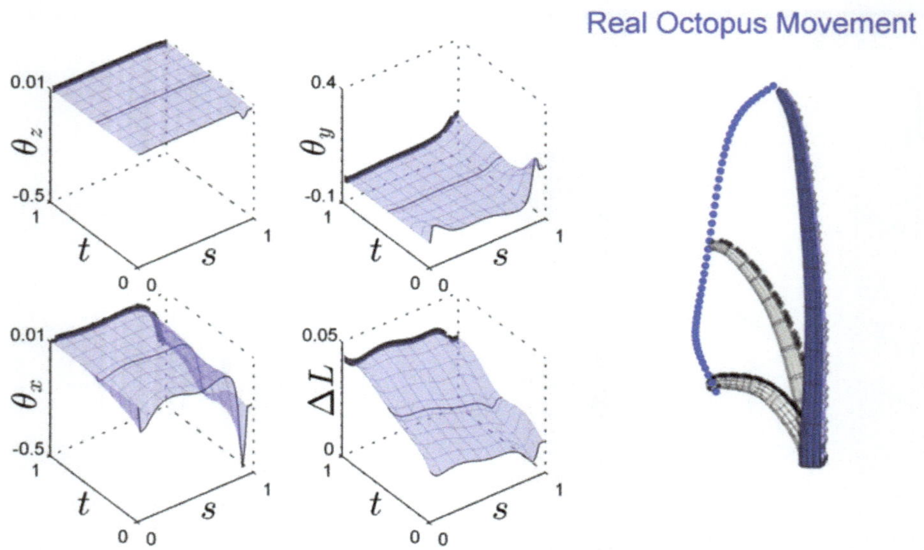

Figure 8 An example of octopus reaching movement after pre-processing of the data. The black lines on the surfaces show configurations at three different time steps (initial, mid, and final configurations), with corresponding octopus arm pose in a 3D Cartesian space depicted in the right graph. The blue dots show the position of the tip during the movement.

The self-refinement approach is thus used to refine this initial estimate, by using as policy parameters the centers of the GMM, as well as the first eigencomponents of ordered eigendecompositions of the covariances (directions of the main axes of the ellipsoids).

Figure 10 shows a GMM model with three Gaussians fitted to the attractor surfaces representing the movement. By considering the centers and the first eigencomponent of the covariance matrix for these three Gaussians, the policy has 36 dimensions. The stiffness and damping

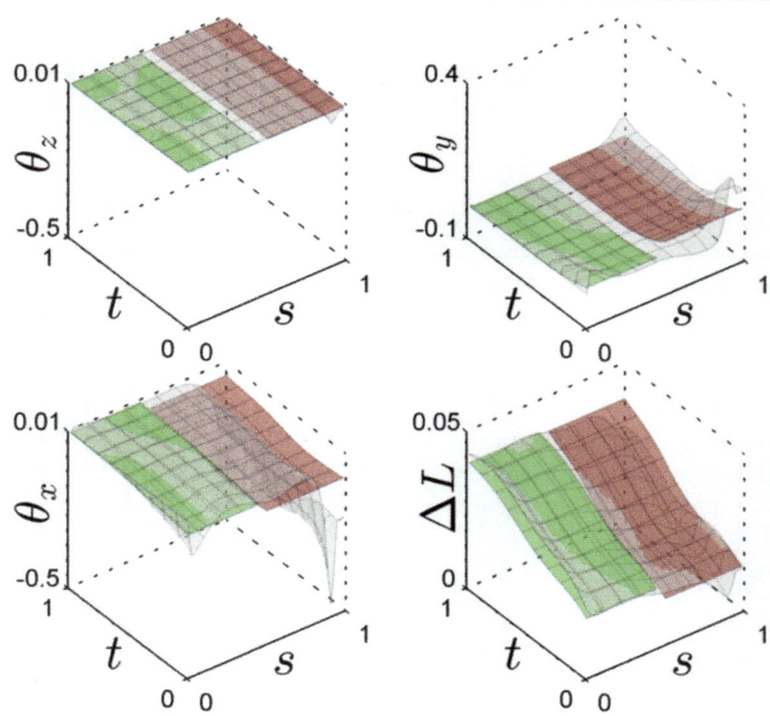

Figure 9 Mapping from the octopus to the STIFF-FLOP robot. The Euler angles and offsets related to the octopus movement (light gray) and the corresponding STIFF-FLOP robot with two modules (green and red surfaces).

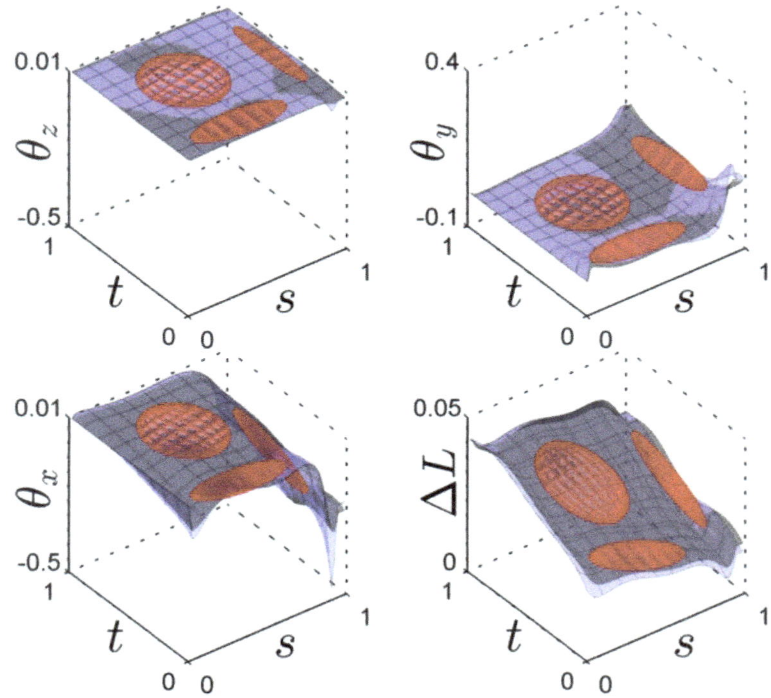

Figure 10 Modeling of surface attractors. The complete set of variable $\boldsymbol{x}$ and the attractor $\boldsymbol{y}$ are plotted in light blue and gray colors, respectively. The Gaussian components of the GMM model are the red ellipsoids fitting the attractor surfaces.

parameters are set empirically with a critical damping constraint.

A reward function based on the Cartesian distance of the STIFF-FLOP end effector ($\boldsymbol{x}^r$) and the octopus tip ($\boldsymbol{x}^o$) is defined at time instances $t = 0.2$, $t = 0.5$, and $t = 1$ as

$$r = \frac{1}{3} \sum_{t=0.2,0.5,1} \exp\left(-\delta||\boldsymbol{x}^r(t) - \boldsymbol{x}^o(t)||\right),$$

where δ is a bandwidth coefficient set experimentally. With three 3D points at different time instances, the goal has nine dimensions.

In order to keep the robot parameters within the real hardware limits, we additionally considered hard constraints in the search process for each of the two modules as

$$L_0 \leq L \leq 1.8L_0 \quad \text{and} \quad \beta \leq 2\pi/3,$$

where L_0, L, and β are the module minimum length, length, and curvature, respectively (see Figure 6).

The results of this experiment are shown in Figures 11 and 12, averaged over 30 runs of the same experiment. Figure 13 shows the convergence of the reward after 10-20 self-refinement iterations, by starting from an initial set of 20 randomly generated policies based on the initial demonstration.

Conclusions

We presented an approach based on statistical dynamical systems to encode movements in biological or robotic systems with a continuous kinematic structure. We showed that the approach could be employed to learn skills from demonstration, and that it could be combined with a self-refinement strategy based on iterative reward-weighted regression.

The aim is to extract relevant motion primitives from biological systems and map those to the STIFF-FLOP robot, while creating a set of natural motion patterns as building blocks. These building blocks could later be combined and reorganized differently to build new types of movements and skills.

The case study with octopus reaching movements, of course, does not target as a final goal to replicate reaching movements in surgery. But such typical motion repertoire allows us to test the efficiency of the proposed encoding approach and to exploit existing databases of biological movements. Even if the same exact movements will not be used in the surgical application, it provides us with a very important starting point to the design of robust models capable of encoding various skills in flexible continuum robots. In other words, the presented work is primarily aimed at exploiting typical behaviors observed in invertebrate systems to test an encoding strategy and guide us toward the design of new modeling techniques that

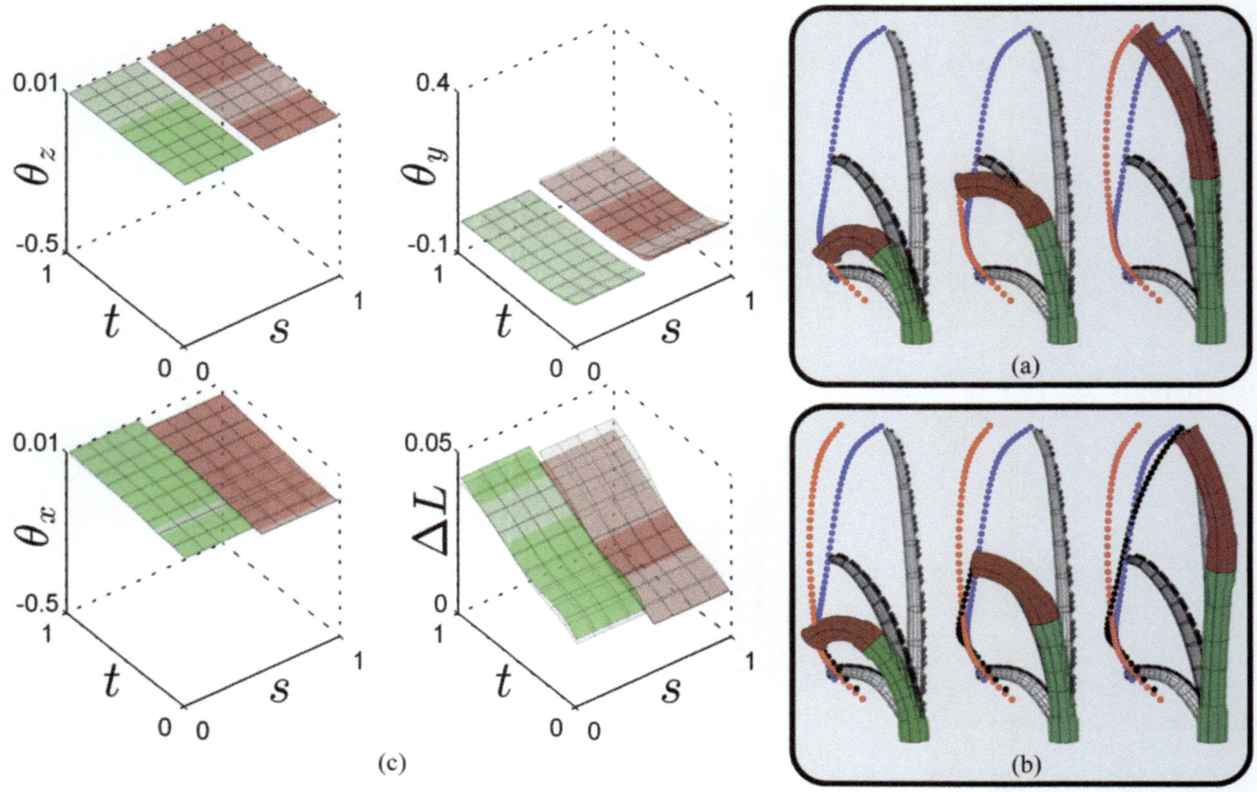

Figure 11 Qualitative comparison of the octopus and STIFF-FLOP robot movements, before and after self-refinement. (a) Comparison of the octopus and STIFF-FLOP robot movements based on the initial learning from demonstration approach (before self-refinement) at time instances $t = 0.2$, $t = 0.5$ and $t = 1$. The red dots represent the tip trajectory. **(b)** Reproduction on the robot after self-refinement. The black dots depict the refined trajectory of the tip. **(c)** The initial (light gray) and refined (green and blue surfaces) control variables.

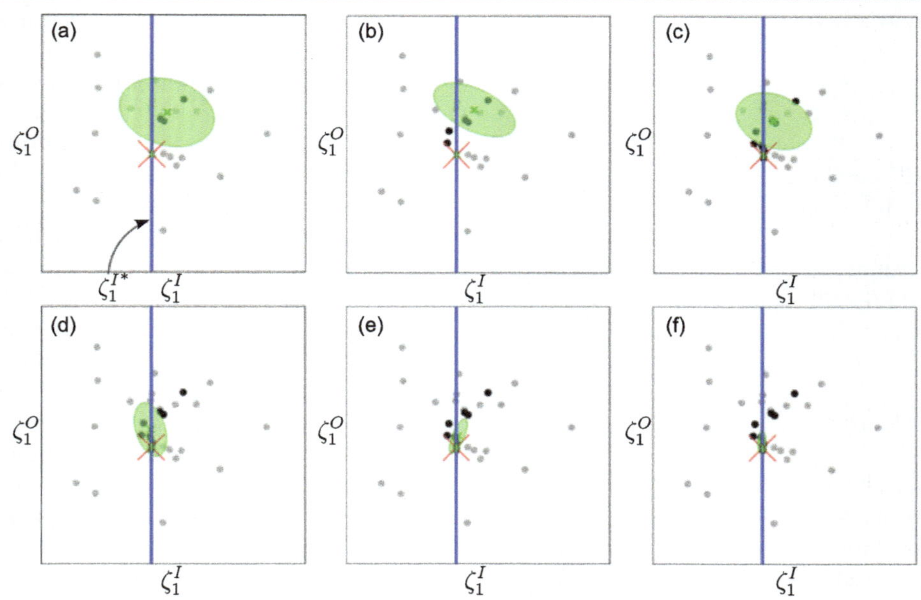

Figure 12 Convergence of the GMR search process. (a-f) A sample search in the augmented goal-policy space $\boldsymbol{\zeta}$ is plotted for six selected self-refinement iterations. Here, only one dimension for $\boldsymbol{\zeta}^O$ and $\boldsymbol{\zeta}^I$ is depicted, but we have $\boldsymbol{\zeta}^O \in \mathbb{R}^{36}$ and $\boldsymbol{\zeta}^I \in \mathbb{R}^9$. The gray dots show the initial set of iterations and the black dots are the new estimated policy. The blue line represents the known goal of the task with highest reward on which the red cross shows the best achieved policy after convergence.

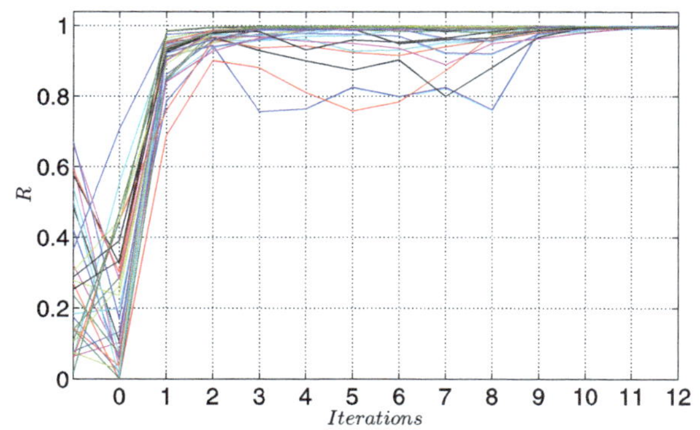

Figure 13 Reward profile for 30 runs, plotted individually.

could be used to encode a broader range of skills and patterns, including new skills that will be relevant to surgical applications.

In future work, we aim at applying this learning approach to the real STIFF-FLOP platform, by considering diverse motion/feedback skill primitives that could be used to assist doctors in surgical operations. We plan to exploit the context-dependent learning interface that we proposed in [18] as a high-level imitation approach to extract and exploit the most relevant cost functions that could explain the observation of octopus movements. In particular, we plan to study how the octopus exploits the degrees of freedom of its flexible arm to perform various tasks, in the form of objective functions that could vary depending on the ongoing situation. The variability information could then be exploited to regulate the stiffness and damping of the robot, with appropriate synergies among curvature, torsion, and elongation variables.

Endnotes

[a] Note that in this implementation, the evolution is only semi bi-dimensional in the sense that the evolution over s with $t = 0$ is first computed, followed by evolutions over t for $s \in [0, 1]$, providing a continuous surface.

[b] k-means is a method of clustering that aims to partition observations into K clusters in which each observation belongs to the cluster with the nearest mean. The center of each cluster after applying k-means, has the minimum sum of distances from all other points in that cluster.

Competing interests

The authors declare that they have no competing interests.

Authors' contributions

MM participated in the development and implementation of the methods and performed the experiments. SC developed the models for the spatiotemporal probabilistic dynamical system representation and the stochastic optimization strategy with GMR and supervised the whole process of implementations and experiments. DB developed the models for the forward and inverse kinematics of the robot, with associated codes, and helped with the implementation of the methods. MM and SC wrote the manuscript and developed the codes for the learning methods and experiments. DC provided advices for the writing of the manuscript. All authors read and approved the final manuscript.

Acknowledgements

We gratefully acknowledge Prof. Binyamin Hochner, Shlomi Hanassy, and Alexander Botvinnik from the Department of Neurobiology, Hebrew University, Jerusalem, Israel, and Prof. Tamar Flash from the Department of Computer Science and Applied Mathematics, Weizmann Institute of Science, Rehovot, Israel, who kindly provided us with the 3D octopus motion data processed from vision sensing and used in this article. We would also like to thank them for their truly inspiring work and for their recommendation concerning the interpretation of the collected octopus behaviors. This work was supported by the STIFF-FLOP European project (FP7-ICT-287728).

Author details

[1] Department of Advanced Robotics, Istituto Italiano di Tecnologia (IIT), Via Morego 30, 16163 Genoa, Italy. [2] Idiap Research Institute, Rue Marconi 19, CH-1920 Martigny, Switzerland.

References

1. Reynolds W (2001) The first laparoscopic cholecystectomy. JSLS 5(1):89–94
2. Vitiello V, Lee S, Cundy TP, Yang G (2013) Emerging robotic platforms for minimally invasive surgery. IEEE Rev Biomed Eng 6:111–126
3. Degani A, Choset H, Wolf A, Zenati MA (2006) Highly articulated robotic probe for minimally invasive surgery. In: Proc. IEEE Intl Conf. on Robotics and Automation (ICRA), Orlando, FL, USA. pp 4167–4172
4. Piccigallo M, Scarfogliero U, Quaglia C, Petroni G, Valdastri P, Menciassi A, Dario P (2010) Design of a novel bimanual robotic system for single-port laparoscopy. IEEE/ASME Trans Mechatronics 15(6):871–878
5. Bajo A, Goldman RE, Wang L, Fowler D, Simaan N (2012) Integration and preliminary evaluation of an insertable robotic effectors platform for single port access surgery. In: Proc. IEEE Intl Conf. on Robotics and Automation (ICRA), Saint Paul, MN, USA. pp 3381–3387
6. Smith KK (1989) Trunks, tongues, and tentacles: moving with skeletons of muscle. Am Sci 77:29–35
7. McMahan W, Chitrakaran V, Csencsits M, Dawson D, Walker ID, Jones BA, Pritts M, Dienno D, Grissom M, Rahn CD (2006) Field trials and testing of the OctArm continuum manipulator. In: Proc. IEEE Intl Conf. on Robotics and Automation (ICRA), Orlando, FL, USA. pp 2336–2341
8. Walker ID (2000) Some issues in creating "invertebrate" robots. In: Proc. intl symp. on adaptive motion of animals and machines, Montreal, Canada
9. McMahan W, Jones BA, Walker ID (2005) Design and implementation of a multi-section continuum robot: air-octor. In: Proc. IEEE/RSJ Intl Conf. on Intelligent Robots and Systems (IROS), Edmonton, Canada. pp 2578–2585
10. Martinez RV, Branch JL, Fish CR, Jin L, Shepherd RF, Nunes R, Suo Z, Whitesides GM (2013) Robotic tentacles with three-dimensional mobility based on flexible elastomers. Adv Mater 25(2):205–212
11. Cianchetti M, Ranzani T, Gerboni G, De Falco I, Laschi C, Menciassi A (2013) STIFF-FLOP surgical manipulator: Mechanical design and experimental characterization of the single module. In: IEEE/RSJ International Conference on Intelligent Robots and Systems (IROS), Tokyo, Japan. pp 3567–3581
12. STIFF-FLOP European project official website, (2014). http://www.stiff-flop.eu/
13. Jiang A, Xynogalas G, Dasgupta P, Althoefer K, Nanayakkara T (2012) Design of a variable stiffness flexible manipulator with composite granular jamming and membrane coupling. In: IEEE/RSJ International Conference on Intelligent Robots and Systems (IROS), Vilamoura, Portugal. pp 2922–2927
14. Kuniyoshi Y, Inaba M, Inoue H (1989) Teaching by showing: Generating robot programs by visual observation of human performance. In: Proc. intl symposium of industrial robots, Tokyo, Japan. pp 119–126
15. Riley M, Ude A, Atkeson C, Cheng G (2006) Coaching: An approach to efficiently and intuitively create humanoid robot behaviors. In: Proc. IEEE-RAS intl conf. on humanoid robots (humanoids), Genova, Italy. pp 567–574

16. Billard A, Calinon S, Dillmann R, Schaal S (2008) Robot programming by demonstration. In: Siciliano B, Khatib O (eds). Handbook of robotics. Springer, Secaucus. pp 1371–1394
17. Argall BD, Chernova S, Veloso M, Browning B (2009) A survey of robot learning from demonstration. Robot Auton Syst 57(5):469–483
18. Calinon S, Bruno D, Malekzadeh MS, Nanayakkara T, Caldwell DG (2014) Human-robot skills transfer interfaces for a flexible surgical robot. Comput Methods Programs Biomed 116(2):81–96. Special issue on new methods of human-robot interaction in medical practice
19. Malekzadeh MS, Bruno D, Calinon S, Nanayakkara T, Caldwell DG (2013) Skills transfer across dissimilar robots by learning context-dependent rewards. In: Proc. IEEE/RSJ Intl Conf. on Intelligent Robots and Systems (IROS), Tokyo, Japan. pp 1746–1751
20. Ng AY, Russell SJ (2000) Algorithms for inverse reinforcement learning. In: Intl conf. on machine learning, Stanford, CA, USA. pp 663–670
21. Zelman I, Titon M, Yekutieli Y, Hanassy S, Hochner B, Flash T (2013) Kinematic decomposition and classification of octopus arm movements. Front Comput Neurosci 7:60
22. Yekutieli Y, Sagiv-Zohar R, Aharonov R, Engel Y, Hochner B, Flash T (2005) Dynamic model of the octopus arm. I. Biomechanics of the octopus reaching movement. J Neurophysiol 94(2):1443–1458
23. Gutfreund Y, Flash T, Fiorito G, Hochner B (1998) Patterns of arm muscle activation involved in octopus reaching movements. J Neurosci 18(15):5976–5987
24. Gutfreund Y, Flash T, Yarom Y, Fiorito G, Segev In, Hochner B (1996) Organization of octopus arm movements: a model system for studying the control of flexible arms. J Neurosci 16(22):7297–7307
25. Flash T, Hochner B (2005) Motor primitives in vertebrates and invertebrates. Curr Opin Neurobiol 15(6):660–666
26. Zelman I, Galun M, Akselrod-Ballin A, Yekutieli Y, Hochner B, Flash T (2009) Nearly automatic motion capture system for tracking octopus arm movements in 3d space. J Neurosci Methods 182(1):97–109
27. Ijspeert A, Nakanishi J, Pastor P, Hoffmann H, Schaal S (2013) Dynamical movement primitives: Learning attractor models for motor behaviors. Neural Comput 25(2):328–373
28. Calinon S, Li Z, Alizadeh T, Tsagarakis NG, Caldwell DG (2012) Statistical dynamical systems for skills acquisition in humanoids. In: Proc. IEEE intl conf. on humanoid robots (humanoids), Osaka, Japan. pp 323–329
29. Dempster AP, Laird NM, Rubin DB (1977) Maximum likelihood from incomplete data via the EM algorithm. J Roy Stat Soc 39(1):1–38
30. Wu C (1983) On the convergence properties of the EM algorithm. Ann Stat 11:95–103
31. MacQueen JB (1967) Some methods for classification and analysis of multivariate observations. In: Proc. of the 5th Berkeley symp. on mathematical statistics and probability, Berkeley, CA, USA. pp 281–297
32. Ghahramani Z, Jordan MI (1994) Supervised learning from incomplete data via an EM approach. In: Cowan JD, Tesauro G, Alspector J (eds). Advances in Neural Information Processing Systems. Morgan Kaufmann Publishers, Inc., Burlington Vol. 6. pp 120–127
33. Dayan P, Hinton GE (1997) Using expectation-maximization for reinforcement learning. Neural Comput 9(2):271–278
34. Peters J, Schaal S (2007) Using reward-weighted regression for reinforcement learning of task space control. In: Proc. IEEE intl symp. on Adaptive Dynamic Programming and Reinforcement Learning (ADPRL), Honolulu, HI, USA. pp 262–267
35. Rueckstiess T, Sehnke F, Schaul T, Wierstra D, Sun Y, Schmidhuber J (2010) Exploring parameter space in reinforcement learning. Paladyn J Behav Robot 1(1):14–24
36. Theodorou E, Buchli J, Schaal S (2010) A generalized path integral control approach to reinforcement learning. J Mach Learn Res 11:3137–3181
37. Kober J, Peters J (2010) Imitation and reinforcement learning: Practical algorithms for motor primitives in robotics. IEEE Robot Autom Mag 17(2):55–62
38. Stulp F, Sigaud O (2012) Path integral policy improvement with covariance matrix adaptation. In: Proc. Intl Conf. on Machine Learning (ICML), Edinburgh, Scotland. pp 1–8
39. Kroese DP, Rubinstein RY (2004) The cross-entropy method: A unified approach to combinatorial optimization. In: Monte-Carlo simulation and machine learning. Springer, New York
40. Hansen N (2006) The CMA evolution strategy: A comparing review. In: Lozano J, Larranaga P, Inza I, Bengoetxea E (eds). Towards a new evolutionary computation. Studies in Fuzziness and Soft Computing. Springer, Berlin-Heidelberg Vol. 192. pp 75–102
41. Calinon S, Kormushev P, Caldwell DG (2013) Compliant skills acquisition and multi-optima policy search with EM-based reinforcement learning. Robot and Autonom Syst, Elsevier 61(4):369–379
42. Kormushev P, Calinon S, Saegusa R Metta G (2010) Learning the skill of archery by a Humanoid robot iCub, Proc. IEEE intl conf. on humanoid robots (humanoids), Nashville. TN, USA. pp 417–423
43. Sumbre G, Fiorito G, Flash T, Hochner B (2005) Neurobiology: Motor control of flexible octopus arms. Nature, Nature Publishing Group, New York City, NY, USA 433(7026):595–596
44. Hanassy S (2008) Reaching movements of the octopus involve both bend propagation and arm elongation. MSc thesis, Medical Neurobiology Department, Jerusalem, Israel: The Hebrew University

Bioinspired non-invasive radial pulse sensor: from biomimetic design, system calibration, to clinic application

Yudong Luo[1], Sanku Niu[1,2], Jose Cordero[1], Hongbin Deng[1,2] and Yantao Shen[1*]

Abstract

The research work aims at developing a real-time non-invasive metabolism and blood circulation surveillance system for monitoring human's health condition by sensing the various bio-signals on the human body. Our goal is to use the developed system to study the functions and characters of organs and tissues that highly relate with the metabolism and blood circulation system, and also, it is expected to help modeling the entire circulation system. At phase I of the research, in this paper we focus on developing a new low-cost, portable, high-accuracy, non-invasive radial pulse sensor. Inspired by touch capability and related biomechanical advantage of human fingertip, the mechanical design of the sensor mimics the physiological structure of human fingertip. The biomimetic sensor is then well calibrated using a high-accuracy force sensor, and the model is accurately identified by the system identification method. Further the calibrated sensor is applied to diagnose the arterial stiffness by measuring the augmentation index (AI) which is the important biomarker of vascular aging. Preliminary results demonstrate the sensor performance that it is capable of non-invasively, accurately, and reliably measuring radial pulse signals at real time, as well as to quantitatively determine the vessel aging.

Keywords: Biomimetic; Non-invasive; Radial pulse sensor

Background

Cardiovascular system is built by central and systemic circulation systems, which generates the blood flow by the heart and transmits the blood to the whole body through the vessel pathway [1]. Associated with the blood flow, the pulse is an important signal containing the pathological information like heart function, blood flow resistance, vessel wall elastic, and blood viscosity. These pathological changes are usually expressed on strength, frequency, altitude, and waveform variation of the pulse signal. Instead of invasive thermodilution methods and complicated non-invasive impedance-cardiography, the pulse wave analysis can thus be an alternative way to monitor the cardiovascular system.

One of these pulses is the radial pulse. A diagnosis method relying on the radial pulse signals detected by physician fingers can be traced back to 500 BC in China [2]. The first mechanical device can measure the arterial pressure waveform at radial called sphygmograph which was developed and improved in the nineteenth century [3]. Later, more and more researchers studied into this area, and different kinds of devices to measure the radial pulse wave were investigated. Nowadays, there are various commercial pressure waveform measurement devices developed by different companies and groups. These devices generally use technologies like RF-signal of an ultrasound, applanation tonometry, photoplethysmography, piezoelectric sensor, oscillometric, mechanotransducer, and volume plethysmography [4]. However, most of these sensing devices are only capable of measuring the systolic blood pressure, diastolic blood pressure, and heart rate [5]. The typical calibration for them is only a static process that relies on several discrete reference values from the cuff-based blood pressure sensors [5,6]. This results in most of them could not achieve radial pulse details and dynamic behaviors that contain rich

*Correspondence: ytshen@unr.edu
[1] Department of Electrical and Biomedical Engineering, University of Nevada at Reno, 1664 N Virginia Street, Reno, NV 89557, USA
Full list of author information is available at the end of the article

physiological and pathological information reflecting cardiovascular disease and vessel aging. In addition, most of devices are not cost effective and are unwearable, and this limits them to be extensively used for continuously monitoring the pulse signals of users. Without continuously monitoring the signals, the history of health condition related to cardiovascular diseases cannot be recovered for helping the diagnosis process and certain important physiological and pathological characteristics may be generally missed. For example, some characteristics from the pulses are significant during the intense exercise or the sleeping time due to the heavy load of the cardiovascular system and increasing of the blood viscosity; however, they may be lost without wearing a portable pulse sensor that has continuous monitoring capability. Recently, some research work also mentioned that continuous monitoring of the pulses will help to prevent the varied cardiovascular diseases at the early stage as well as to alarm users when the pulse signals become singular so as to avoid the situations like stroke, myocardial infarction, and sudden death [2,7]. In prospect of the increase in demand, developing a low-cost, wearable pulse sensor that can offer real-time continuous monitoring, immunizes mechanical and electronic noises, as well as displays some basic diagnostic results is becoming one of the most popular research topics in this area [5,6].

This paper aims at developing a real-time non-invasive metabolism and blood circulation surveillance system for monitoring human's health condition by sensing the various bio-signals on the human body; as its phase 1, we firstly address developing a high-accuracy, high-sensitivity, low-cost, non-invasive, and wearable radial pulse sensor for the system. More importantly, inspired by touch capability and associated biomechanical advantage of human fingertip, the mechanical design of the sensor mimics is the human fingertip physiological structure. In addition, to pursue the accuracy, the biomimetic sensor is well calibrated using high-accuracy force sensor and the model is accurately identified by the system identification method. The extended applications of the sensor include monitoring human's health condition in real time to support medical diagnosis process and physiological and pathological studies. It can be also expected to prevent and alarm the varied cardiovascular diseases at the early stage and boosts related clinical studies. Extensive calibration results show the good performance such as high reliability, high accuracy, and high sensitivity of the developed biomimetic sensor. In the first application, the sensor was successfully used to measure one critical factor called augmentation index (AI) that is related to arterial stiffness and cardiovascular diseases [7].

The paper is structured as follows. In Section 'Methods', Subsection 'Biomimetic radial pulse sensor design' reviews the human fingertip physiological structure and describes the mechanical and electrical designs of the biomimetic radial pulse sensor. Subsection 'Sensor calibration, identification, and verification' presents the calibration and identification processes of the developed sensor and validates the biomimetic performance of the sensing structure. Section 'Results and discussion' depicts the radial pulse measurement performance of the sensor, the clinical application and its preliminary results. Finally, Section 'Conclusions' concludes the work.

Methods

Biomimetic radial pulse sensor design

Human fingertip physiological structure

Finger palpation methods have been performed many centuries, and they are still widely applied in the clinic process. The mechanism of the palpation is that the physical properties containing the physiological and pathological information of the object that is contacted with the fingertip can be purely transferred to mechanoreceptors inside of the fingertip through its fine physiological structure with mechanical filtering functions [8]. This indicates that human fingertip owns the biomechanical advantages for achieving high-performance touch/force sensing and inspires us to mimic it for designing our radial pulse sensing structure. To start the mimicking, we firstly review the physiological structure of the human fingertip. Figure 1 shows the anatomy of the fingertip and nail, and there are several layers for the whole fingertip structure constituted with different organisms. The epidermis is the first layer from the bottom to the top, which works as a protective layer isolating the interior of the fingertip from outside. Dermis is the interlayer next to epidermis and is distributed with various types of mechanoreceptors that can sense touch, pressure, and vibrations from 0.5 to

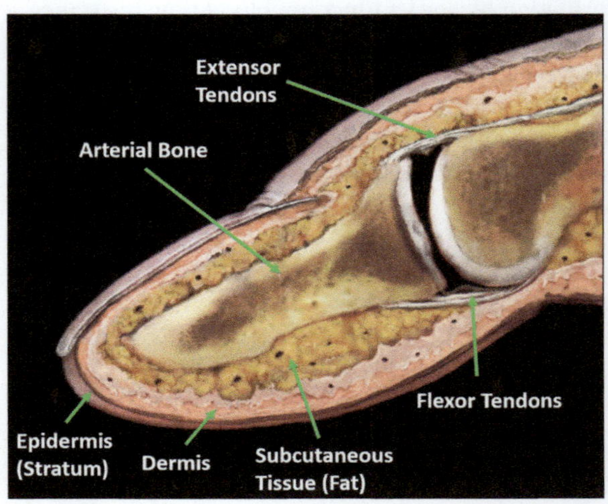

Figure 1 Illustration of human fingertip and nail anatomy.
Adapted from [11].

500 Hz [8]. Subcutaneous tissue is mainly a fat layer with 60% to 72% fluid in volume [9] and works like a damper to remove the unwanted information to the mechanoreceptors. The bone controlled by extensor and flexor tendons can move up and down to provide both passive and initiative movements for the effective touch and supports the fingertip as well. Due to the complex mechanical behavior and structures of the skin and subcutaneous, two simplified biomechanical models of the fingertip were reported [10]. These models include mechanical modeling of combined bio-components such as the bone, fat, soft tissues, and skin membrane. The results validating biomechanical advantages inspire us to design a radial pulse sensor that mimics the physiological structure of the human fingertip. As shown in Figure 1, several bio-components are arrowed and they are mimicked in designing our sensor components as shown in Figure 2b.

Mechanical design of biomimetic sensor

Figure 2 illustrates the mechanical design of the biomimetic radial pulse sensor using SolidWorks. Figure 2a displays the three-view diagram of the sensor structure. The length of the sensor is 50 mm, the width is 30 mm, and the height is 25 mm (without the adjustable screw); two grooves at two sides of the sensor can attach an adjustable belt to help wear it at the radial, or other body locations such as brachial, carotid, popliteal, and even superficial temporal comfortably. Figure 2b details the sensor components that functionally mimic the fingertip bio-components and their functions. All components except the sensing film and polydimethylsiloxane (PDMS) damper were made by the 3D printer using the solid plastics. The diameter of the skin contact ball is 10 mm, back side of the ball is glued with the sensing film, and the open side directly contacts with the radial

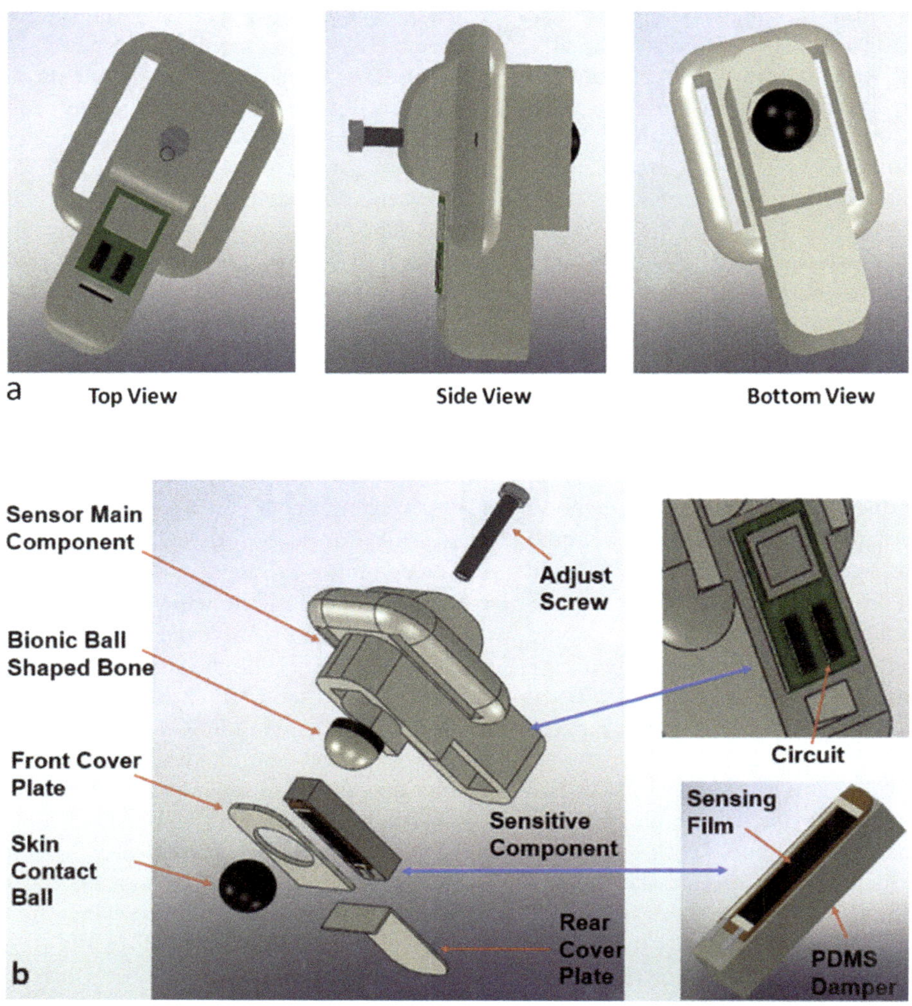

Figure 2 Illustration of mechanical design of the biomimetic radial pulse on SolidWorks. (a) Three-view diagram of the radial pulse sensor and **(b)** 3D assembly diagrams and components of the pulse sensor.

pulse location of wrist. The rigidity of the ball is carefully chosen so that it can transfer the radial pulse signal to the sensing film without significant mechanical loss. The rigid contact ball design can also help user to quickly find the exact pulse points. The sensing film is one of the key components that mimic the mechanoreceptors of fingertip for the developed pulse sensor. It is an ultrasensitive hybrid carbon/polymer-based piezoresistive (HCP) film with thickness of 0.127 mm, width of 3 mm, and length of 12 mm. The film has high sensitivity, very low thermal drift, and low hysteresis that were reported in our previous study [12]. The sensing film is glued on the elastic PDMS damper, the film will be deformed, and its resistance will change following the strength of transferred radial pulse signals .

The elastic PDMS damper is another key component in the sensor. It mimics subcutaneous soft tissues and fat in the human fingertip. To reach or close to the function of the soft tissues and fat, the hardness of the damper is an important factor. In this work, a digital shore hardness tester (TYPE A, range: 0 to 100 HA, resolution: 0.5 HA, accuracy: $\leq \pm 1\%$ HA) was firstly used to estimate the hardness of the human fingertips. The average hardness of the fingertips measured by the digital shore hardness tester is around 30 HA. To make an elastic damper with the similar hardness, Sylgard 184 Silicone Elastomer (Dow Corning Corp., Midland, MI, USA) (temperature range −45°C to 200°C) was selected. Five different samples with different mixing ratios between base liquid (part A) and curing agent (part B) were tested in order to achieve the hardness that is close to the hardness of the soft fingertip. Vacuumed Thermo Scientific Nalgene Chambers (Thermo Fisher Scientific, Waltham, MA, USA) was then used to store all the mixed silicone elastomer liquids for 48-hour curing process at the room temperature. After cured, five different hardness PDMS elastic dampers were made using different mixing ratios between base liquid and curing agent such as 1:2, 2:3, 1:1, 3:2, and 2:1. Through a long-time aging test, we found that all hardness of all damper samples gradually reaches the constant values after 33 days, as illustrated in Figure 3a. Especially for the sample with mixing ratio 1:2 (base liquid 1: curing agent 2), it can be found in Figure 3b, its hardness was measured around 30 HA that is the most close to the average hardness of the human fingertips. This PDMS sample was then chosen to be an elastic damper for the biomimetic sensor. In our design, the damper is a cuboid with a thickness of 3 mm (matches the thickness between the skin and the bone of fingertip), width of 3 mm, and length of 15 mm. In addition, the ball-shaped bone in the sensor mimics the arterial bone in the fingertip. The adjustable screw serves as the regulating unit of the bone. It functions as the extensor and flexor tendons in the fingertip. As shown in Figure 2b, all centers of the screw, bionic ball-shaped bone, and skin contact ball are aligned on a line in order to reduce the shear force and twist distortion during the operation.

Electronics design

The electronic circuit interfacing with the HCP sensing film was designed and custom-built for the sensor. It is used to process and to transform the signals that are from the film resistance changes caused by the radial pulse deformation. Since a preload force is applied on the sensing film when the sensor is worn at the radial, the gain and sensitivity of the circuit needs to be carefully adjusted to guarantee high sensitivity and high accuracy of the sensor, as well as to avoid saturation of the measured signals. To meet the requirement, as shown in Figure 4, a custom-built Wheatstone quarter bridge circuit with an AD620 amplifier is employed.

In the schematic, R_6 is the resistor that represents the value of the resistive layer on the HCP sensing film. Its resistance is changed following the micro deformation caused by the radial pulse. At the top of R_6, a potentiometer is used to trim the bridge circuit through removing the zero offset. R_1, R_2, and R_5 are the constant resistors with the same resistance as the undeformed resistive layer. R_3 is the resistor used to modify the gain of the amplifier AD620, and the maximum gain of the amplifier can reach 10,000. The circuit output V_o is the analog voltage signal that is input to a PC for further display and processing through a 16-bit and 200 kS/s DAQ system PCI-DAS6013 made by Measurement Computing Corp.

Sensor calibration, identification, and verification

Sensor calibration setup

As shown in Figure 5, all calibration and verification processes were implemented on a high-performance Newport vibration-isolated workstation. The calibration and verification procedures were automatically performed by the computer program in order to minimize the environmental noises and human effects.

Two high-performance sensors, a Shimpo FGV-1XY digital force gauge (Shimpo Instruments, Itasca, IL, USA) (capacity: 5 N, resolution: 1 mN, high accuracy: ± 0.2% FS) and a Baumer OADM 20I6X41/S14F laser distance sensor (Baumer Electric, Southington, CT, USA) (range: 70 mm, resolution: 20 μm), were employed in the calibration process. The laser distance sensor OADM 20I6X41/S14F was set up on the vibration-isolated workstation and its laser beam perpendicularly shoot on the moving stage of the micromanipulator for tracking the displacement change of the skin contact ball in the developed radial pulse sensor, when the digital force gauge is controlled to push/release the contact ball during the calibration. The pushing/releasing process was controlled by a Sutter MPC 385 3-D micromanipulator (Sutter Instruments, Novato,

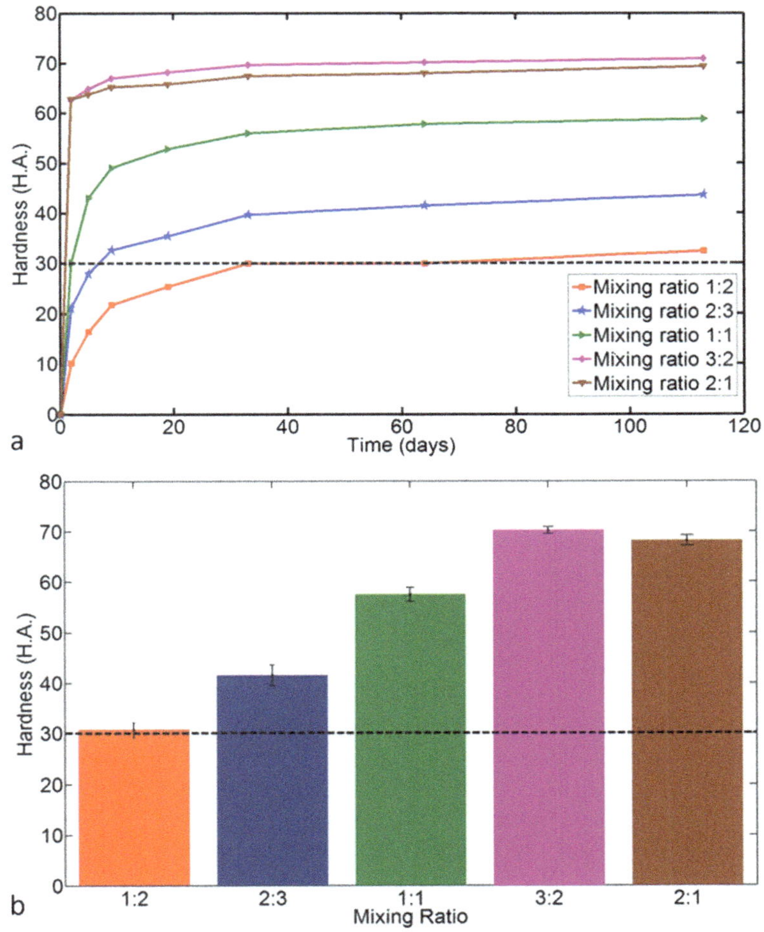

Figure 3 Long-time aging test result of multiple PDMS samples. (a) Five different hardness PDMS elastic dampers made by five different mixing ratios. **(b)** The damper made by the mixing ratio 1:2 has the hardness of 30 (after 33-day curing process) that is close to the average hardness of human fingertips.

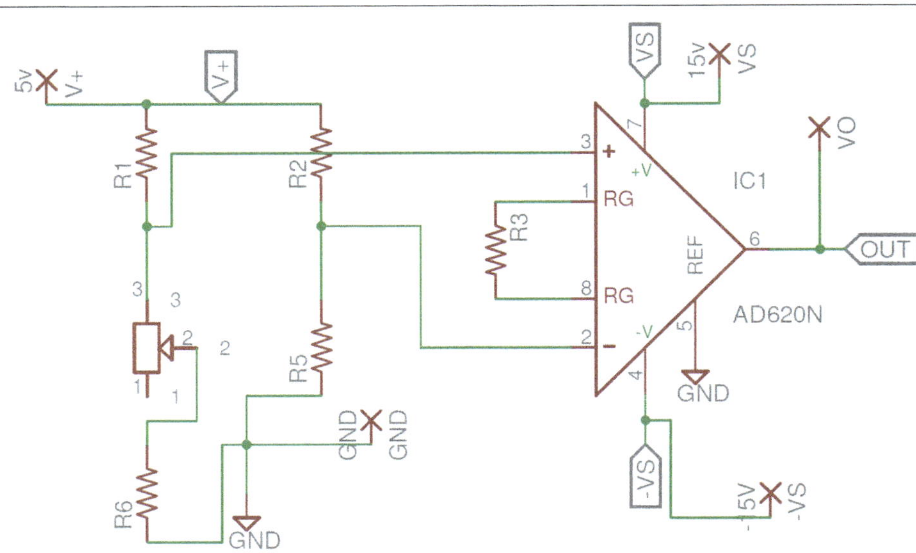

Figure 4 Circuit schematic of the pulse sensor.

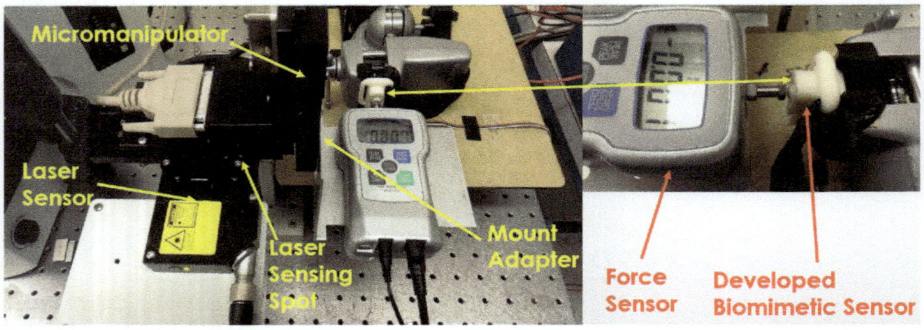

Figure 5 The setup for calibrating the biomimetic radial pulse sensor.

CA, USA) (range: 25,000 μm, resolution: 1 μm) where the digital force gauge is attached on the micromanipulator. The process can generate 150-μm height square wave pushing/releasing movement by the program-controlled micromanipulator. Meanwhile, the force information was recorded by the force gauge. In the calibration process, the developed pulse sensor was fixed on the vibration-isolated workstation through a clamping stander. Note that, to meet the real condition that the preload force is applied to the developed sensor during wearing it, an average 1.6 N preload force to the contact ball was initially set by the force gauge FGV-1XY before and during the calibration. In addition, the designed electronic circuit with the gain of 50 times was interfaced with the developed sensor. All signals were recorded to a PC through the DAQ board PCI-DAS6013 (Measurement Computing Corporation, Norton, MA, USA) (resolution: 16 bits, range: ± 10 V, number of channels: 16, accuracy: ± 8.984 mV, speed: 200 kS/s) with sampling rate of 1 kS/s.

Identification and calibration

Figure 6 plots all experimental data including pushing/releasing force, the developed sensor output, and pushing/releasing displacement on the skin contact ball of the sensor. The first row is the force curve measured by the digital force gauge FGV-1XY. The second row plots the voltage output from our developed pulse sensor, and the last row shows the displacement detected by the laser distance sensor OADM 20I6X41/S14F. The data demonstrates the developed sensor which has the fast response speed and good dynamic behavior.

Based on the experimental data shown above, the transfer function model between the developed pulse sensor voltage output and the force measurement by the force gauge FGV-1XY can be found through the system identification method developed in MATLAB. Here, the sensor output is used as the input, and the force measurement is used as the output (reference). The Laplace

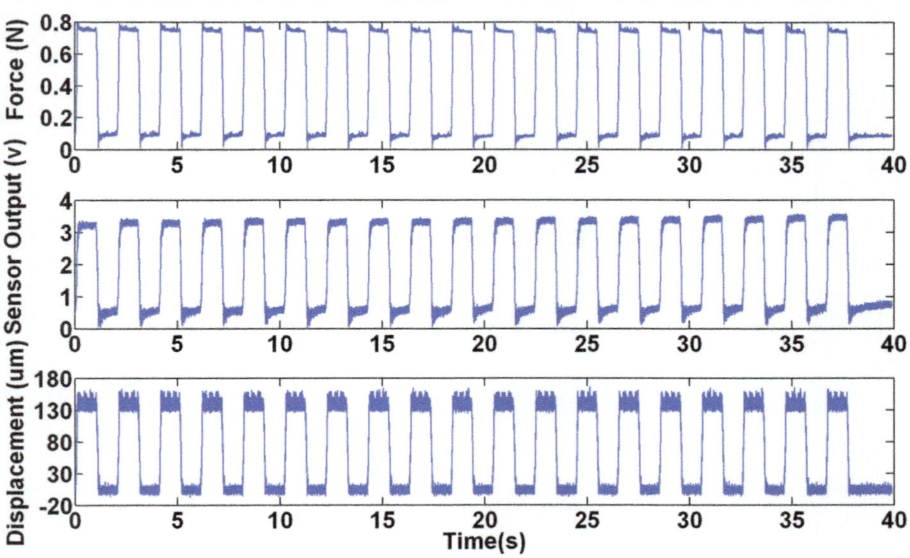

Figure 6 Raw data of force, the pulse sensor output, and displacement.

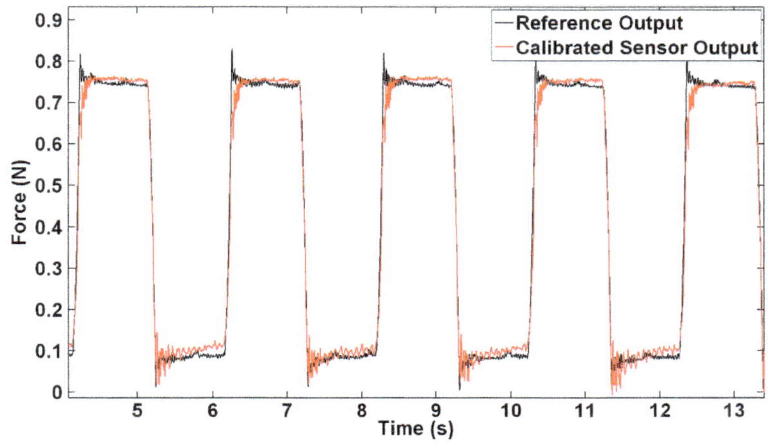

Figure 7 The calibrated sensor output and the reference measurement by force gauge FGV-1XY.

transfer function is identified as shown in the following equation:

$$G(s) = K\frac{1 + T_z s}{(1 + T_{p1}s)(1 + T_{p2}s)(1 + T_{p3}s)} \quad (1)$$

where the gain $K = 0.20998$, the coefficients $T_{p1} = 0.001$, $T_{p2} = 9.7751$, $T_{p3} = 0.0023316$, and $T_z = 11.063$. By applying the identified model in Equation 1 to the raw sensor voltage output, as shown in Figure 7, we can see that the fitting rate between the reference force measurement (black line) and the calibrated sensor force output (red line) is 90.75% high and the maximum error at peaks is 3.25%.

Further, Figure 8 plots the comparison results on frequency-amplitude relationship between the calibrated sensor output signals and the reference signals. The consistence between the two results further indicates the calibration performance of the sensor, that is, the sensor is well calibrated and can be used for the radial pulse measurement.

Biomimetic performance verification of sensing structure

The calibration results of the sensor have demonstrated the good mechanical to electronic conversion performance of the developed sensor. To further verify the biomimetic performance of the sensing structure that is close to the structure advantage of the human fingertip, an experiment was conducted to investigate the force-displacement relationship of serially connected components in the sensing structure consisting of a rigid skin contact ball, a sensing film, a silicone elastic damper, and a ball-shaped bone structure. In this experiment, the setup is similar to the calibration one described in Subsection 'Sensor calibration, identification, and verification'. Instead of using 150-μm height square wave motion, 150-μm height triangle wave movement was generated by the computer program to push

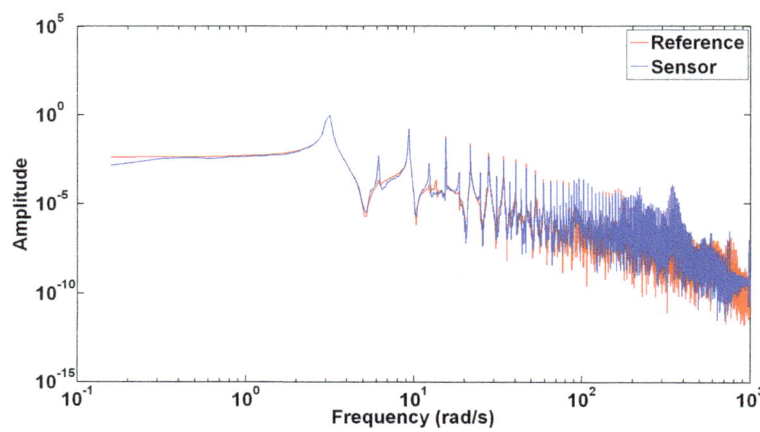

Figure 8 Comparison on spectral analysis results between the calibrated sensor output and reference signals.

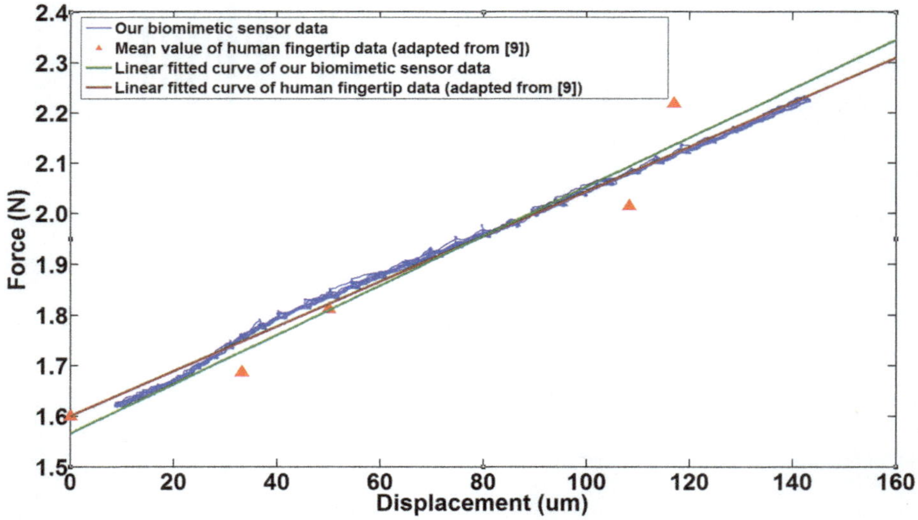

Figure 9 Relationship/comparison between force and deformation of both the sensing structure and human fingertip structure.

and release the sensing structure through the contact ball. The laser distance sensor OADM 20I6X41/S14F was set to a higher resolution but with a short range (range: 30 mm, resolution: 4 μm) to accurately detect displacement/deformation of the sensing structure caused by pushing and releasing operations.

Figure 9 plots the experimental force-displacement data points of the sensing structure, the mean value data points from real human fingertips conducted by Serina et al. in [9], and two curves fitted using the above two data point groups. Note that, a preload force of 1.6 N was applied to the sensing structure before conducting experiments. Some small spines in our experimental force-displacement data curves were induced by the movement of the step driven motors of the micromanipulator. It is clear that our sensing structure force-displacement data points demonstrate a good linear relationship between the force and the displacement in several pushing and releasing experimental loops. The small overlap between these data points (curves) in several experimental loops indicates a very low hysteresis effect of the designed sensing structure. More importantly, our experimental result agrees well with the force-displacement behavior of the human fingertips conducted by Serina et al. [9]. In the figure, the brown linear fitted curve from the data points of our biomimetic sensor is very close to the green linear fitted curve from the human fingertip data measured by Serina et al. [9] and it experimentally verifies the successful biomimetic design of our sensor.

Results and discussion

Radial pulse measurement and results

After the calibration process, the sensor can be used to measure the radial pulses. As shown in Figure 10, one user worn the developed biomimetic pulse sensor and placed his arm on the flat table where the user sits in front of. The developed acquisition and signal processing program in a PC is started, and it records the processed radial pulse signals from the sensor and the circuit and displays the results on the screen in real time.

Figure 11 shows the raw radial pulse signal (in red) and the calibrated radial pulse signal (in blue) of the user.

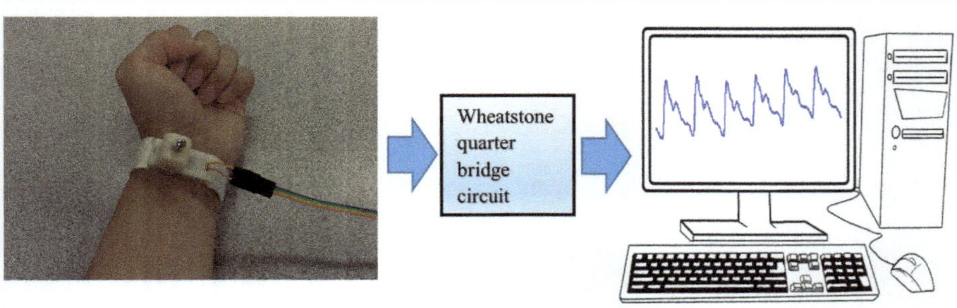

Figure 10 Illustration of the radial pulse measurement setup.

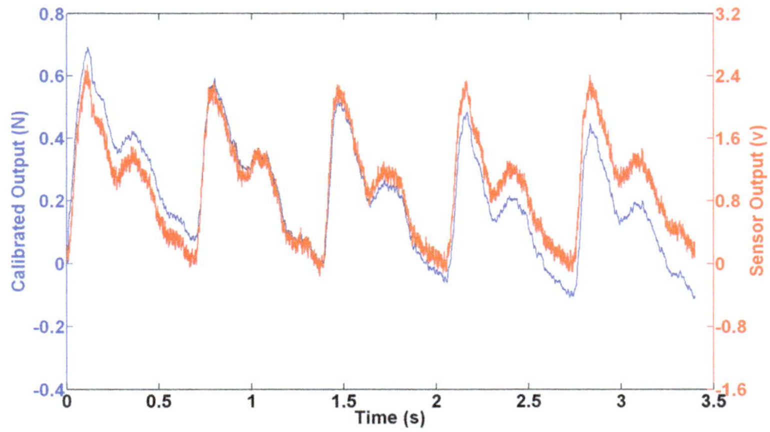

Figure 11 Radial pulse signal measurement and calibration results.

The results demonstrate the good measurement performance of the developed sensor as its raw data is very close to the calibrated data, as well as the detail of the pulse information is well revealed.

Clinic application and results

Next to the achieved accurate radial pulse signals using the sensor, in this section, one clinical application based on the radial pulse measurement is extensionally presented. The application is related to determining the arterial stiffness by computing the AI using the measured radial pulse signals of the sensor.

In the medical area, arterial stiffness is seen as an important biomarker of the cardiovascular risk [7]. The central augmentation index (cAI) is widely acknowledged as an evaluation factor for this biomarker. cAI is usually defined as the ratio of difference between the first and second systolic peaks with the pulse pressure. As shown in Figure 12, the complex pulse waveform (in black) at carotid (central) artery contains a forward wave (the first one is in blue) generated by the heart and a backward wave (in red) reflected from the peripheral artery. The central augmentation index is then defined as $cAI = (Pi-PS)/(PS-PD)$ that presents the ratio difference at carotid (central). This index was first reported by Murgo et al. [13], and then, Kelly et al. [14] discovered the way to find the inflection point using the fourth derivative of the pulse waveform. Later, Takazawa et al. [15] researched the radial (peripheral) pulse augmentation index (rAI) and found that rAI is the ratio of the height of late systolic pressure to the early systolic pressure. They also demonstrated the automatic way to find the late systolic pressure point at radial artery by fourth derivative of the radial pulse waveform [15].

Recent research shows that the central augmentation index cAI can be estimated by applying a transfer function to the radial pulse augmentation index (rAI) [16,17], or can be obtained directly from the radial pulses [18] due to

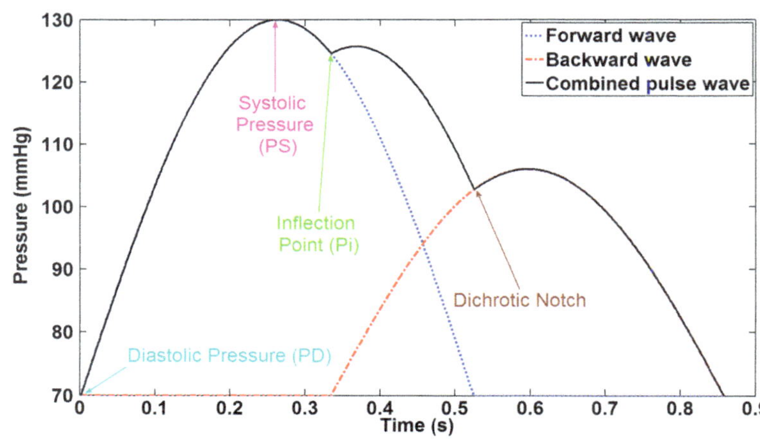

Figure 12 Illustration of C-type waveform at carotid artery.

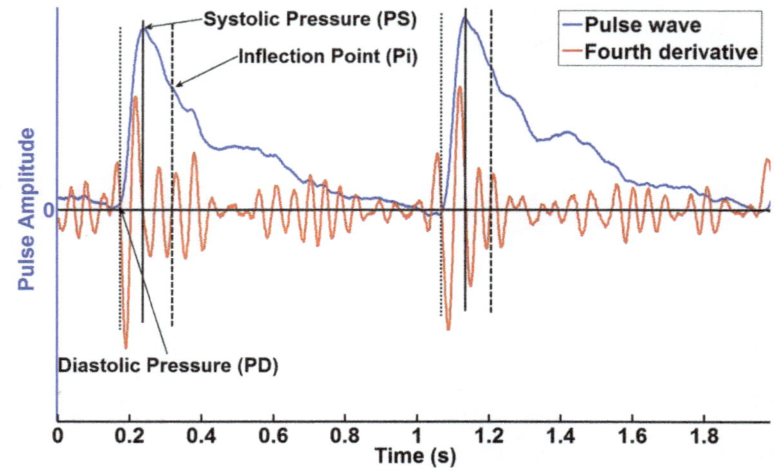

Figure 13 Measured radial pulse signal and its processing for achieving the augmentation index.

high correlation. This indicates rAI can be also used as the evaluation factor like cAI. In this paper, based on the radial pulse measurement using the developed sensor, we preliminarily start to investigate radial augmentation index rAI as well to correlate it into the cAI for determining the arterial stiffness.

As shown in Figure 13, the blue curve presents the pulse wave measured by our calibrated sensor and the red curve is the fourth derivative of the pulse wave after filtering using Savitzky-Golay FIR smoothing filter. In addition, the dot, solid, and dash vertical lines indicate the diastolic pressure (PD) point, systolic pressure (PS) point, and inflection point (Pi), respectively. These points in time axis are determined when the fourth derivatives of the pulse wave cross to zero [15]. After determining these points, radial augmentation index, rAI = (Pi−PD)/(PS−PD) can be computed. The rAI is strongly correlated with the carotid (central) augmentation index (cAI), and the corrected results can be used to determined arterial stiffness and cardiovascular risk directly [19,20].

To validate the augmentation index achieved by the pulse signals from the developed pulse sensor, a photoplethysmography (PPG) (Embedded Lab, HRM-2511-E) sensor was synchronously employed for additional measurements during measuring the radial pulse. PPG signals can be used to compute the index of SI. The SI index is an equal factor as the augmentation index and can be used as an important value that evaluates the artery stiffness. The index is computed based on the height of the subject/user divided by the time delay between the subject's systolic and diastolic peaks [21], as illustrated in Figure 14.

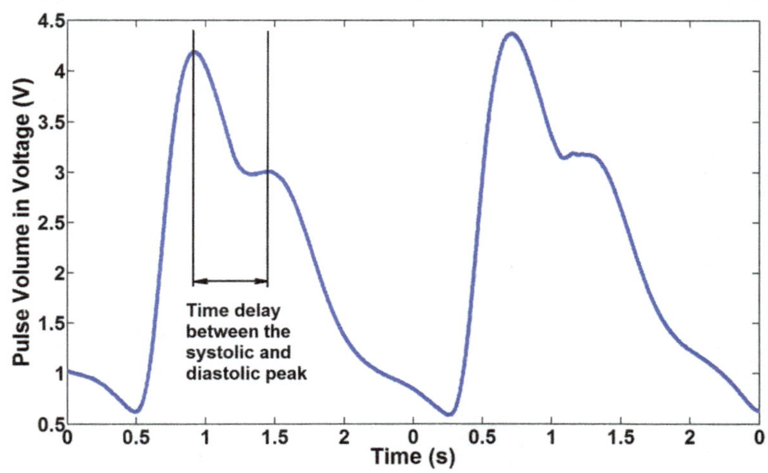

Figure 14 Measured PPG signals for achieving the SI index.

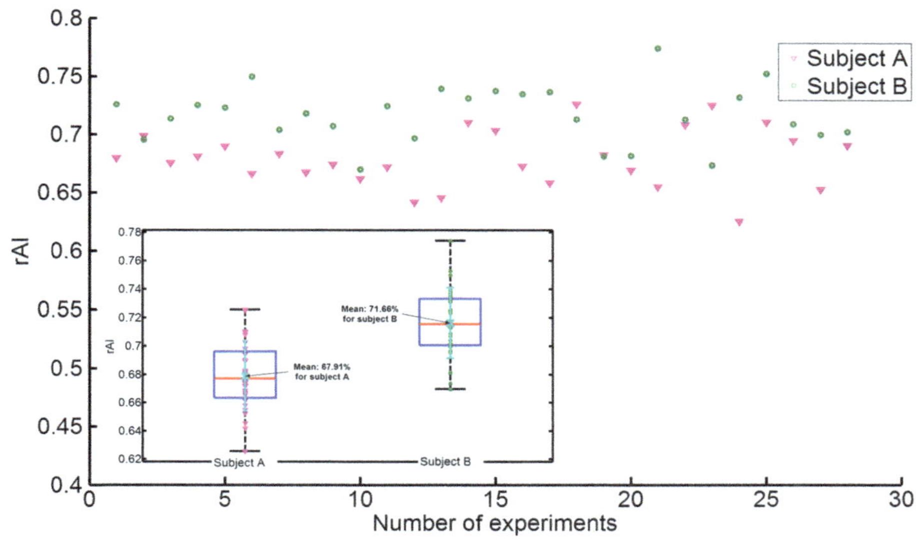

Figure 15 Experimental results and statistical analysis boxplot of two-subject's rAIs (measured by our pulse sensor).

Note that, in this clinic test, two non-smoking health subjects without cardiovascular disease history, subjects A and B, with the same height and similar weight were chosen to measure after 30-min rest in a quiet room with more than 3-h fasted, also abstained from caffeine and alcohol. The developed pulse sensor was worn at the radial on the left arm of each subject. The PPG sensor was clamped at the index finger of the right hand of each subject. Both signals from our pulse sensor and the PPG sensor were collected by the DAQ board PCI-DAS6013 (sample rate: 1 kS/s) in the same time.

Corresponding to Figure 13, Figure 15 shows two-subject experimental data of the rAI measured by our sensor and their statistical analysis results. Note that, experiments were conducted 28 times for each subject. In the figure, the pink triangle data points indicate the rAI values of subject A and the green circle data points represent rAI values of subject B. Embedded small figure at the left bottom of Figure 15 displays the statistical analysis results of two-subject experimental data points by the boxplot method with the mean and error bar. Both boxplot graphs represent the range between the 25th percentiles and 75th percentiles using the blue squares,

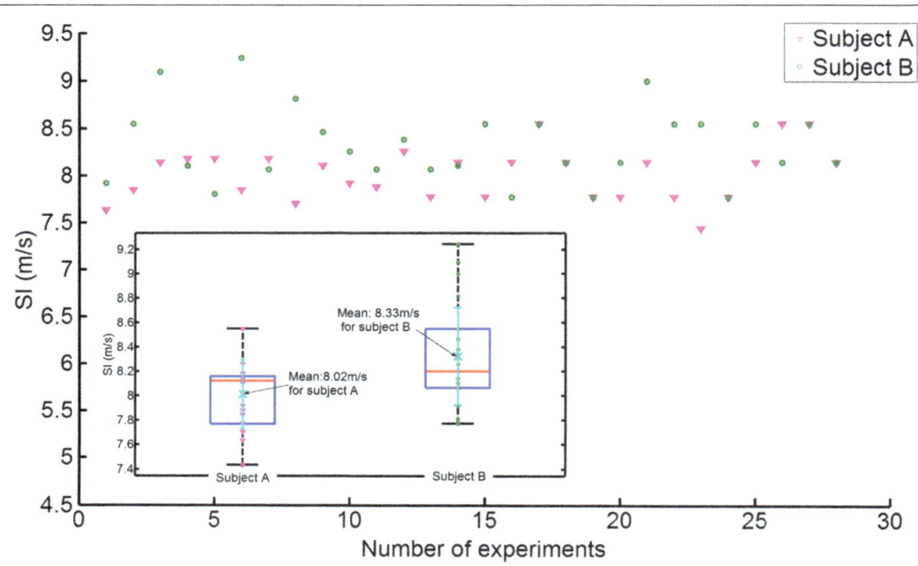

Figure 16 Experimental results and statistical analysis boxplot of two-subject's SIs (measured by the PPG sensor).

Table 1 Physiological characteristics of the subjects and measured AI and SI indexes

Characteristic	Subject A	Subject B
Height	1.71 m	1.71 m
Weight	73 kg	75 kg
Body mass index	25	25.6
Age	25	30
Sex	Male	Male
Cardiovascular disease history	No	No
AI (our pulse sensor)	67.91%	71.66%
SI (PPG)	8.02 m/s	8.33 m/s

and the red lines in the blue squares are the medians of two data sets, respectively. The celeste blue cross and the vertical celeste blue line in the middle of each square show the mean and the standard deviation of the experimental data of each subject, respectively. Similarly, corresponding to Figure 14, 28 experimental results of two subjects on the SI index measured by the PPG sensor and their boxplot statistical analysis results are plotted in Figure 16. Both results in Figures 15 and 16 demonstrate that the vessel aging evaluated by both mean values of rAI and SI of subject B is higher than the value of subject A. This result agrees well with age truth of two subjects and indicates our pulse sensor can achieve highly accurate and precision results for evaluating vessel aging that is highly related to subject age.

Table 1 summarizes all experimental tests. As shown in the table, the physiological characteristics of the two subjects are very close except for age. Two biomarkers, the mean value of rAI measured by the developed pulse sensor and the mean value of SI detected by the commercial PPG, which are related to the arterial stiffness, are listed. The AI index for the younger subject (subject A) is 67.91% while for the 30 years old subject (subject B) is 71.66% that is slightly stiffer than the value of the younger subject. In the normal condition, the augmentation index will increase as age increases [14,22]. The achieved AI results reasonably reflect the truth of arterial stiffness of the two subjects. In addition, the truth reflected by the measured AI index results is well validated by the SI index results from the PPG. These results clearly verify the sensor performance that the developed biomimetic sensor can be successfully used in determining the arterial stiffness and vessel aging and related clinic applications.

Conclusions

In this paper, inspired by touch capability and biomechanical advantage of human fingertip, we developed a high-performance radial pulse sensor with the highly sensitive, durable, low-cost HCP film, and biomimetic sensing structure. Calibration and verification prove the sensor performance that it could extract the accurate and reliable pulse signals non-invasively. One clinic application of the developed sensor is to determine the arterial stiffness based on the calculated augmentation index at radial. The results indicate that the developed sensor can be used in clinic applications and can be potentially extended to use in our real-time non-invasive metabolism and blood circulation surveillance system in the future.

Competing interests

The authors declare that they have no competing interests.

Authors' contributions

YL participated in all stages of the sensor design, carried out all calibration and experiments, performed data processing and analysis, and drafted the manuscript. SN and JC helped in conducting the experiments and data processing. HD participated in the design of the sensor. YS conceived the research, participated in the sensor design, data analysis and the research coordination, and revised the manuscript. All authors read and approved the final manuscript.

Acknowledgements

This research work is partially supported under the NSF Career Award CBET 1352006 and the Research Enhancement Grants (REG) from UNR office of the vice president for research and innovation.

Author details

[1]Department of Electrical and Biomedical Engineering, University of Nevada at Reno, 1664 N Virginia Street, Reno, NV 89557, USA. [2]Department of Mechatronic Engineering, Beijing Institute of Technology, 5 South Zhongguancun Street, Haidian, Beijing 100081, China.

References

1. Green JF (1987) Mechanical concepts in cardiovascular and pulmonary physiology. 2nd edn. Philadelphia, Lea & Febiger
2. Fan ZP, Zhang G, Liao S (2011) Chap. 2, Advanced biomedical engineering, Edited by Gargiulo GD. InTech, http://www.intechopen.com/books/advanced-biomedical-engineering
3. Kim DH, Braam B (2013) Assessment of arterial stiffness using applanation tonometry. Can J Physiol Pharmacol 91:999–1008
4. Avest ET, Stalenhoef AF, Graaf JD (2007) What is the role of non-invasive measurements of atherosclerosis in individual cardiovascular risk prediction? Clin Sci 112:507–516
5. Tu T, Chao PCP, Lee Y (2013) A new non-invasive cuff-less blood pressure sensor. In: Proceedings of the 2013 IEEE International Conference on Sensors, 857-860, Baltimore, MA, USA
6. Hsu YP, Young DJ (2013) Skin-surface-coupled personal health monitoring system. In: Proceedings of the 2013 IEEE International Conference on Sensors, 210-213, Baltimore, MA, USA
7. Mitchell GF (2009) Arterial stiffness and wave reflection: biomarkers of cardiovascular. Artery Res 3(2):56–64
8. Johansson RS, Flanagan JR (2009) Coding and use of tactile signals from the fingertips in object manipulation tasks. Nature 10:345–359
9. Serina ER, Mockensturm E, Mote CDJ, Rempel D (1998) A structural model of the forced compression of the fingertip pulp. J Biomech 31(7):639–646
10. Wu JZ, Dong RG, Rakheja S, Schopper AW, Smutz WP (2004) A structural fingertip model for simulating of the biomechanics of tactile sensation. Med Eng Phys 26:165–175
11. Zimmer SM (2008) Assessment and initial care of fingertip and nailbed injuries. JUCM 3(2):11–17
12. Luo YD, Shen YT, Mohan N (2013) Durable and cost-effective 3-D microforce sensor for musical tuning enhanced micro palpation of

biological entities. In: Proceedings of the 2013 IEEE international conference on Sensors, 928-931, Baltimore, MA, USA
13. Murgo JP, Westerhof N, Giolma JP, Altobelli SA (1980) Aortic input impedance in normal man: relationship to pressure wave forms. Circulation 62:105–116
14. Kelly R, Hayward C, Avolio A, O'Rourke M (1989) Non-invasive determination of age-related changes in the human arterial pulse. Circulation 80:1652–1659
15. Takazawa K, Tanaka N, Takeda K, Kurosu F, Ibukiyama C (2005) Underestimation of vasodilator effects of nitroglycerin by upper limb blood pressure. Hypertension 26(3):520–523
16. Fetics B, Nevo E, Chen CH, Kass DA (1999) Parametric model derivation of transfer function for noninvasive estimation of aortic pressure by radial tonometry. IEEE Trans Biomed Eng 46:698–706
17. Chen CH, Nevo E, Fetics B, Pak PH, Yin FC, Maughan WL, Kass DA (1997) Estimation of central aortic pressure waveform by mathematical transformation of radial tonometry pressure: validation of generalized transfer function. Circulation 95:1827–1836
18. Millasseau SC, Patel SJ, Redwood SR, Ritter JM, Chowienczyk PJ (2003) Pressure wave reflection assessed from the peripheral pulse is a transfer function necessary? Hypertension 41:1016–1020
19. Sugawara J, Komine H, Hayashi K, Maeda S, Matsuda M (2007) Relationship between augmentation index obtained from carotid and radial artery pressure waveforms. J Hypertens February (2):375–381
20. Kohara K, Tabara Y, Oshiumi A, Miyawaki Y, Kobayashi T, Miki T (2005) Radial augmentation index: a useful and easily obtainable parameter for vascular aging. America J Hypertens 18(S1):11–14
21. Millasseau SC, Kelly RP, Ritter JM, Chowienczyk PJ (2002) Determination of age-related increases in large artery stiffness by digital pulse contour analysis. Clin Sci 103:371–377
22. Salvi P (2012) Pulse waves: how vascular hermodynamics affects blood pressure. Springer, Milan

A novel RGB Fourier transform-based color space for optical microscopic image processing

Tao Guan[1*], Dongxiang Zhou[1], Chao Xu[1] and Yunhui Liu[2]

Abstract

This paper presents an IF_1F_2 color space for describing colors in microscopic images. Unlike the classical color spaces that treat the R, G, and B (RGB) components of a pixel's color as independent elements, the proposed color space treats RGB as a sampled spectral signal, called a RGB signal. Then, based on the Fourier spectrum analysis of the RGB signal, the 1-D discrete Fourier transform is introduced to describe the color features of microscopic images. *K*-means clustering experiments on two microscopic image datasets validate the superiority of the proposed IF_1F_2 color space compared with the classical RGB and HSV (hue, saturation, and value) color spaces.

Keywords: RGB color space; HSV color space; Fourier transform; *K*-means

Background

Color image processing is one of the hot topics in image processing and computer vision. To effectively represent colors, various color spaces have been presented in the literature, such as the RGB (R, G, and B) color space, the HSV (hue, saturation, and value) color space, the CIELAB color space, and the YCbCr color space. A detailed survey of these color spaces can be found in [1]. In recent years, a number of new methods have also been proposed for describing colors more effectively. Chengjun Liu and Jian Yang [2] introduced the independent component analysis (ICA) into the analysis of color features and proposed an ICA-based color space. Margarita Bratkova et al. developed an oRGB color space [3] based on the opponent color theory. Ingmar Lissner and Philipp Urban derived a perception-based color space [4] that has high perceptual uniformity. For color image processing algorithms, it is of vital importance to select the optimal color space [1,2,5-7]. However, most existing color spaces, such as the RGB color space and the HSV color space, are application dependent. Once the image varies, the used color space may decrease its performance. In microscopic image processing, such as the cervical cell image segmentation, most algorithms are still based on classical color spaces, which cannot best describe the color features of microscopic images [8,9,7,10-13]. In the existing studies of how to effectively represent colors, we found that few methods analyze colors from the frequency domain. Therefore, we investigate colors in microscopic images based on the frequency analysis of the R, G, and B components of pixels' colors in this paper.

Color images are generally stored in the format of the RGB color space. Through linear or nonlinear transforms, the RGB color space can be transformed to other classical color spaces. However, these classical color spaces mostly treat the R, G, and B components as three independent elements. In this paper, on contrary to classical color spaces, we treat the R, G, and B components as a sequence of a spectral signal, called a RGB signal. Then, based on the methodology of frequency spectrum analysis in the signal processing area, we combine the characteristics of 1-D discrete Fourier transform [14] with the color features of microscopic images and propose a RGB Fourier transform-based color space, namely an IF_1F_2 color space, to describe colors of microscopic images. In this color space, similar colors have uniform data distributions whereas different colors have obviously different data distributions.

Methods

Color can be represented by a combination of tristimuli R (red), G (green), and B (blue) [1]. Actually, the R, G, and B components are obtained from three separate color filters,

* Correspondence: gt_mike2003@126.com
[1]College of Electronic Science and Engineering, National University of Defense Technology, Kaifu, Changsha 410073, Hunan, China
Full list of author information is available at the end of the article

namely red filter S_R, green filter S_G, and blue filter S_B, according to the equations below:

$$\begin{aligned} R &= \int P(\lambda) S_R(\lambda) d\lambda \\ G &= \int P(\lambda) S_G(\lambda) d\lambda \\ B &= \int P(\lambda) S_B(\lambda) d\lambda \end{aligned} \tag{1}$$

where $P(\lambda)$ is the light radiance with the wavelength λ.

From Equation 1, we may find that the R, G, and B components are not independent but actually correlated with each other. Unlike many existing classical color spaces that treat the R, G, and B as independent elements, we treat them as a sequence of a sampled spectral signal that has three samples with the value of R, G, and B in order. In the area of signal processing, we know that similar signals have a similar Fourier frequency spectrum [14]. Consequently, if we treat the R, G, and B as signal samples, similar colors in the image may have a similar signal frequency spectrum, whereas different colors have a different signal frequency spectrum. Then, the next question is how to describe the color features (the intensity and the chromatic information) based on the Fourier frequency spectrum? We know that in the RGB color space, the gray level (without chromatic information) from black to white is represented by $R = G = B = \{0 \sim 255\}$ [1]. To some extent, the chromatic information is caused by the difference between R, G, and B. When the R, G, and B components are arranged into a sequence and become a signal, the high frequency component of the signal's Fourier transform can describe the difference between R, G, and B. Inspired by this idea, we introduce the 1-D discrete Fourier transform into the color feature analysis of microscopic images as follows.

Assume that a pixel's RGB value is (r, g, b), then (r, g, b) is arranged into a sequence s, where $s(0) = r$, $s(1) = g$, $s(2) = b$. We assume s to be a sampled signal, which has three samples with the value r, g, and b, respectively. We name this signal a RGB signal. According to the definition of 1-D discrete Fourier transform (DFT) [14], the DFT of s is:

$$X(k) = \mathrm{DFT}\{s\} = \sum_{n=0}^{2} s(n) \cdot \exp\left(-j\frac{2\pi}{3}nk\right), \; k = 0, 1, 2 \tag{2}$$

where X is called the RGB Fourier transform of the pixel's color.

Then, the proposed color space is constituted of three components:

$$\begin{aligned} I &= \mu \cdot \mathrm{abs}(X(0)) \\ F_1 &= \mathrm{real}(X(2)) \\ F_2 &= \mathrm{imag}(X(2)) \end{aligned} \tag{3}$$

where $X(0)$ is the low frequency (zero frequency) component of X and $X(2)$ is the high frequency component of X. μ is a scaling parameter used to scale I into a proper numerical range for specific image processing applications, abs($X(0)$) is the complex magnitude of $X(0)$, real($X(2)$) is the real part of $X(2)$, and imag($X(2)$) is the imaginary part of $X(2)$. Since the RGB signal only has three samples, we use 3-point DFT to calculate its discrete Fourier spectrum. For a 3-point DFT, $X(1)$ and $X(2)$ both belong to the high frequency components. The magnitudes of $X(1)$ and $X(2)$ are equivalent, while their phases are different. Therefore, we can either choose $X(1)$ or $X(2)$ as the chromatic component of the proposed color space. If two colors are slightly different, the major difference between them may be caused by how the RGB values of each color vary. In the methodology of frequency spectrum analysis, the high frequency component of a signal's Fourier spectrum represents the differences between the signal's samples, and these differences, to some extent,

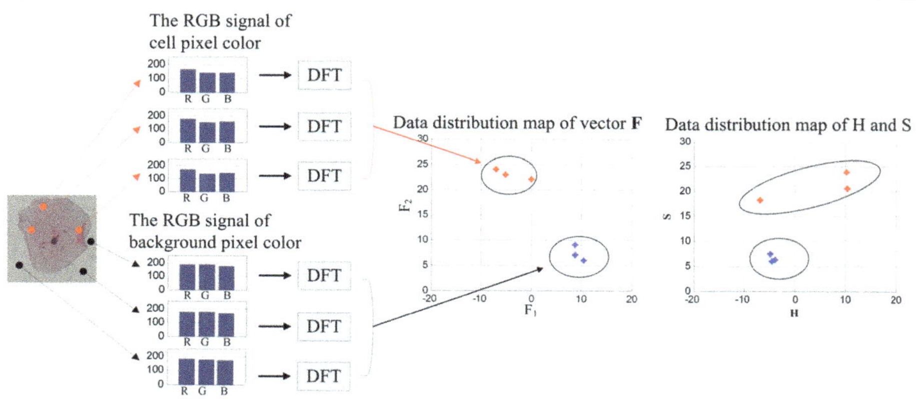

Figure 1 RGB signals of different colors. Their corresponding data distribution map of vector **F** as well as the H and S components in the HSV color space is also shown.

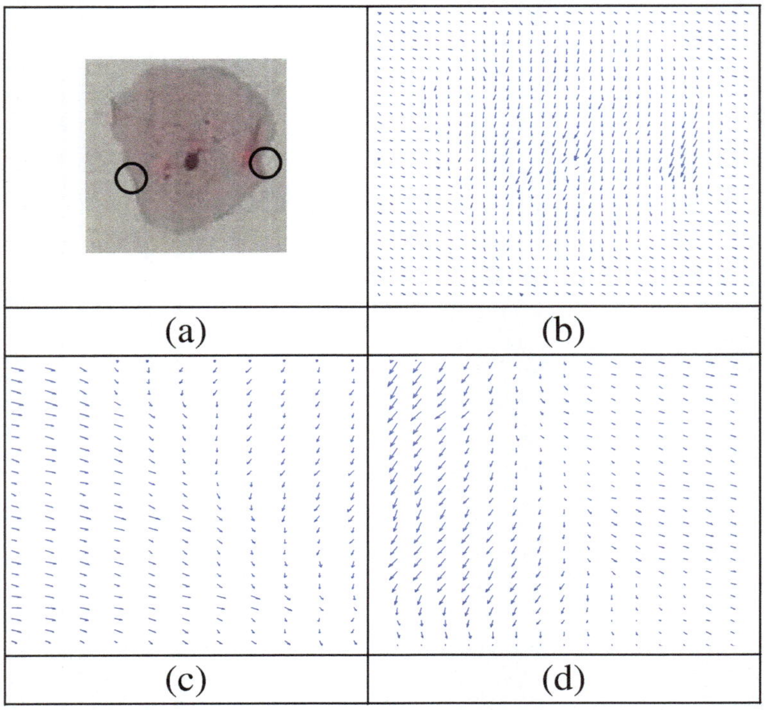

Figure 2 Color vector field of a cervical cell image. (a) Original color cell image. **(b)** Color vector field of **(a)**. **(c, d)** Color vector fields in high resolution corresponding to the left and right circles in **(a)**, respectively.

describe how the signal's samples vary. Therefore, we use the high frequency component of the RGB signal to describe the color features of microscopic images in this paper.

Then, we construct a vector $\mathbf{F} = [F_1, F_2]$, and then I and F are the intensity component and the chromatic component of the proposed color space, respectively. This new color space is called an IF_1F_2 color space. We use a color cervical cell image as an example to analyze the characteristics of different colors in the IF_1F_2 color space. In Figure 1, the second column shows the RGB signals. Each RGB signal corresponds to a sampled pixel's color (as the arrow shows in Figure 1) in the cell image. The data distribution map in the middle part of Figure 1 shows the data distribution of vector **F** corresponding to each RGB signal in the second column. We can find that, for similar colors, the data distribution of vector **F** is concentrated, whereas for different colors, the data distribution of vector **F** is dispersed. In addition, we also provide the data distribution map of the chromatic components (the H and S components) in the HSV color space concerning the same pixels in the left cell image, as shown in the right part of Figure 1. To make fair comparisons, we rescaled the H and S values to the same range as that of the values in vector **F**. We can see that the data distributions of H and S are not as concentrated as that of vector **F** for the red points in Figure 1.

Table 1 The ZSI mean μ_{ZSI} and the deviation σ_{ZSI} of the clustering results in different color spaces

	$\mu_{ZSI} \pm \sigma_{ZSI}$		
	RGB	HSV	IF_1F_2
Dataset 1	0.8090 ± 0.0880	0.8237 ± 0.0838	0.8801 ± 0.0543
Dataset 2	0.8121 ± 0.0856	0.8103 ± 0.0809	0.8910 ± 0.0508

In addition, we provide the color vector field of a cervical cell image to further demonstrate the effectiveness of the IF_1F_2 color space, as shown in Figure 2. In the color vector field of the image, each point corresponds to one image pixel and the vector on that point corresponds to the color vector **F** of that pixel. From Figure 2, we may observe that the vectors within the cell region are similar, whereas the vectors in the cell region are obviously different from those in the background region.

Results and discussion

Our experiments focused on validating the superiority of the proposed IF_1F_2 color space over the classical RGB and HSV color spaces for describing the color features of microscopic images. Because K-means algorithm is a widely used clustering algorithm, which can merge data points with high similarity into the same class and split data points with low similarity into different classes, we

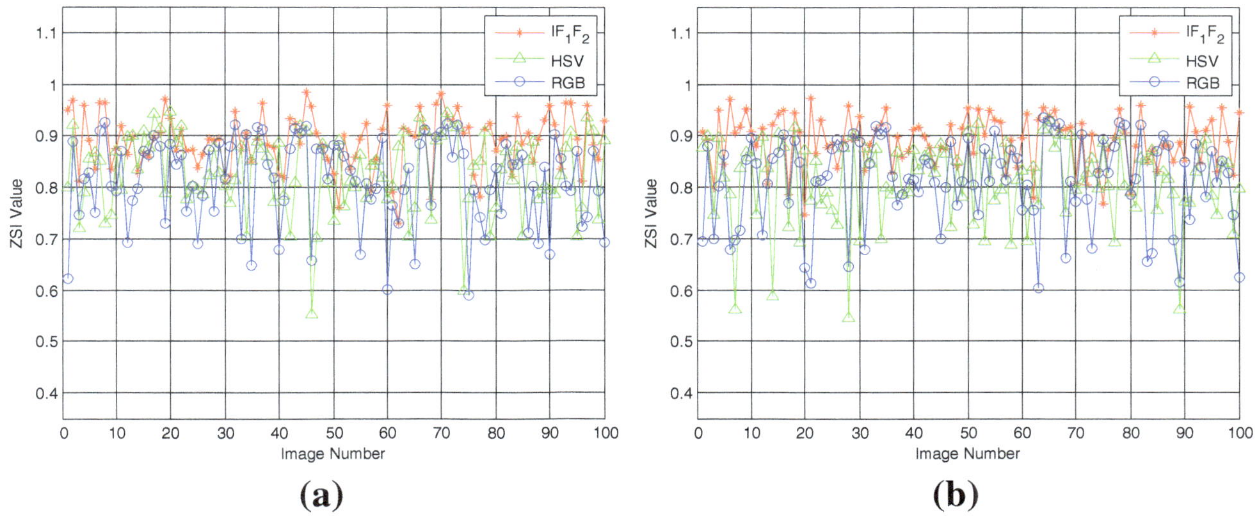

Figure 3 The statistical ZSI results. **(a)** Dataset 1. **(b)** Dataset 2.

adopted *K*-means clustering algorithm to make rough segmentation of the cell images and test the discriminating power of different color spaces.

Two optical microscopic image datasets including the Herlev dataset [15] containing 100 normal cervical cell images (Dataset 1) and the single-cell cervical smear image dataset of our lab containing 100 images (Dataset 2) have been tested. The cell images of our lab were acquired by an automatic microscopic image acquisition system with a CCD camera (DA-HENG, Beijing, China), an optical microscope (Olympus CX31, Olympus Corporation, Shinjuku-ku, Japan), and a 40× magnification lens.

In each cell image, there are typically two kinds of pixels: the pixels in the cell region (with approximately red or blue color) and the pixels in the background (with approximately white color). The aim of the rough segmentation is to split the pixels into two classes: the cell (foreground) and the background. The colors of the pixels within the same region (the cell or the background) are similar, whereas the major difference between the cell and the background pixels is caused by the chromatic components of pixels' colors. Therefore, we chose the chromatic components of the HSV and IF_1F_2 color spaces as the input feature vector to the *K*-means algorithm. For the RGB color space, because there are no chromatic components of the RGB space, we chose all the R, G, and B components. Since the color difference between the nuclei and the cytoplasm is mainly caused by the intensity contrast, whereas their chromatic components are similar, they are supposed to be within the same class.

We employed the same metric, ZSI (Zijdenbos similarity index) [16], as used in [8] to test the accuracy of the rough segmentation results in different color spaces. The ZSI is defined as

$$\mathrm{ZSI} = 2\frac{\#\{A_1 \cap A_2\}}{\#\{A_1\} + \#\{A_2\}} \tag{4}$$

where A_1 is the set of ground truth foreground pixels, A_2 is the set of segmented foreground pixels, and #{} is the number of pixels in the set. This index is similar to the Dice [17] similarity coefficient in the literature. The ground truth cell regions are manually segmented by an expert cytopathologist. The ZSI results of the IF_1F_2 color space and the other two color spaces are provided in Table 1 and Figure 3.

Figure 4 shows several rough segmentation results. The first three cell images are from Dataset 1, and the other images are from Dataset 2. We can find that the foreground regions (black) of the clustering results in the IF_1F_2 color space are more approximated to the real cell regions than the RGB and HSV color spaces. Many pixels are classified into the wrong class in the RGB and HSV color spaces. In Table 1, we can find that the ZSI for our method has a mean larger than 0.89 and standard deviation smaller than 0.06, which indicates that the accuracy of the rough segmentation results in the IF_1F_2 color space is higher than that in the other two color spaces. This demonstrates that the discriminating power of the IF_1F_2 color space is higher than that of the other two classical color spaces. Because the high frequency component of the RGB signal's Fourier spectrum is, to some extent, the reflection of the difference between the R, G, and B, it may be more suitable for describing colors of the microscopic images than classical color spaces. From Table 1 and Figure 4, it can be found that the performance of the proposed IF_1F_2 color space is better than that of the other two color spaces.

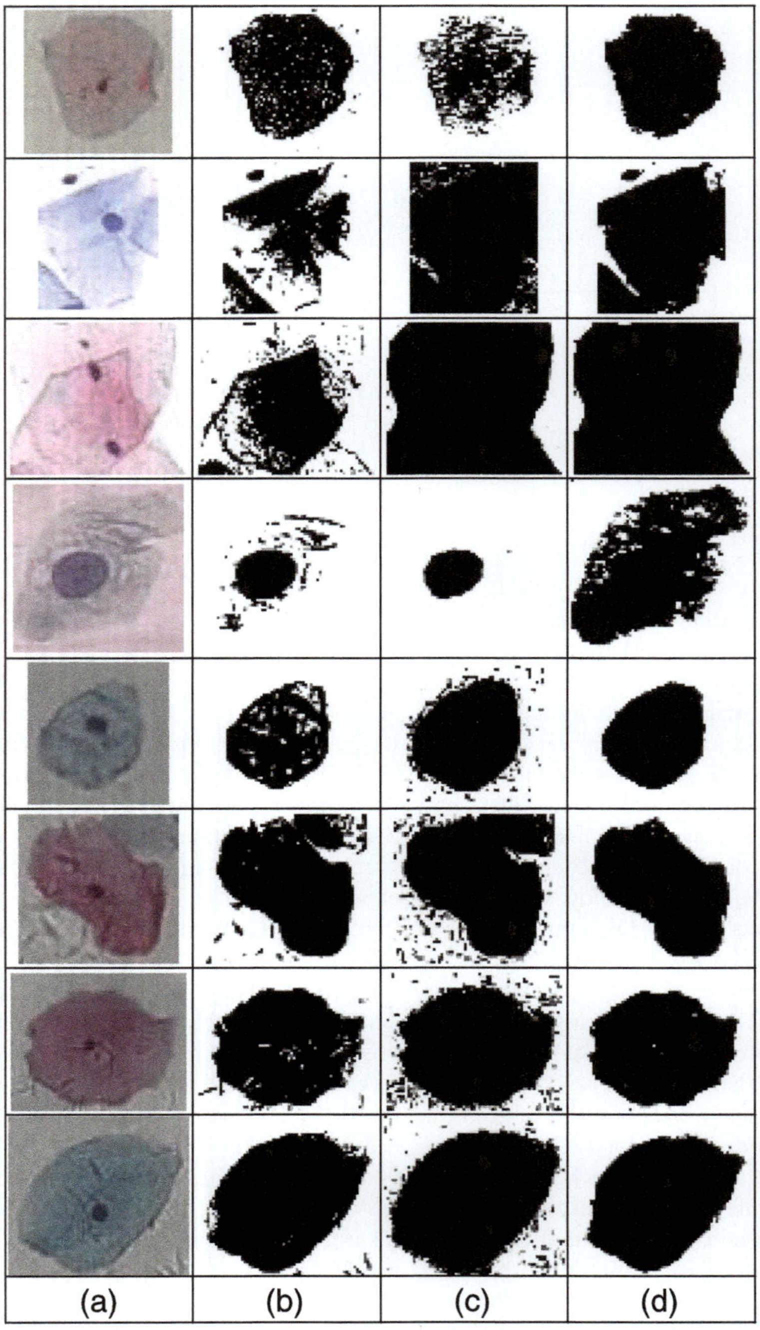

Figure 4 ***K*-means clustering results. (a)** Original microscopic images. **(b)** Results of the RGB space. **(c)** Results of the HSV space. **(d)** Results of the proposed color space.

Conclusions

In conclusion, we proposed a novel color space for describing colors of the microscopic images in this paper. On contrary to classical color spaces, we treat the R, G, and B components as a sampled spectral signal. Based on the Fourier spectrum analysis of the signal, we propose a RGB Fourier transform-based color space, namely IF_1F_2 color space. *K*-means clustering results of two microscopic image datasets show that the performance of the proposed IF_1F_2 color space is better than that of the classical RGB and HSV color spaces. The proposed IF_1F_2 color space may be also useful for other kinds of color images and the relative automatic medical systems.

In this paper, we treat the R, G, and B components of a pixel's color as a sampled signal. Since this signal only has three samples, we use 3-point DFT to calculate its discrete Fourier transform. For a 3-point DFT, the Fourier transform is just expended to the second order. However, we think that higher orders of the Fourier transform may be useful for specific applications, and we will investigate this topic in our future work.

Competing interests

The authors declare that they have no competing interests.

Authors' contributions

TG carried out the study of the proposed color space, participated in the programming of the experiments, and drafted the manuscript. DZ carried out the study of using color vector field for analyzing the effectiveness of the proposed color space. CX performed the statistical analysis in the experiments. YL conceived of the study and helped to draft the manuscript. All authors read and approved the final manuscript.

Authors' information

TG is a Ph.D. student at National University of Defense Technology, China. He received his master's degree in information and communication engineering from National University of Defense Technology in 2009. His research interests include computer vision and image processing.
DZ received his Ph.D. degree in information and communication engineering from National University of Defense Technology, China, in 2000. He is currently a professor at National University of Defense Technology. His major research interests include computer vision and image processing.
CX is a Ph.D. student at National University of Defense Technology, China. He received his master's degree in information and communication engineering from National University of Defense Technology in 2011. His research interests include computer vision and image processing.
YL received his Ph.D. degree in mathematical engineering and information physics from the University of Tokyo, Japan, in 1992. He is currently a professor at the Chinese University of Hong Kong, China. His research interests include visual servoing, robotics, and automatic control. He is a fellow of IEEE.

Author details

[1]College of Electronic Science and Engineering, National University of Defense Technology, Kaifu, Changsha 410073, Hunan, China. [2]Department of Mechanical and Automation Engineering, The Chinese University of Hong Kong, Shatin, Hong Kong, China.

References

1. Cheng H, Jing X, Sun Y, Wang J (2001) Color image segmentation: advances and prospects. Pattern Recogn 34(12):2259–2281
2. Liu C, Yang J (2009) ICA color space for pattern recognition. IEEE Trans Neural Netw 20(2):248–257
3. Bratkova M, Boulos S, Shirley P (2009) oRGB: a practical opponent color space for computer graphics. IEEE Comput Graph Appl 29(1):42–55
4. Lissner I, Urban P (2012) Toward a unified color space for perception-based image processing. IEEE Trans Image Process 21(3):1153–1168
5. Asmare MH, Asirvadam VS, Iznita L (2009) Color space selection for color image enhancement applications. Paper presented at the 2009 international conference on signal acquisition and processing, Kuala Lumpur, Malaysia, 3–5 Apr 2009
6. Jiao C, Gao M, Shi Y (2008) Color image segmentation in a novel dynamic color space. Paper presented at the proceedings of the 7th world congress on intelligent control and automation, Chongqing, China, 25–27 June 2008
7. Zhang C, Wang P (2000) A new method of color image segmentation based on intensity and Hue clustering. Paper presented at the proceedings of the 15th international conference on pattern recognition, Barcelona, Spain, 3–7 Sept 2000
8. Li K, Lu Z, Liu W, Yin J (2012) Cytoplasm and nucleus segmentation in cervical smear images using radiating GVF snake. Pattern Recogn 45(4):1255–1264
9. Plissiti ME, Nikou C, Charchanti A (2011) Automated detection of cell nuclei in Pap smear images using morphological reconstruction and clustering. IEEE Trans Inf Technol Biomed 15(2):233–241
10. Genctav A, Aksoy S, Onder S (2012) Unsupervised segmentation and classification of cervical cell images. Pattern Recogn 45(12):4151–4168
11. Plissiti ME, Nikou C, Charchanti A (2010) Accurate localization of cell nuclei in Pap smear images using gradient vector flow deformable models. Paper presented at the proceedings of 3rd international conference on bio-inspired signals and systems (BIOSIGNALS 2010), Valencia, Spain, 20–23 Jan 2010
12. Plissiti ME, Nikou C, Charchanti A (2010) Watershed-based segmentation of cell nuclei boundaries in Pap smear images. Paper presented at the proceedings of the IEEE/EMBS Region 8 international conference on information technology applications in biomedicine, Corfu, Greece, 3–5 Nov 2010
13. Plissiti ME, Nikou C, Charchanti A (2011) Combining shape, texture and intensity features for cell nuclei extraction in Pap smear images. Pattern Recogn Lett 32(6):838–853
14. Lathi BP (1998) Signal processing linear systems. Oxford University Press, New York
15. Jantzen J, Dounias G (2006) Analysis of Pap-smear image data. Paper presented at the proceedings of the nature-inspired smart information systems 2nd annual symposium, Tenerife, Spain, 29 Nov–1 Dec 2006
16. Zijdenbos A, Dawant B, Margolin R, Palmer A (1994) Morphometric analysis of white matter lesions in MR images: method and validation. IEEE Trans Med Imaging 13(4):716–724
17. Dice LR (1945) Measures of the amount of ecologic association between species. Ecology 26(3):297–302

11

Shape of mole nose providing minimum axial resistance

Yi Shen, Xuyan Hou*, Yiwei Qin, Shengyuan Jiang and Zongquan Deng

Abstract

Introduction: As a carrier of different sensors, moles can penetrate into the regolith automatically and keep investigating the subsurface environment continuously. In this section, features of several moles with different applications are introduced to explain why we choose a hammer-driven mole to study.

Mole driven by a hammer: In this section, the penetrating principle of a hammer-driven mole is illustrated and a circular arc shape for the front nose is proposed. Moreover, applying the penetrating principle, experiments of the mole with an arc-shaped nose are performed to observe the penetration phenomena in a simulated lunar regolith.

Mechanics analysis: According to soil mechanics theory, regions of soil failure are divided and a mechanics model is established between soil and mole with an arc-shaped nose. The work is done to get approximate axial resistance equations which are analyzed with the defined geometric parameters caliber-radius-head.

EDEM simulations: EDEM is leading global software based on discrete element method, whose main function is to analyze and observe the movement of particles. Lunar soil simulacrum is established to simulate axial resistance. Eventually, the theoretical results are validated by simulation.

Keywords: Mole; Circular arc shape; EDEM

Background

As a novel technique of *in situ* investigation, a mole device employs a self-penetrating mechanism, which is considered to be a promising direction for space missions to get information about the geological structure, evolution, and physical and chemical properties of the material if we want to know whether there is life on Mars or the conditions were ever suitable for humans. It is more compact, is lightweight, and has low power consumption. Once it has penetrated into a certain depth, it can acquire geological information of the medium constantly, using various sensors.

In order to conduct deep space exploration, many countries have been involved in the development of a low-speed, unmanned subsurface investigation device. Several prototypes of the mole have been designed and tested in laboratory conditions even though it started late. The mole was first developed by the Russian Federal Space Agency for the Mars-96 mission, called Mars 96 Penetrator, utilizing high speed to penetrate into the regolith, which can penetrate a deeper distance but cannot be used repeatedly and causes great damage to the detecting instruments, leading to data errors [1]. To overcome the disadvantages above, a hammer-driven mechanism was applied to the mole, which will be described in the next sections. In 2011, Japanese researchers proposed a robotic screw explorer which can excavate into soil and transport it backward automatically [2,3]. Inspired by animals, such as mouse and earthworm, more and more researchers focus on the bionics design, e.g., an earthworm-type robot which can make use of the reactive force caused by pushing the discharged regolith above the robot [4,5]. As for China, apart from the development of a thermal drill which is a combination of a rotary drill and a melting probe in Hong Kong Polytechnic University [6], it is seldom studied so far. In addition, it is worth mentioning that all those prototypes are still in the exploratory pilot phase and none is implemented in space mission successfully, which provides a great space for China to develop the automatic penetrating device.

* Correspondence: houxuyan@hit.edu.cn
School of Mechatronics Engineering, Harbin Institute of Technology, West Dazhi Street, 15001 Harbin, China

Although discharging soil backward has great advantage compared with squeezing soil, the hammer-driven mole penetrates better on current technology. So this paper still focuses on optimization of a mole driven by a hammer. As opposed to that of the conventional rotary drill, forward motion of the mole is done by displacement and compression of the soil. Penetration depth depends on the matching of three qualities and shape of the front nose. So it is crucial to study the shape of the front nose. Up to now, the hammer-driven mole has several types of front nose, e.g., MMUM [7], derived from PLUTO developed by DLR for the Beagle 2 lander on the ESA Mars Express mission with added capability of sampling at the 60° front cone that can open during further penetration [8,9]. Another interesting device is MUPUS on the Philae lander by PAS, whose front nose was designed with sharp and elastic barbs to get anchoring property [10]. Following the successful MUPUS development, a nonlinear conical shape (ogive-shaped tip angle starts from 45° at the base and 30° at the end) was applied to the KRET [11].

The remainder of this paper is organized as follows. The principle of operation of the hammer-driven mole is illustrated and an arc-shaped front nose is proposed in the 'Methods' section. Besides, experiment results of a convex arc-shaped nose in simulated lunar are also presented in the 'Methods' section. A mechanical model between the front nose and soil is built in the 'Results and discussion' section. What's more, EDEM simulations of the mole with different geometric parameters are given in the 'Results and discussion' section to further prove the relationship between axial resistance and caliber-radius-head, followed by the 'Conclusions' section.

Methods

Principle of penetration

As for the hammer-driven, low-speed type of mole, the basic principle of operation is to make full use of impacting movement to get energy needed for penetration. As shown in Figure 1, the mole mainly consists of seven parts: hammer, servo driving unit, casing, driving spring, buffer spring, escapement mechanism, and tether.

Therefore, the work cycle can be divided into four stages:

(a) Accumulation of energy in the driving spring by the movement of the hammer upward relatively to the casing via the servo driving unit.
(b) Escapement mechanism separates suddenly when the hammer reaches a certain distance and the driven hammer accelerates and hits the bottom of the casing, thus contributing the displacement x_1 by overcoming the restraint of the soil around.
(c) At the instant of the release of the hammer, the servo driving unit moves backward compared with the hammer, and then the buffer spring is compressed to avoid reverse movement of the mole.
(d) Under the effect of the buffer spring and its own gravity, the servo driving unit hits the casing, thus forcing the mole to move down at x_2. The movement of the mole repeats just like that, which is suitable for the regolith that can be compressed for making space. After the mole penetrates to a certain depth, a hole with a higher density appears at the back.

An arc-shaped front nose

The shape and geometric parameters of the front nose, which contacts soil first, must have a great influence on the penetrating feature of the mole. Consider the axisymmetric mole as a rigid body, remaining undeformed during dynamic penetration. A convex arc shape of the front nose and notations are shown in Figure 2.

The shape is arc of a circle with radius S that is tangent to the casing of the mole. Caliber-radius-head is defined as $\psi = S/2R$, where R is the radius of the casing. The length of the front nose is l.

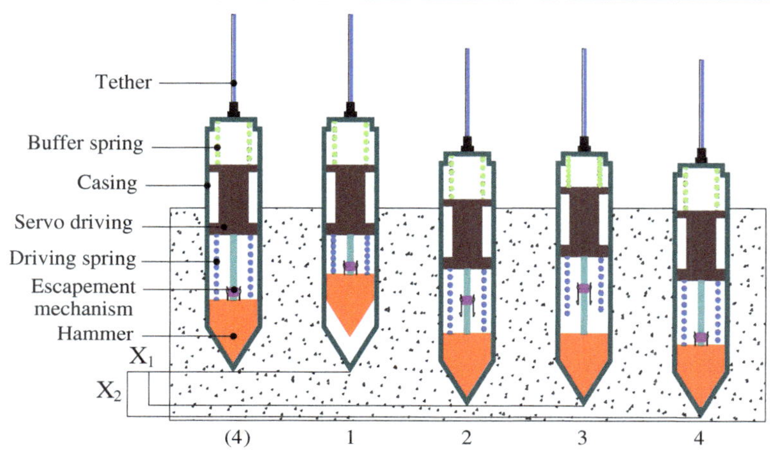

Figure 1 Schematic principle of the operation of the mole.

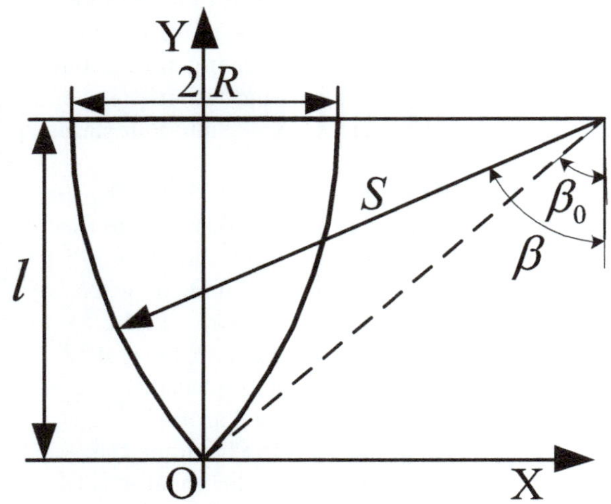

Figure 2 Cross section of a convex arc-shaped nose and notations.

Figure 3 Experiments of the arc-shaped nose.

According to the geometric relationship, the following equations can be obtained:

$$\begin{cases} \sin\beta_0 = \dfrac{S-R}{S} \\ l = \sqrt{S^2-(S-R)^2} = R\sqrt{4\psi-1} \end{cases} \tag{1}$$

Penetrating experiments

Automatic penetrating process in space is complicated due to the ultrahigh vacuum, micro-gravity conditions, very low temperature, and huge uncertainties about the surface properties. But in order to further observe the penetration phenomenon of an arc-shaped front nose, penetration experiments have been performed in the laboratory. As shown in Figure 3, experiments have been conducted in the following conditions: (1) target: simulated lunar, (2) caliber-radius-head: 3.7, (3) stroke energy: 1.2 J, (4) mass of the hammer: 1.2 kg, (5) stroke height: 83.3 mm. Facts show that progress of the mole driven by a hammer in a granular medium largely depends on the density of the soil. To ensure the same penetrating conditions, experiments are performed with the same compaction level of simulated lunar.

The arc-shaped nose has been pounded into a 150-mm-diameter and 500-mm-deep barrel containing dry sand which represents a cohesionless material. Recording the penetrating data after ten hammer strokes, each experiment and the result can be seen in Figure 4. The convex arc-shaped nose travels a distance of some 210 mm under 600 strokes in the sand. The performed experiments demonstrate that the nose is capable of good penetration performance in sand. Even though the penetration progress becomes more and more difficult with the increase of depth, the nose is able to move downward under sufficient strokes.

Results and discussion

Theory of mechanics analysis

Soil failure

As mentioned previously, there is a great deal of complexity about the regolith of space bodies. Currently, it is accepted that the surface and subsurface materials of space bodies consist of granulated matter with grain size in a range from several micrometers to several millimeters, like terrestrial sand [12]. In addition, the mole device is driven by a hammer, which means that it is capable of working in the regolith in the dynamic range.

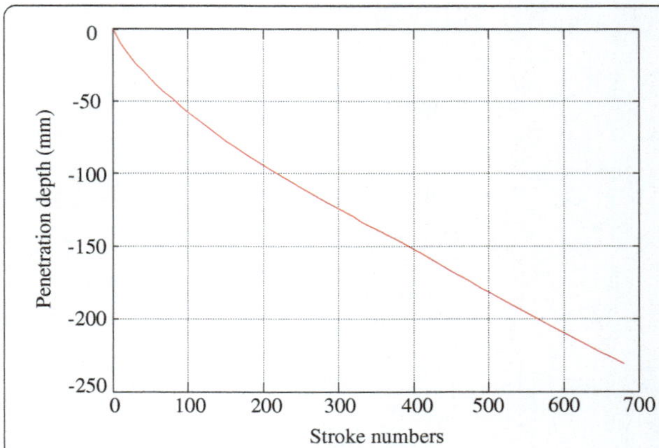

Figure 4 Penetration depth vs stroke numbers.

Thus, the mechanical properties and failure model of the regolith are crucial to correctly calculate resistance force. Based on the study of soil on the earth, we assume the mole penetrates a semi-infinite shield.

As shown in Figure 5, soil around the front nose can be divided into three regions called compacted region I, transition region II, and passive region III during the penetration. Due to the effect of large friction on the arc-shaped segments AO and BO, soil in region I moves downward as a rigid body with the mole, and the adjacent region II is forced sideways and region III is displaced upwards. Under positive pressure P, the mole penetrates into the geological layer and soil moves to region III initially. When positive pressure reaches a certain value, soil in region III reaches the limited state. And soil starts to fail if the pressure continues to increase. Region II has two sets of slip lines. One set is two rays, AC from point A and BE from point B. The other set is logarithmic spiral curves CG and EF, expressed as

$$r = r_0 e^{\theta \tan\varphi} \tag{2}$$

where φ is the internal friction angle [13].

When region III is in the limited state, mechanics analysis of block BEFK is shown in Figure 6. The angle between the slip line and horizontal plane is $\delta = \pi/4 - \phi/2$. After the mole with a convex arc shape penetrates into soil, the initial length of the logarithmic spiral curves is:

$$r_0 = \overline{BE} = 2\psi R\left(\frac{\sqrt{4\psi-1}}{2\psi-1}\sin\beta - \cos\beta\right) \tag{3}$$

According to geometric relationship, it is easy to see:

$$\begin{cases} \overline{BF} = r_0 e^{\left(\frac{\pi}{4}+\frac{\varphi}{2}\right)\tan\varphi} \\ \overline{BK} = \overline{BF}\cos\delta = r_0 e^{\left(\frac{\pi}{4}+\frac{\varphi}{2}\right)\tan\varphi}\cos\left(\frac{\pi}{4}-\frac{\varphi}{2}\right) \\ \overline{FK} = \overline{BF}\sin\delta = r_0 e^{\left(\frac{\pi}{4}+\frac{\varphi}{2}\right)\tan\varphi}\sin\left(\frac{\pi}{4}-\frac{\varphi}{2}\right) \end{cases} \tag{4}$$

Considering the gravity of BEFK at this time and turning it to pressure acting on the surface of BK:

$$q = \gamma H \tag{5}$$

where γ is the bulk density of soil. H is given by

$$H = \frac{r_0}{3} N_\varphi \tag{6}$$

where

$$N_\varphi = 1 + \frac{e^{\left(\frac{\pi}{2}+\varphi\right)tg\varphi}\sin^2\left(\frac{\pi}{4}-\frac{\varphi}{2}\right)}{1+e^{\left(\frac{\pi}{4}+\frac{\varphi}{2}\right)tg\varphi}\sin\left(\frac{\pi}{4}-\frac{\varphi}{2}\right)}$$

According to the failure criterion of Mohr-Coulomb, the principal stress satisfying the relationship can be expressed as:

$$\sigma_1 = \sigma_3 \tan^2\left(\frac{\pi}{4}+\frac{\varphi}{2}\right) + 2c\cot\left(\frac{\pi}{4}+\frac{\varphi}{2}\right) \tag{7}$$

When soil in region III is in the limited state, longitudinal stress σ_z is the minimum principal stress σ_3 and horizontal stress σ_x is the maximum main stress σ_1. So we can get:

$$p_P = q\cot^2\left(\frac{\pi}{4}-\frac{\varphi}{2}\right) + 2c\cot\left(\frac{\pi}{4}-\frac{\varphi}{2}\right) \tag{8}$$

where c is cohesion.

Stresses on surface EF are cohesion of the soil c (distribution uniform), normal stress p_n, and friction force $p_n\tan\varphi$ (distribution non-uniform). p_r is the joint stress of p_n and $p_n\tan\varphi$, and it has an included angle of φ with p_n. According to the characters of the logarithmic spiral curve, point B should be on the force line of p_r.

Soil in region I moves downward integrally with the mole. Therefore, BE can be regarded as a retaining wall and the wall back is upright. Suppose the normal stress on surface BE is p_B, namely Rankine's active earth pressure. Taking point B as centroid, according to

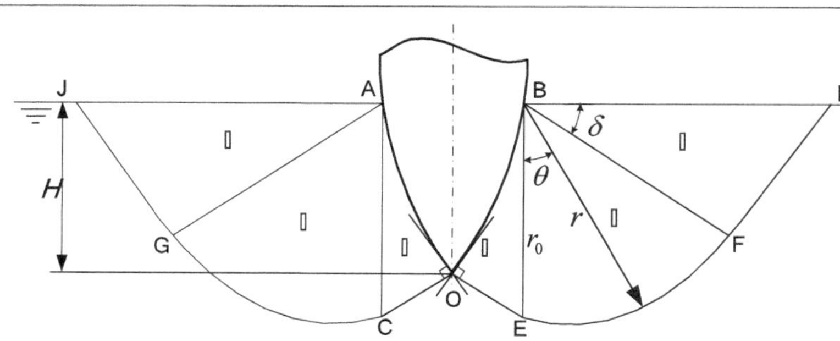

Figure 5 Regions of soil failure around the arc-shaped nose.

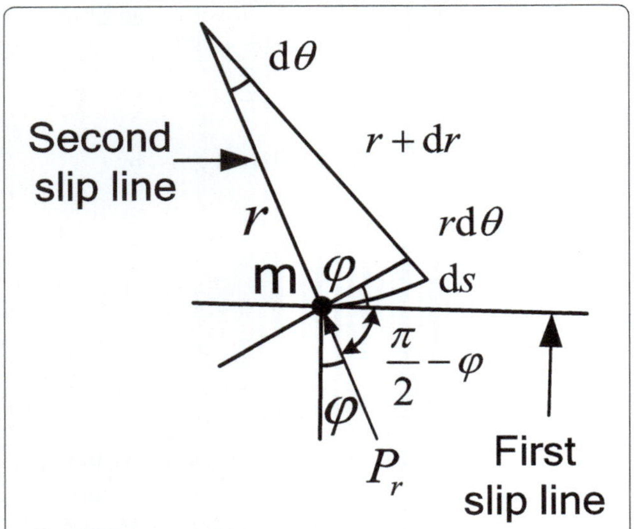

Figure 6 Mechanics analysis of block BEFK.

the condition of the static equilibrium of block BEFK, we have:

$$\sum M_B = M_{p_F} - M_{p_P} - M_q - M_c = 0 \tag{9}$$

The moment of p_r to point B is zero and the rest of the moment to point B are:

$$\begin{cases} M_{p_F} = \frac{1}{2} p_F \left(\overline{\mathrm{BF}}\right)^2 = \frac{1}{2} p_F (r_0)^2 \\ M_{p_P} = \frac{1}{2} p_P \left(\overline{\mathrm{HK}}\right)^2 \\ \quad = \frac{1}{2} (r_0)^2 e^{\left(\frac{\pi}{2}+\varphi\right)\tan\varphi} \left[\gamma H \cos^2\left(\frac{\pi}{4}-\frac{\varphi}{2}\right) + c\cos\varphi\right] \\ M_q = \frac{1}{2} q \left(\overline{\mathrm{BK}}\right)^2 = \frac{1}{2} \gamma H (r_0)^2 e^{\left(\frac{\pi}{2}+\varphi\right)\tan\varphi} \cos^2\left(\frac{\pi}{4}-\frac{\varphi}{2}\right) \\ M_c = \int c\,ds\cos\varphi r = \int_0^{\left(\frac{\pi}{4}+\frac{\varphi}{2}\right)} c \frac{r d\theta}{\cos\varphi} \cot\varphi r \\ \quad = \frac{1}{2} c (r_0)^2 \cot\varphi \left[e^{\left(\frac{\pi}{2}+\varphi\right)\tan\varphi} - 1\right] \end{cases} \tag{10}$$

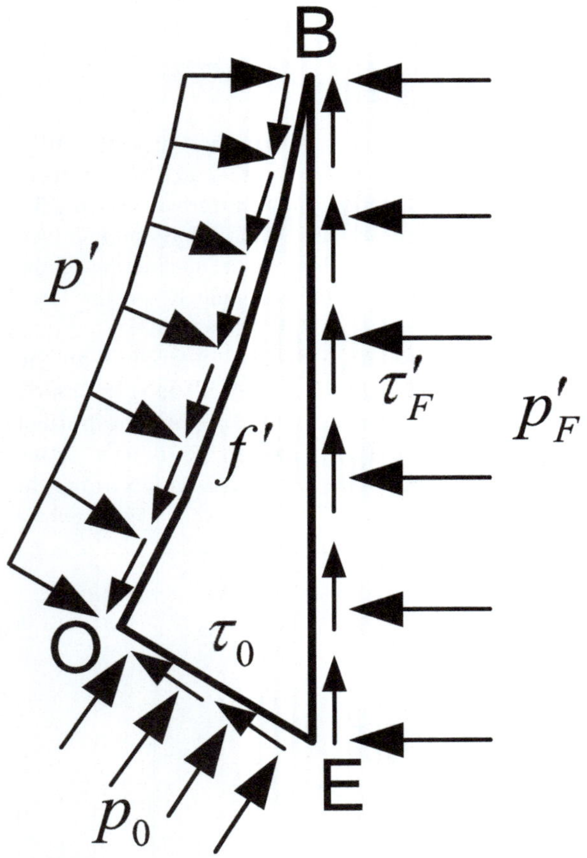

Figure 7 Mechanics analysis of block BOE.

Substituting Equation 10 into Equation 9, we can get

$$p_P = qN_q + cN_c \tag{11}$$

where

$$N_q = 2e^{\left(\frac{\pi}{2}+\varphi\right)\tan\varphi} \cos^2\left(\frac{\pi}{4}-\frac{\varphi}{2}\right)$$

$$N_c = (\cos\varphi + \cot\varphi) e^{\left(\frac{\pi}{2}+\varphi\right)\tan\varphi} - \cot\varphi$$

N_q and N_c are the bearing capacity coefficients of soil, which are functions of internal friction angle φ.

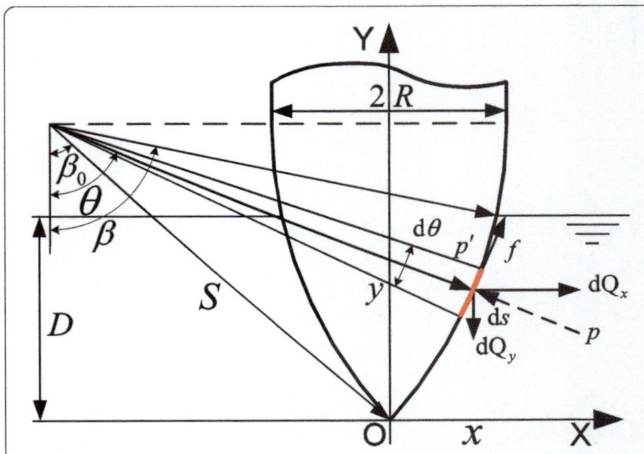

Figure 8 Mechanics analysis of the front nose.

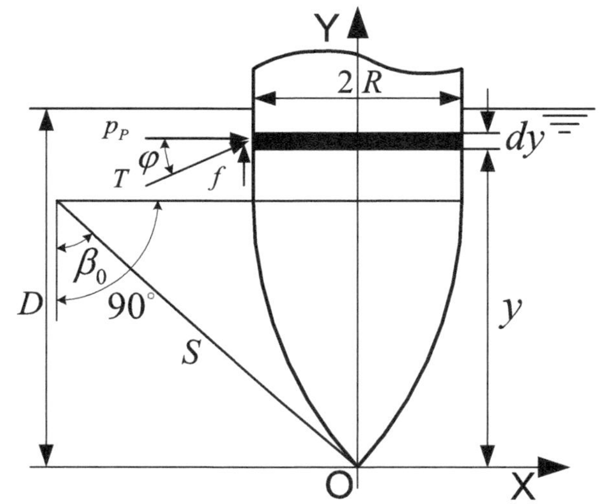

Figure 9 Mechanics analysis of the lateral surface.

Mechanics analysis of block BOE in the compacted region is shown in Figure 7, neglecting the gravity of the soil. Increments of horizontal force and vertical force are:

$$\begin{cases} dQ_x = (p' \sin\theta - f' \cos\theta)ds \\ dQ_y = (p' \sin\theta + f' \cos\theta)ds \end{cases} \tag{12}$$

where ds is the increment of the arc length on the surface of the nose and p' is the radial stress acting on the cambered surface BO. Take the coefficient of friction to be μ and

$$f' = \mu p' \tag{13}$$

Substituting Equation 13 into Equation 12 and integraling, horizontal force and vertical force on the cambered surface BO can be expressed as follows:

$$\begin{cases} Q_x = Sp'\left[\dfrac{\sqrt{4\psi-1}+\mu(2\psi-1)}{2\psi} - \cos\beta - \mu\sin\beta\right] \\ Q_y = Sp'\left[\dfrac{\mu\sqrt{4\psi-1}-(2\psi-1)}{2\psi} + \sin\beta - \mu\cos\beta\right] \end{cases} \tag{14}$$

According to the condition of the static equilibrium of block BOE, we find:

$$\begin{cases} Q_x + p_0\cos\beta_0\overline{OE} - \tau_0\sin\beta_0\overline{OE} - p_F'\overline{BE} = 0 \\ p_0\sin\beta_0\overline{OE} + \tau_0\cos\beta_0\overline{OE} + \tau_F'\overline{BE} - Q_y = 0 \end{cases} \tag{15}$$

where τ_0 and τ_F' can be expressed as:

$$\begin{cases} \tau_0 = c + p_0\tan\varphi \\ \tau_F' = c + p_F'\tan\varphi \end{cases} \tag{16}$$

Considering the interaction with block BEFK, we can find:

$$p_F = p_F' \tag{17}$$

Combining Equation 14 to Equation 17, radial stress on surface BO is obtained:

$$p' = a\psi + b \tag{18}$$

where

$$a = \frac{2\gamma R N_\varphi N_q(1-\tan^2\varphi)\sin^2(\beta-\beta_0)}{3\sin\beta_0 e}$$

$$b = c\frac{\sin\beta_0\sin(\beta-\beta_0)(N_c - \tan\varphi - \cot\beta_0) - \cos2\beta_0\left(\sin\beta - \frac{2\psi-1}{2\psi}\right)}{\sin\beta_0 e}$$

$$e = (\sin2\beta_0 - \mu\cos2\beta_0) - (\mu\tan\varphi - 1)\cos\beta_0\cos\beta \\ -\mu(\mu\tan\varphi + 1)\sin\beta_0\sin\beta - \tan\varphi\cos(\beta-\beta_0) + \tan\varphi$$

Axial resistance

Assume the mole is always along a vertical direction. Then the penetrating process can be divided into three stages:

$D < l$

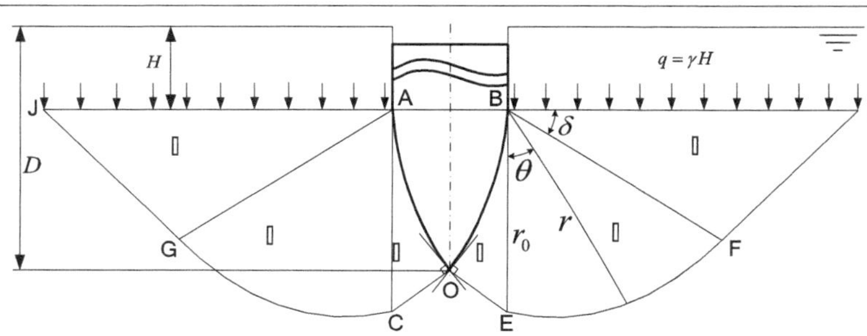

Figure 10 Soil failure when the mole penetrates totally.

During the initial penetration, soil resistance acts on the front nose only. Mechanical analysis of the front nose and infinitesimal circle is shown in Figure 8. Then the infinitesimal force of the infinitesimal area can be expressed as:

$$dF = 2\pi x(p\cos\theta + f\sin\theta)ds \tag{19}$$

where $x = S\sin\theta - (S - R)$.

Using a formula from differential geometry, the following expression for axial resistance acting on the convex arc-shaped nose can be derived:

$$F_1 = \int dF = \int_{\beta_0}^{\beta} 2\pi x p_1 S(\cos\theta + \mu\sin\theta)d\theta = K_1\psi^3 + K_2\psi \tag{20}$$

where

$$K_1 = 4\pi R^2 a[\mu\cos(\beta_0+\beta) + \sin(\beta_0+\beta)]\sin(\beta_0-\beta) + 4\pi R^2 a[2(\sin\beta_0 - \sin\beta) - 2\mu(\cos\beta_0 - \cos\beta)]$$

$$K_2 = 4\pi R^2 b[\mu(\cos\beta_0 - \cos\beta) + (\sin\beta - \sin\beta_0) + 4\mu(\beta_0 - \beta)]$$

$$l < D < l + L$$

With increase of penetration depth, the front nose penetrates into soil totally. Substituting $\beta = \pi/2$ into Equation 20, we can get the axial resistance $F_1^{'}$ on the total convex arc-shaped front nose. What's more, the lateral surface of the mole also contacts soil, whose mechanical analysis is shown in Figure 9. Regarding the lateral surface as a retaining wall, according to Rankine's earth pressure theory, passive earth pressure p_{P1} has the following expression:

$$P_p = \gamma(D-y)K_p + 2c\sqrt{K_P} \tag{21}$$

where K_P is the passive earth pressure coefficients and $K_P = tg^2(45° + \varphi/2)$.

Therefore, horizontal positive pressure on the infinitesimal surface can be expressed as:

$$dN = p_P dA = 2\pi p_P R dy \tag{22}$$

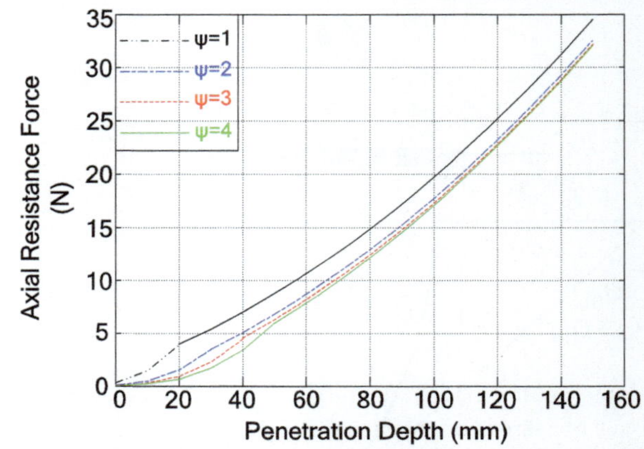

Figure 11 Axial resistance vs penetration depth for different ψ.

which can be integrated between $y = l$ and $(D - l)$ to give the net positive force N:

$$N = \int dN = \pi R^2\left[\gamma\frac{K_P}{R}\left(D - R\sqrt{4\psi-1}\right)^2 + 4c\frac{\sqrt{K_P}}{R}\left(D - R\sqrt{4\psi-1}\right)\right] \tag{23}$$

Then we can get the resistance on the lateral surface:

$$F_c = \mu N = \mu\pi R^2\left[\gamma\frac{K_P}{R}\left(D - R\sqrt{4\psi-1}\right)^2 + 4c\frac{\sqrt{K_P}}{R}\left(D - R\sqrt{4\psi-1}\right)\right] \tag{24}$$

Combining $F_1^{'}$ on the convex arc-shaped front nose, the axial force resisting motion of the mole can be obtained:

$$F_1 = F_1^{'} + F_c \tag{25}$$

$$D > l + L$$

When the mole penetrates into the regolith totally, the length of the front nose is smaller than the penetration depth relatively and we can get the region of soil failure in Figure 10.

Neglect the gravity of soil above the front nose and stress on surfaces AJ and BI can be regarded as the gravity of soil above the front nose. Then H is the distance

Table 1 Physical parameters of the mole and soil

	Parameters					
	R (mm)	L (mm)	c	φ (°)	γ (kN/m³)	μ
Value	15	200	0	30	18	0.3

Table 2 Material parameters

Material	Poisson's ratio	Shear modulus (Pa)	Density (kg/m³)
Particle	0.3	2×10^7	1,750
Mole	0.3	2×10^{10}	7,850

Table 3 Interaction parameters

Interaction	Particle-particle	Particle-mole
Coefficient of restitution	0.1	0.2
Coefficient of static friction	0.574	0.3
Coefficient of rolling friction	0.01	0.01

from the ground to surfaces AJ and BI, which can be expressed as:

$$H = D - l = D - R\sqrt{4\psi - 1} \tag{26}$$

Applying the mechanics analysis of soil block before, we can get the stress on surface BE:

$$p_F = \gamma N_q\left(D - R\sqrt{4\psi - 1}\right) + cN_c \tag{27}$$

Afterwards, radical stress on the contact stress between soil and nose is:

$$p'' = a''\sqrt{4\psi - 1}\left(\frac{D}{R} - \sqrt{4\psi - 1}\right) + cb'' \tag{28}$$

where

$$a'' = \frac{\gamma R N_q\left[2\tan\varphi\frac{\sqrt{4\psi-1}}{2\psi-1} + (1-\tan^2\varphi)\right]}{(1-\mu\tan\varphi)\sqrt{4\psi-1} + (\mu + \tan\varphi)}$$

$$b'' = \frac{2(N_c\tan\varphi + 1)\frac{4\psi-1}{2\psi-1} - [\tan\varphi - N_c(1-\tan^2\varphi)]\sqrt{4\psi-1} - 1}{(1-\mu\tan\varphi)\sqrt{4\psi-1} + (\mu + \tan\varphi)}$$

According to Equation 19, integrate dF from $\theta = \beta_0$ and $\theta = \pi/2$. So axial resistance on the front nose has the following expression:

$$\begin{aligned} F_1'' &= \pi R^2\left[a''\sqrt{4\psi-1}\left(\frac{D}{R} - \sqrt{4\psi-1}\right) + cb''\right] \\ &= \left[4\mu\left(\frac{\pi}{2} - \beta_0\right)\psi^2 - \mu(2\psi-1)\sqrt{4\psi-1} + 1\right] \end{aligned} \tag{29}$$

Moreover, the lateral surface contacts soil totally in this stage. Combining $F_c = \mu N$, integrate Equation 22 between $y = l$ and $y = l + L$ to get the net resistance on the lateral surface:

$$F_c'' = \mu N = K_1' - K_2'\sqrt{4\psi - 1} \tag{30}$$

where

$$K_1' = 2\mu L\sqrt{K_P}\left[\pi R\gamma\sqrt{K_P}(2D - L) + c\right]$$

$$K_2' = 4\pi R^2\mu\gamma K_P L$$

Then axial resistance of the mole is obtained as follows:

$$F_1 = F_1'' + F_c' \tag{31}$$

Axial resistance vs caliber-radius-head

Through mechanics analysis, it is known that for a convex arc-shaped front nose, axial resistance is identified with penetration depth and geometric parameters caliber-radius-head. In order to get the relationship between axial resistance and caliber-radius-head directly, other physical parameters which will be used are listed in Table 1 according to the acquaintance of the mole and soil.

Based on axial resistance equations, curves relating axial resistance to penetration depth for various values of ψ are drawn (shown in Figure 11). As can be seen from Figure 11, the curve of axial resistance is similar to a parabola changing with penetration depth. Moreover, axial resistance decreases with the increase of ψ. However, the advantage is not obvious when ψ increases to 3 and a very large value of ψ causes an increase in total length of the front nose which is not advisable.

EDEM simulation

Generally, granular matter behaves like a compressible non-Newtonian complex fluid including fluid solid transition and can be simulated using EDEM simulation. An obvious advantage of EDEM simulation is that it provides the possibility of obtaining movement, forces, and other dynamical properties of the system at any time.

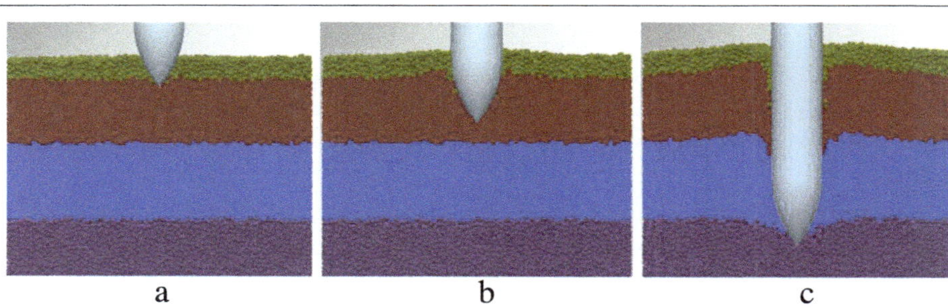

Figure 12 EDEM simulation of a mole with an arc-shaped front nose (a-c). The shape parameter of the mole is 2(ψ =2), and particles are set with different colors by section to observe the flow directly during the penetration.

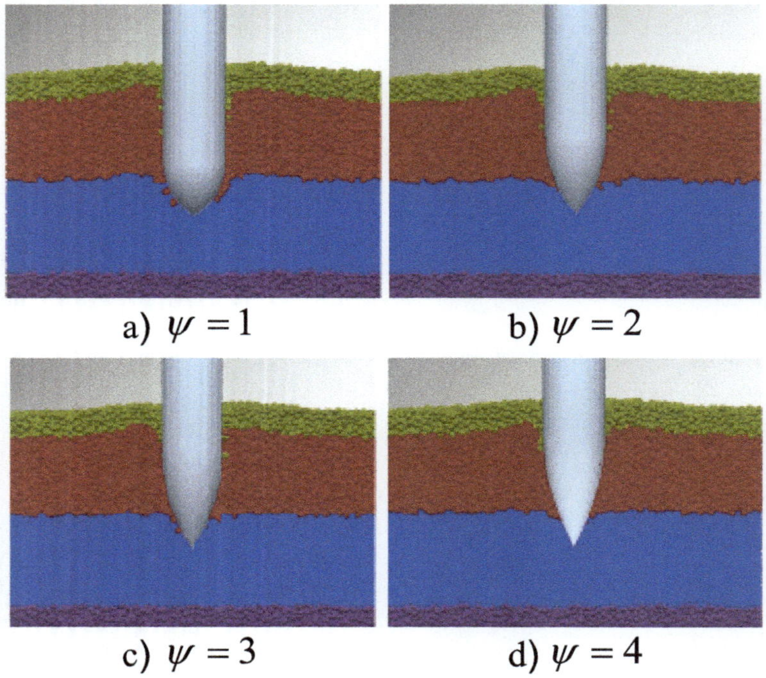

Figure 13 Penetration simulations of mole with an arc-shaped front nose for different ψ. **(a)** ψ =1. **(b)** ψ =2. **(c)** ψ =3. **(d)** ψ =4.

For ensuring establishment of the simulation model consistent with theoretical analysis, the model of the mole is built in SolidWorks and has the same parameter setting as shown Table 1. Besides, the Hertz-Mindlin (no slip) interaction model is employed in EDEM simulations, and the setting of the other variables is shown in Tables 2 and 3. Simulation of the mole whose shape parameter is 2 is shown in Figure 12 to observe the flow of the lunar soil.

It can be seen that the mole works properly in granular matter and the failure of soil increases along with the increase of penetration depth. Change of the values of the shape parameter ψ from 1 to 4 and all simulations are shown in Figure 13.

According to Figure 13, it can been seen that when the mole penetrates into the same depth, there is vertical displacement of the value of ψ. But the variation is not obvious when ψ increases from 3 to 4, which can explain the relationship between axial resistance and ψ in the certain conditions. Exporting the EDEM simulation data, curves of axial resistance vs penetration depth for various values of ψ are shown in Figure 14, which is consistent with theoretical analysis.

Conclusions

In this paper, we focus on the front nose of a hammer-driven mole. Distinguishing with the conical one, an arc-shaped front nose has been proposed, which is tested in laboratory conditions. The contributions of this paper are the establishment of a mechanics model, derivation of axial resistance based on soil mechanics, and application of the discrete element method to observe the flow of the lunar soil. Through theoretical analysis, it is found that the axial resistance of the mole shows a negative correlation with the shape parameter ψ. Besides, parameters like penetration depth D, friction coefficient μ, internal friction angle φ, and cohesion c also affect the axial resistance to some extent. Such theoretical approach can be used to optimize the geometric design of the front nose. What's

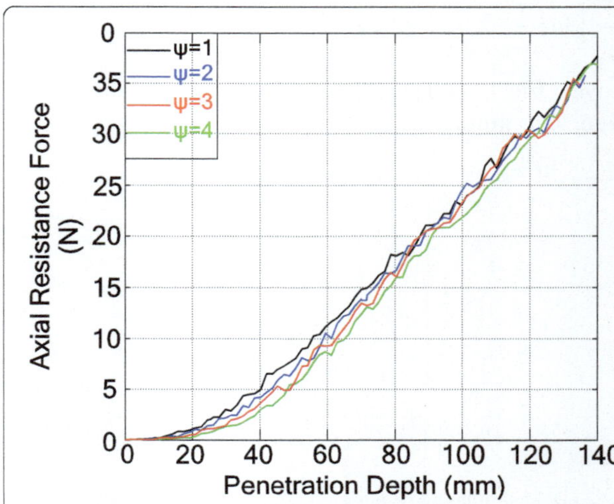

Figure 14 Results of EDEM simulation for different ψ.

more, a series of EDEM simulations show excellent accordance with the result. However, the length of the front nose increases with a large value of ψ, which decreases the stiffness of the front nose seriously. Therefore, we compromise to balance the two factors to get an appropriate value depending on demand. The study of optimization of the front nose is still ongoing. Other parameters and other shapes will be taken into consideration in order that the mole can penetrate into a deeper subsurface. Besides, several noses have already been tested and more experiments will be carried out in the future.

Competing interests

The authors declare that they have no competing interests.

Authors' contributions

YS carried out the derivation of the theoretical study, organized and completed the penetration experiments, participated in the EDEM simulations, and drafted the manuscript. XH participated in the design of the study and reviewed the manuscript. YQ participated in the theoretical analysis and was involved in drafting the manuscript. SJ conceived of the study and helped to review the manuscript. ZD participated in its design and coordination. All authors read and approved the final manuscript.

Acknowledgements

The project is financially supported by the National Nature Science Foundation of China (Grant No. 51105092) and College Discipline Innovation Wisdom Plan of China (111 Project, Grant No. B07018).

References

1. Surkov YA, Kremnev R (1998) Mars-96 mission: Mars exploration with the use of penetrators. Planetary Space Sci 46(11):1689–1696
2. Nagaoka K, Kubota T, Otsuki M, Tanaka S (2008) Experimental study on autonomous burrowing screw robot for subsurface exploration on the moon. In: IEEE/RSJ international conference on intelligent robots and systems, 2008. IROS 2008, IEEE, Piscataway, pp 4104–4109
3. Nagaoka K, Kubota T, Otsuki M, Tanaka S (2009) Robotic screw explorer for lunar subsurface investigation: dynamics modeling and experimental validation. In: International conference on advanced robotics. IEEE, Piscataway, pp 1–6
4. Kubota T, Nagaoka K, Tanaka S, Nakamura T (2007) Earth-worm typed drilling robot for subsurface planetary exploration. In: IEEE international conference on robotics and biomimetics, 2007. ROBIO 2007, IEEE, Piscataway, pp 1394–1399
5. Omori H, Nakamura T, Iwanaga I, Hayakawa T (2010) Development of mobile robots based on peristaltic crawling of an earthworm. In: Robotics 2010 current and future challenges. InTech, Rijeka, pp 299–320
6. Weiss P, Yung K, Kaufmann E, Kargl G (2011) Thermal drill sampling system onboard high-velocity impactors for exploring the subsurface of Europa. Adv Space Res 48(4):743–754
7. Stoker C, Richter L, Smith W, Lemke L, Hammer P, Dalton J, Glass B, Zent A (2003) The Mars underground mole (MUM): a subsurface penetration device with in situ infrared reflectance and Raman spectroscopic sensing capability. In: Sixth international conference on Mars, Pasadena, 20–25 July 2003, Lunar and Planetary Institute science conference abstracts, vol 34., p 1201
8. Richter L, Coste P, Gromov V, Kochan H, Nadalini R, Ng T, Pinna S, Richter H-E, Yung K (2002) Development and testing of subsurface sampling devices for the Beagle 2 lander. Planetary Space Sci 50(9):903–913
9. Richter L, Kroemer O (2009) Application of a remote controlled hammering drill from space to deep sea. In: OCEANS 2009-EUROPE. IEEE, Piscataway, pp 1–4
10. Grygorczuk J, Banaszkiewica M, Cichocki A, Ciesielska M, Dobrowolski M, Kedziora B, Krasowski J, Kucinski T, Marczewski M, Morawski M, Rickman H, Rybus T, Seweryn K, Skocki K, Spohn T, Szewczyk T, Wawrzaszek R, Wiśniewski Ł (2011) Advanced penetrators and hammering sampling devices for planetary body exploration. In: Proceedings of ASTRA 11th symposium on advanced space technologies in robotics and automation, Noordwijk, pp 12–14
11. Grygorczuk J, Seweryn K, Wawrzaszek R, Banaszkiewicz M (2009) Technological features in the new mole penetrator KRET. In: Proceedings of the 13th European space mechanisms and tribology symposium. ESMATS (2009), Vienna, 23–25 Sept 2009
12. Carrier WD III (1973) Lunar soil grain size distribution. The Moon 6(3–4):250–263
13. Chen Z, Zhou J, Wang H (2007) Soil mechanics. Tsinghua University Press, Beijing

12

Path planning of mechanical polishing process for freeform surface with a small polishing tool

Weiyang Lin[1,2,3], Peng Xu[1,4], Bing Li[1*] and Xiaojun Yang[1]

Abstract

Products with freeform surface are widely applied in industries, and the surface quality plays an important role in order to fulfill the targeted functions. As polishing path of small polishing tool affects the polishing removal function considerably, it is highly necessary to study the polishing path of freeform surface for obtaining good polishing efficiency and well-proportioned surface quality. By combining the Preston polishing removal function, the material removal model of small polishing tool under the control of constant polishing force and pressure is established. Based on this model, the material removal functions of scan line, Archimedean spiral, and Hilbert fractal polishing path are derived. The simulation results show that the Hilbert fractal polishing path has the best comprehensive performance. By using the projection relation of differential geometry, the optimal path generation algorithm of the Bézier surface based on Hilbert fractal polishing path is established. The polishing experiments are conducted on a self-developed polishing machine which is based on a parallel manipulator. The experimental results demonstrate that the surface roughness is improved from level 9 to level 11.

Keywords: Mechanical polishing; Small polishing tool; Freeform surface; Polishing path; Removal function

Background

Product structure can be optimized by using freeform surfaces, which opens the door for solutions with improved performance, reduced complexity, lower mass, and smaller size. However, due to the geometric particularity, freeform surfaces face more challenges and difficulties in precision manufacturing. Polishing is usually one of the final processing steps of precision manufacturing, and the results directly affect the appearance and longevity of parts [1]. It is fundamentally different from other precision manufacturing technologies. Removal of polishing does not only depend on the position of the tool orthogonal to the workpiece, as for grinding and cutting processes but also proportional to the product of local pressure and relative-speed between tool and workpiece and the dwell time. In recent years, there were efforts to develop versatile polishing processes in order to achieve high accuracy. The current polishing methods mainly include electrochemical polishing [2], magnetorheological fluid polishing [3,4], plasma polishing [5,6], ultrasonic polishing [7,8] and computer-controlled mechanical polishing [9,10]. Nowadays, a lot of freeform components are still produced by final manual polishing. It not only heavily relies on the know-how and experience of technicians but also needs much attention for processing and testing. To achieve a given level of precision with high efficiency and reliability, process automation is clearly the way forward. As the computer-controlled mechanical polishing has high efficiency and can be controlled easily [11], it is the focus of this study. The mechanical polishing is a statistical 'rubbing' process that the microscopic loose-abrasive particles in the polishing liquid which is driven by high-speed rotational polishing tool could produce friction with the part surface. Protruding portions of the surface are removed to meet the roughness requirement. However, due to the various factors, the polishing process heavily relies on trial and experience, which leads to a slow development in the mechanical polishing process.

Compared with the large polishing tool, the small polishing tool used in this study has many advantages [12]. The small tool can follow the freeform surface with relatively large curvature, while large tool cannot polish the freeform surface with curvature smaller than its radius. The small tool can redress the error of local surface, while

*Correspondence: libing.sgs@hit.edu.cn
[1] Shenzhen Graduate School, Harbin Institute of Technology, Shenzhen University Town, Xili, 518055 Shenzhen, China
Full list of author information is available at the end of the article

a large tool may polish the nearby surface when it polishes a local surface. The small can also operate at greater pressure and velocity than a large tool; therefore, it can remove the material in a rapid manner. In the case of manufacturing one single piece, small polishing tool has a higher polishing efficiency.

As the polishing surface is highly nondeterministic, one of the reasons is that the polishing path of the tool affects the removal considerably. A basic requirement for polishing paths is that the surface can be completely and uniformly covered during a polishing cycle. Therefore, the study of removal function of different polishing path has a very important significance. A lots of polishing paths in mechanical polishing process can be found in literatures [13-18]. The ultimate goal of the research about different polishing removal function is to select an optimal polishing path so as to avoid any under-polishing and over-polishing phenomenon and ensure the uniformity of the polishing process. Meanwhile, it can also present a clear understanding of the various factors that influence the polishing quality. The core issue of this research is to study the polishing removal function and the polishing performance in a given surface shape and a specified path, so that it can facilitate the subsequent polishing path control.

The rest of the paper is organized as follows. In the 'Theoretical background' section, some basic polishing assumptions and fundamental theories are introduced. In the 'Methods' section, the scan line path, Archimedean spiral path, and Hilbert fractal path are described and analyzed in details, while the modeling method of polishing path from plane surface to freeform surface is also given. The simulation and experiment are conducted, and the results are depicted in the 'Results and discussion' section. Conclusions are drawn in the last section.

Theoretical background

Removal rate of polishing

Removal rate of polishing, which depicts the properties of part surface's removal amount at a local polishing position during every unit of time, is an important part of study about polishing process [19,20]. Many factors affect the polishing removal rate, such as polishing tool and polishing environment [21-23]. For the polishing tool, the main factors are material property, shape, and surface roughness. While for the polishing part, the results are relevant to material property and friction coefficient between the part and the polishing tool. Additionally, for the polishing environment, the main factors include the particle size and viscosity of polishing paste and polishing liquid, the temperature and pressure during polishing and the chemical correction of part in the polishing liquid.

Preston presented theoretical basis for the prediction of material removal for mechanical polishing [18]. Although there are many other polishing formulations in polishing literatures, most of them are based on the Preston equation. Tseng and Wang [24] proposed a modification equation based on the Preston equation in chemical-mechanical polishing; Nanz [25] considered the bending of pad and flow of slurry into the Preston equation to obtain a new polishing formula. The formula of Preston polishing removal function is displayed as below:

$$f(\vec{q},t) = \frac{\mathrm{d}z}{\mathrm{d}t} = k \cdot p(\vec{q},t) \cdot v(\vec{q},t) \tag{1}$$

where $f(\vec{q},t)$ is the material removal in unit time; $\vec{q}$ is the position vector of polishing point; t is the time; k is the Preston coefficient which is related to the material, polishing liquid, and environment; $p(\vec{q},t)$ is the pressure at the polishing point; and $v(\vec{q},t)$ is the instantaneous polishing velocity at the polishing point.

Actually, the problem of polishing removal rate is a very complex nonlinear problem. Preston polishing removal function is a linearization assumption, but it is approximately valid in many practical polishing experiments.

Removal amount of polishing path

In terms of mechanical polishing, the polishing velocity and polishing pressure can be controlled by computer numerical control (CNC) machine precisely. Therefore, during the polishing process, the Preston coefficient k is regarded as constant. Moreover, the polishing tool is usually made of relatively soft materials. At the micro level, the soft tool is adaptive to the shape of surface. It can keep the polishing tool and word-piece contact with each other throughout the polishing process. Thus, the polishing pressure applied by small polishing tool on freeform surface is assumed to be uniform; that is to say, the pressure of contact surface is equal everywhere at the same time.

At time τ and point $\vec{q}$ in the part coordinate system, the polishing velocity and pressure are set as $v(\vec{q},\tau)$ and $n(\tau)$, respectively. When the radius of polishing tool is R, the polishing removal amount is defined as follow:

$$z(x,y,t) = \int_0^t k \cdot \frac{n(\tau)}{2\pi \cdot R^2} \cdot v(\vec{q},\tau)\mathrm{d}\tau \tag{2}$$

The polishing pressure during the polishing process can be divided into the constant polishing pressure n_s and the time-varying polishing pressure $n_v(\tau)$, namely $n(\tau) = n_s + n_v(\tau)$. However, in order to simplify the subsequent discussion, we assume that the polishing force remains constant during the polishing process. Then, the polishing removal function can be rewritten as:

$$z(x,y,t) = \frac{k \cdot n_s}{2\pi \cdot R^2} \cdot \int_0^t v(\vec{q},\tau)\mathrm{d}\tau \tag{3}$$

Equation 3 is a general polishing removal function. Given different polishing paths, the polishing velocity in Equation 3 will change correspondingly, and this will result in different integration results.

Removal function at a fixed point

Figure 1 is the diagram of polishing model at a fixed point. When the polishing tool is stationary and rotates with angular velocity ω_t, the polishing removal function of the point with radius r to the center of polishing tool is described as:

$$z(r,t) = \frac{k \cdot n_s}{2\pi \cdot R^2} \cdot r \cdot \omega_t \cdot t \tag{4}$$

The Equation 4 is a conical surface as shown in Figure 2. The velocity of polishing tool center equals zero which results in zero removal amount at this point.

In practice, the polishing velocity is the sum of feed velocity and rotation velocity; that is to say, $v = v_t + r \cdot \omega_t$. However, the angular velocity ω_t of the polishing tool is very large which can reach about thousands of revolutions per minute. As a result, v_t is very small and $r \cdot \omega_t \gg v_t$, so it can be considered that $v \approx r \cdot \omega_t$.

Methods

Tool paths are one of the key factors for the automatic mechanical surface polishing. Three polishing paths are covered in this study: scan line path, Archimedean spiral path, and Hilbert fractal path. How they affect the polishing results are investigated here.

Scan line polishing path

The scan line path, a conventional machining path, is frequently used in CNC machining. In the scan line polishing process, the polishing tool offsets an interval at the boundary and continues the machining until the processing is done.

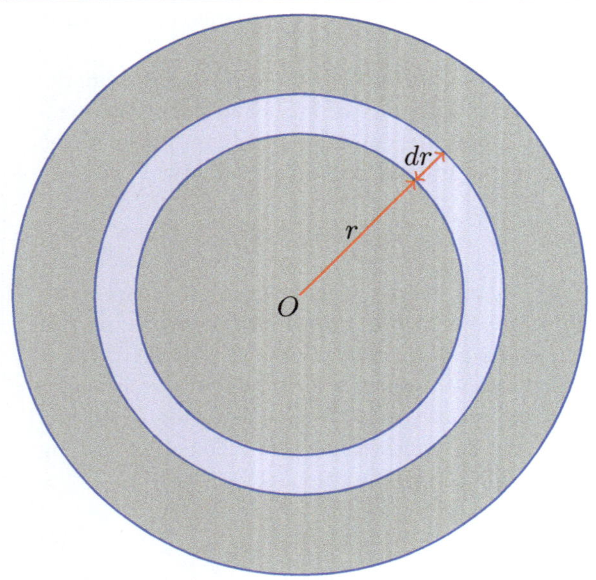

Figure 1 Polishing model of a fixed point.

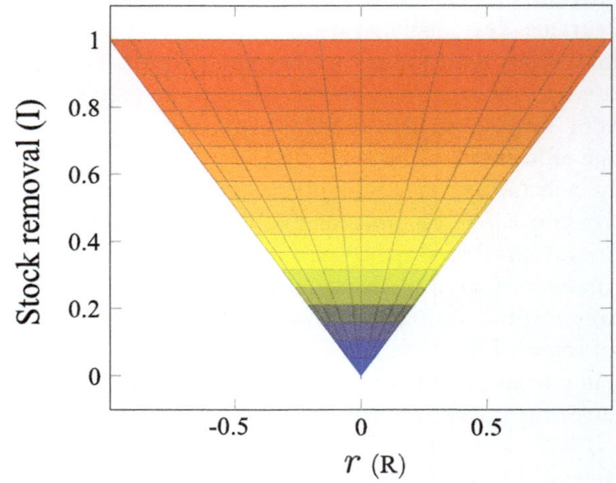

Figure 2 Removal function of a fixed point.

Removal function of single line polishing path

In the modeling of single line polishing path along the X-axis, the removal function Equation 3 at point (x_p, y_p) on plane can be rewritten as:

$$\begin{aligned} z(x_p, y_p) &= \frac{k \cdot n_s}{2\pi \cdot R^2} \cdot \int_0^{\infty} v(\vec{q}(x(\tau), y_p), \tau) d\tau \\ &= \frac{k \cdot n_s}{2\pi \cdot R^2} \cdot \int_{-\frac{\sqrt{r^2 - y_p^2}}{v_t}}^{\frac{\sqrt{r^2 - y_p^2}}{v_t}} \omega_t \cdot \sqrt{(v_t \tau)^2 + y_p^2}\, d\tau \\ &= \frac{k \cdot n_s}{2\pi \cdot R^2} \cdot \frac{\omega_t}{v_t} \cdot \Big[R\sqrt{R^2 - y_p^2} \\ &\quad + y_p^2 \left(\ln\left(R + \sqrt{R^2 - y_p^2}\right) - \ln(|y_p|)\right)\Big] \end{aligned} \tag{5}$$

where $y_p \in [-R, R]$.

The coefficient term of Equation 5 is proportional to the removal function, and it does not impact the shape of the removal function curve. Thus, let I represent the coefficient term:

$$I = \frac{k \cdot n_s}{2\pi \cdot R^2} \cdot \frac{\omega_t}{v_t} \tag{6}$$

Figure 3 is the removal function of single line polishing path in the XZ plane. The maximum removal value is not at the polishing tool center, although the center point has a longest valid polishing length.

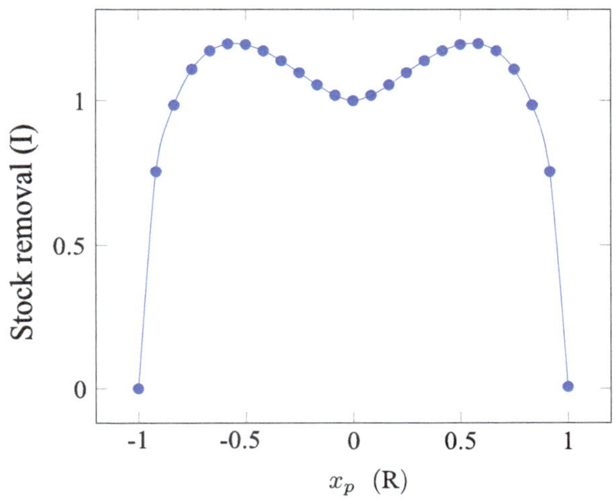

Figure 3 Removal function of single line polishing path.

Removal function of scan line polishing path

Figure 4 is the polishing model of scan line path. L denotes the interval of two adjacent lines. Equation 5 can be rewritten as the following piecewise function:

$$Z_s(x,y) = \begin{cases} I \cdot \left(R\sqrt{R^2-y^2} + y^2\left(\ln(R+\sqrt{R^2-y^2}) - \ln(|y|)\right)\right), & y \in [-R, R] \\ 0, & y \notin [-R, R] \end{cases} \tag{7}$$

Let $Z_{y_0}(x,y) = Z_s(x, y - y_0)$ denote removal function of the polishing line $y = y_0$. Calculating the sum of all the polishing lines with interval L, the removal function of scan line is:

$$Z(x,y) = \sum_{i=-n}^{n} Z_{(i \cdot L)}(x,y) \tag{8}$$

Archimedean spiral polishing path

Archimedean spiral is an arithmetic spiral, which is one of the most useful machining paths. Due to the rotational motion of workpiece, with no need of reversing, the polishing machine can achieve smooth and steady motion.

Modeling of Archimedean spiral polishing path

Figure 5 is the model of Archimedean spiral polishing path. Let L be the constant separation distance and v_t be the feed velocity of the polishing tool. The Archimedean spiral formula in the polar coordinate system is $\rho = (L \cdot \theta)/2\pi$. The feed velocity of polishing tool is $v_t = \sqrt{\dot{\rho}^2 + \rho^2\omega^2}$. Therefore, the angular velocity of the Archimedean spiral $\omega = \dot{\theta}$ can be obtained:

$$\dot{\theta} = \frac{2\pi v_t}{L} \cdot \frac{1}{\sqrt{1+\theta^2}} \tag{9}$$

With the initial condition $\theta(0) = 0$, the integral for Equation 9 is:

$$\theta\sqrt{1+\theta^2} + \operatorname{arcsinh}(\theta) = \frac{4\pi}{L} \cdot v_t t \tag{10}$$

where $\operatorname{arcsinh}(\theta) = \ln\left(\theta + \sqrt{1+\theta^2}\right)$ is the inverse hyperbolic function. There is no analytical expression for θ in Equation 10. The value of θ should be calculated by numerical method and then the value of ρ can be obtained.

Figure 6 is the model of Archimedean spiral polishing path. R is the radius of the polishing tool and r is the distance between point (ρ_0, θ_0). (ρ, θ) denotes the polishing tool center. According to the cosine law $r(\rho, \theta, \rho_0, \theta_0) =$

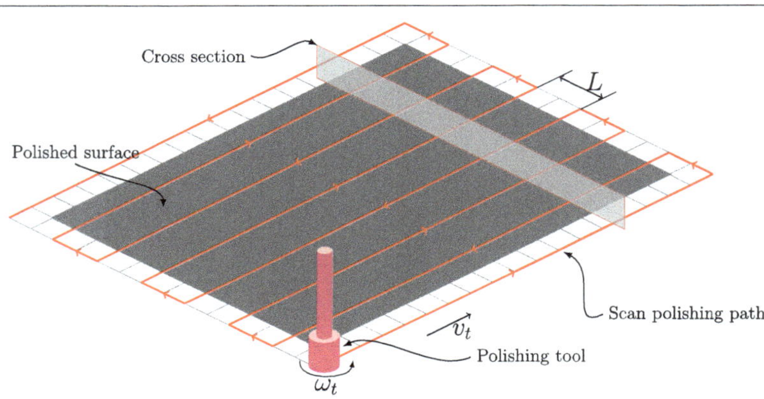

Figure 4 Scan line polishing path.

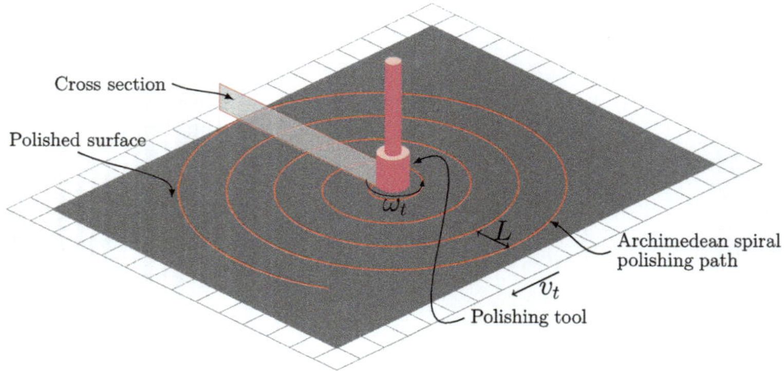

Figure 5 **Archimedean spiral polishing path.**

$\sqrt{\rho^2+\rho_0^2-2\rho\rho_0\cos(\theta-\theta_0)}$, the polishing velocity v_R of the polishing point (ρ,θ) can be expressed as:

$$v_R(\rho,\theta,\rho_0,\theta_0)=\sigma_R\left[r(\rho,\theta,\rho_0,\theta_0)\right]\cdot\omega_t \tag{11}$$

where the piecewise function σ_R is defined as follows:

$$\sigma_R=\begin{cases} r, & |r|\le R\\ 0, & |r|>R\end{cases} \tag{12}$$

For a given velocity v_t and a separation distance L, the variables ρ and θ are the function of t. With Equations 3 and 11, the integral of Archimedean spiral path can be obtained as:

$$Z(\rho_0,\theta_0)=I\cdot v_t\cdot\int_0^{\infty}\sigma_R\times\left(\sqrt{\rho(\tau)^2+\rho_0^2-2\rho(\tau)\rho_0\cos(\theta(t)-\theta_0)}\right)d\tau \tag{13}$$

The polishing process of Archimedean spiral is the uniform motion polishing process, the natural coordinate is $s=v_t\cdot t$. Thus, Equations 10 and 13 can be rewritten as:

$$\theta\sqrt{1+\theta^2}+\operatorname{arcsinh}(\theta)=\frac{4\pi}{L}\cdot s \tag{14}$$

$$Z(\rho_0,\theta_0)=I\cdot\int_0^{\infty}\sigma_R\times\left(\sqrt{\rho(s)^2+\rho_0^2-2\rho(s)\rho_0\cos(\theta(s)-\theta_0)}\right)ds \tag{15}$$

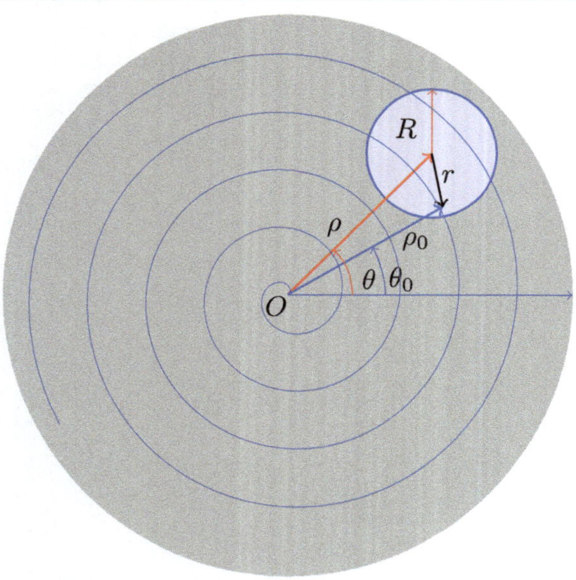

Figure 6 **Polishing model of Archimedean spiral path.**

In the natural coordinate system, Archimedean spiral path is expressed by natural coordinate s rather than time t.

Removal function of Archimedean spiral polishing path

The expression of natural coordinate s can be obtained from Equation 14:

$$s=\frac{L}{4\pi}\left[\theta\sqrt{1+\theta^2}+\operatorname{arcsinh}(\theta)\right] \tag{16}$$

Set the radius of the polishing tool R as a reference unit 1. The polar coordinate of polishing tool center is (ρ,θ). Let $s_e=s(\rho_0+R)$; $s_o=s(\rho_0-R)$ in condition $\rho_0>R$, and $s_o=0$ in condition $\rho_0\le R$. s_o and s_e are the new lower and upper limits of Equation 15. Thus, the removal function can be obtained with new limits:

$$Z(\rho_0, \theta_0) = I \cdot \int_{s_0}^{s_e} \sigma_R \times \left(\sqrt{\rho(s)^2 + \rho_0^2 - 2\rho(s)\rho_0 \cos(\theta(s) - \theta_0)} \right) ds \tag{17}$$

Using numerical method and discrete method, the removal function of Equation 17 can be rewritten approximately as:

$$Z(\rho_0, \theta_0) = I \cdot \sum_{s_0}^{s_e} \sigma_R \times \left(\sqrt{\rho(s)^2 + \rho_0^2 - 2\rho(s)\rho_0 \cos[\,\theta(s) - \theta_0]} \right) \Delta s \tag{18}$$

The value of Equation 18 at polishing point (ρ_0, θ_0) can be calculated by Newton's method.

Hilbert fractal polishing path

The scan line and Archimedean spiral polishing path discussed above are very directional. Therefore, the removal functions of them have strong anisotropy property. The anisotropy property destroys the polishing isotropy of the polished surface. In order to obtain high polishing quality, the removal function of designed polishing path with isotropy property is expected to apply into freeform surface polishing task.

Fractal curve

The fractal curve refers to the curve with fractal dimension. The local simple structure of fractal curve is suitable for CNC programming and controlling, while the global randomness property shows a global isotropy characteristic to remove the streaks of polished freeform surface and improves the polishing quality.

Fractal curves include snowflake fractal curve, triangle fractal curve, Hilbert fractal curve, and so on, but not every fractal curve is qualified for surface polishing. The qualified fractal curve needs to satisfy the following conditions: surface ergodic property, easy motion control for CNC, and strong isotropy property.

The surface ergodic property requires the fractal path with dimension greater than one, and the fractal curve can cover the polished surface when the order of the fractal curve is large enough. In mathematics, it means that the closure of the fractal curve is homeomorphism to two-dimensional plane.

The surface ergodic property requires that fractal curve can cover the whole higher-order polished surface. The snowflake fractal curve does not fulfill the surface ergodic property. For example, the cubic Koch snowflake cannot cover the two-dimensional surface. On the other hand, the easy motion control property requires the simple curve. The isometric segment is a good choice for this requirement because of easy planning, programming, and CNC machining. The snowflake fractal curve does not have well isotropic property globally, and it has a strong global orderliness with the high orders. Hence, it is not suitable for polishing purpose. Fulfilling the three requirements for freeform surface polishing, the Hilbert fractal curve is selected for freeform surface polishing here.

As shown in Figure 7, the Hilbert fractal path is organized with segments and traverses the whole two-dimensional plane. The segments of the Hilbert curve are parallel to X axis or Y axis. They can be easily controlled by CNC machine. Also, Hilbert fractal path has a good isotropy and no orderliness in global.

Modeling of Hilbert fractal polishing path

Figure 8 is the polishing model of the Hilbert curve. Similar to the scan line and Archimedean spiral polishing path, let ω_t and v_t denote the rotational velocity and the feed velocity in polishing process, respectively.

The basic unit of the Hilbert curve is segment. Therefore, Equation 3 with segment path can be applied for Hilbert polishing path. The segments of Hilbert curve are all parallel to X axis or Y axis.

For a given segment along X axis direction, the motion of polishing tool center is uniform linear motion in segment $C_x\left[(x_s, y) - (x_e, y)\right]$. The removal function is non-negative. Hence, changing the integral variable from $d\tau$ to dx needs to use absolute function to guarantee the non-negative value of Equation 3. The removal function on point (x_p, y_p) generated by the polishing segment C_x can be obtained:

$$Z_{C_x}(x_p, y_p, y) = I \cdot \left| \int_{x_s}^{x_e} \sigma_R \left(\sqrt{(x - x_p)^2 + (y - y_p)^2} \right) dx \right| \tag{19}$$

Similarly, $C_y\,[(x, y_s) - (x, y_e)]$ denotes the uniform linear segment of polishing tool center along Y axis direction, the removal function on point (x_p, y_p) generated by the polishing segment C_y can be obtained as:

$$Z_{C_y}(x_p, y_p, x) = I \cdot \left| \int_{y_s}^{y_e} \sigma_R \left(\sqrt{(x - x_p)^2 + (y - y_p)^2} \right) dy \right| \tag{20}$$

The σ_R in Equations 19 and 20 has a mutation at the point where $|r| = R$. Therefore, the mutation is not convenient for solving the integration of Equations 19 and 20. x_- and x_+ are the x-coordinate of intersection points

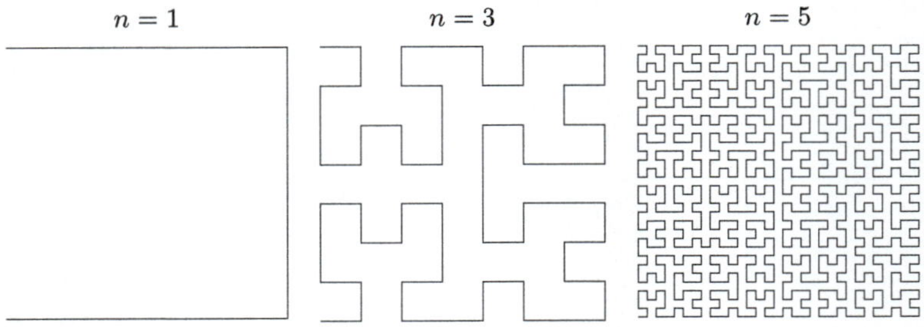

Figure 7 **Hilbert fractal path.**

between the line and the circle. Let $X_{\min} = \min(x_s, x_e)$ and $X_{\max} = \max(x_s, x_e)$; let $x_- = x_p - \sqrt{R^2 - (y - y_p)^2}$ and $x_+ = x_p + \sqrt{R^2 - (y - y_p)^2}$ under condition $|y - y_p| \leq R$. The valid lower and upper limits $x'_s \leq x'_e$ of integral function along X axis direction the polishing segment can be defined as:

$$x'_s = \begin{cases} x_- , & X_{\min} \leq x_- \leq X_{\max}, \text{ and } |y - y_p| \leq R \\ X_{\min} , & x_- \leq X_{\min} \leq x_+, \text{ and } |y - y_p| \leq R \\ \text{not defined} , & \text{other} \end{cases} \tag{21}$$

$$x'_e = \begin{cases} x_+ , & X_{\min} \leq x_+ \leq X_{\max}, \text{ and } |y - y_p| \leq R \\ X_{\max} , & x_- \leq X_{\max} \leq x_+, \text{ and } |y - y_p| \leq R \\ \text{not defined} , & \text{other} \end{cases} \tag{22}$$

Similarly, let $Y_{\min} = \min(y_s, y_e)$ and $Y_{\max} = \max(y_s, y_e)$; let $y_- = y_p - \sqrt{R^2 - (x - x_p)^2}$ and $y_+ = y_p + \sqrt{R^2 - (x - x_p)^2}$ under condition $|x - x_p| \leq R$. The valid lower and upper limits $y'_s \leq y'_e$ of integral function along Y axis direction polishing segment can be defined as:

$$y'_s = \begin{cases} y_- , & Y_{\min} \leq y_- \leq Y_{\max}, \text{ and } |x - x_p| \leq R \\ Y_{\min} , & y_- \leq Y_{\min} \leq y_+, \text{ and } |x - x_p| \leq R \\ \text{not defined} , & \text{other} \end{cases} \tag{23}$$

$$y'_e = \begin{cases} y_+ , & Y_{\min} \leq y_+ \leq Y_{\max}, \text{ and } |x - x_p| \leq R \\ Y_{\max} , & y_- \leq Y_{\max} \leq y_+, \text{ and } |x - x_p| \leq R \\ \text{not defined}, & \text{other} \end{cases} \tag{24}$$

The integrands of Equations 19 and 20 are given, respectively:

$$f_{x_p,y_p}(x, y) = \frac{\delta x\sqrt{\delta x^2 + \delta y^2} + \delta y^2 \ln\left(\delta x + \sqrt{\delta x^2 + \delta y^2}\right)}{2} \tag{25}$$

$$g_{x_p,y_p}(x, y) = \frac{\delta y\sqrt{\delta x^2 + \delta y^2} + \delta x^2 \ln\left(\delta y + \sqrt{\delta x^2 + \delta y^2}\right)}{2} \tag{26}$$

where $\delta x = x - x_p$, $\delta y = y - y_p$.

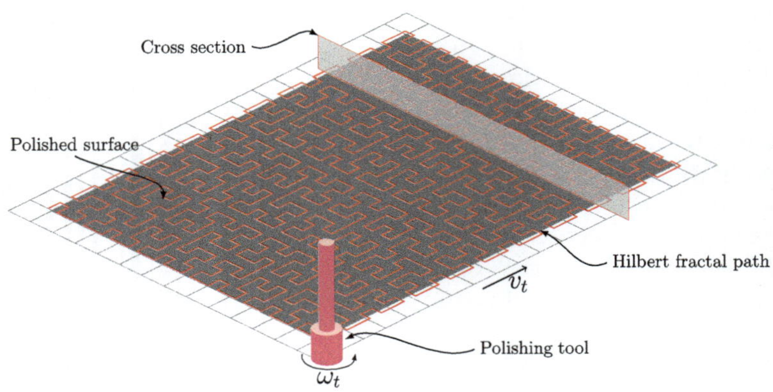

Figure 8 **Hilbert fractal polishing path.**

When x in $f_{x_p,y_p}(x,y)$ is not defined, $f_{x_p,y_p}(x,y) = 0$; when y in $g_{x_p,y_p}(x,y)$ is not defined, $g_{x_p,y_p}(x,y) = 0$. With new defined lower and upper limits, the absolute function can be removed from the integration. The single uniform linear polishing segment can be rewritten as:

$$Z_X\left(x_p, y_p, x_s, x_e, y\right) = I \cdot \left|f_{x_p,y_p}\left(x'_e, y\right) - f_{x_p,y_p}\left(x'_s, y\right)\right| \tag{27}$$

$$Z_Y\left(x_p, y_p, x, y_s, y_e\right) = I \cdot \left|g_{x_p,y_p}\left(x, y'_s\right) - g_{x_p,y_p}\left(x, y'_s\right)\right| \tag{28}$$

where x'_s, x'_e, y'_s, and y'_e are defined in Equations 21, 22, 23, and 24.

Applying uniform linear motion segment, the removal functions of Equations 27 and 28 with the Hilbert fractal curve are the segments set $\mathbf{H}$, the removal function of Hilbert fractal curve at point (x_p, y_p) is the sum of all the segments removal functions:

$$Z\left(x_p, y_p\right) = \sum_{C_{X_i} \in \mathbf{H}} Z_{X_i}\left(x_p, y_p, x_{si}, x_{ei}, y_i\right) + \sum_{C_{Y_j} \in \mathbf{H}} Z_{Y_j}\left(x_p, y_p, x_j, y_{sj}, y_{ej}\right) \tag{29}$$

Freeform surface polishing modeling

The previous polishing paths are all discussed in the planer plane. Referring to differential geometry method, the planer plane can be mapped on a freeform surface. The following discusses the general mapping method.

Natural coordinate expression of freeform surface path

The freeform curve expressed by natural coordinate s can be directly used in path planning. Freeform surface refers to a smooth two-dimensional surface with finite or infinite degrees of freedom. Because the order of freeform

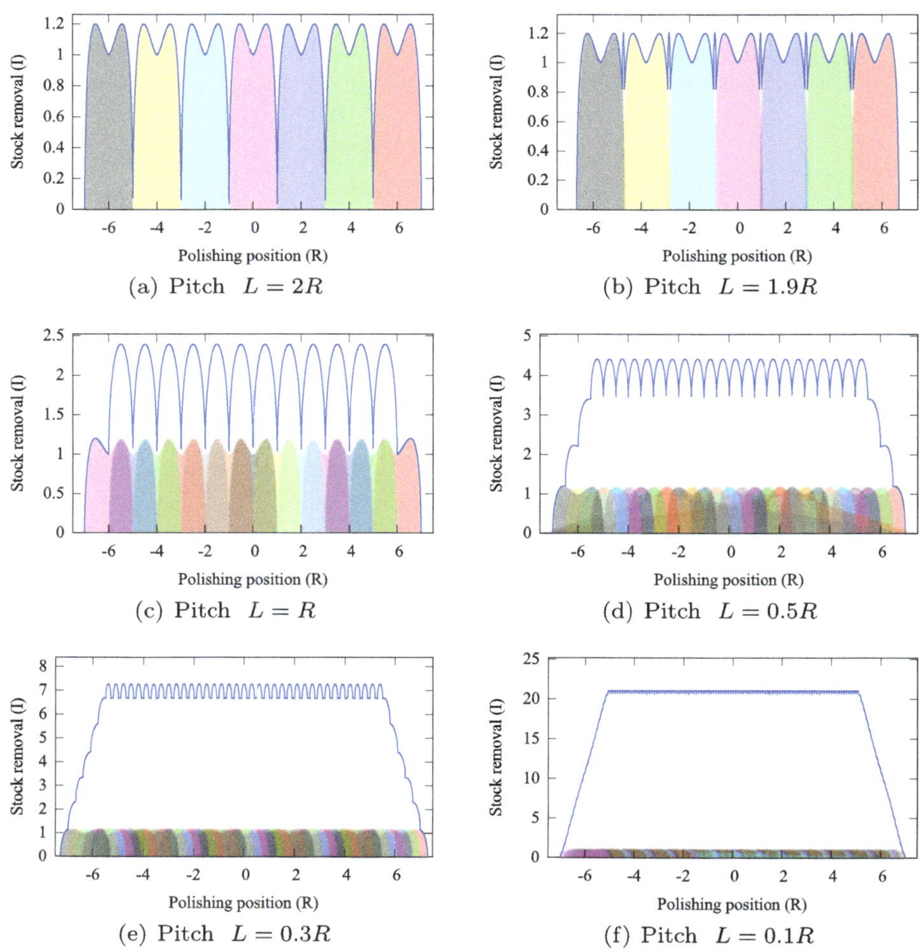

Figure 9 Removal function of scan line path with different spaces. **(a)** Pitch $L = 2R$, **(b)** $L = 1.9R$, **(c)** $L = R$, **(d)** $L = 0.5R$, **(e)** $L = 0.3R$, **(f)** $L = 0.1R$.

surface is unknown, it can hardly use an unified expression to describe the surface exactly. The freeform surface is expressed by linear combining of several basic functions to approach the freeform surface itself. The polynomial is one of the most useful basic functions due to its simpleness and easy calculation. According to the approximation theory, piecewise polynomial surfaces can approximate a smooth freeform surface in arbitrary precision.

Let C_{uv} be polishing path in parameter u-v plane and t be the parameter of path C_{uv}. Mapping the plane to the freeform surface obtains the project path C_f. The space path C_f can be expressed as follows:

$$x = x(u(t), v(t)),\ y = y(u(t), v(t)),\ z = z(u(t), v(t)) \tag{30}$$

In application, $x(t)$ and $y(t)$ are selected as the parameter variables of $u(t)$ and $v(t)$. The path C_f on freeform surface is described as $z = z\,(x(t),\ y(t))$.

The natural coordinate parameter s which is the length of curve is used to replace the parameter variable t to describe the space path C_f. Referring to formula $\mathrm{d}s = \sqrt{\mathrm{d}x^2 + \mathrm{d}y^2 + \mathrm{d}z^2}$, natural coordinate parameter s can be written as:

$$s = \int_0^t \sqrt{\left(1 + z_x^2\right) x_t^2 + 2 z_x z_y x_t y_t + \left(1 + z_y^2\right) y_t^2} \cdot \mathrm{d}\tau \tag{31}$$

where $z_x = \mathrm{d}z/\mathrm{d}x$.

With Equation 31, the interpolation algorithm of the polishing path on freeform surface can be easily designed.

Modeling of polishing path on the Bézier freeform surface

The traditional surface fitting methods include the Hermite method, Bézier method, B-spline method, and NURBS method. The Bézier surface can approximate any freeform surface and has fewer parameters which make it easier to use in practice. Thus, the Bézier method is studied in this paper for freeform surface polishing.

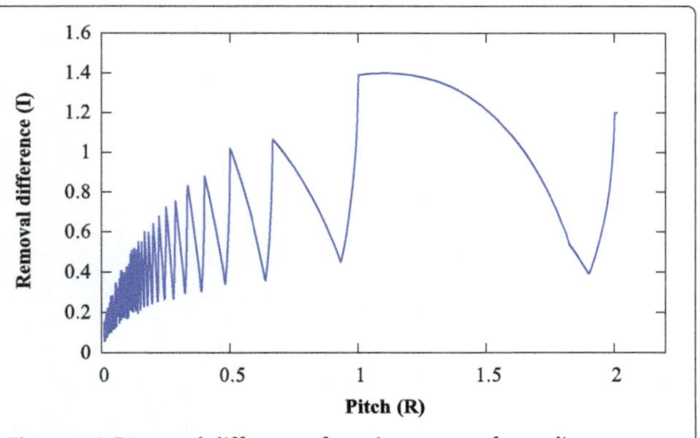

Figure 10 Removal difference function curve of scan line.

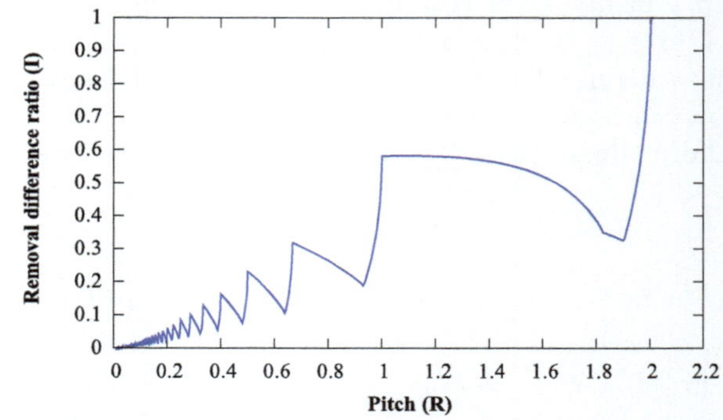

Figure 11 Removal difference ratio function curve of scan line.

Bézier method is polynomial-based approximation algorithm. The degree n of the polynomial is related to the number of the knots. Polynomial can approximate a smooth function in arbitrary precision. Hence, the polynomial-based Bézier surface can approximate the smooth freeform surface as well in arbitrary precision.

Given m rows and n columns Bézier control points $P_{ij} = \left[P_{ij}^x, P_{ij}^y, P_{ij}^z\right]$. Let $\mathbf{B}_{k-1}(\alpha) = \left[B_{k-1}^0(\alpha), B_{k-1}^1(\alpha), \cdots, B_{k-1}^{k-1}(\alpha)\right]^{\mathrm{T}}$, $\mathbf{P}_{mn}^x = \left\{P_{ij}^x\right\}$, $\mathbf{P}_{mn}^y = \left\{P_{ij}^y\right\}$, and $\mathbf{P}_{mn}^z = \left\{P_{ij}^z\right\}$. Thus, the Bézier surface can be expressed as:

$$\begin{cases} x(u,v) = \mathbf{B}_{m-1}^{\mathrm{T}}(u) \cdot \mathbf{P}_{mn}^x \cdot \mathbf{B}_{n-1}(v) \\ y(u,v) = \mathbf{B}_{m-1}^{\mathrm{T}}(u) \cdot \mathbf{P}_{mn}^y \cdot \mathbf{B}_{n-1}(v) \\ z(u,v) = \mathbf{B}_{m-1}^{\mathrm{T}}(u) \cdot \mathbf{P}_{mn}^z \cdot \mathbf{B}_{n-1}(v) \end{cases} \tag{32}$$

where $u \in [0,1]$, $v \in [0,1]$.

The Bernstein basic function $B_n^i(\alpha)$ is defined as:

$$B_n^i(\alpha) = \sum_{i=0}^{n} \frac{n!}{i!\,(n-i)!} (1-\alpha)^{n-i} \alpha^i \tag{33}$$

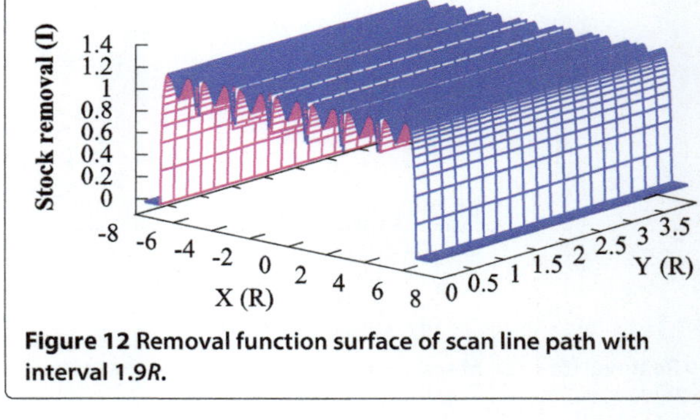

Figure 12 Removal function surface of scan line path with interval 1.9*R*.

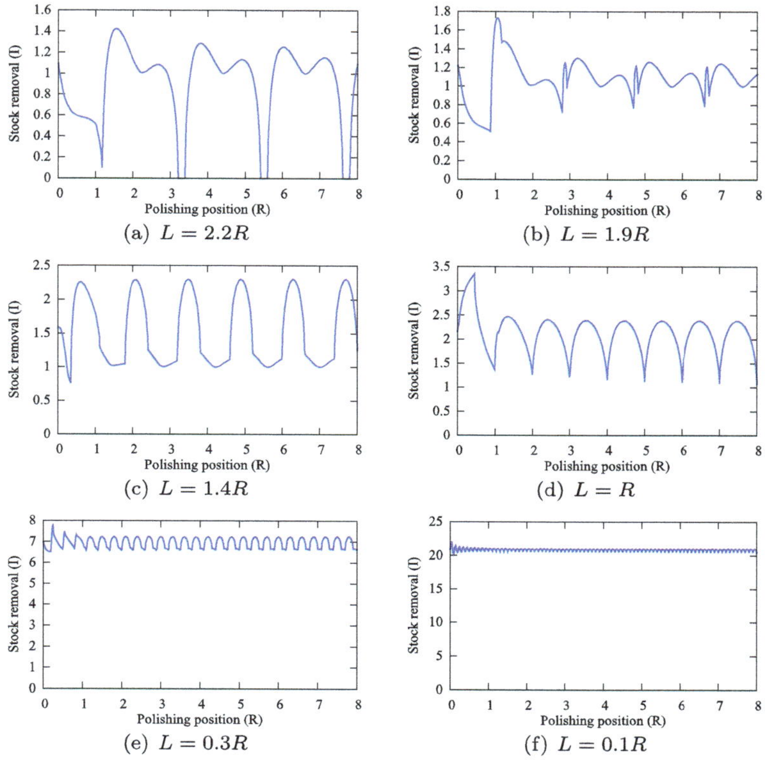

Figure 13 Removal function of Archimedean spiral path with different spaces at $\theta = 0$. (a) $L = 2.2R$, **(b)** $L = 1.9R$, **(c)** $L = 1.4R$, **(d)** $L = R$, **(e)** $L = 0.3R$, **(f)** $L = 0.1R$.

Set the Bézier surface with the range of $[X_s, X_e] \times [Y_s, Y_e]$. The affine transformation from u-v to x-y is defined as:

$$u = (x - X_s)/(X_e - X_s) \ , \ v = (y - Y_s)/(Y_e - Y_s) \quad (34)$$

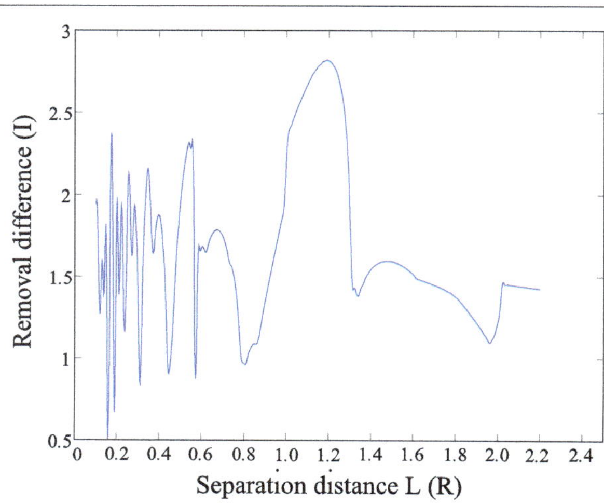

Figure 14 Removal difference function of the Archimedean spiral.

The surface is evenly divided into $m \times n$ matrix points. Therefore, $P^x_{ij} = X_s + \frac{i-1}{m-1} \cdot (X_e - X_s)$ and $P^y_{ij} = Y_s + \frac{j-1}{n-1} \cdot (Y_e - Y_s)$. Bernstein polynomial is endowed with a binomial expansion $\sum_{i=0}^{n} B^i_n(\alpha) = 1$. The identical relation for the $x(u, v)$ and $y(u, v)$ in Equation 32 is:

$$\begin{cases} x\left(\dfrac{x - X_s}{X_e - X_s}, t\right) & = x \\ y\left(t, \dfrac{y - Y_s}{Y_e - Y_s}\right) & = y \end{cases} \quad (35)$$

The Bézier surface can be simplified as follows:

$$z(x, y) = \mathbf{B}^{\mathrm{T}}_{m-1}\left(\frac{x - X_s}{X_e - X_s}\right) \cdot \mathbf{P}^z_{mn} \cdot \mathbf{B}_{n-1}\left(\frac{y - Y_s}{Y_e - Y_s}\right) \quad (36)$$

Thus, z_x and z_y in Equation 31 are expressed as:

$$z_x = \mathbf{B}'^{\mathrm{T}}_{m-1}\left(\frac{x - X_s}{X_e - X_s}\right) \cdot \mathbf{P}^z_{mn} \cdot \mathbf{B}_{n-1}\left(\frac{y - Y_s}{Y_e - Y_s}\right) \quad (37)$$

$$z_y = \mathbf{B}^{\mathrm{T}}_{m-1}\left(\frac{x - X_s}{X_e - X_s}\right) \cdot \mathbf{P}^z_{mn} \cdot \mathbf{B}'_{n-1}\left(\frac{y - Y_s}{Y_e - Y_s}\right) \quad (38)$$

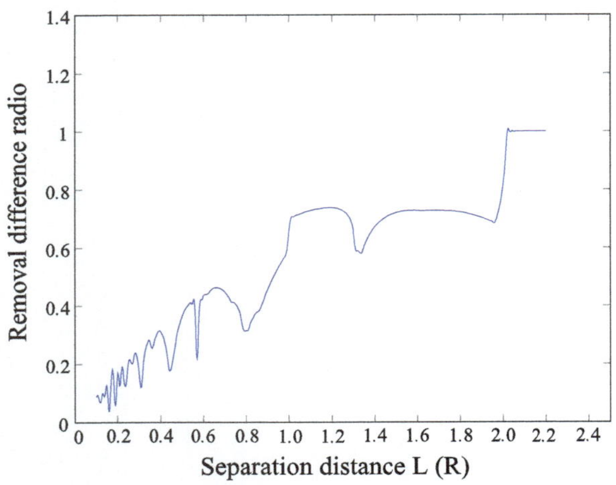

Figure 15 Removal difference ratio function curve of Archimedean spiral.

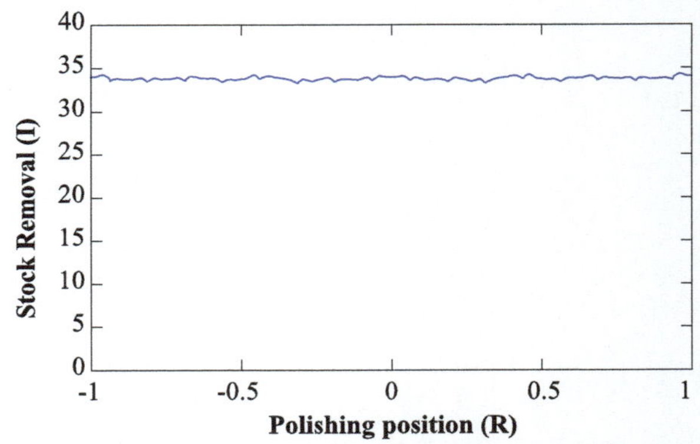

Figure 17 Removal function graph of five-order Hilbert curve along the X axis.

The first order derivative of Bernstein function is $B_n^{i\,\prime}(t) = n\left[B_{n-1}^{i-1}(t) - B_{n-1}^{i}(t)\right]$. Hence, $\mathbf{B}'^{\,\mathrm{T}}_{m-1}\left(\frac{x-X_s}{X_e-X_s}\right)$ and $\mathbf{B}'_{n-1}\left(\frac{y-Y_s}{Y_e-Y_s}\right)$ can be calculated by a series of simple algebraic operation. With the natural coordinate parameter s, the interpolation algorithm of freeform surface curve can be directly obtained.

Results and discussion

Different parameters for the same type of polishing path may cause different quality. The following will discuss removal function of the different polishing paths and optimize the parameters of the polishing path.

Simulation

Scan line polishing path

Figure 9 shows the removal function curves of the scan line polishing with different spaces L. Ignoring the boundary effect, the points on the surface have no more than two times of repeatedly polishing along the feed direction when $R < L < 2R$. However, when $L < R$, there are several times of repeatedly polishing of the surface. If the equivalent coefficient and the ratio between the rotational speed and the feed speed of polishing tool is fixed, the polishing removal amount and the uniformity will increase with the decrease of the polishing space $L < R$. As the main role of polishing is to remove the microscopic irregularities of the surface, the smaller removal amounts the better the surface uniformity based on the condition that the accuracy of polishing is ensured.

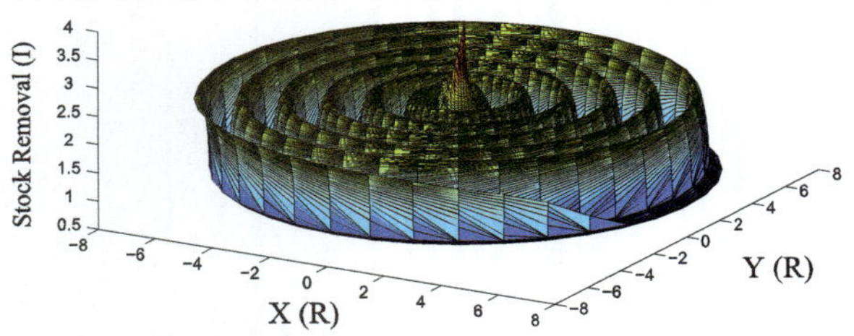

Figure 16 Removal function of the Archimedean spiral.

Set the maximum removal amount and the minimum removal amount at a specific polishing point during a polishing cycle as $Z_{\max}$ and $Z_{\min}$, respectively. The difference of the removal amount $\Delta Z = Z_{\max} - Z_{\min}$ which is varied with polishing space is studied. By numerical calculation, the curve of the ΔZ changed with polishing space can be obtained as shown in Figure 10. It displays that the removal amount increases with the decrease of the polishing space according to a zigzag trend. By only considering the difference of removal amount and eliminating the effect of the removal amount, it can be seen that the smaller polishing space has a better polishing results. Of

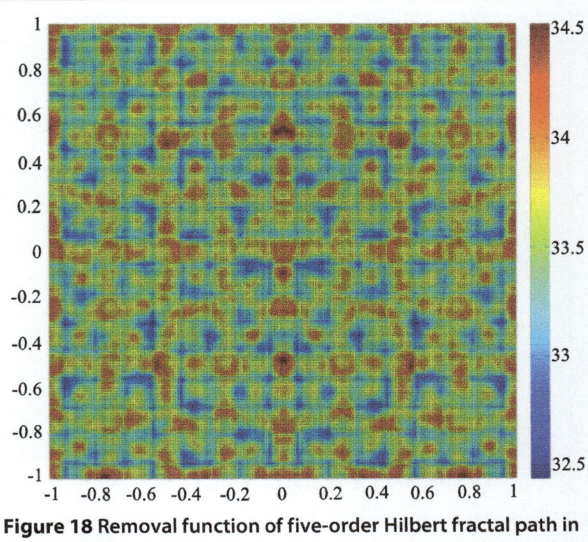

Figure 18 Removal function of five-order Hilbert fractal path in the XY plane.

Table 1 Feature comparison of the polishing path

Polishing type	Uniformity	Isotropy	Motion control	Comprehensive review
Scan line	Bad	Bad	Easy	Easy control, low polishing quality
Archimedean spiral	Bad	General	Hard	Hard control, general polishing quality
Hilbert curve	Good	Good	General	Easy control, high polishing quality

course, this value needs to be selected from the trough point on the zigzag curve.

The uniformity of the polishing cannot be fully seen if only the difference of the polishing removal amount ΔZ is discussed, because different polishing space has different difference maximum polishing amount Z_{max}. Thus, the difference ratio coefficient $R_{\Delta Z} = \Delta Z/Z_{max}$ is introduced to eliminate the effect of Z_{max} on ΔZ. Figure 11 shows that the ratio coefficient $R_{\Delta Z}$ varies with the polishing space. Although the trend of the results is similar to the curve line of the polishing removal amount during a cycle, Figure 11 can better reflect the polishing amount uniformity of the polishing path.

In order to ensure the polishing efficiency and avoid excessive repeatedly polishing of the surface, the polishing space should be set within the interval of $[R, 2R]$. From Figures 10 and 11, it can be found that the optimal value appear in the vicinity of $1.9R$. The planar graph and the three-dimensional graph of the removal function at this value is shown in Figure 9b and Figure 12, respectively.

Archimedean spiral polishing path

Figure 13 shows that the curves of the polishing removal function vary with ρ_0 when $\theta_0 = 0$. When $L > 2R$, the spiral polishing path do not interfere with each other. Thus, the curve of removal function is an independent waveform. Similar to the scan line polishing path, the polishing amount increased with the decrease of the polishing space. What different to the scan line polishing path is that the Archimedean spiral polishing path displays quasi-periodicity characteristic. The periodicity becomes increasingly stronger with the distance from the center of Archimedean spiral path and the removal amount near the center of the Archimedean spiral path varies considerably.

In order to depict the uniformity of the Archimedean spiral polishing, Z_{max} is set as the maximum removal amount and Z_{min} is set as the minimum removal amount. The difference of the removal amount is represented as $\Delta Z = Z_{max} - Z_{min}$. By numerical calculation, ΔZ changing with the space can be derived as shown in Figure 14. Different from the scan line polishing path, the smaller polishing space does not display a better polishing results.

The difference ratio coefficient $R_{\Delta Z} = \Delta Z/Z_{max}$ is introduced, and the curve line that it varies with the polishing space is shown in Figure 15. It can be seen from the figure that with the decrease of the polishing space, the uniformity of the surface becomes better and better.

In practical, the polishing space of the Archimedean spiral cannot tend to zero. In order to improve the polishing efficiency, the polishing space usually set within the interval of $[R, 2R]$. Figure 16 demonstrates the three-dimensional graph of removal function at the optimal polishing space $L = 1.3R$ which can be obtained from

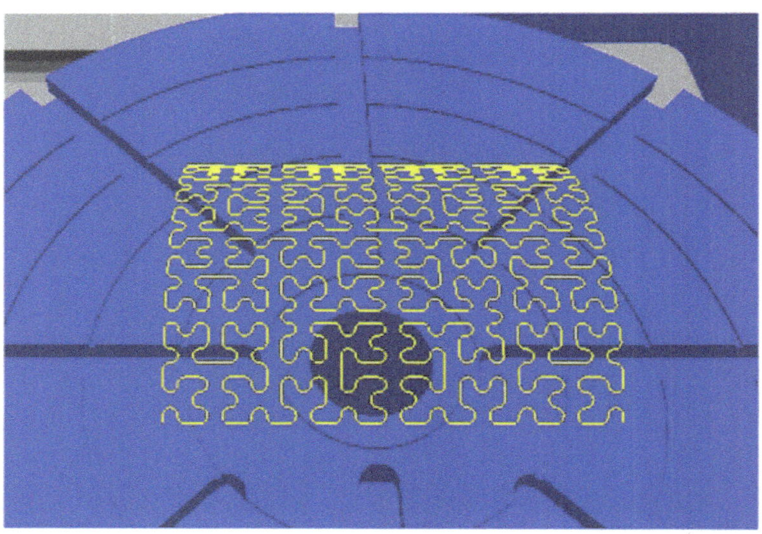

Figure 19 Hilbert fractal polishing path.

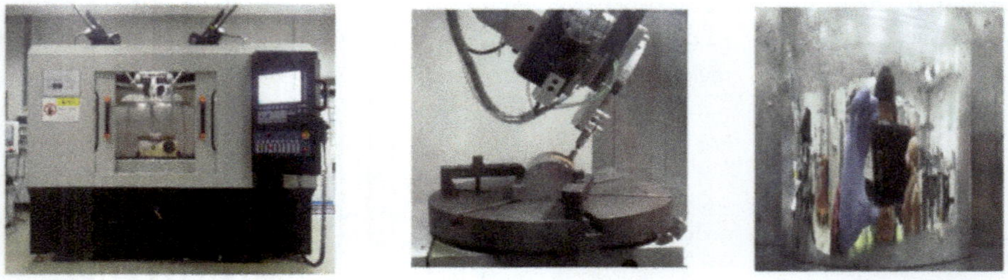

(a) Polishing machine (b) Polishing process (c) Polished surface

Figure 20 Automatic mechanical polishing of Bézier freeform surface. (a) Polishing machine. **(b)** Polishing process. **(c)** Polishing surface.

the Figure 14. Although this is the optimized result, the removed amount near the center of Archimedean spiral is still highly asymmetric.

Hilbert fractal polishing path

The quintic Hilbert fractal curve is projected on a plane with the range $[-2R, 2R] \times [-2R, 2R]$. In order to ignore the boundary effect, only the results in the area within the range $[-R, R] \times [-R, R]$ are studied. The polishing space of small segment line in the quintic Hilbert fractal curve is $0.125R$. Figure 17 is the polishing removal function curve of the quintic Hilbert fractal curve along the X-axis. It can be seen from the figure that the removal amount is symmetrical to the Y-axis. In addition, the periodic of the polishing removal function is not obvious, which means it satisfies the overall randomness requirement of polishing.

The overall randomness characteristic of the fractal curve polishing path can be seen from the three-dimensional graph of the removal function which is shown in Figure 18. This kind of characteristic is very suitable for surface polishing, and it can make the polishing results achieve good performance.

Experiment

The features for the three kinds of polishing path are listed in Table 1. Thus, the polishing path selected in this study is the Hilbert fractal polishing path which has geometric ergodicity and affine projection invariance characteristics.

In order to verify the performance of the presented polishing path, a part with Bézier freeform surface is selected. First, the discrete point positions and their normal vectors of the surface are saved. Next, by importing them into the fractal path generation module developed with Equation 31, the Hilbert fractal path is obtained. Then, the data file of the Hilbert fractal path is imported into the polishing simulation software of the self-developed polishing machine and the path on the freeform surface is obtained, which is shown in Figure 19.

Based on the freeform surface polishing requirement, a self-developed automatic polishing machine is proposed, which mainly includes a five-DOF parallel manipulator, a single DOF rotary table, and a linear motion platform with a force feedback system. It has been described extensively elsewhere in the literatures [26-28]. The parallel manipulator is a closed mechanical structure, and it has high stiffness, low inertia, and high dynamic performance. In the polishing process, the high stiffness helps to reduce the vibration generated by high speed rotation of polishing spindle, and the low inertia feature allows high polishing speed. A redundant translation is provided by

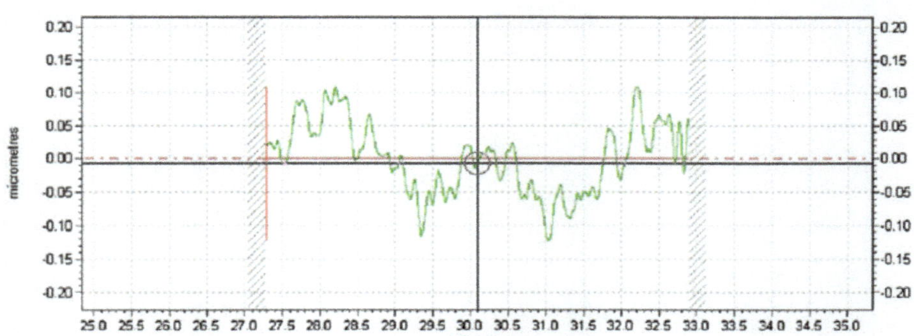

Figure 21 Roughness of polished surface.

a linear motion actuator on the moving platform to realize a constant polishing force. The rotary table is used to enlarge the polishing workspace of the automatic polishing machine. Moreover, with the help of rotary table, the path generation module could generate a singular-free and optimized dexterity polishing path for the parallel manipulator. Wool wheel polishing tool and polishing paste are used to polish the freeform surface part. During the experiment, the objective constant control force is set as 6N, and the polishing process is shown in Figure 20.

The roughness of the part surface is detected by the Taylor Hobson roughmeter. The original surface is only dealt with polishing cloth, and the roughness S_a equals 173.5 nm and the maximum height of the profile S_z equals 383.9 nm. This belongs to level 9 precision. After polishing, the roughness S_a decrease to 43.1 nm and the maximum height of the profile S_z decrease to 119.4 nm. This belongs to level 11 precision. The measuring roughness of the freeform surface after polishing is demonstrated in Figure 21.

Conclusions

Path planning is one of the key issues of the polishing process for freeform surface. Using the classical Preston polishing function, the mathematical model of small polishing tool is derived. With the assumption of a constant polishing pressure, the removal functions of scan line polishing path, Archimedean spiral polishing path, and Hilbert fractal polishing path are derived. The modeling method for polishing paths on a planar surface projected on a freeform surface is also discussed.

As different parameters for the same type polishing path may cause different polishing quality, the optimize parameters of polishing path are analyzed. The optimal polishing space for the scan line path and the Archimedean spiral polishing path is 1.9R and 1.3R, respectively. By comparing the removal functions of different polishing path, the reasonable polishing path and parameters are chosen. In order to verify the effectiveness of the selected Hilbert fractal polishing path, a polishing experiment is conducted on a self-developed mechanical polishing machine and the sequence of polishing process can be automatically scheduled. The results demonstrate that the roughness of freeform surface is improved from level 9 to level 11.

Competing interests

The authors declare that they have no competing interests.

Authors' contributions

WL carried out all the simulations and developed the polishing machine. PX drafted the manuscript and participated in the polishing experiment. BL conceived the study and participated in its design and coordination. XY participated in the polishing experiment. All authors read and approved the final manuscript.

Acknowledgements

This work was supported in part by the National Science Foundation under grant 51175105 and in part by the Research Projects of Shenzhen under grant numbers JCYJ20140417172417129 and SGLH20131010144128266.

Author details

[1]Shenzhen Graduate School, Harbin Institute of Technology, Shenzhen University Town, Xili, 518055 Shenzhen, China. [2]School of Astronautics, Harbin Institute of Technology, Xidazhi Street, 150001 Harbin, China. [3]Mechanical Automation of Engineering, the Chinese University of Hong Kong, NT Road, 999077 Shatin, Hong Kong. [4]Department of Industrial and Systems Engineering, the Hong Kong Polytechnic University, Yuk Choi Road, 999077 Kowloon, Hong Kong.

References

1. Tsai MJ, Chang J-L, Haung J-F (2005) Development of an automatic mold polishing system. Automation Sci Eng IEEE Trans 2(4):393–397
2. Kao P, Hocheng H (2003) Optimization of electrochemical polishing of stainless steel by grey relational analysis. J Mater Process Technol 140(1):255–259
3. Kordonski WI, Jacobs S (1996) Magnetorheological finishing. Int J Modern Phys B 10(23n24):2837–2848
4. Sidpara A, Jain V (2011) Experimental investigations into forces during magnetorheological fluid based finishing process. Int J Mach Tools Manufacture 51(4):358–362
5. Wang B, Zhang J, Dong S (2009) New development of atmospheric pressure plasma polishing. Chinese Optics Lett 7(6):537–538
6. Yamamura K, Takiguchi T, Ueda M, Deng H, Hattori A, Zettsu N (2011) Plasma assisted polishing of single crystal SiC for obtaining atomically flat strain-free surface. CIRP Annals-Manufacturing Technol 60(1):571–574
7. Zhao J, Zhan J, Jin R, Tao M (2000) An oblique ultrasonic polishing method by robot for free-formsurfaces. Int J Mach Tools Manufacture 40(6):795–808
8. Zhao M, Li Y, Kao HT (2012) Design and dynamic analysis of horn for ultrasonic polishing. Appl Mech Mater 121:2619–2623
9. Cheung C, Kong L, Ho L, To S (2011) Modelling and simulation of structure surface generation using computer controlled ultra-precision polishing. Precision Eng 35(4):574–590
10. Shu Y, Nie X, Shi F, Li S (2014) Smoothing evolution model for computer controlled optical surfacing. J Optical Technol 81(3):164–167
11. Zhan J, Zhao J, Xu S, Zhu P (2002) Study of the contact force in free-form-surfaces compliant EDM polishing by robot. J Mater Process Technol 129(1):186–189
12. Jones RA (1977) Optimization of computer controlled polishing. Appl Optics 16(1):218–224
13. Tam H-y, Cheng H (2010) An investigation of the effects of the tool path on the removal of material in polishing. J Mater Process Technol 210(5):807–818
14. Wang W, Yun C (2011) A path planning method for robotic belt surface grinding. Chin J Aeronaut 24(4):520–526
15. Zheng D, Lu F, Zhang LZ, Shi YJ (2012) The effect of polishing tool path on polishing parameters. Appl Mech Mater 101:1043–1046
16. Pan R, Yang W, Guo YB, Yang F, Zhang DX (2012) Research on the optimized tool-path planning of computer controlled optical surfacing. Adv Mater Res 399:1763–1767
17. Rososhansky M, Xi FJ (2011) Coverage based tool-path planning for automated polishing using contact mechanics theory. J Manufacturing Syst 30(3):144–153
18. Marinescu ID, Uhlmann E, Doi T (2006) Handbook of lapping and polishing. CRC Press, New York
19. Tsai M-J, Fang J, Chang J-L (2006) Robotic path planning for an automatic mold polishing system. Int J Robot Automation 19(2):81–90
20. Li H, Walker D, Yu G, Zhang W (2013) Modeling and validation of polishing tool influence functions for manufacturing segments for an extremely large telescope. Appl Optics 52(23):5781–5787
21. Tam H-y, Cheng H (2010) An investigation of the effects of the tool path on the removal of material in polishing. J Mater Process Technol 210(5):807–818

22. Lambropoulos JC, Jacobs SD, Ruckman J (1999) Material removal mechanisms from grinding to polishing. Ceram Trans 102:113–128
23. Lin S-C, Wu M-L (2002) A study of the effects of polishing parameters on material removal rate and non-uniformity. Int J Mach Tools Manufacture 42(1):99–103
24. Tseng W-T, Wang Y-L (1997) Re-examination of pressure and speed dependences of removal rate during chemical-mechanical polishing processes. J Electrochemical Soc 144(2):15–17
25. Nanz G, Camilletti LE (1995) Modeling of chemical-mechanical polishing: a review. Semiconductor Manufacturing IEEE Trans 8(4):382–389
26. Li B, Lin W, Yang X, Deng Z (2010) Kinematic analysis and design for a 4URHU-1URHR parallel kinematic polishing machine. In: Proceedings of CLAWAR. World Scientific, Nagoya, Japan
27. Lin W, Li B, Yang X, Zhang D (2013) Modelling and control of inverse dynamics for a 5-DOF parallel kinematic polishing machine. Int J Advanced Robot Syst 10:1–21
28. Li B, Li G, Lin W, Xu P (2014) Design and constant force control of a parallel polishing machine. In: Information Science and Technology (ICIST), 2014 4th IEEE International Conference On, IEEE, Shenzhen, China, pp 324–328

13

Experimental design and signal selection for construction of a robot control system based on EEG signals

Masataka Yoshioka[1], Chi Zhu[1*], Kazuyuki Imamura[1], Feng Wang[1], Haoyong Yu[2], Feng Duan[3] and Yuling Yan[4]

Abstract

Aiming to develop an noninvasive BMI control system with EEG (electroencephalogram) signals to control external devices such as prostheses and robots for rehabilitation and/or power support, four different tasks corresponding to different brain excitation degrees are designed. Their EEG spectra are analyzed with short-time fast Fourier transform (STFFT), and their features of mu and beta rhythms corresponding to the different tasks are extracted. The extracted features are used to control a one-joint robot arm and their corresponding results are compared. The results show that the EEG signal when a subject is holding a weight is comparatively more stable than the EEG signals in other tasks such as motor imagery. This implies an effective way for power assist by EEG signals.

Keywords: EEG; BMI; Motor imagery; Robot control

Background

Brain-machine interfaces (BMIs) are a technology that allow interaction between humans and artificial devices. They rely on continuous, real-time interaction between living neuronal tissue and artificial effectors. The past decade has seen a fast growing interest to develop various effective BMIs that can be invasive or noninvasive. Invasive BMIs, which derive the subject's intent from neuronal action potentials or local field potentials recorded within the brain, are being studied mainly in nonhuman primates [1-7]. Recently, prosthetic devices controlled by a human subject with invasive BMI is reported [8]. Though these invasive BMIs have made great success, they face substantial technical difficulties and entail significant clinical risks: they require that recording electrodes be implanted in the cortex and function well for long periods, and they risk infection and other damages to the brain. The efforts to develop them, despite these disadvantages, are based on the widespread belief that only invasive BMIs will be able to provide users with real-time multidimensional control of a robotic arm or a neuroprosthesis.

On the other hand, noninvasive BMIs for human users derive the user's intent from scalp-recorded electroencephalogram (EEG) activity. Because these systems do not require invasive surgical implants, they can be used for wide range of applications. Up so far, they have been mainly used for cursor control and communication by means of selection of letters or items on a computer screen [9-11]. Recently, an electrical powered wheelchair is controlled by noninvasive BMI [12]. However, despite having the great advantage of not exposing the subject to the risks of brain surgery, EEG-based techniques provide communication channels of limited capacity. In this study, we not only use EEG signals as communication tool but also use them as control signals of external artificial devices. Therefore, one of our main purposes here is to find and extract stable EEG signals by task design and use the signals to control an external robot for power assist.

In awake people, primary sensory or motor cortical areas often display 8 to 12 [Hz] EEG activity when they are not engaged in processing sensory input or producing motor output. This idling activity is called μ rhythm when focused over the somatosensory or motor cortex. These μ rhythms are usually associated with 13 to 30 [Hz]

*Correspondence: zhu@maebashi-it.ac.jp
[1] Department of Systems Life Engineering, Maebashi Institute of Technology, Kamisadori 460-1, 371-0816 Maebashi, Gunma, Japan
Full list of author information is available at the end of the article

β rhythms. While some β rhythms are harmonics of μ rhythms, some are separable from them by topography and/or timing and thus are independent EEG features. Note that they do not require actual movement, and they occur also with motor imagery [9]. A voluntary movement results in a circumscribed desynchronization (decrease) in the μ and lower β bands. This decrease is called event-related desynchronization (ERD) and begins in the contralateral rolandic region about 2 s prior to the onset of a movement and become bilaterally symmetrical immediately before execution of movement. After a voluntary movement and with relaxation, the power in the brain rhythms increases. This phenomenon called event-related synchronization (ERS) is dominant over the contralateral sensorimotor area and reaches a maximum around 600 ms after movement offset [13].

Because BMIs do not depend on nerves or muscles, BMIs are conventionally and mainly used to provide communication and control to people with severe neuromuscular disorders such as amyotrophic lateral sclerosis (ALS), brainstem stroke, cerebral palsy, and spinal cord injury. Besides, to achieve the above purposes, this study also aims to construct a BMI that can be used to control robotic devices to help healthy people (e.g. caregivers) with power assistance.

In this study, aiming to develop a BMI control system with EEG signals to control external devices such as prostheses and robots for rehabilitation and/or power support, four different tasks corresponding to different brain excitation degrees are designed. Their EEG spectra are analyzed with short-time fast Fourier transform (STFFT), and their features of μ and β rhythms corresponding to the different tasks are extracted. The extracted features are used to control a one-joint robot arm and their corresponding results are compared. The results show that the EEG signal when a subject is holding a weight is comparatively much stabler than the EEG in other cases such as motor imagery. This implies the possibility of power assistance using EEG signals.

Overview of the research

Several factors suggest that μ and/or β rhythms from the sensorimotor cortex could be good signal features for EEG-based communication and control. In this study, these two sensorimotor μ and β rhythms, generated in motor imagery or motion, are used to control a robotic arm.

In order to steadily and effectively get EEG signals, in this study, four different tasks corresponding to the different brain excitation degrees are designed. Their features of μ and β rhythms corresponding to the different tasks are extracted with their EEG spectra by STFFT, and a one-joint robot arm is controlled by the extracted features. Then, the control results using the EEG signals from these four tasks are compared to determine which task is more appropriate for robot control. The procedures of the research are as follows.

1. Design four different experimental tasks: idling task, gazing task, motion imagery task, and motion task. In these four tasks, the brain excitation degree gradually increase task by task. Hence, their corresponding μ and/or β rhythms should have different changes.
2. Acquire the EEG signals in each task into computer for analyzing.
3. Divide the obtained EEG signals at a time interval of 256 ms with 206-ms overlap to get a series of the short-time windows.
4. Perform STFFT to the EEG signals in each divided short-time window repetitively to get their power spectra. In this way, the power spectra of EEG signals are obtained while their time information is also kept. The time information of EEG signals is very important for control.
5. Take the obtained power spectrum magnitudes of μ and β rhythms as the features of the motion imagery or motion.
6. Convert the power spectrum magnitudes of μ and β rhythms into the target angles of the robot arm and control the robot arm.

With the above steps, a one-joint robot arm is controlled with EEG signals and the control results are obtained. These results are compared to determine which of the tasks is the most suitable for robotic control and power assist.

Methods

Measurement and signal processing of EEG

Acquisition of EEG signals

In this study, the EEG signals of one of two subjects (two healthy 20's young men) are measured while the subject is sitting in a chair as shown in Figure 1. The electrodes embedded in a head cap are positioned in the International 10-20 system of electrode placement. The cap is so worn that the position C_z is right at the top of the subject's scalp, and the cap is firmly held on the head with a belt passed through the subject's chest. The embedded electrodes do not directly contact to the scalp of the subject. Instead, electrolytic gel is poured into the electrode holes with a special syringe to suppress the contact impedance between the electrodes and the scalp. The EEG signals over the contralateral sensorimotor area (C_4 to A_2) are measured. The signals are assumed to be corresponding to the movement of the subject's left arm. Position C_z (the top of the scalp) is used as ground, and the right wrist is used as body earth. The sampling frequency is 1 kHz. The EEG signals are amplified 10,000

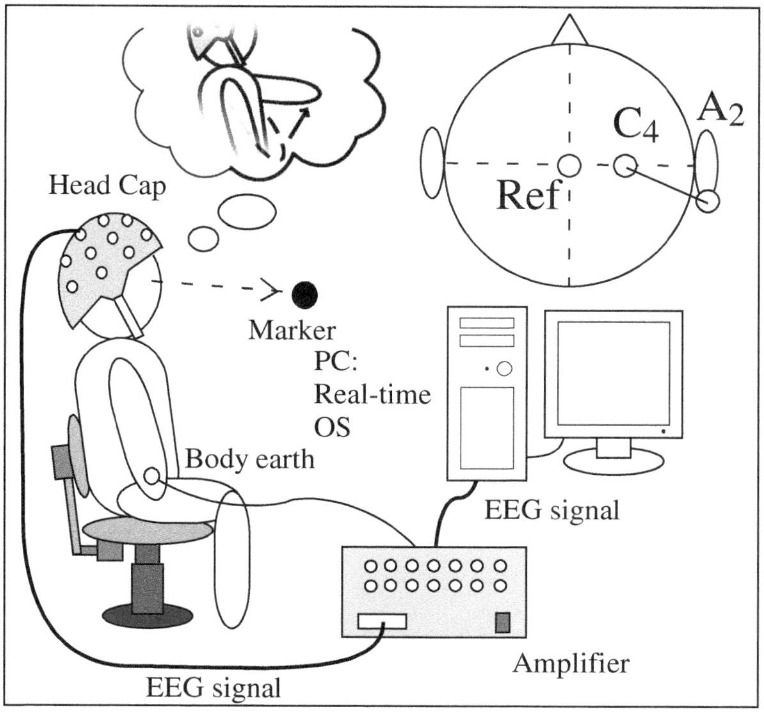

Figure 1 Measurement of EEG signals.

times by an amplifier (g.BSamp, g.tec Co., Austria). Then, the signals are acquired through 12-bit A/D converters in a multi-functional interface board and further processed online on a PC running real-time operation system. In our case, shield room is not used in consideration of our BMI system which would be used in daily life. The extracted EEG signals are used to drive a one-joint robot arm like the subject flexing his elbow joint. The robotic arm is also controlled with real-time operation system.

Extraction of μ and β rhythms

Sensorimotor rhythms (SMRs) include an arch-shaped μ rhythm, usually with a frequency of 10 Hz (range 8 to 12 Hz), often mixed with a β rhythm (around 20 Hz, range 13 to 30 Hz) and a γ component (around 40 Hz) recorded over somatosensory cortices, most preferably over C_3 and C_4. The SMR is related to the motor cortex with contributions of somatosensory areas such as that the β component arises from the motor and μ component from the sensory cortex. SMR is blocked by movements, movement imagery, and movement preparation; thus, it is seen as 'idling' rhythm of the cortical sensory region. In this study, the μ and β rhythms in movement imagery and/or movement are used as the EEG features to control a robot arm. To extract such μ and β rhythms and further to control a robot arm with the extracted EEG features, STFFT is employed so that both of the spectra of the μ and β rhythms and the time information are obtained.

Here, we briefly describe the concept of STFFT. A complex Fourier series of function $x(t)$ is defined as

$$x(t) = \sum_{k=-\infty}^{\infty} C_k e^{jk\omega_0 t} \tag{1}$$

By inner product over the interval $[-T/2, T/2]$, the Fourier coefficients C_k are defined by the formula

$$C_k = \left\langle x(t), e^{jk\omega_0 t} \right\rangle = \frac{1}{T}\int_{-T/2}^{T/2} x(t) \cdot e^{-jk\omega_0 t} dt \tag{2}$$

Further, the discrete Fourier transform about the discrete signal x_i with sampling interval T_s is expressed as

$$X(\omega) = \sum_{i=-\infty}^{\infty} x_i e^{j\omega i T_s}, \quad \omega = -\frac{\pi}{T_s} \text{ to } \frac{\pi}{\mathrm{T_s}} \tag{3}$$

To the amplified and sampled sequence of N EEG signals x_n $(n = 0, 1, \cdots, N-1)$, its discrete FFT (DFFT) is given as

$$X_k = \sum_{n=0}^{N-1} x_n W^{kn} \tag{4}$$

where $W = e^{-j\frac{2\pi}{N}}$, X_k is the obtained spectrum of x_n, and k is the number of frequencies.

As aforementioned, STFFT is used in this study. The data x_n to be transformed is divided into N blocks (saying, N windows, corresponding to time shift t) with an overlap to reduce artifacts at the boundary. Each block is discrete Fourier transformed, then the complex result is added to a matrix, which records magnitude and phase for each point in time and frequency. This process is repeated so that both the spectra and time information of the EEG signals are obtained. For example, let $t = 0, 10, 20, \cdots$, [ms], then

$$X_k = \{X_{(k,10)}, X_{(k,20)}, \cdots, X_{(k,t)}\} \tag{5}$$

Shortening the window width will increase the time resolution, but it will increase the computation cost. The concept of the STFFT is shown in Figure 2 and one example of obtained spectra of μ rhythm by the STFFT is shown in Figure 3.

Task design for motion imagery and motion

In order to extract the features of the μ and β rhythms in motion imagery and/or motion, four different tasks are designed, in which the subject is asked to image to flex his nondominant arm (for example, left arm) but without real movement (only motion imagery) or to perform a real actually motion of flexing his elbow joint for holding up a dumbbell. The four tasks are so designed to generate the different EEG signals from relaxed idling task to excited motion task. When the subject is performing each task of motion imagery or actual motion, he is sitting in a chair wearing the head cap for EEG measurement. The chair height is adjusted to fit the subject's height and let him feel comfortable.

The detail of designed each task is explained as follows.

Task 1: idling task

In this task, the subject just closes his eyes in relaxation state, and he is not doing and not thinking anything as shown in Figure 4 task 1.

This task is designed to get the subject's EEG in idling (relaxed) state. It is expected to get the highest μ rhythm in all of four tasks. The results of STFFT for this task are shown in Figure 5, in which we find that μ rhythm is greatly appeared.

Task 2: gazing task

The subject is opening his eyes and continuously gazing at a marker at the endpoint of the robot arm as shown in Figure 4 task 2.

In this task, the subject just concentrates himself to gaze at the marker and is not doing and not thinking anything else. This task is designed to get the subject's EEG when he opens his eyes and gazes at an object. The brain excitation degree of the subject in this task is higher than that in task 1. It is expected that his μ rhythms would be decreased or saying, suppressed, and the β rhythms would have changes accompanying the changes of μ rhythms in this task. Figure 6 shows the results of STFFT for this task 2. We find that μ rhythms are suppressed and there are a lot of noises as indicated by 'A' in other frequency bands that is probably caused by eye-blinking artifacts.

Task 3: motion imagery task

In this task, the subject is asked to continuously image to flex his left arm but not really perform an actual motion while he is gazing at the marker at the endpoint of the robot arm as shown in Figure 4 task 3.

This task is designed to get the subject's EEG in motion imagery. It is expected that his μ rhythm would be further decreased in this task. Same as the purpose of task 2, the second purpose of this task is to see what happens to the subject's β rhythm in motion imagery.

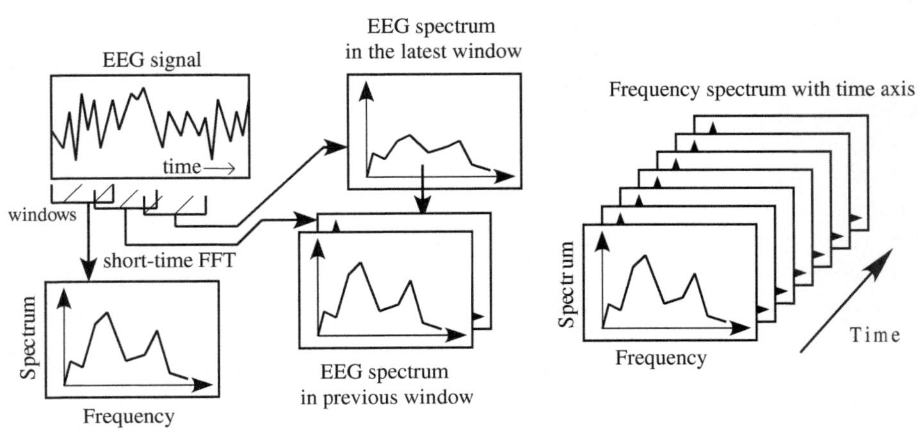

Figure 2 Concept of discrete short-time FFT for EEG signals.

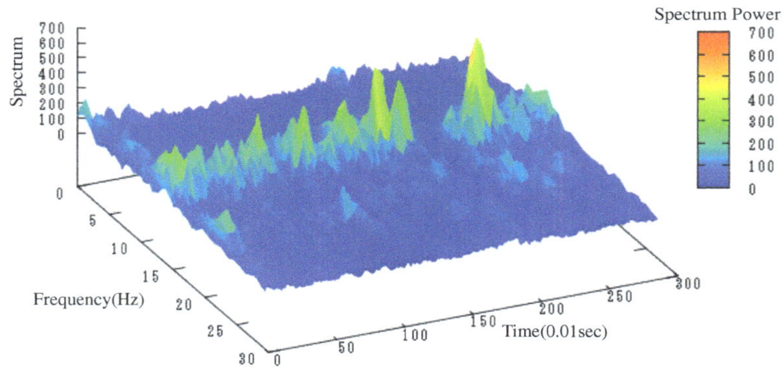

Figure 3 Spectra of μ rhythm obtained by short-time FFT.

Figure 7 shows the results of task 3 with STFFT. We find that μ rhythms are suppressed but β rhythms are generated. This motion imagery task confirmed such a fact that motion imagery can really result in the changes of μ and β rhythms.

Task 4: motion task

In this task, the subject is actually holding a 4 kg dumbbell in 90° angle of his left elbow joint while he is still gazing at the marker at the endpoint of the robot arm as shown in Figure 4 task 4. Of course, to conduct the holding task of

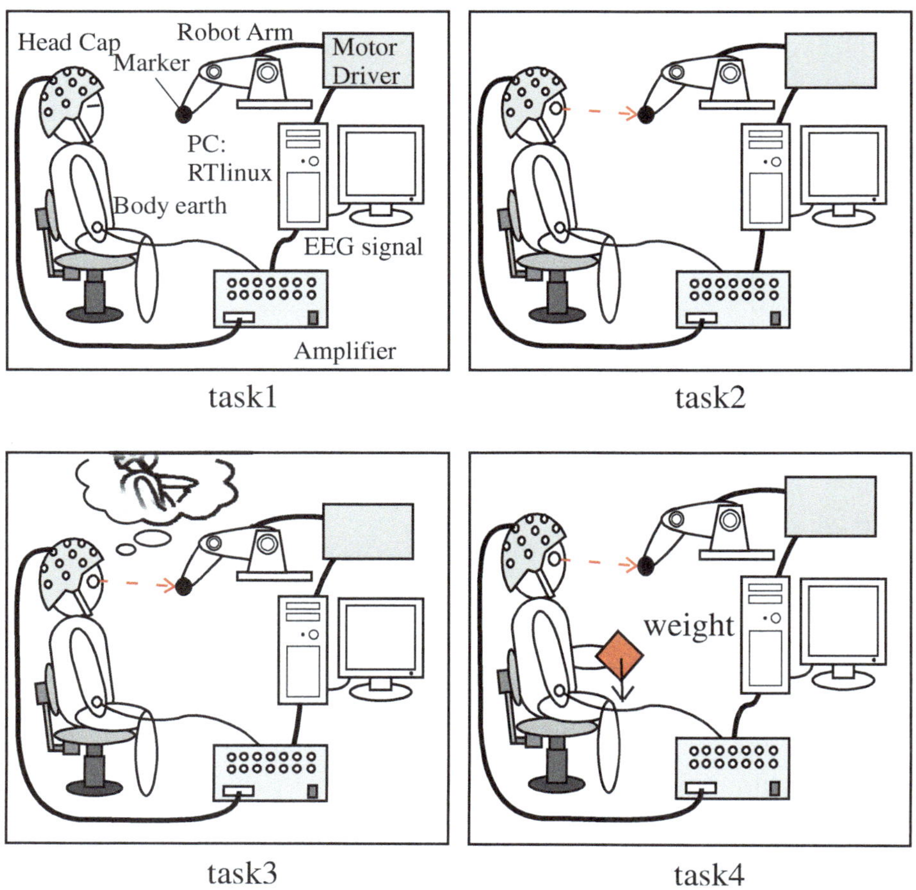

Figure 4 Four tasks designed to obtain the μ and β rhythms in motion imagery and/or motion.

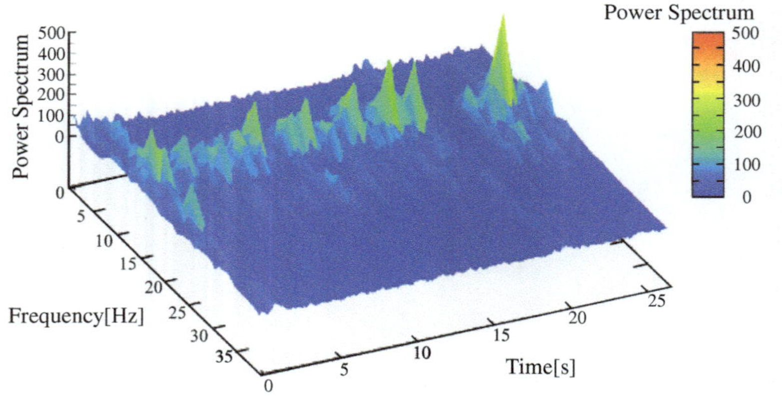

Figure 5 Results of short-time FFT for task 1. μ rhythm is greatly generated but almost no β rhythm.

a dumbbell, the subject will first have an intent, or saying, image, to flex his left arm. This task is in fact a combination of motion imagery and real motion. In this paper, we simply define it as motion task.

This task is designed to get the subject's EEG in motion task. It is expected that his μ rhythm would be suppressed to minimum and his β rhythm would be further increased. The results of this task with STFFT are shown in Figure 8. We find that μ rhythm is really suppressed and β rhythm is greatly generated as we expected.

Spectrum characteristics of each task and some remarks

As described above, tasks 1 to 4 are designed according to the degree of brain excitation with/without the existence of motion. It is expected that the spectrum of μ rhythm will decline and the spectrum of β rhythm will change, probably will increase, task by task .

Since the spectrum of β rhythm is much less than that of μ rhythm, the spectrum of EEG is normalized with the ratio $X_r = X_{(t)}/X_{(\max,t)}$, where $X_{(t)}$ and $X_{(\max,t)}$ are the spectrum and the maximum spectrum in the frequency band 1 to 30 Hz during the time interval t, respectively.

The EEG between C_4 and A_2 in each task is measured in 24 trials. The measurement time is 30 s, and the window width of STFFT (i.e. the analytical period) is 100 ms. The averages of the spectrum ratios (here, we sometimes still call it as power spectrum) in 24 trials for 30 s are shown in Figure 9. From this figure, we can find that μ rhythm greatly occurs in task 1 and it is really suppressed in other three tasks as we expected. On the other hand, the suppression extent of μ rhythms in other three tasks is not so gradually intensified as the degree of brain excitation is increased task by task. The differences among the spectra of the other three tasks indeed exist, but they are not obvious. This is different from our expectation, in which we expected that the suppression of μ rhythm would be gradually intensified as the increases of the brain excitation degree.

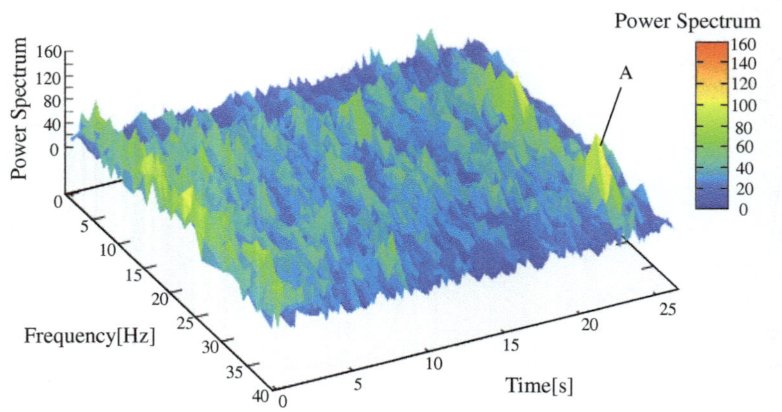

Figure 6 Results of short-time FFT for task 2. μ rhythm is suppressed and there are many mixed noises as indicated by 'A' probably caused by eye-blinking artifacts.

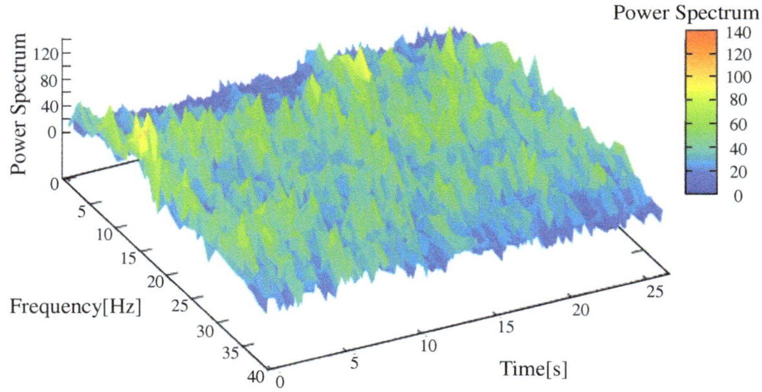

Figure 7 Results of short-time FFT for task 3. μ rhythm is suppressed; meanwhile, β rhythm is generated in motor imagery.

Meanwhile, we can also find that the β rhythm basically increases from tasks 1 to 4. This shows the fact that the spectrum in β band increases according to the excitation degree of the brain. In this study, these two features of μ and β rhythms are used to control the robot.

Experiments

The subjects are two young men. Before experiments, they gave the consent for the study, which was reviewed and approved by the Ethics Committee of our university. In order to get the EEG signals in motion imagery and/or motion, each subject who is wearing a head cap with embedded electrodes is asked to perform the defined four tasks, and his μ rhythm and β rhythm extracted from the EEG are used to control a robot arm.

Control approach

The flowchart of controlling a robot arm using EEG signals is shown in Figure 10. The EEG signals from the electrodes are amplified by an amplifier and input to a personal computer for A/D conversion via an interface board. Then, STFFT is implemented to the converted data series to obtain their spectra $X_{(k,t)}$ (k means frequency band and t indicates time). Further, μ rhythm ($X_{(\mu,t)}$) and β rhythm $X_{(\beta,t)}$ are extracted from the computed $X_{(k,t)}$. The target joint angle θ_d of the robot arm is calculated according to the obtained spectra of μ rhythm($X_{(\mu,t)}$) and β rhythm $X_{(\beta,t)}$ with the following formula,

$$\theta_d = K_\mu \left(1 - \frac{X_{(\mu,t)}}{X_{(max,t)}}\right) + K_\beta \frac{X_{(\beta,t)}}{X_{(max,t)}} \tag{6}$$

where $X_{(max,t)}$ is the maximum spectrum in the frequency band 1 to 30 Hz during the time duration t, and K_μ and K_β are the weight coefficients for μ and β rhythms, respectively. Since the spectrum magnitudes of μ rhythm and β rhythm are different and the spectrum magnitude of β rhythm is much less than that of μ rhythm, the two different weight coefficients respectively for μ rhythm and β rhythm are necessary. Further, in the measured EEG signals, there are a lot of noises such as artifacts. These contaminated EEG signals in μ and β bands may lead to a sharp change of the target angle θ_d. This would result in

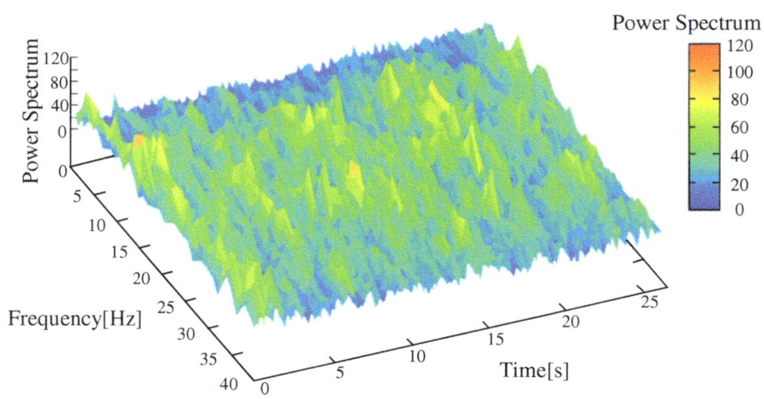

Figure 8 Results of short-time FFT for task 4. μ rhythm is suppressed and β rhythm is greatly produced in motion and motion imagery.

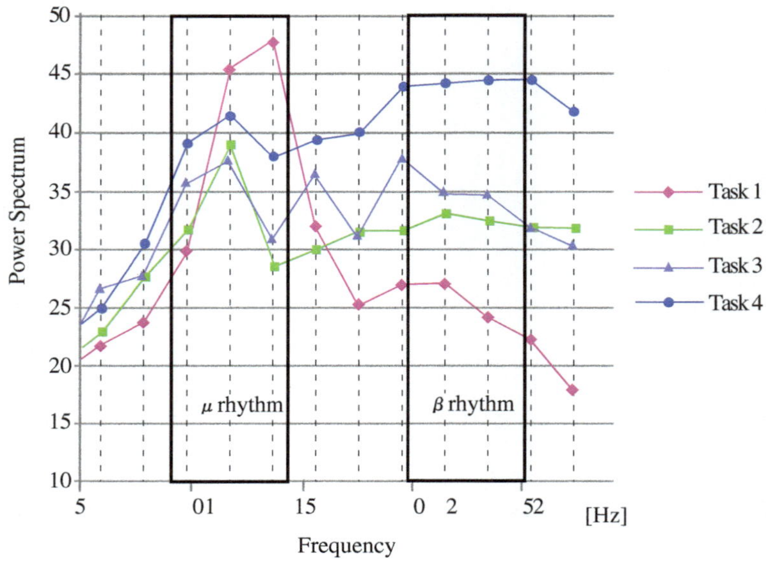

Figure 9 Averages of normalized power spectrum of each task in 24 measurement times for 30 s.

an abrupt movement of the robot arm. To prevent this, a 5-Hz low-pass filter is employed to smooth the output of the target angle θ_d. Moreover, a threshold is set to the filtered target angle θ_d to avoid frequently trivial motions of the robot arm. Finally, this target angle θ_d is used as the command for position control of the robot arm.

Experimental procedures

The experimental procedures are basically same as the steps for four tasks described in Section 'Task design for motion imagery and motion', in which the subjects are asked to perform each task for 38 s. The EEG is measured between $C_4$to A_2 and the reference point is C_z. The ground of the measurement system is right wrist. This ground is separated from the ground of the motor driving system to prevent the subject from the possible electrical shock. The angle of the robot arm is controlled to investigate the performance of each task. The experiment system is shown in Figure 11.

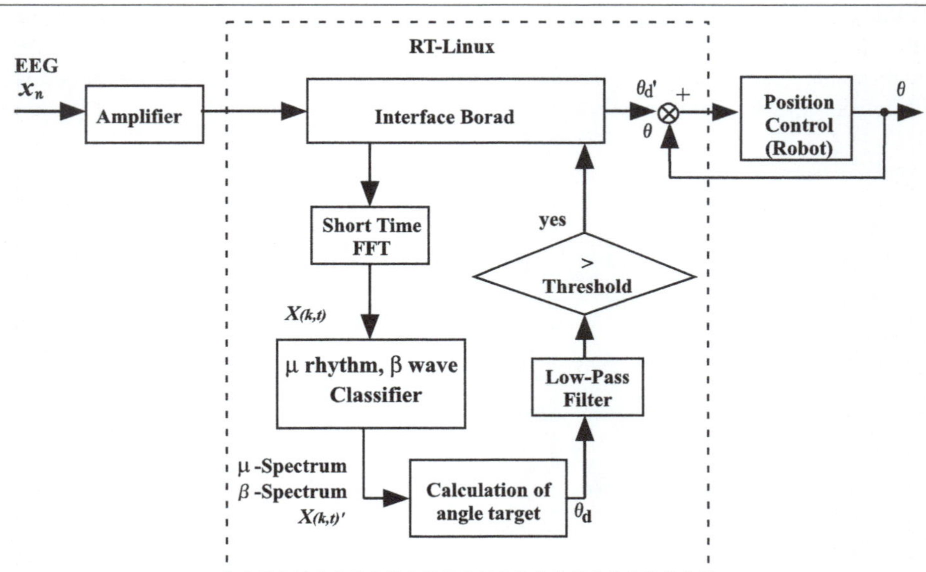

Figure 10 Control flowchart.

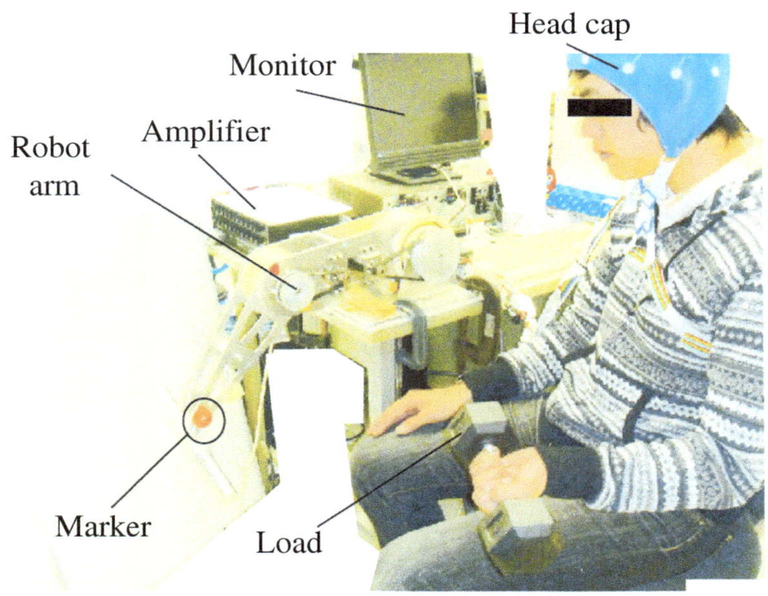

Figure 11 Experimental system.

Results and discussion

Experimental results

The experimental results of subjects A and B are shown in Figures 12 and 13, respectively. In the two figures, the horizontal axes indicate time and the vertical axes represent the controlled angle of the robot arm. The experimental time from the start is 38 s for all of four tasks.

Remarks on experimental results

Figures 12 and 13 show the experimental results of controlling the robot arm using extracted EEG features from the designed four tasks. We find:

- In task 2 and task 3 of subject A (Figure 12), from start to about 12 s, the angle of the robot arm in task 3 is larger than that in task 2. However, after that, this difference has disappeared. Further, after 21 s, the situation is reversed, that is, the angle of the robot arm in task 3 decreases to be less than the angle in task 2. For subject B (Figure 13), from start to about 9 s, the angles of the controlled robot arm in task 2 are almost the same as the angles in task 3. But from 9 to 23 s, the angle of the robot arm in task 3 is larger than that in task 2. However, after 17 s, the angle in task 3 decreases and then has a slight rise.
- The reason for such changes could be interpreted as follows. Task 3 is mainly related to the motion imagery and such kind of action of the subjects to perform motion imagery is greatly dependent on his concentration. It could be postulated that after 12 s

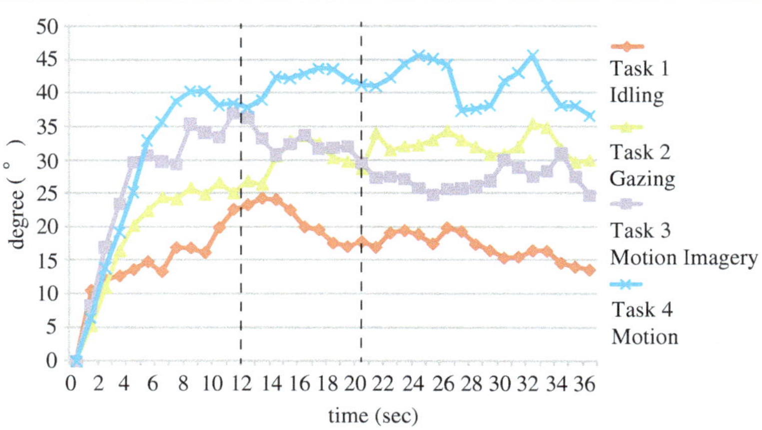

Figure 12 Experimental results of subject A.

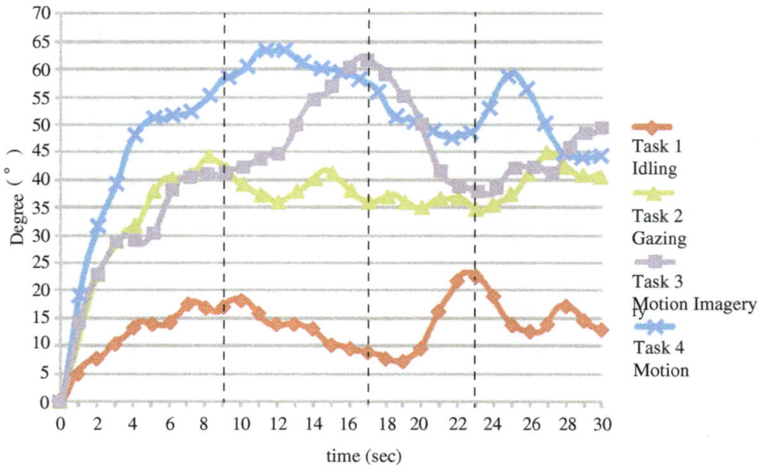

Figure 13 Experimental results of subject B.

(subject A) or 17 s (subject B), the concentration of the subjects starts to decline and after 21 s, the subject A's concentration further recedes. Though we could not observe the similar phenomenon for subject B, it could be considered as an individual difference.

- The experimental results tell us two facts. One is that the subject training for motion imagery is necessary to keep his/her concentration. On the other hand, long-time motion imagery may not be a good way for robot control or external device control since the subject could not keep his concentration over dozen seconds.
- For both subjects A and B, different from the results of tasks 2 and 3, during the entire experiment duration, the angles of the robot arm in the task 1 are almost always the minimum in all of four tasks. Contrarily, the angles corresponding to task 4 are basically always the maximum in all of four tasks. In other words, compared with the results in tasks 2 and 3, the angles of the robot arm in tasks 1 and 4 are more stable. This phenomenon is meaningful.
- The result of the task 1 is natural because the brain excitation of the subject is the lowest in all of the four tasks. Though it is comparatively stable, this task has no practical application value since the subjects in this task have not done and thought anything.
- However, in task 4, the subjects are holding a 4 kg dumbbell in 90° angle of his left elbow joint while gazing at the marker at the endpoint of the robot arm. Since their brain excitation is the highest, the angles of the robot arm are basically always the maximum in all of four tasks. This result tells us such a fact that the EEG signals of the subjects are comparatively stable when the subjects are holding or touching something. Furthermore, this fact implies that EEG signal can be used for robotic power assist to help a subject holding or moving a heavy load.

Conclusions

In this study, aimed to construct a robot control system using EEG signals, four different tasks with different human's brain excitation degree are designed, and their spectra of μ and β rhythms of EEG signals in the four tasks are extracted with STFFT. These spectra are used to control a one-joint robot arm. The experimental results show that the classic motion imagery cannot be simply used to real robot control since the subject's concentration cannot be kept over dozen seconds. Contrarily, since the EEG signals in the task of holding a load are comparatively stable, we conclude that this holding task has a great application potential to be used to construct a robotic power assist system to help person to hold or move a heavy load. This is one of our future work.

Competing interests

The authors declare that they have no competing interests.

Authors' contributions

MY and CZ proposed the basic idea of this study and designed the experiments. Moreover, MY implemented the experiments, analyzed and interpreted the experiment data. He also wrote the draft of this manuscript. CZ revised and complete the manuscript. KI and FW advised the experiment design and the measurement of EEG signals. HY, FD and YY advised the method of signal processing. All authors read and approved the final manuscript.

Author details

[1]Department of Systems Life Engineering, Maebashi Institute of Technology, Kamisadori 460-1, 371-0816 Maebashi, Gunma, Japan. [2]Department of Bioengineering, Faculty of Engineering, National University of Singapore, 9, Engineering Drive 1, 117576 Singapore City, Singapore. [3]Department of Automation, College of Information Technical Science, Nankai University, No. 94 Weijin Road, 300071 Tianjin, People's Republic of China. [4]Department of Bioengineering, School of Engineering, Santa Clara University, 500, El Camino Real, Santa Clara, CA 95053, USA.

References

1. Chapin JK, Moxon KA, Markowitz RS, Nicolelic MAL (1999) Real-time control of a robot arm using simulataneously recorded neurons in the motor cortex. Nat Neurosci 2:664–670
2. Wessberg J, Stambaugh CR, Kralik JD, Beck PD, Laubach M, Chapin JK, Kim J, Biggs AJ, Srinivasan MA, Nicolelic MAL (2000) Real-time prediction of hand trajectory by ensembles of cortical neurons in primates. Nature 408:361–365
3. Nicolelic MAL (2001) Action from thoughts. Nature 409:403–407
4. Taylor DM, Tillery SIH, Schwartz AB (2002) Direct cortical control of 3d neuroprosthetic. Science 296:1829–1832
5. Carmena JM, Lebeddev MA, Crist RE, O'Doherty JE, Santucci DM, Dimitrov DF, Patil PG, Henriquez CS, Nicolelic MAL (2003) Learning to control a brain-machine interface for reaching and grasping by primates. PLOS Biol 1:193–208
6. Lebeddev MA, Nicolelic MAL (2006) Brain-machine interfaces: past, present and future. Trends Neurosci 29:536–546
7. Velliste M, Perel S, Spalding MC, Whitford AS, Schwartz AB (2008) Cortical control of a prosthetic arm for self-feeding. Nature 453:1098–1101
8. Hochberg LR, Serruya MD, Friehs GM, Mukand JA, Saleh M, Caplan AH, Branner A, Chen D, Penn RD, Donoghue JP (2006) Neuronal ensemble control of prosthetic devices by a human with tetraplegia. Nature 442:164–171
9. Wolpaw JR, Birbraumer N, McFarland DJ, Pfurtscheller G, Vaughan TM (2002) Brain-computer interfaces for communication and control. Clin Neurophysiol 113:767–791
10. Wolpaw JR, McFarland DJ (2004) Control of a two-dimensional movements signal by a noninvasive brain-computer interface in humans. Proc Natl Acad Sci U S A 101:17849–17854
11. Dornhege G, Millan J, Hinterberger T, McFarland JD, Muller KR (2007) Towards Brain-Computer Interfacing. The MIT Press, Cambridge
12. Iturrate I, Antelis JM, Kubler A, Minguez J (2009) A noninvasive brain-actuated wheelchair based on a P300 neurophysiological protocol and automated navigation. IEEE Trans. Robot 25:614–627
13. Bashashati A, Fatourechi M, Ward RK, Brich GE (2007) A survey of signal processing algorithms in brain-computer interfaces based on electrical brain signals. J Neural Eng 4:32–57

14

Cell segmentation and pipette identification for automated patch clamp recording

Runhuai Yang[1,2], Chi Ho Tam[1], Ka Lun Cheung[1], Kai Chun Wong[1], Ning Xi[3], Jie Yang[2] and King Wai Chiu Lai[1*]

Abstract

A visual-based approach for identifying living cells and performing the automated patch clamp recording was reported. Based on the image processing and blob detection algorithm, the vision-based method was developed for the detection and identification of biological cells and micropipette. The method was implemented in a micromanipulation system that enabled the identification of the boundary and the center of the target cell and separation from its neighboring cells. The method successfully identified a batch of neuroblastoma cells with the highest yield of 90%. The results demonstrated that the visual-based approach can be integrated to the micromanipulation system to automatically manipulate the patch pipette tip to the center of the target cell, and as a result, the whole-cell recording can be performed precisely and effectively.

Background

Patch clamp technique is one of the promising intracellular bio-electrical signal recording methods for measuring the electric current flowing through ion channels in a cell membrane [1]. It is well understood that the ion channels played important roles in regulating the flow of ions across the cell membrane, which are involved critically in various physiological processes [2]. Therefore, patch clamp recording was widely used for studying the electrophysiological properties of cells [3] and applied for investigating the effect of different drugs on the cell [4]. In addition, patch clamp was specially suitable for high-resolution current measurements under low noise by using the patch clamp pipette and applying a known voltage across the membrane [5]. Therefore, the adoption of the patch clamp technique has led to significant advances in the life sciences and biomedical studies, including the characterization of the properties of ion channels in various types of cells and the pharmacological treatment of diseases implicated with ion channels such as diabetes and heart disease [6] and Parkinson's disease [7].

However, the current patch clamp method required a skilled technician to manually manipulate the pipette, apply the appropriate pressure, and observe the electrical resistance [8]. This complicated operation normally requires several months or even 1 year of training before the technician is able to reliably record from the cells [9] because of the difficulty in obtaining successful recording. This also significantly increases the cost of the whole patch clamp recording process. Due to these challenging issues, several groups started incorporating automated systems to current patch clamping so as to reduce the operation time, complexity, and cost. The automated patch clamp technology was previously reported for the integrative analysis of molecular, anatomical, and electrophysiological properties of a single cell [10]. The planar patch clamp is another kind of patch clamp systems that provides the automatic recording of cells [11]. More recently, the lateral patch clamp device was demonstrated to provide the automatic whole-cell recording [12]. The micro glass pipette has also been used for automated patch clamp [9,13]. Compared to the methods of planar and lateral patch clamp, the micro glass pipette provided better seal formation [14].

The current automated patch clamp system normally applies a negative pressure to catch the suspended cell in a buffer medium. This method causes some problems, for example, the tip of the pipette is blocked by the contaminant in the buffer. Moreover, the automated patch clamp system is currently difficult to fully determine the condition of the cell before catching and selecting the cell for patch clamp recording. Additionally, most of the

* Correspondence: kinglai@cityu.edu.hk
[1]Department of Mechanical and Biomedical Engineering, City University of Hong Kong, Tat Chee Ave, Kowloon, Hong Kong
Full list of author information is available at the end of the article

current automated patch clamps were performed on the suspended cells, and the cells were fixed in the fluid medium by either mechanical method, e.g., cell scrapers or chemical method using trypsin [11], which would change the property of the cells [15,16]. An automated patch clamp is still difficult to be performed on those adherent cells that are grown directly on the substrate. Therefore, it is necessary to develop an automated micromanipulation system for manipulating the glass pipette to perform patch clamp recording on adherent cells.

To overcome the above limitations, we aim at designing a visual-based approach together with an image processing method for the automated patch clamp system. The system will be demonstrated to identify cells and pipettes automatically and perform patch clamp recording non-blindly on adherent cells. The typical process of the patch clamp recording on the adherent cells is shown in Figure 1. In the traditional manual operation, the target cell should be chosen manually. By observing and identifying the cells and glass pipette under the microscope (Figure 1a), the technician can manipulate the glass pipette to the cells as shown in Figure 1b. Here, the visual-based approach is designed to perform this task by the system automatically.

Several methods based on computer vision were reported which have potential to be used for automated patch clamp. The method based on template matching was used to successfully identify sperm cells [17] and other suspended cells [18]. Its concept is to find the sub-region in the input image that best matches the pre-defined template. However, the template of adherent cells is hardly described because the pattern of adherent cells is quite different with each other. It is difficult to use the template-matching method to identify adherent cells with high yield. Feature matching is another widely used matching algorithm to identify cells by matching the nuclei of the cells because the nuclei of same type of cells have similar features [19]. However, the limitation of this method is that the height of the nuclei of the cells should be much higher than other regions of the cells. The yield of identification is strongly affected by the condition of the cells. In this paper, the blob detection will be used as the main method to detect cells and pipettes for patch clamp. Blobs are binary objects or points which are in the same state [20]. Blobs can be classified as different objects based on color, shape, area, perimeter, etc. [21]. This method can thus be applied to identify the target cell based on its own shape and feature and separate the cell from other neighboring cells in the same image. Suspended cells such as red blood cells and white blood cells have been successfully detected and distinguished by using the blob detection method [22]. For detecting adherent cells, the situation is more complicated than the suspended cells because the pattern of adherent cells is varied. Moreover, the adherent cells after several days of culture are grown quite close to each other. Therefore, it is necessary to separate the neighboring cells during the identification process so as to avoid the wrong identification. For pipette detection, some methods were used to identify the pipettes for cell injection [23] or micromanipulation [24]. However, these methods cannot be directly used for the identification of the patch pipette in the patch clamp recording. It is because the pipette and cells are located closely.

To overcome these challenges, a method for detection and identification of a single cell and pipette for automated patch clamp is developed in this paper. More

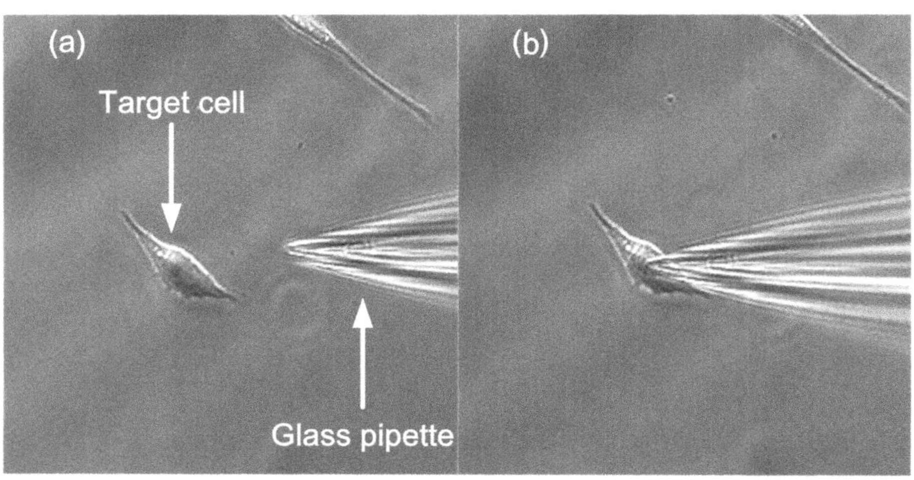

Figure 1 Optical images showing two states of the patch clamp recording process. (a) The initial state before the patch clamp recording, which indicates the initial position of the target cell and patch pipette and **(b)** the pipette is successfully moved to the center of the cell and patched on the cell.

importantly, we report the novel use of the visual-based approach for automatically identifying the target adhered cells and pipettes. Based on the position and feature of the cells and pipette, the condition of the cell can be revealed more clearly, and the cell can be selected for automated patch clamp recording. The method is further combined with the automatic control techniques for performing automated patch clamp on adherent cells, which is different with the current non-visual-based automated patch clamp system.

Methods

Cell preparation

Neuroblastoma cells were employed as the sample cells for the identification and patch clamp recording. The neuroblastoma cells were cultured in Dulbecco's modified eagle medium containing nutrient mixture F-12 (Gibco, Thermo Fisher Scientific Inc., Waltham, MA, USA) with 10% fetal bovine serum (Gibco, USA) and 1% penicillin-streptomycin-neomycin antibiotic mixture (Gibco, USA) and maintained at 37°C in 5% CO_2 humidified air. Stock cultures were passaged at 1:6 weekly and fed twice weekly. The cells were subcultured with the cell density of 100,000 in a culture flask. During the process of subculture, the cells were detached from the culture flask by the treatment of 1 ml trypsin-EDTA (Gibco, USA). Around 5,000 cells were then counted and seeded onto the cover-slip coated with Poly-D-lysine hydrobromide (Sigma-Aldrich, St. Louis, MO, USA) and cultured for 6 days before patch clamp recording. These cells were induced to differentiate by the treatment with 10 μM retinoic acid. All the experiments were conducted at 37°C.

Experimental setup and control strategy

The illustration of the experimental setup for implementing the patch clamp recording based on the visual-based identification is shown in Figure 2. The cells and pipette were first observed under an inverted microscope (Olympus, Shinjuku-ku, Japan). The microscope was equipped with four different objectives with the magnification of × 4, ×10, ×20, and × 40. The live video and digital real-time image were then captured using a CCD camera. The resolution of the image was 1,280 × 768 pixels. Based on the image processing and identification method as described in Figure 2, the coordinates of the cell and the pipette can be obtained from the image, and the movement of the manipulator was then calculated.

To achieve the automated patch clamp, the patch pipette was mounted to a computer-controllable micromanipulator (Sutter MP-285; Sutter Instrument Company, Novato, CA, USA) connected to a controller (Sutter MPC-200, USA). The micromanipulator was controlled to

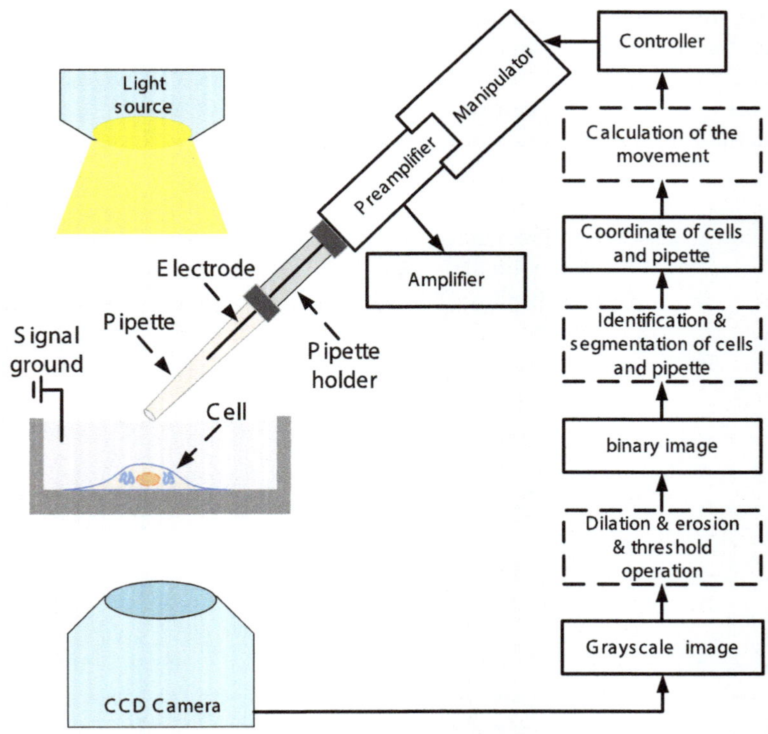

Figure 2 Experimental setup for performing the automated patch clamp based on the developed identification method.

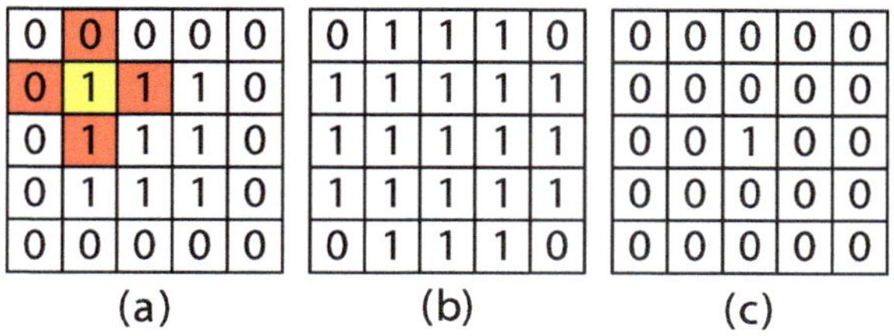

Figure 3 Processes of dilation and erosion operations. (a) A 3 × 3 cross shape structural element applied to a 5 × 5 binary image. Each pixel value in the element represented the intensity. **(b)** Output image obtained from dilation. **(c)** Output image obtained from erosion.

move along the x-, y-, or z-axis with the range from 0 to 25,000 μm. The micromanipulator was then moved to the desired position based on the command received from the controller. The pipette was connected to the manipulator via a pipette holder, and the position of pipette tip was then controlled by the manipulator. The amplifier was used to perform the patch clamp recording. The signal ground of the patch clamp amplifier was connected to the extracellular buffer in the dish. The patch clamp amplifier applied a voltage pulse to the cell and the corresponding current between the electrode inside the pipette and the signal ground was then recorded. By monitoring the current, the condition of the pipette and the cell membrane was determined.

The automatic control strategy can be divided into four steps. In the first step, a digital image containing cells and pipette is captured by the CCD camera, and the system transforms the digital image to the grayscale image. In the second step, the dilation, erosion, and threshold operation are applied to the grayscale image, and the binary image is then automatically obtained. In the third step, the boundary of the cell is identified, and the center of the cell is calculated from the binary image. Besides, the tip of the pipette is also displayed and identified in the binary image. Finally, the coordinate of the center of the cell and the tip of the pipette are obtained. The movement from the tip of the pipette to the center of the cell is calculated based on the updated coordinates, and the command is sent to the controller for moving the manipulator and the pipette tip to the center of the cell. The movement of the pipette tip can be adjusted based on the updated coordinates of the cell and the pipette.

Identification of cells

To perform the detection of the cell and pipette for patch clamp, the digital image of the sample was first transformed to a grayscale image. In the grayscale image, every pixel was represented by an intensity value ranging from 0 to 255. The image was then processed by applying two basic morphological operations, namely dilation and erosion [25]. The concept of the dilation operation was to increase the area of bright regions in the image as the erosion was to increase the area of dark regions in the image. In order to perform the appropriate morphological operation to the image, a structural element, a matrix consisting of only 0 and 1 values, was defined and used. As shown in Figure 3a, a 3 × 3 cross shape structural element was applied to a 5 × 5 binary image, and each pixel value in the element represented the intensity. When the dilation was applied, the maximal pixel value overlapped by the structural element was found to replace the pixel value at the center of the structural element. Therefore, the resulting bright region was increased after performing the dilation as shown in Figure 3b. When the erosion was applied, the minimal pixel value overlapped by the structural element was found, and the pixel value at the center of the structural element was then replaced by this minimal pixel value. Therefore, the resulting bright region was decreased after performing the erosion as shown in Figure 3c.

The process flow of applying the morphological operation to the cell image is shown in Figure 4. As shown in the grayscale image of the cell in Figure 4a, the cell exhibited higher intensity in the region of the cell membrane as compared with that was displayed in other areas of the cell. In order to allow the system to detect and obtain the cell

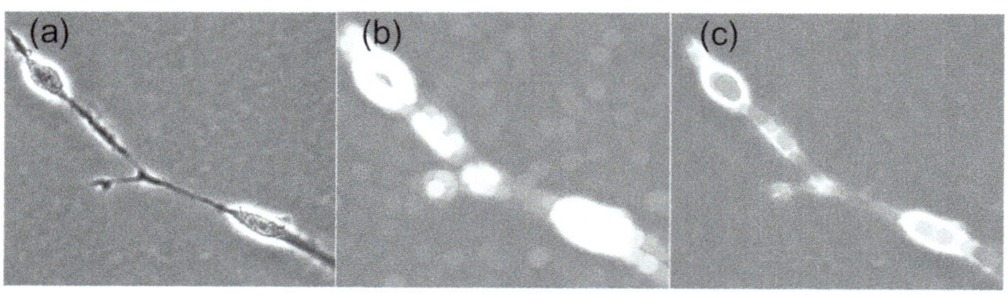

Figure 4 Images obtained before and after dilation and erosion operations. (a) Grayscale image of cells; **(b)** cell image obtained from dilation; and **(c)** cell image obtained from dilation followed by erosion.

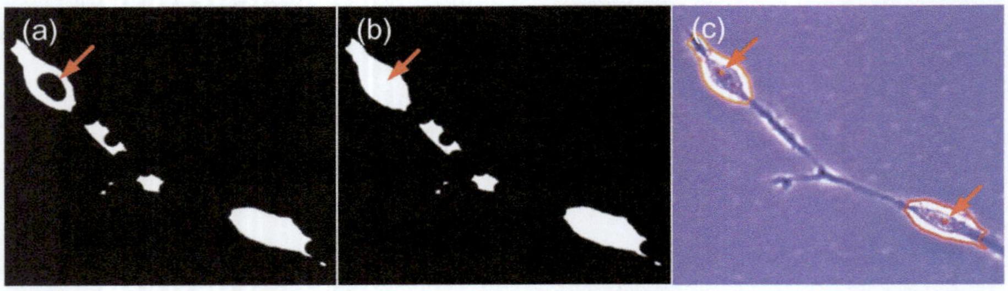

Figure 5 Image obtained by using threshold operations and centers obtained by calculation. **(a)** Cell image after applying threshold operations. **(b)** The black pixels inside the white pixels (indicated by red arrow in **(a)**) are filled. **(c)** Image showing the detected cells (the boundary is indicated by red lines) and their centers are indicated by red arrows.

outline, the dilation followed by erosion was applied to the image as shown in the output image of the cells in Figure 4c, the cell boundary with higher intensity than the background was generated.

After performing the dilation and erosion of the cell image, the cells were further segmented from the background by applying threshold operations to the image. For a grayscale image with intensity value ranging from 0 to 255, pixels with an intensity value that is higher than the threshold were changed to the maximum intensity value, i.e., 255, while pixels with the intensity value equal or lower than the threshold were

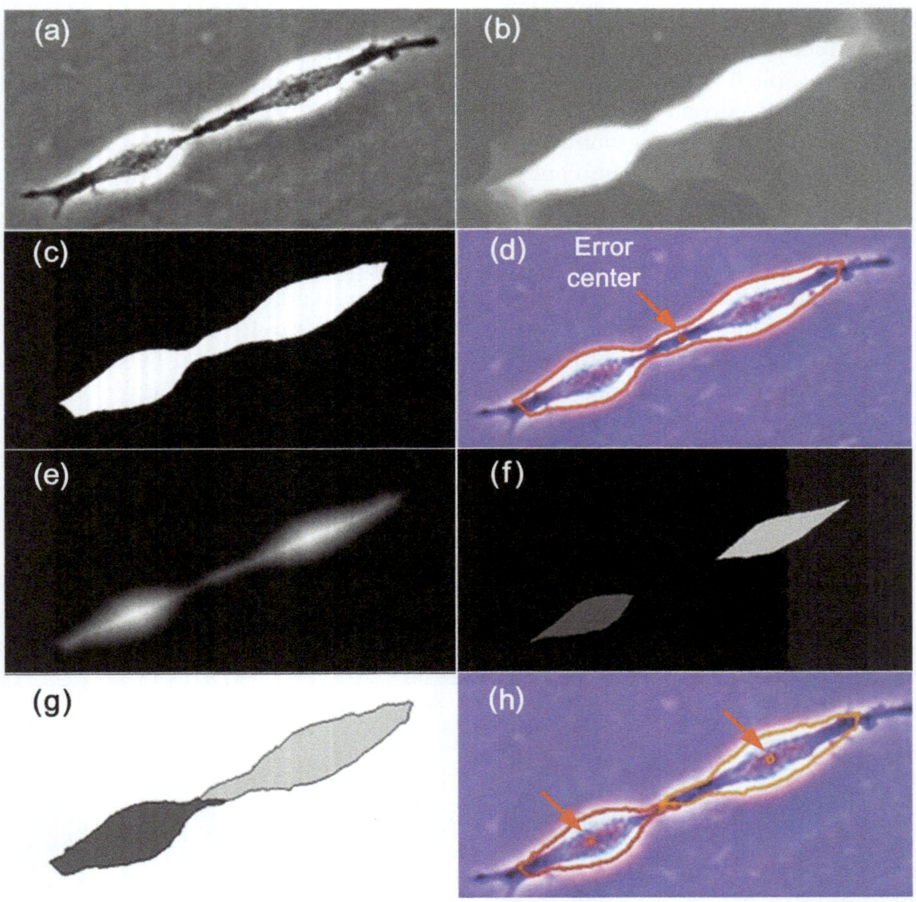

Figure 6 Processes of segmentation of neighboring cells. **(a)** Grayscale image of two adjacent cells which are closed to each other; **(b)** grayscale image, **(c)** binary image, and **(d)** the matching result obtained from morphological operation (dilation followed by erosion); **(e)** distance transform image; **(f)** markers of the cells and background in different gray level values; **(g)** result of applying watershed transformation using the markers; **(h)** corrected identification of the two cells (the boundaries are indicated by red lines and their centers are indicated by red arrows).

0	0	0	0	0	0	0
0	1	1	1	1	1	0
0	1	1	1	1	1	0
0	1	1	1	1	1	0
0	1	1	1	1	1	0
0	1	1	1	1	1	0
0	0	0	0	0	0	0

(a)

0	0	0	0	0	0	0
0	1	1	1	1	1	0
0	1	2	2	2	1	0
0	1	2	3	2	1	0
0	1	2	2	2	1	0
0	1	1	1	1	1	0
0	0	0	0	0	0	0

(b)

Figure 7 Distance transform. (a) Original binary image and **(b)** its distance transform.

changed to 0. This operating process can be represented by the following equation:

$$dst(x,y) = \begin{cases} \text{MaxVal if } src(\text{x, y}) > \text{threshold} \\ 0 \text{ if } src(\text{x, y}) \leq \text{threshold} \end{cases} \quad (1)$$

where *dst*(x, y) represents the intensity value of each pixel in the output image, *src*(x, y) represents the intensity value of each pixel in the input image, and threshold is the threshold value.

After the threshold operation, a binary image was obtained in which the white areas indicated the cell features and the black areas indicated the background as shown in Figure 5a. In the next step, the black areas inside the white area were filled with white color as shown in Figure 5b. The cells were then identified by detecting the white areas.

After identifying the cells, the center of the cell was calculated by using the following equations:

$$C_x = \frac{\text{Sum of x-coordinate within the blob}}{\text{Number of pixels within the blob}} \quad (2)$$

$$C_y = \frac{\text{Sum of y-coordinate within the blob}}{\text{Number of pixels within the blob}} \quad (3)$$

where C_x is the x-coordinate of the center of the cell, C_y is the y-coordinate of the center of the cell. Finally, the membrane and the center of the cells were identified as shown in Figure 5c.

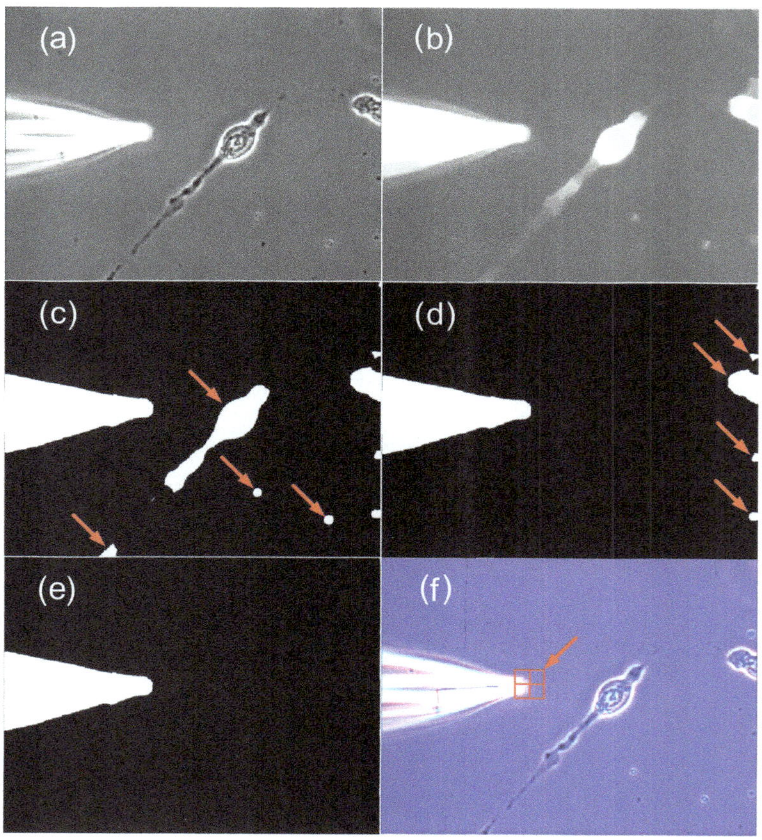

Figure 8 Processes of identification of pipette. (a) Grayscale image of a pipette and cell; **(b)** image obtained after dilation followed by erosion; **(c)** binary image of the pipette; **(d)** the blobs which are not connected to the boundary (indicated by red arrows in **(c)**) are filled with black pixels; **(e)** only the blob which has the longest distance in the x-direction is left; **(f)** the tip of the pipette is successfully identified.

Segmentation of neighboring cells by using distance transform and watershed transformation

Further methods were used to segment the cell from neighboring cells when the cells were too close to each other. As shown in the grayscale image of two adjacent cells (Figure 6a), the two cells were considered as one blob after the morphological operation (dilation followed by erosion (Figure 6b,c), and therefore, the center of the cells was found at the wrong position as shown in Figure 6d.

Two algorithms, distance transform and watershed transformation, were used to further process the image obtained from the morphological operation and segment the connected cells accordingly. The distance transform was first applied to the binary image, and the basic model for applying the distance transform is shown in Figure 7. Consider the binary image as shown in Figure 7a, the foreground was extracted, and distance transform was applied from the boundary of the foreground region in the binary image. Each pixel in the foreground regions in the binary image was replaced by a pixel with gray value that was determined from the computed distance between that pixel to the closest boundary in its foreground region [26] (Figure 7b). Based on this method, the binary image of the cells after distance transform was obtained as shown in Figure 6e.

Threshold operations were then applied to the distance transform image. After the distance transform image was obtained (shown in Figure 6e), another binary image can be generated by the threshold operations. The pixels having intensity value equal or lower than the threshold were changed to 0. Therefore, the area of the cells in the binary image generated by the threshold operations (shown in Figure 6f) was smaller than the area of the cells in the distance transform image (shown in Figure 6e). These smaller foreground regions represent the markers of the cells as shown in Figure 6f. Moreover, after the binary image was generated by the threshold operations, markers can be generated from the different blobs in the binary image. After obtaining the markers of the cells and the background as well as the gradient image of the cells as shown in Figure 6f, the watershed transformation [27] was used, and the markers were colored with different gray levels, i.e. white color for background, lighter gray color for the upper cell, and darker gray color for the bottom cell. In the watershed transformation, the lighter gray maker spreads over all pixels of the upper cell while the darker gray marker spreads over all pixels of the bottom cell. When different markers merged together, they can be recognized from different sources, and thus watershed lines were drawn to separate them as shown in Figure 6g. The cells were finally segmented and identified as shown in Figure 6h.

Identification of pipettes

The method for detection of pipette was developed when the pipette and cells appeared in the same image. The pipette was first calibrated to achieve two conditions

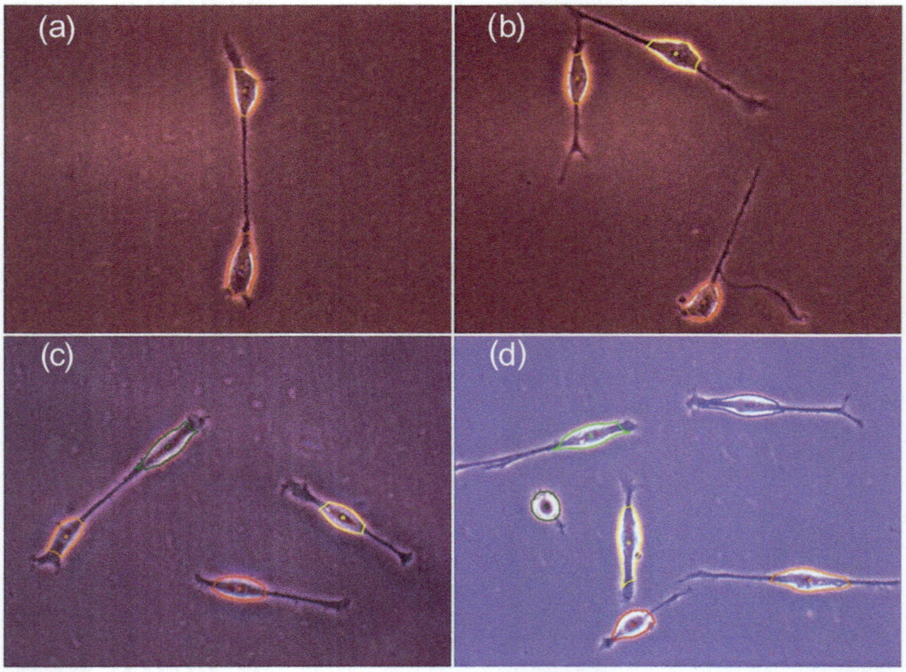

Figure 9 Different shapes and numbers of cells were successfully identified with the marked cell centers and boundaries (a-d). The boundary of each cell is indicated by the line in different colors.

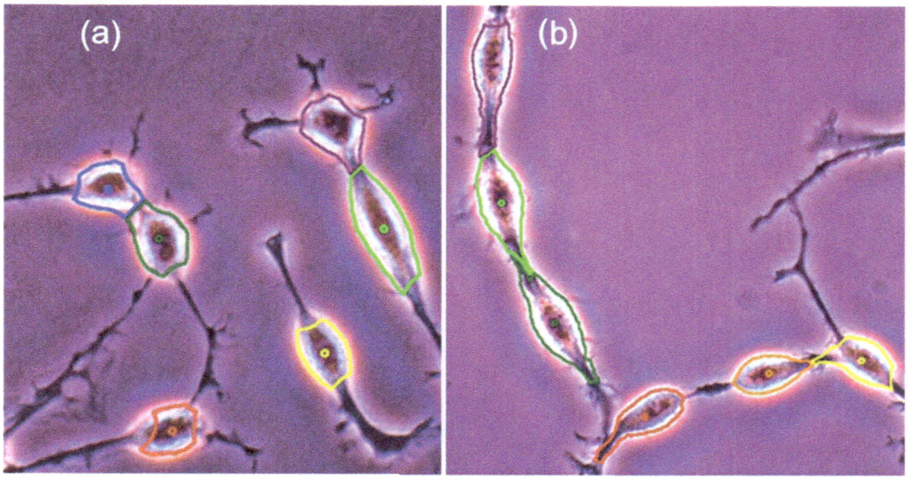

Figure 10 Segmentation of different groups of adjacent cells (a-b). The boundary of each cell is indicated by the line in different colors.

for the identification. Firstly, the pipette should only be moved along the x-direction. Secondly, the length of the pipette on the x-direction in the image should be longer than one-third of the image. The grayscale image of the pipette and the cells is shown in Figure 8a. As similar to the method for identification of cells, the dilation followed by erosion was used. The threshold operation was then used to obtain the binary image. After the binary image was obtained as shown in Figure 8c, all blobs of white pixels that were not connected to the left and right edge of the image were filled with black pixels as shown in Figure 8d. After this operation, the cells which are connected to the edge still appeared in the image. Afterwards, the blobs were filled by black pixels except the block that has the longest distance in the x-direction as shown in Figure 8e. As from the prior calibration, the length of the pipette is longer than the length of the cell. The situation of some neighboring cells was also avoided by the operation of distance transform. The boundary of the pipette was identified, and the position of the tip of the pipette was determined.

Results and discussion

Identification of different shapes of cells

To further verify the identification method, cells in different shapes and sizes were tested and after obtaining the image of the cells, the detection and identification methods were applied to the images accordingly. As shown in Figure 9, different numbers and shapes of the cells were successfully identified, where their corresponding center and boundary were located and marked.

The identification and segmentation of the neighboring cells were also tested. The combination of the distance transform and the marker-based watershed transformation were applied to the images. The adjacent cells that were close to each other were successfully segmented, and the resulting images are shown in Figure 10.

Yield of cell identification

As the structural element and threshold value are the key parameters for the successful identification of cells, a batch of cells was tested independently, and a certain range of size of structural element and threshold values were applied. Experiments were conducted on 95 cells with the applied threshold value ranging from 130 to 250. The experiment was repeated for the changing applied size of structural element ranging from 31 to 61. A cell was counted as successfully identified when its center and boundary were accurately located from the output image. The yield of cell identification is shown in

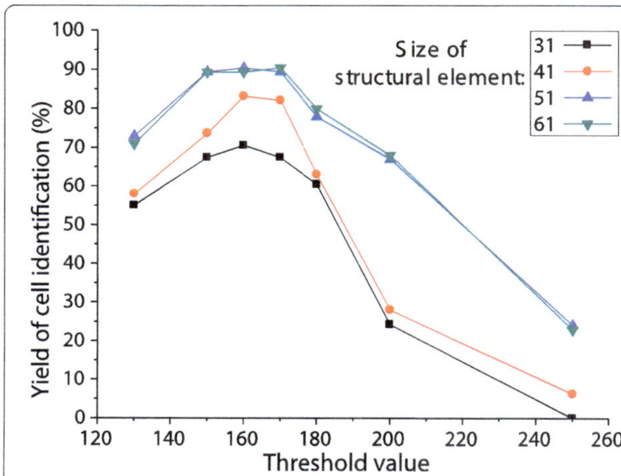

Figure 11 The yield of identification of cells on different sizes of structural element and threshold values.

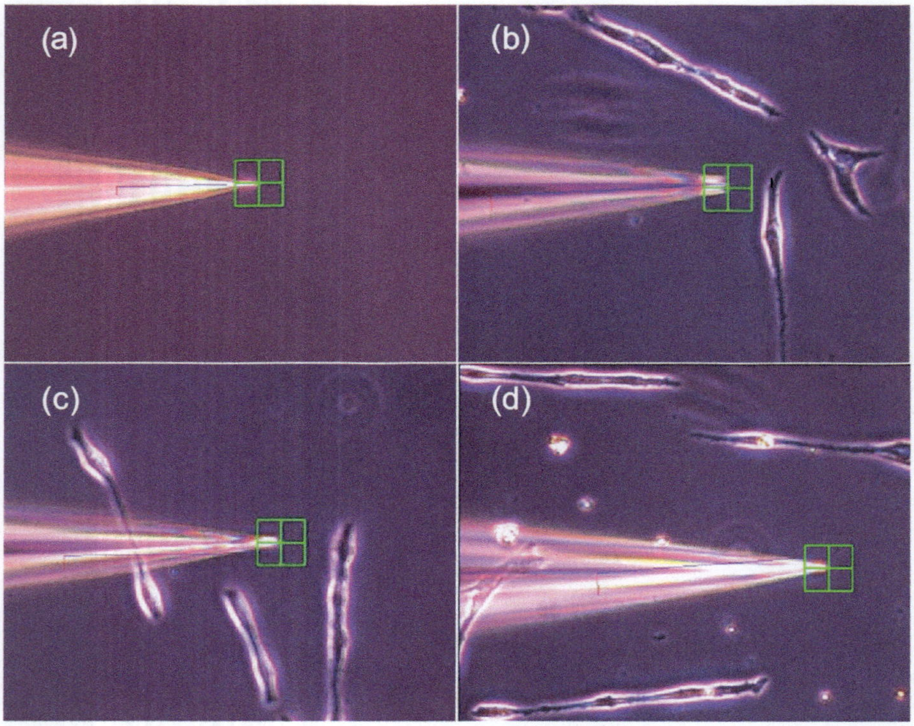

Figure 12 The images showing the successful identification of the pipette under different situations. The detected pipette was marked by the green box. **(a)** Identification of pipette without any cells. **(b)** Identification of pipette when couple cells are located around the pipette. **(c)** Identification of the pipette when couple cells are overlapped with the pipette. **(d)** Identification of the pipette when a batch of cells connected to each other.

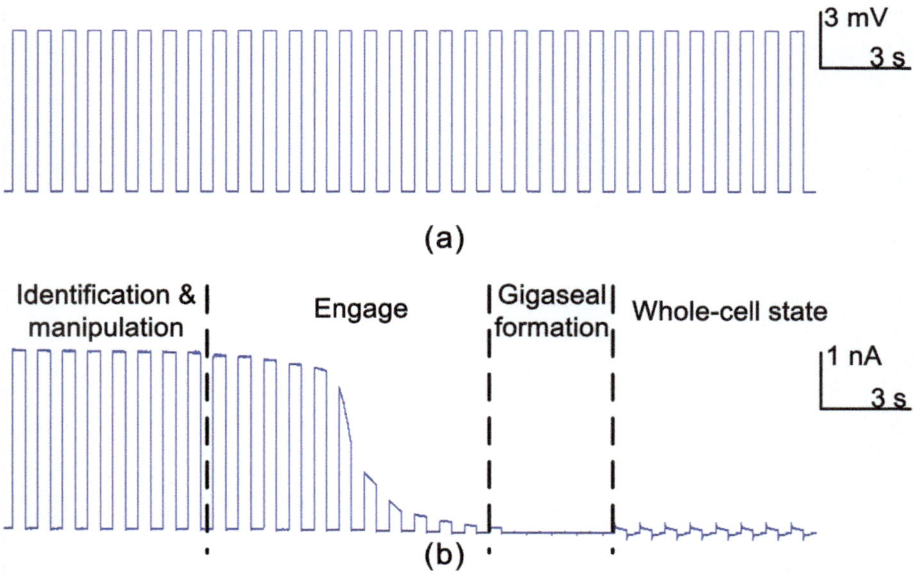

Figure 13 **Patch clamp recording: (a) applied voltage pulse and (b) the induced transmembrane current.**

Table 1 Comparison of the gigaseal formation in manual and automated patch clamp process

	Conditions of identification of the cells	Yield of gigaseal formation (%)
Manual process	One cell in the image can be identified at a time	69
Automated process	Ninety percent of the cells in the image can be identified at a time	72.4

Figure 11. The results indicated that when the threshold value was 160 to 180 and the size of structural element was higher than 51, the highest yield of identification which was equal to 90% was obtained.

Identification of pipettes

Experiments were conducted for testing the identification of pipettes under different test conditions. As seen from Figure 12a, the individual pipette was successfully identified. When the cells were near the pipette, the system was also able to detect the pipette as shown in Figure 12b. As shown in Figure 12c, a more complex situation was tested in which the pipette was overlapped by the cells. When a batch of cells was connected and located near the pipette, the pipette was also successfully identified as shown in Figure 12d.

Validation

Based on the experimental results showing that the cells and pipettes were successfully identified by using the developed approach, the method was integrated to a micromanipulation system. After detecting and locating the cells and pipette, the manipulator of the system was controlled to move the tip of the pipette to the center of the cell for patch clamping. An automated patch clamp program that was previously developed by our group was used [28], and the whole-cell patch clamp recording was then performed. A voltage pulse with the peak voltage of 10 mV (Figure 13a) was applied and the corresponding transmembrane current was monitored (Figure 13b), and thus the state of patch clamp was indicated. The results indicated that different values of transmembrane current were obtained at different state of the patch clamp process, from identification, engaging, and gigaseal formation to the whole-cell state.

The yield of gigaseal formation in manual process and automated process were analyzed and compared. In the manual process, only one cell was chosen at a time. By using the visual-based approach, around 90% of cells in a digital image were identified at a time. The yield of the gigaseal formation was studied in both manual and automated patch clamp process (29 selected cells for each process). In the manual process, 69% of cells achieved

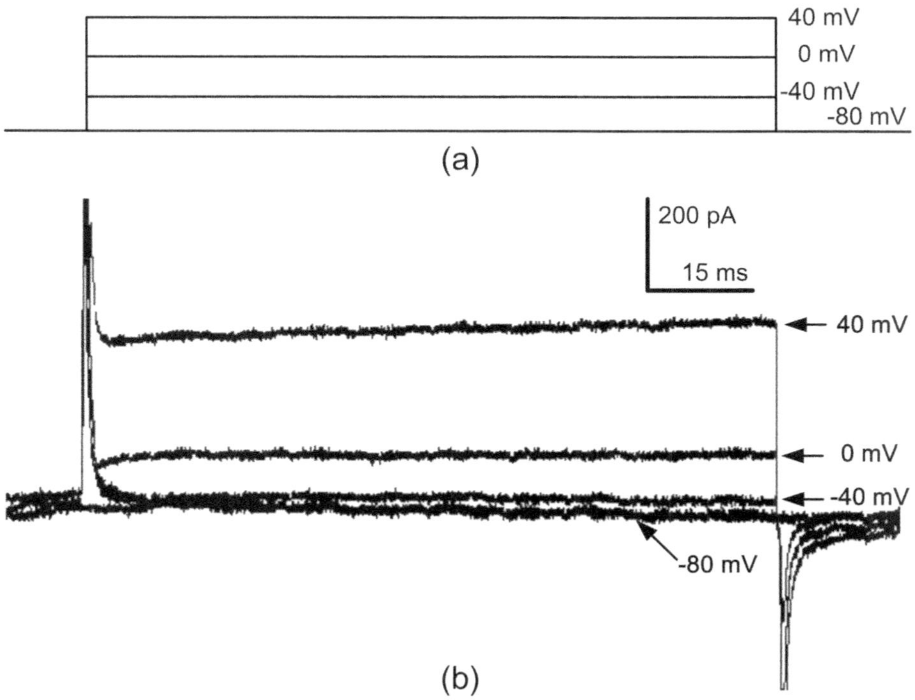

Figure 14 Whole-cell recording of the neuroblastoma cell. (a) Depolarizing voltage pulse with different amplitudes applied to the cells. **(b)** Transmembrane current induced by different depolarizing voltage pulses.

the gigaseal formation, and in the automated process, 72.4% of cells the achieved gigaseal formation. Therefore, the seal condition of the cell by using the automated patch clamp was comparable with seal condition which was obtained by using the manual process (Table 1).

The whole-cell recording was further performed by changing the voltage pulse amplitudes. The cell consists of various voltage-gated ion channels on cell membrane and the transmembrane currents, which indicated the ion channel response, were obtained and compared at different voltage pulse amplitudes. By applying the depolarizing voltage pulse to the cell, the voltage-gated ion channels can be activated and transmembrane current can be recorded. The transmembrane currents were induced using 100 ms depolarizing voltage pulse (−40, 0, 40 mV while holding potential = −80 mV) as shown in Figure 14. A 20 to 30 pA increment of transmembrane currents was obtained between −40 and −80 mV, which showed that the voltage-gated ion channels were not activated at −40 mV. The transmembrane current increased significantly when the voltage pulse was changed to 0 and 40 mV, which showed that the voltage-gated ion channels were activated by these voltage pulses. The curve of transmembrane currents is consistent with the related studies reported in [29], and this further validates the successful patch clamp recording using the method.

Conclusions

The identification method based on blob detection for the detection of cells and pipette was developed and applied for the automated patch clamp. The capability of identifying different shapes of adherent cells and pipettes was demonstrated. By combining the blob detection to distance transform and watershed algorithm, the automatic cell segmentation was realized. The automated patch clamp was implemented with the identification approach successfully. The method presented here provides a visual-based approach to the current automated patch clamp system that further facilitates the patch clamp recording.

Competing interests

The authors declare that they have no competing interests.

Authors' contributions

RY developed the system, carried out the patch clamp experiments, and wrote the manuscript. RY, CHT, KLC, and KCW designed and implemented the visual based method. NX and JY participated in the design of the study. KWCL designed the study, developed the methodology, analyzed the data, and wrote the manuscript. All authors read and approved the final manuscript.

Acknowledgement

The project described is supported in part by GRF grant from the Research Grant Council of the Hong Kong Special Administrative Region Government (CityU139313 and CityU124213).

Author details

[1]Department of Mechanical and Biomedical Engineering, City University of Hong Kong, Tat Chee Ave, Kowloon, Hong Kong. [2]University of Science and Technology of China, Jinzhai RdShushan, Hefei, China. [3]Michigan State University, Trowbridge Rd, East Lansing, MI 48824, USA.

References

1. Jurkat-Rott K, Lehmann-Horn F (2004) The patch clamp technique in ion channel research. Curr Pharm Biotechnol 5(4):387–395
2. Ward JM, Mäser P, Schroeder JI (2009) Plant ion channels: gene families, physiology, and functional genomics analyses. Annu Rev Physiol 71:59–82
3. Kornreich BG (2007) The patch clamp technique: Principles and technical considerations. J Vet Cardiol 9(1):25–37
4. Kaczorowski GJ, McManus OB, Priest BT, Garcia ML (2008) Ion channels as drug targets: the next GPCRs. J Gen Physiol 131(5):399–405
5. Cang J, Isaacson J (2003) In vivo whole-cell recording of odor-evoked synaptic transmission in the rat olfactory bulb. J Neurosci 23:4108–4117
6. Kass R (2005) The channelopathies: novel insights into molecular and genetic mechanisms of human disease. J Clin Investig 115(8):1986–1989
7. Ahmed I, Bose SK, Pavese N, Ramlackhansingh A, Turkheimer F, Hotton G, Hammers A, Brooks DJ (2011) Glutamate NMDA receptor dysregulation in Parkinson's disease with dyskinesias. Brain 134(Pt 4):979–986
8. Zhao Y, Inayat S, Dikin D (2008) Patch clamp technique: review of the current state of the art and potential contributions from nanoengineering. J Nanoeng Nanosyst 222:1–11
9. Kodandaramaiah SB, Franzesi GT, Chow BY, Boyden ES, Forest CR (2012) Automated whole-cell patch-clamp electrophysiology of neurons in vivo. Nat Methods 9(6):1–6
10. Brueggemann A, George M, Klau M, Beckler M, Steindl J, Behrends JC, Fertig N (2004) Ion channel drug discovery and research: the automated Nano-Patch-Clamp technology. Curr Drug Discov Technol 1(1):91–96
11. Fertig N, Blick R, Behrends J (2002) Whole cell patch clamp recording performed on a planar glass chip. Biophys J 82(6):3056–3062
12. Tang KC, Reboud J, Kwok YL, Peng SL, Yobas L (2010) Lateral patch-clamping in a standard 1536-well microplate format. Lab Chip 10(8):1044–1050
13. Vasilyev D, Merrill T, Iwanow A, Dunlop J, Bowlby M (2006) A novel method for patch-clamp automation. Pflugers Arch - Eur J Physiol 452(2):240–247
14. Fejtl M, Czubayko U, Hümmer A, Krauter T, Lepple-Wienhues A (2007) Automated glass pipette-based patch-clamp techniques. Neuromethods 38:435–450
15. Vukicevic M, Weder G, Boillat A, Boesch A, Kellenberger S (2006) Trypsin cleaves acid-sensing ion channel 1a in a domain that is critical for channel gating. J Biol Chem 281(2):714–722
16. Wai Wong C, Dye DE, Coombe DR (2012) The role of immunoglobulin superfamily cell adhesion molecules in cancer metastasis. Int J Cell Biol 2012:340296
17. Ouyang WZPR, Zhang WJ, Gupta MM (2007) Overview of the development of a visual based automated bio-micromanipulation. Int Conf Comput Vision 17(10):147–151
18. Priyankara GPM, Seneviratne OW, Silva RKOH, Soysa WVD, De Silva CR (2011) An extensible computer vision application for blood cell recognition and analysis. IEEE Conf Comput Vision Pattern Recognit 805–902
19. Dewan MAA, Ahmad MO, Swamy MNS (2011) Tracking biological cells in time-lapse microscopy: an adaptive technique combining motion and topological features. IEEE Trans Biomed Eng 58(6):1637–1647
20. Damerval C, Meignen S (2007) Blob detection with wavelet maxima lines. IEEE Signal Processing Letters 14(1):39–42
21. Moon WK, Shen Y-W, Bae MS, Huang C-S, Chen J-H, Chang R-F (2013) Computer-aided tumor detection based on multi-scale blob detection algorithm in automated breast ultrasound images. IEEE Trans Med Imaging 32(7):1191–1200
22. Peng H, Zhou X, Li F, Xia X, Wong STC (2009) Integrating multi-scale blob/curvilinear detector techniques and multilevel sets for automated segmentation of stem cell images. In: Proceedings - 2009 IEEE International Symposium on Biomedical Imaging: From Nano to Macro, ISBI 2009., pp 1362–1365
23. Ouyang PR, Zhang WJ, Gupta MM, Zhao W (2007) Overview of the development of a visual based automated bio-micromanipulation system. Mechatronics 17(10):578–588

24. Becker BC, Yang S, MacLachlan RA, Riviere CN (2012) Towards vision-based control of a handheld micromanipulator for retinal cannulation in an eyeball phantom. In: Proceedings of the IEEE RAS and EMBS International Conference on Biomedical Robotics and Biomechatronics., pp 44–49
25. Mason S (2003) Computer vision and image processing. ISPRS J Photogramm Remote Sens 48(2):24–27
26. Holzer S, Hinterstoisser S, Ilic S, Navab N (2009) Distance transform templates for object detection and pose estimation. In: 2009 IEEE Computer Society Conference on Computer Vision and Pattern Recognition Workshops, CVPR Workshops 2009., pp 1177–1184
27. Couprie M, Bertrand G, Psi L, Esiee G (2003) Topological grayscale watershed transformation. SPIE Vision Geometry VI Proceedings 3168:136–146
28. Yang R, Lai KWC, Xi N, Yang J (2013) Development of automated patch clamp system for electrophysiology. IEEE Int Conf Robotics Biomim ROBIO 2013:2185–2190
29. Reeve HL, Vaughan PF, Peers C (1992) Glibenclamide inhibits a voltage-gated K+ current in the human neuroblastoma cell line SH-SY5Y. Neuroscience Letter 135(1):37–40

Terrain perception for a reconfigurable biomimetic robot using monocular vision

Arnab Sinha*, Ning Tan and Rajesh Elara Mohan

Abstract

This paper presents a reconfigurable biomimetic robot which is able to crawl and roll. The robot mimics the morphology of a huntsman spider that can transform between crawling and rolling by reconfiguring its legs. Terrain perception for reconfigurable biomimetic robots has not been studied in literature. This work tends to perceive and segment the terrain when the robot is crawling or in the steady state between rollings. A remote control system is designed with a server-client mechanism which can perform real-time image processing with GPGPU coding and develop a probabilistic framework for terrain perception. For validation, we test the system in both an indoor lab environment and a more uncontrolled outdoor environment. The results suggest that the system provides a trustable performance.

Keywords: Reconfigurable robot; Biomimetics; Terrain perception; Unsupervised terrain classification; Mean shift

Background

Reconfigurable robots capable of performing multi-state locomotion offer enormous potential with their versatility, fault tolerance, and efficiency for a variety of rugged missions in real world. There have been a lot of works on reconfigurable robotics [1-5].

The design of reconfigurable robots can be inspired by nature. Some attempts tried to develop reconfigurable biomimetic robotic platforms with rolling and crawling capabilities in the cases of BiLBIQ [6]. However, these efforts had been focused completely on mechanism design with almost no effort associated with perception or autonomous features. However, real-world deployments of these reconfigurable robots often require some intelligent capabilities, such as terrain perception and understanding, that go beyond purely mechanism design. Unfortunately, integrating complex reconfigurable design mechanisms with perception introduces a lot of new research challenges.

In terrain perception field, researchers choose in general two directions, namely, predictive and reactive terrain perceptions. Predictive terrain perception is done in a pre-entry manner, that is, the robot starts to perceive the terrain in front of itself. On the other hand, reactive terrain perception considers post-entry classification or recognition of terrain on which the robot perceives. In the case of reactive terrain perception, usually, researchers use inertial measurement unit (IMU) like sensors as described in [7]. The authors have modeled the vibration to classify terrain. Amongst other important works, [8-12] are notable. There are multiple sensors used in all these works, such as gyros, accelerometers, encoders, motor current and voltage sensors, multi-axis force sensor, tachometer etc. The basic disadvantage within reactive terrain perception methologies is that before entering a terrain, the robot cannot do the processing.

On the other hand, for path or trajectory planning, predictive terrain perception is required and usually an elevation map is constructed. Some important works using this approach are [10,13-15], etc. In predictive terrain perception, usually, the sensors used are stereo-camera, laser, IR sensor, etc. These sensors can provide a 3D point cloud or a 2D point cloud along with (or not) a camera color information.

Iagnemma et al. in [10] had used both reactive sensors and predictive sensors for the estimation of two key terrain parameters, cohesion, and internal friction angle. Whereas, Ojeda et al. in [11] used both types of sensors for both terrain characterization and classification. Every sensor modality was trained/learned with respect to different

*Correspondence: arnab_sinha@sutd.edu.sg
Singapore University of Technology and Design, 20 Dover Drive, Singapore, Singapore

kinds of terrains with a neural network and henceforth can be used with some error in terrain classification or characterization.

A particular work which is worth mentioning was done by Fukuoka et al. [16]. The authors have tried to emulate the muscle structure through a bio-inspired mechanical design and to provide flexibility with respect to the terrain characteristics.

In the case of predictive terrain perception algorithms, there are again two broad classes, namely, supervised and unsupervised. Mostly, the works, such as [17-19], etc., are supervised. Like our work, [19] also uses only one camera. However, we try to move our algorithm from supervised texture classification to unsupervised terrain perception. Moreover, we concentrate our terrain perception algorithm on a single camera-based algorithm, which do not have any localization information, no mapping information, and no geometry information (i.e., no 3D/2D point cloud).

Recently, a family of reconfigurable robotic platforms (i.e., Scorpio) capable of crawling and rolling locomotion have been developed. These robots mimic the morphology of a huntsman spider that can transform between crawling and rolling by reconfiguring their legs. In this paper, we present our effort to achieve terrain perception in the Scorpio robot using one monocular camera.

The rest of the paper is organized as follows. The 'Design of the robot' section describes the design specification of our robot. Thereafter, in the 'Terrain perception' section, we describe our methodology, whose results are described in the 'Results and discussion' section. Finally, the 'Conclusions' section concludes the paper with discussions and future research directions.

Huntsman spider

The robot to be discussed in this paper is inspired by a species of huntsman spider, *Cebrennus rechenbergi*. As shown in Figure 1, other than crawling, this kind of spider is also able to roll. The rolling locomotion of such spider was discovered by Ingo Rechenberg from TU Berlin [6]. The habitat of *C. rechenbergi* is the sand dunes of the Erg Chebbi desert in Southern Morocco, boundary to the Sahara Desert.

Normally, the spider crawls with eight legs. However, if provoked or threatened by an external stimulus, the spider can escape by doubling its normal crawling speed using forward or backward flips similar to acrobatic flic-flac movements used by gymnasts with the use of its eight legs simultaneously. What is the most curious thing is that the spider turns somersaults to move independently from surrounding conditions, which means that it does not need a slope to initiate the rolling process by using the gravitational force or does not need to walk a little first or perform a startup gesture to trigger the rolling locomotion.

So far, it could also be observed that only if very certain situations occur, the spider starts to switch from a normal crawling locomotion to this unique rolling locomotion in a somersaulting manner. Such situations might be the appearance of a predator, for example, the fennec

Figure 1 Huntsman spider performing crawling and rolling locomotion.

fox and sand cat, or meeting a conspecific. But it has not been researched that whether the sex of the conspecific plays a role or not. Besides, the circumstances that the spider makes use of the rolling locomotion, for instance, to change positions, to hunt down its prey, or to search for its tunnel, could not be observed either unfortunately.

Methods

Design of the robot

The section presents the mechanical design and system architecture of the Scorpio robot.

Mechanical design

The design of Scorpio robot is based on the real huntsman spider introduced above, which is capable of crawling and rolling. The huntsman spider has eight legs, but we simplify the design to four legs which are sufficient to perform crawling and rolling.

Figure 2 contains a part-by-part view of Scorpio robot showing the assemblies. It is observed that the Scorpio robot consists of four legs (tibia), four servo covers and joints (femur), four main joints (coxa), and a body. The processor, controller, and sensors are placed inside

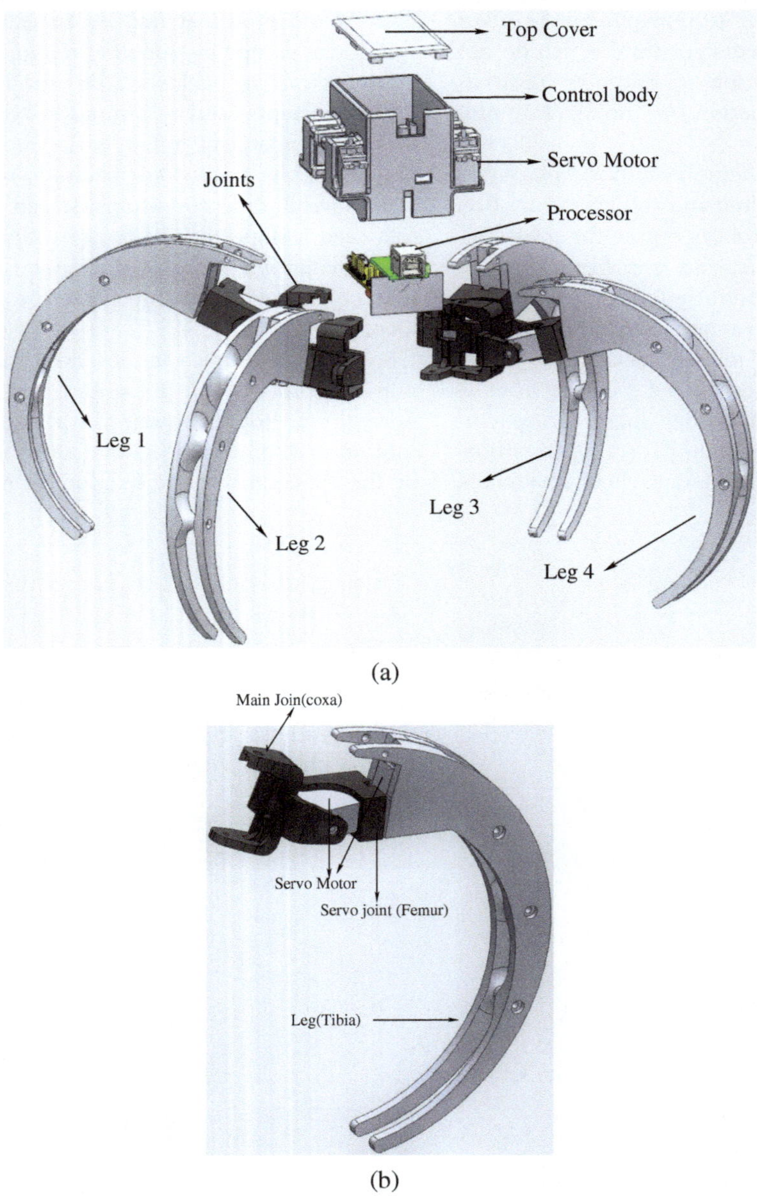

Figure 2 Design of Scorpio robot in exploded view with assembly parts. (a) Robot parts. **(b)** Robot leg parts.

the body which is made from PLA plastic. Twelve servo motors are used in this Scorpio robot to generate locomotion. Each leg is mounted with three servos, so it has 3 degrees of freedom. These legs are able to rotate and transform from crawling to rolling gaits. The specifications of the Scorpio robot are listed in Table 1.

For crawling motion, the Scorpio robot opens up its four legs as shown in Figure 3a. The crawling involves 2 degrees of freedom. Transformation from crawling pose to cylindrical exoskeleton for rolling requires a motion of 3 degrees of freedom. The Scorpio robot uses its legs to push from the ground and shift the center of gravity to achieve the rolling motion with 1 degree of freedom. The rolling speed of the Scorpio robot doubles the rate of crawling speed.

System architecture

For the autonomous movement and reconfiguration, we need to perceive a terrain. Since the platform in this work is a lightweight and small-sized (around 15 cm) robot, and the robot should reconfigure during movement, one cannot place a laser sensor or any kind of range sensors because usually they are too heavy. Moreover, ground-level sensors provide some good data with respect to physical interactions. Based on these perspectives, we choose to build up a vision system, which can provide reliable data for terrain perception. The vision system should be lightweight and small-sized to minimize the influence on the locomotion and control of the robot. Thus, we further alleviate this problem by restricting the vision system with only one camera situated on top of the robot.

In this work, we have chosen a wireless network camera to stream video to a local network computer. Thereafter, we process the frames on this remote processor. The Ai-Ball camera [20] that we use is shown in Figure 4 with the robot. When the robot is crawling (Figure 4a), the camera can always be looking ahead. Whereas in the case of rolling gait (Figure 4b), the view might be blocked when the camera is rotated underneath. The camera may rotate along yaw and pitch rotation axis when the robot is crawling and rolling. Therefore, we cannot have any *a priori* geometry assumption relating the image to the world. After one roll of 360°, the robot gets back to the *stand up* position and checks (for now the decision is coming from human operator) the terrain for further rolling or crawling decision.

In Figure 5, we show the overall architechture of the system. XBee [21] is used for communication between the robot and the remote computer for message passing. The robot has a Arduino Mini [22] controller to drive the servo motors for the creation of the gaits, for different motions and reconfigurations. This controller also sends its current state (according to its movement) to the remote desktop.

The remote desktop receives information from the robot and camera, respectively, as shown in green arrows in Figure 5. These two information are then managed within ROS [23], for time synchronization, and software stability. After considering all the inputs, if the system tries to take feedback, it asks the human operator and, thereafter, transfers the required morphological gait commands to the robot, as shown in blue arrow in Figure 5. The time synchronization within different inputs and related decision-making process of our semi-autonomous system (developed within ROS) is required for stable robot performance.

Table 1 Specifications of the Scorpio robot

	Specifications
Controller	Arduino Mini Pro 328
Servo motor	JR ES 376
Servo controller	Pololu-Micro Maestro 18-channel USB servo controller
Camera	WiFi Ai-Ball Cam
Battery	LiPo 1,200 mah 7.4 v
ZigBee	XBee Pro S1, Digi International
Full body material	Polylactic acid or polyclactide (PLA)
Dia (while rolling) in mm	168 mm
L × W × H (while walking) in mm	230 mm × 230 mm × 175 mm
Weight (full weight) in g	430 g

Terrain perception

In this section, we describe the image perception/segmentation based on the mean shift algorithm.

Mean shift image segmentation

In this work, the algorithm for semi-autonomy depends on image segmentation, and therefore, we describe earlier works with respect to mean shift-based image segmentation in the following.

Mean shift has been introduced by Yijong Cheng in 1995 [24]. Thereafter, Comaniciu and Meer in [25] popularized the mean shift algorithm by providing a theory for robust feature space analysis. Since then, this algorithm has been used, studied, and further developed by a number of researchers, such as in [26], Han et al. have used it for object tracking. Yang et al. further improved its performance by introducing a new similarity measure in [27]. Some other important works involve image segmentation [28], visual tracing [29], stereo matching [30], etc.

Here, we first describe the mean shift algorithm. Given n data points (or vectors), $\mathbf{x}_i; i = 1, 2, \ldots, n$, where each data vector $\mathbf{x_i}$ lies in a d-dimensional space say R^d. Here, R represents state space of each dimension, i.e., in case of a color channel of a RGB image, $R = \{0, 1, 2, \ldots, 255\}$. The

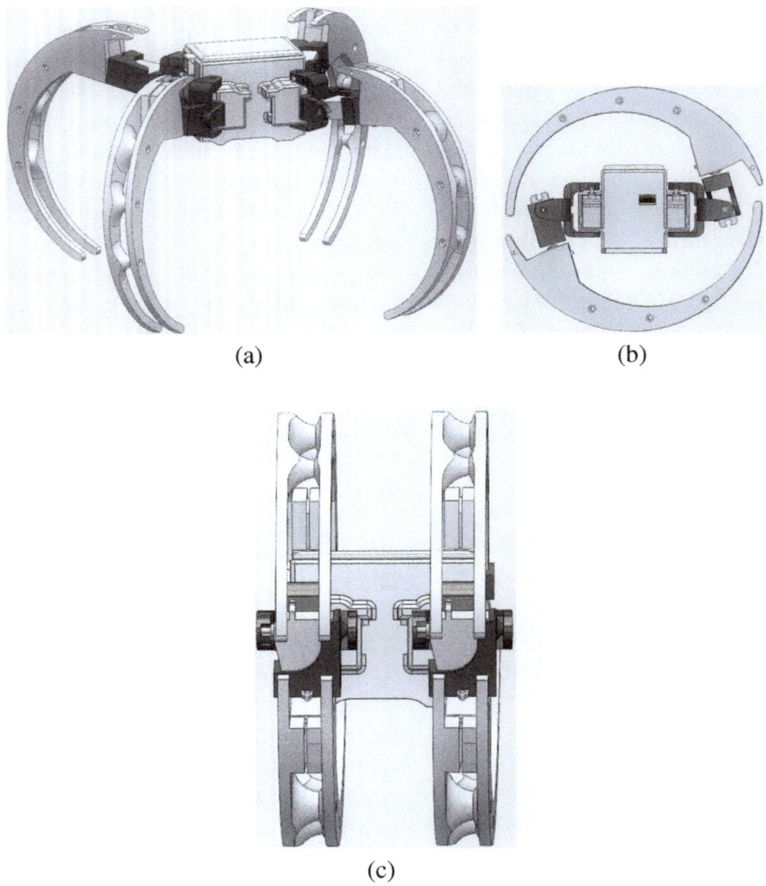

Figure 3 Design of Scorpio robot in crawling and rolling gestures. (a) Crawling configuration. **(b)** Rolling configuration - side view. **(c)** Rolling configuration - front view.

(a) (b)

Figure 4 Scorpio robot with wireless network camera Ai-Ball in different gaits. (a) Crawling gait. **(b)** Rolling gait.

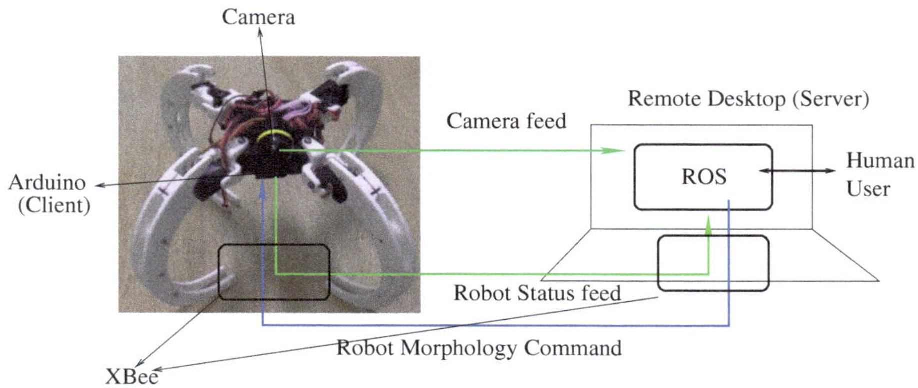

Figure 5 Overall system architecture. The arrows show the flow of information.

multi-variate kernel density estimator with kernel function $K(\mathbf{x})$ and a $d \times d$ bandwidth matrix H, at a point $\mathbf{x}$ can be computed as:

$$\hat{f(\mathbf{x})} = \frac{1}{n}\sum_{i=1}^{n} K_H(\mathbf{x} - \mathbf{x}_i) \tag{1}$$

where, $K_H(\mathbf{x}) = |H|^{-1/2}K(H^{-1/2}\mathbf{x})$. The kernel function $K(\mathbf{x})$ should satisfy the following conditions:

$$\int_{R^d} K(\mathbf{x})d\mathbf{x} = 1 \quad \lim_{||\mathbf{x}||\to \inf} ||\mathbf{x}||^d K(\mathbf{x}) = 0 \tag{2}$$

$$\int_{R^d} \mathbf{x}K(\mathbf{x})d\mathbf{x} = 0 \quad \int_{R^d} \mathbf{x}\mathbf{x}^T K(\mathbf{x})d\mathbf{x} = c_k \mathbf{I} \tag{3}$$

where, c_k is a constant, $\mathbf{I}$ is the identity matrix, and $||.||$ is the norm defined on R^d space. In terms of popularity, product kernel are mostly used due to its mathematical and practical simplicity, it is defined in the following:

$$K^P(\mathbf{x}) = \prod_{j=1}^{d} K_1(x_j). \tag{4}$$

Moreover, we have used Gaussian kernel here, i.e.:

$$K_1(x_j) = \frac{1}{\sqrt{2\pi h_j^2}} \exp \frac{(x_j - x_{i,j})^2}{2\pi h_j^2}, \tag{5}$$

where $x_{i,j}$ is the jth element of data vector $\mathbf{x}_i$, and x_j is the jth element of data vector x, where we are estimating the density. Now, we define function $g(x) = -\frac{dk(x)}{dx}$ as the derivative of the single variable kernel profile. Here, we assume that $g(x)$ exists for all values of $x \in R$. Given the definition of $g(x)$, we define the mean shift in the following equation:

$$\mathbf{m}_{h,g}(\mathbf{x}) = \frac{\sum_{i=1}^{n} \mathbf{x}_i g\left(||\frac{(\mathbf{x}-\mathbf{x_i})}{h}||^2\right.}{\sum_{i=1}^{n} g\left(||\frac{(\mathbf{x}-\mathbf{x_i})}{h}||^2\right.} - \mathbf{x}. \tag{6}$$

In mean shift segmentation, a new data vector is created according to the following equation:

$$\mathbf{x}_i = \mathbf{x}_i + \mathbf{m}_{h,g}. \tag{7}$$

Through iterations of Equation 7, the data vectors converge to its mode, i.e., $\mathbf{x}_i \to M_i$, within the original density function, which signifies a cluster. The convergence depends solely upon the bandwidth parameter h, which is also known as the smoothness parameter. Since, as one increases the value of this smoothness parameter, the convergent mode becomes more global, that is, it smoothes out local perturbations. In this work, we have considered three color channels (RGB) and two spatial channels (XY). Therefore, in our case, $d = 5$, and the state space of color channels is $L = \{0, 1, \ldots 255\}$ and spatial channel is $\Re$. The smoothness parameters for color channels are equal to each other and controlled by one parameter h_C, similarly, the smoothness parameter for each spatial channels is h_S. After convergence, we color the given image according to the average color value of the local mode or cluster. In the next section, we use these identified modes for final terrain perception.

Computational complexity of mean shift algorithm is quite extensive, and hence, it is not so popular within real-time algorithms. Recently, researchers have identified the parallel power of mean shift algorithm, and with the advance of general purpose graphics processing unit (GPGPU) architecture, one can have almost real-time performance with CUDA coding, [31]. In this work, we use this GPGPU mode for computing. The actual computational complexity of this approach comes from the calculation of kernel weight according to given input, i.e., Equation 1. Here, the computational complexity is $O(n)$, for each data $\mathbf{x}$. Therefore at each iteration, Equation 7 also has a computational complexity of $O(n)$ for each data point $\mathbf{x}$. In case of a serial computational algorithm, this computational complexity is scaled up by the number of

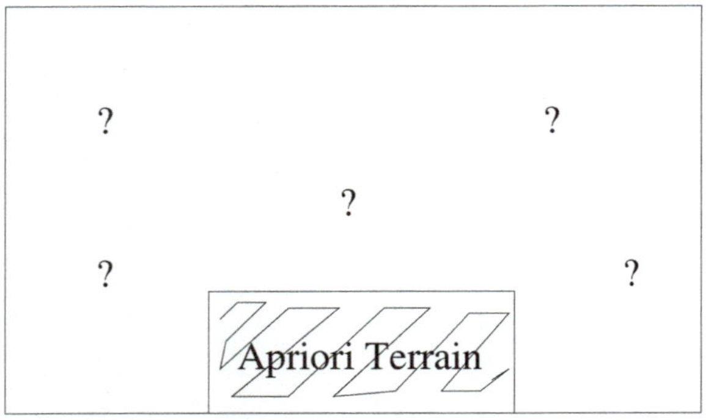

Figure 6 Camera image with a priori terrain region assumption.

data points, whereas in the case of parallel computing, it remains the same. Within an image, the number of pixels represent the number of data points, and henceforth with parallel computing architecture, one can reduce the huge computational burden.

Terrain perception

Unsupervised terrain perception is one of the fundamental requirements of future robotics. For the perception of terrain, we need some *a priori* assumptions on which the algorithm could be based. The two assumptions about terrain are listed in the following:

- The terrain is mostly flat at least locally.
- The robot is standing on a terrain, that is in its immediate front, there is no obstacle.

Given these two *a priori* assumption, we build up a probabilistic framework for the terrain estimation. As shown in Figure 6, we describe our *a priori* assumption about the terrain. Moreover, from sub-section, we know the modes of each pixel, say M_s, where $s = (i,j) \in \Omega$ is a pixel site. Now, we define our *a priori* terrain assumption region as $\Omega_T \subset \Omega$. We are now able to define probability of mode $M_s; s \in \Omega$, according to all the modes defined within Ω_T, as described in Equation 8. Thereafter, we use this likelihood measure to make a decision about whether a particular pixel is representing terrain or not. For this, one can use a hard threshold, i.e., if $P(M_s) \geq T$th, where Tth is the terrain likelihood threshold, we say $s = (i,j)$ represents a terrain pixel. This threshold is estimated according to Equation 9.

$$P(M_s) = \frac{1}{|\Omega_T|} \sum_{s_t \in \Omega_T} K_H(M_{s_t} - M_s) \tag{8}$$

$$T_{th} = \min P(M_{s_t}); s_t \in \Omega_T \tag{9}$$

Once we segment the image pixels in terrain and non-terrain classes, we can further segment non-terrain class into two classes using the above approach. It is assumed that there are regions which are probable terrain next to the terrain class. It is worth mentioning that the robot is not a wheeled robot; hence, the camera can rotate along yaw, pitch, and roll axis. That is, we cannot use vertical lines in the image as those in real world.

Semi-autonomous motion transformation

After terrain classification, we have three classes, namely terrain, probable terrain, and obstacle. Then, we can proceed to semi-autonomous motion transformation. Some terrains which have less friction best suit to robot rolling, whereas there are some rugged terrains where the robot cannot rotate and crawl. Such ruggedness is difficult to estimate just from an image; therefore, we leave this decision to the human operator. As the segmented terrain area in front of the robot becomes smaller, we notify the human operator for suggestions. The human operator then takes a decision to either change crawling direction or roll.

Results and discussion

In our experiments, we have considered two types of flat terrains, since our robot is not capable to move on a sloppy terrain or stair-like terrain or big pebble terrain. One terrain is in indoor environment, using normal carpet, and the other terrain is outside, where a cemented small pebble terrain and a much smoother marble terrain is situated.

Figure 7 Terrain segmentation/perception in indoor environment. Green is a terrain, blue defines a probable terrain, and red signifies an obstacle.

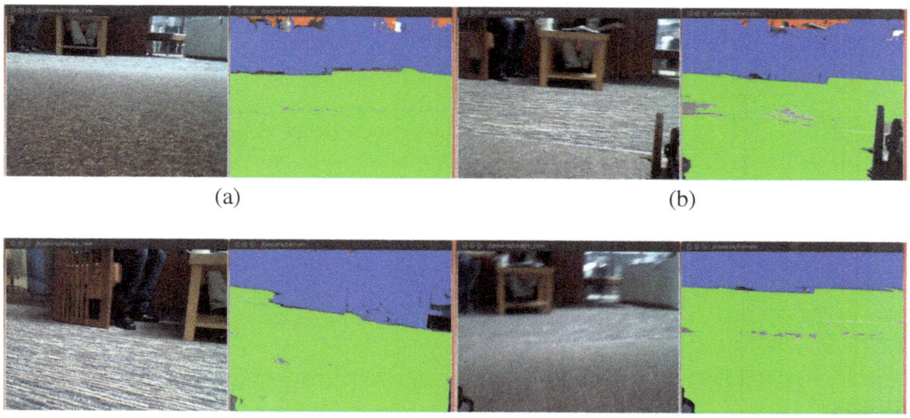

(a) (b)

(c) (d)

Figure 8 Terrain segmentation/perception results of a serial of motions. Green is a terrain, blue defines a probable terrain, and red signifies an obstacle. **(a)** Moving straight. **(b)** Robot leg is not detected as terrain. **(c)** Turned motion. **(d)** Again straight.

There are two main parameters within mean shift segmentation algorithm as mentioned in the 'Mean shift image segmentation' section, i.e., color smoothness parameter h_C and spatial smoothness parameter h_S. For our case of implementation, $h_C = 40$ and $h_S = 40$ are chosen. The image processing speed is less than 20 ms which can be considered to be real time.

The terrain perception can be performed during the crawling motion. When the robot is rolling, the camera is also in rolling state. Thus, the feedback from the camera cannot be considered. What we treat is that when the rolling motion is finished, the robot and camera are in a standing state, so the terrain perception is performed again.

With a single camera and without localization algorithm, it would be very difficult to have any geometrical information of the terrain. Moreover, the selection of the terrain also depends on the robot's capability. In this work,

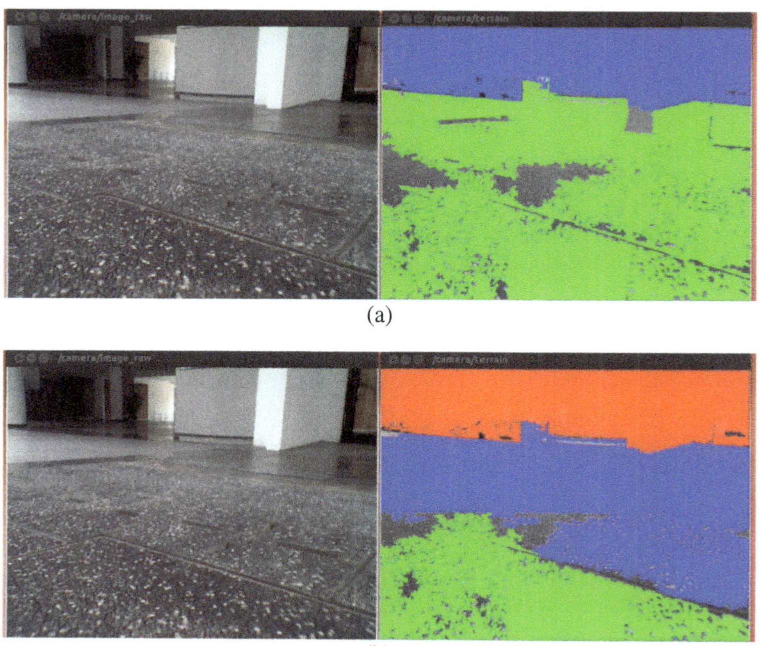

(a)

(b)

Figure 9 Initial results with outdoor terrain perception. Green is a terrain, blue defines a probable terrain, and red signifies an sobstacle. **(a)** Initial result - different terrain texture not segmented. **(b)** Initial result - different terrain texture not well segmented.

for the sake of simplification, the perception experiments are performed on flat terrains, with a rugged one and a smooth one.

Indoor experiment

In an indoor environment, Figure 7 provides an example of the terrain perception result. We colorize the pixels in four classes: green signifies an identified terrain, blue represents a possible terrain, and red is an obstacle. Other pixels are shown with their mode color, since we could not decide whether they belong to any class or not. The same color representation is followed for all other results. As can be seen from the figure, the two different carpets are well segmented. Figure 8 shows the results as the robot is crawling in the indoor environment. Sometimes, the robot legs go in front of the camera. But they are not recognized as the terrain class. From the experiment results, we can observe that in indoor carpet case, the terrain perception works quite good.

Outdoor experiment

Outdoor experiments are more difficult due to a lot of uncontrolled variations and disturbances. Figure 9 shows the results as the robot starts crawling on a more rugged terrain and approaching to a smoother terrain. Here, the segmentation, in between two terrains, are not so good, as depicted in the figure.

In Figure 10, we show the results of our algorithm as it is moving from rugged terrain to a smoother terrain from different angles. Figure 10a shows the motion blur, and most notably, our perception algorithm works quite well here. Figure 10b shows the effect of light reflection on

(a)

(b)

(c)

Figure 10 Effect of motion blur and reflection on terrain perception. Green is a terrain, blue defines a probable terrain, and red signifies an obstacle. **(a)** Effect of motion blur on terrain perception. **(b)** Effect of sunlight reflection on terrain perception. **(c)** Effect of sunlight on terrain perception - best result.

smoother surface. With respect to surface reflection, we show our best result in Figure 10c.

Normal execution time per frame of the video is less than 20 ms, which is real-time. One can also decrease the time of the algorithm, by changing the bandwidth parameter, required for mean shift-based image segmentation, but the results are not very good.

Conclusions

This work reports the study of real-time terrain perception for the reconfigurable biomimetic robot. The robot can mimic the huntsman spider to crawl and roll by its legs. Using the single camera situated on the robot, the robot can perceive different terrains. There are some demerits of our system, which are, our algorithm is not perfect all the time; but with a consensus over multiple test on same scenario, we can stabilize it. The advantages of our algorithm are it is fast, real-time, and especially un-supervised. That is, it needs not any training *a priori* for a particular unknown terrain. The experimental results show that in both indoor or outdoor condition, if the light is not reflected, our algorithm achieves good terrain perception, even when there is a motion blur.

This work is a fundamental step to build more advanced and lightweight reconfigurable biomimetic robots. In the future, we will try to include IMU sensor to extend the perception module with both reactive and predictive sensors.

Competing interests

The authors declare that they have no competing interests.

Authors' contributions

AS carried out the development of the 'Results and discussion' section, and NT has contributed in developing the section 'Mechanical design'. AS and NT has jointly written the 'Background' section. Finally, RME has guided the original motivation and sought through the development of the overall work presented within this paper. All authors read and approved the final manuscript.

Acknowledgement

We hereby acknowledge that this work is supported by the Temasek Laboratory and SUTD-MIT International Design Centre at Singapore University of Technology and Design.

References

1. Nansai S, Rojas N, Elara MR, Sosa R (2013) Exploration of adaptive gait patterns with a reconfigurable linkage mechanism. In: Intelligent Robots and Systems (IROS) 2013 IEEE/RSJ International Conference On. IEEE, Japan. pp 4661–4668
2. Yim M, Shen W-M, Salemi B, Rus D, Moll M, Lipson H, Klavins E, Chirikjian GS (2007) Modular self-reconfigurable robot systems [grand challenges of robotics]. IEEE Robot Autom Mag 14(1):43–52
3. Murata S, Kurokawa H (2007) Self-reconfigurable robots. IEEE Robot Autom Mag 14(1):71–78
4. Moubarak P, Ben-Tzvi P (2012) Modular and reconfigurable mobile robotics. Robot Autonom Syst 60(12):1648–1663
5. Gilpin K, Kotay K, Rus D (2008) Miche: modular shape formation by self-disassembly. Int J Robot Res 27:345–372
6. King RS (2013) BiLBIQ: a biologically inspired robot with walking and rolling locomotion, Biosystems & Biorobotics, vol. 2. Springer. http://www.springer.com/engineering/robotics/book/978-3-642-34681-1
7. Tick D, Rahman T, Busso C, Gans N (2012) Indoor robotic terrain classification via angular velocity based hierarchical classifier selection. In: Robotics and Automation (ICRA), 2012 IEEE International Conference On. IEEE, Minnesota, USA. pp 3594–3600
8. Garcia Bermudez FL, Julian RC, Haldane DW, Abbeel P, Fearing RS (2012) Performance analysis and terrain classification for a legged robot over rough terrain. In: Intelligent Robots and Systems (IROS), 2012 IEEE/RSJ International Conference On. IEEE, Portugal. pp 513–519
9. Rebula JR, Neuhaus PD, Bonnlander BV, Johnson MJ, Pratt JE (2007) A controller for the littledog quadruped walking on rough terrain. In: Robotics and Automation, 2007 IEEE International Conference On. IEEE, Roma, Italy. pp 1467–1473
10. Iagnemma K, Kang S, Shibly H, Dubowsky S (2004) Online terrain parameter estimation for wheeled mobile robots with application to planetary rovers. IEEE Trans Robot 20(5):921–927
11. Ojeda L, Borenstein J, Witus G, Karlsen R (2006) Terrain characterization and classification with a mobile robot. J Field Robot 23:103–122
12. Hoepflinger MA, Remy CD, Hutter M, Spinello L, Siegwart R (2010) Haptic terrain classification for legged robots. In: Robotics and Automation (ICRA), 2010 IEEE International Conference On. IEEE, Alaska, USA. pp 2828–2833
13. Belter D, Skrzypczynski P (2010) Rough terrain mapping and classification for foothold selection in a walking robot. In: Safety Security and Rescue Robotics (SSRR), 2010 IEEE International Workshop On. IEEE, Bremen, Germany. pp 1–6
14. Poppinga J, Birk A, Pathak K (2008) Hough based terrain classification for realtime detection of drivable ground: Research articles. J Field Robot 25(1-2):67–88
15. Manduchi R, Castano A, Talukder A, Matthies L (2005) Obstacle detection and terrain classification for autonomous off-road navigation. Autonomous Robots 18(1):81–102
16. Fukuoka Y, Kimura H, Hada Y, Takase K (2003) Adaptive dynamic walking of a quadruped robot on irregular terrain based on biological concepts. Int J Robot Res 22(3-4):187–202
17. Best G, Moghadam P, Kottege N, Kleeman L (2013) Terrain classification using a hexapod robot. In: The Australasian Conference on Robotics and Automation (ACRA). ARAA, Sydney, Australia
18. Zenker S, Aksoy EE, Goldschmidt D, Worgotter F, Manoonpong P (2013) Visual terrain classification for selecting energy efficient gaits of a hexapod robot. In: Advanced Intelligent Mechatronics (AIM), 2013 IEEE/ASME international conference on. IEEE, Wollongong, Australia. pp 577–584
19. Filitchkin P, Byl K (2012) Feature-based terrain classification for littledog. In: Intelligent Robots and Systems (IROS), 2012 IEEE/RSJ International Conference On. IEEE, Vilamoura-Algarve, Portugal. pp 1387–1392
20. LTD. T..I: Ai-Ball (2010). http://www.thumbdrive.com/aiball/index.html
21. http://www.digi.com: XBee®802.15.4. http://www.digi.com/products/wireless-wired-embedded-solutions/zigbee-rf-modules/point-multipoint-rfmodules/xbee-series1-module#overview
22. Banzi M (2008) Getting Started with Arduino, Ill edn. Make Books - Imprint of: O'Reilly Media, Sebastopol
23. Quigley M, Conley K, Gerkey BP, Faust J, Foote T, Leibs J, Wheeler R, Ng AY (2009) ROS: an open-source robot operating system. In: ICRA workshop on open source software. IEEE, Kobe, Japan
24. Cheng Y (1995) Mean shift, mode seeking, and clustering. IEEE Trans Pattern Anal Mach Intell 17(8):790–799. doi:10.1109/34.400568
25. Comaniciu D, Meer P (2002) Mean shift: a robust approach toward feature space analysis. IEEE Trans Pattern Anal Mach Intell 24(5):603–619
26. Han B, Comaniciu D, Zhu Y, Davis L (2004) Incremental density approximation and kernel-based bayesian filtering for object tracking. In: Computer Vision and Pattern Recognition, 2004. CVPR 2004. Proceedings of the 2004 IEEE computer society conference on, vol. 1. IEEE, Washington DC, USA. pp 638–6441. doi:10.1109/CVPR.2004.1315092,
27. Yang C, Duraiswami R, Davis L (2005) Efficient mean-shift tracking via a new similarity measure. In: Computer Vision and Pattern Recognition, 2005. CVPR 2005. IEEE computer society conference on, vol. 1. IEEE, San Diego, CA, USA. pp 176–1831. doi:10.1109/CVPR.2005.139

28. Sfikas G, Nikou C, Galatsanos N, Heinrich C (2011) Majorization-minimization mixture model determination in image segmentation. In: Computer Vision and Pattern Recognition (CVPR), 2011 IEEE conference on. IEEE, Colorado, USA. pp 2169–2176. doi:10.1109/CVPR.2011.5995349
29. Lu L, Hager GD (2007) A nonparametric treatment for location/segmentation based visual tracking. In: Computer Vision and Pattern Recognition, 2007. CVPR '07. IEEE conference on. IEEE, Minnesota, USA. pp 1–8. doi:10.1109/CVPR.2007.382976
30. Mei X, Sun X, Dong W, Wang H, Zhang X (2013) Computer Vision and Pattern Recognition (CVPR), 2013 IEEE conference on. IEEE, Ohio, USA. pp 313–320
31. Corporation N (2014) NVIDIA CUDA: Parallel Programming and Computing Platform. http://www.nvidia.com/object/cuda_home_new.html

16

Path following control of planar snake robots using virtual holonomic constraints: theory and experiments

Ehsan Rezapour[1*], Kristin Y Pettersen[1], Pål Liljebäck[1,2], Jan T Gravdahl[1] and Eleni Kelasidi[1]

Abstract

This paper considers path following control of planar snake robots using virtual holonomic constraints. In order to present a model-based path following control design for the snake robot, we first derive the Euler-Lagrange equations of motion of the system. Subsequently, we define geometric relations among the generalized coordinates of the system, using the method of virtual holonomic constraints. These appropriately defined constraints shape the geometry of a constraint manifold for the system, which is a submanifold of the configuration space of the robot. Furthermore, we show that the constraint manifold can be made invariant by a suitable choice of feedback. In particular, we analytically design a smooth feedback control law to exponentially stabilize the constraint manifold. We show that enforcing the appropriately defined virtual holonomic constraints for the configuration variables implies that the robot converges to and follows a desired geometric path. Numerical simulations and experimental results are presented to validate the theoretical approach.

Keywords: Path following; Snake robots; Biologically inspired robots; Virtual holonomic constraints

Introduction

Although wheels and legs are extensively used in mobile robots and are known as the conventional locomotion tools, we sometimes need more adaptable and flexible locomotion systems in order to carry out tasks in complex, narrow, and unstructured environments. In such situations, snake robots, which have significant adaptability and structural flexibility properties, are a potentially useful alternative to conventional types of mobile robots. Snake robots are relevant for applications where restriction of human involvement is important due to safety (e.g. in firefighting operations [1]) and in applications where human presence is impossible (e.g. in narrow pipe inspection tasks [2,3]).

This paper considers path following control of snake robots. In the path following problem, the goal is to ensure that the error between the system output and a desired geometric path is asymptotically less than any prespecified constant, while guaranteeing a forward motion along the path and boundedness of the states of the controlled system [4]. This control problem is particularly challenging for snake robots. This is due to the fact that such mechanisms are generally hyper-redundant, i.e. they have a large degree of kinematic redundancy, and this gives rise to a complicated dynamical behaviour for the system. Moreover, snake robots are underactuated, i.e. they have fewer independent control inputs than degrees of freedom (DOF), and this complicates the control design for these robots.

Background and literature review

In general, snake robots can be categorized into two classes: snake robots which are subject to nonholonomic velocity constraints and snake robots without nonholonomic velocity constraints. Path following control of both classes of snake robots has been considered in several previous works. The majority of these works consider snake robots with nonholonomic velocity constraints, which is inspired by the world's first snake robot developed in 1972 [5]. Nonholonomic constraints are in the form of sideslip constraints on the links of the robot, i.e. where each link is constrained from moving sideways. These constraints

*Correspondence: Ehsan.Rezapout@itk.ntnu.no
[1] Department of Engineering Cybernetics, Norwegian University of Science and Technology (NTNU), Trondheim 7491, Norway
Full list of author information is available at the end of the article

allow the control input to be specified directly in terms of the desired propulsion of the snake robot, which is employed in [6-8] for computed torque control of the position and heading of snake robots with nonholonomic velocity constraints. In [9], position and path following controllers are proposed for the case where some, but not all, of the snake robot links are subject to sideslip constraints. These constrained links can be lifted from the ground, which give the system more DOF that can be utilized to follow a trajectory while simultaneously maintaining a high manipulability. Similar approaches are considered in [10], where strategies for sinus lifting during the lateral undulatory motion are proposed. In [11], a path following controller for a snake robot with nonholonomic velocity constraints is proposed, and Lyapunov analysis is employed in order to analyse the controller.

Path following control of snake robots without nonholonomic velocity constraints is only considered in a few previous works. In [12], path following control of swimming snake robots is achieved by moving the joints according to a predetermined gait pattern while introducing an angular offset in each joint to steer the robot to some desired path. Methods based on numerical optimal control are considered in [13] for determining optimal gaits during positional control of snake robots. In [14], a control strategy is proposed for sinus lifting during lateral undulation by solving a quadratic optimization problem. In [15], numerical simulations are used to study the properties of lateral undulation that are related to the optimality of motion of the snake robot. In [16,17], cascaded systems theory is employed to achieve path following control of a snake robot described by a simplified model. In this simplified model of the snake robot, the motion of the links is approximated as translational motion instead of rotational motion, which is valid for small joint angles. In [18], a dynamic feedback control law is proposed which controls the body shape and orientation of the snake robot. Controllability analysis of planar snake locomotion is presented in [19], and a controller for straight line path following control of snake robots is proposed and a Poincare map is investigated to prove that the resulting state variables of the snake robot, except for the position in the forward direction, trace out an exponentially stable periodic orbit.

Research design and methodology

The contribution of this paper is a solution to the path following control problem for a snake robot without nonholonomic velocity constraints, by using virtual holonomic constraints, which is a particularly useful concept for control of oscillations (see, e.g. [20-24]). Using this approach, we constrain the state evolution of the system to an appropriately defined submanifold of the configuration space, which is called the constraint manifold. This manifold is defined based on the specified geometric relations among the generalized coordinates of the system which are called virtual holonomic constraints. The proposed feedback control law is designed to exponentially stabilize the constraint manifold, i.e. to enforce the virtual holonomic constraints, which allows the convergence of the snake robot to the desired path.

To our best knowledge, the application of model-based motion control approaches which rely on formal stability proofs for snake robots is very restricted in the previous literature. In particular, the only previous works which formally prove the stability of a path following controller for a snake robot without nonholonomic velocity constraints are presented in [17,18]. In [17], cascaded systems theory is used to stabilize a desired straight path for the position of the center of mass (CM) of a snake robot which is described based on simplified kinematic and dynamic models. In contrast with [17], in this paper, we consider a more complex and accurate model of the snake robot, where the motion of the links is not modelled based on the simplifying assumptions of [17]. Preliminary results of this paper are presented in [18], where the method of virtual holonomic constraints is used for path following control of a planar snake robot. In this paper, these results are extended with a new simulation study of a 11-link snake robot. Furthermore, we extend the results presented in [18] by an experimental investigation of the performance of the control approach. A back-to-back comparison between simulations and experimental results is given, in order to contribute to bridge the gap between advanced control theory and practice.

The paper is organized as follows. First, we derive the Euler-Lagrange equations of motion of the robot. Subsequently, we state the control design objectives. Afterwards, we use the virtual holonomic constraints approach for path following control design for the snake robot. Finally, we present the results of numerical simulations and real-time experiments which illustrate the performance of the theoretical control design.

Methods

In this section, we derive the kinematic model along with the dynamic equations of motion of the snake robot in a Lagrangian framework. Moreover, we use partial feedback linearization to write the model in a simpler form for model-based control design.

In order to perform control design, we need to write the governing equations of the system in an implementable way. This is often done by choosing a local coordinate chart and writing the system equations with respect to (w.r.t.) these coordinates. According to the illustration of the snake robot in Figure 1, we choose the vector of the generalized coordinates of the N-link snake robot as

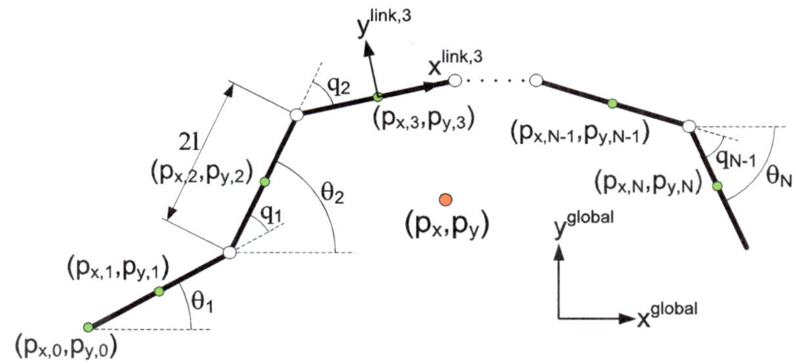

Figure 1 An illustration of the *N*-link snake robot. Kinematic parameters of the snake robot.

$x = \left[q_1, q_2, \ldots, q_{N-1}, \theta_N, p_x, p_y\right]^T \in \mathbb{R}^{N+2}$, where q_i with $i \in \{1, \ldots, N-1\}$ denotes the ith joint angle, θ_N denotes the head angle, and the pair (p_x, p_y) describes the position of the CM of the robot w.r.t. the global $x-y$ axes. Since the robot is not subject to nonholonomic velocity constraints, the vector of the generalized velocities is defined as $\dot{x} = \left[\dot{q}_1, \dot{q}_2, \ldots, \dot{q}_{N-1}, \dot{\theta}_N, \dot{p}_x, \dot{p}_y\right]^T \in \mathbb{R}^{N+2}$. Using these coordinates, it is possible to specify the kinematic map of the robot. In this paper, we denote the first N elements of the vector x, i.e. $(q_1, \ldots, q_{N-1}, \theta_N)$, as the angular coordinates, and the corresponding dynamics as the angular dynamics of the system.

The geometry of the problem

The $(N+2)$-dimensional configuration space of the snake robot is denoted as $\mathcal{Q} = \mathcal{S} \times \mathcal{G}$, which is composed of the shape space $\mathcal{S}$ and a Lie group $\mathcal{G}$ which is freely and properly acting on the configuration space. In particular, the shape variables, i.e. $q_a = (q_1, \ldots, q_{N-1})$, which define the internal configuration of the robot and which we have direct control on, take values in $\mathcal{S}$. Moreover, the position variables, i.e. $q_u = (\theta_N, p_x, p_y)$, which are passive DOF of the system, lie in $\mathcal{G}$. The velocity space of the robot is defined as the differentiable $(2N+4)$-dimensional tangent bundle of $\mathcal{Q}$ as $T\mathcal{Q} = \mathbb{T}^N \times \mathbb{R}^{N+4}$, where $\mathbb{T}^N$ denotes the N-torus in which the angular coordinates live. The free Lagrangian function of the robot $\mathcal{L} : T\mathcal{Q} \to \mathbb{R}$ is invariant under the given action of $\mathcal{G}$ on $\mathcal{Q}$. The coupling between the shape and the position variables causes the net displacement of the position variables, according to the cyclic motion of the shape variables, i.e. the *gait pattern*. Note that for simplicity of presentation, in this paper, we consider local representation of $T\mathcal{Q}$ embedded in an $(2N+4)$-dimensional open subset of the Euclidean space $\mathbb{R}^{2N+4}$.

The forward kinematic map of the snake robot

Based on the kinematic parameters of the snake robot given in Figure 1, it is possible to write the coordinate representation of the forward kinematic map. The map between the absolute link angles θ_i and the relative joint angles q_i is given by

$$\theta_i = \sum_{n=i}^{N-1} q_n + \theta_N \tag{1}$$

The position of the CM of the ith link w.r.t. the global x-y axes can be, respectively, given as

$$p_{x,i} = p_{x,0} + 2l \sum_{j=1}^{i-1} \cos\theta_j + l\cos\theta_i \tag{2}$$

$$p_{y,i} = p_{y,0} + 2l \sum_{j=1}^{i-1} \sin\theta_j + l\sin\theta_i \tag{3}$$

where $2l$ denotes the length of each link, and $(p_{x,0}, p_{y,0})$ denotes the tail position (*cf.* Figure 1). The linear velocities of the CM of the ith link w.r.t. the global x-y axes can be found by taking the time derivative of (2)-(3) which gives

$$\dot{p}_{x,i} = \dot{p}_{x,0} - 2l \sum_{j=1}^{i-1} \sin\theta_j\dot{\theta}_j - l\sin\theta_i\dot{\theta}_i \tag{4}$$

$$\dot{p}_{y,i} = \dot{p}_{y,0} + 2l \sum_{j=1}^{i-1} \cos\theta_j\dot{\theta}_j + l\cos\theta_i\dot{\theta}_i \tag{5}$$

Since all the links have equal length and mass, the position of the CM for the whole structure of the robot is defined as

$$(p_x, p_y) = \left(\frac{1}{N}\sum_{i=1}^{N} p_{x,i}, \frac{1}{N}\sum_{i=1}^{N} p_{y,i}\right) \tag{6}$$

To facilitate path following control of the CM of the snake robot, we replace the tail position $(p_{x,0}, p_{y,0})$ in (2)-(3) with the position of the CM of the robot (p_x, p_y) using the following change of coordinates:

$$p_{x,0} = p_x - \frac{1}{N}\sum_{i=1}^{N}\left(2l\sum_{j=1}^{i-1}\cos\theta_j + l\cos\theta_i\right) \tag{7}$$

$$p_{y,0} = p_y - \frac{1}{N}\sum_{i=1}^{N}\left(2l\sum_{j=1}^{i-1}\sin\theta_j + l\sin\theta_i\right) \tag{8}$$

Substituting (7)-(8) along with their time derivatives into (2)-(5) completes the derivation of the forward kinematic map of the snake robot w.r.t. the desired specified coordinate chart $(x, \dot{x})$.

Equations of motion of the snake robot

The majority of the previous literature on snake robots and similar mobile multi-body robotic structures, such as eel-like robots, have derived the equations of motion of these robots with a Newton-Euler formulation, i.e. where the equations describing the linear and angular motion of individual links are written separately (see, e.g. [15,16]). This is due to the fact that it is usually not straightforward to integrate the anisotropic external dissipative forces, i.e. ground friction forces, acting on these complex robotic structures into their Euler-Lagrange equations of motion. However, ground friction forces have been proved to play a fundamental role in snake robot locomotion (see, e.g. [16]). In this paper, we derive the equations of motion of the snake robot in a Lagrangian framework, i.e. treating the robot as a whole and performing the analysis using a Lagrangian function, which is simple to follow and better suited for studying advanced mechanical phenomena such as elastic link deformations [25], which might be insightful for future research challenges on snake robots. Moreover, we integrate the anisotropic friction forces into these equations using the Jacobian matrices of the links, which gives a straightforward mapping of these forces for the equations of motion.

Snake robots are a class of *simple* mechanical systems, where the Lagrangian $\mathcal{L}(q_a, \dot{x})$ is defined as the difference between the kinetic energy $\mathcal{K}(q_a, \dot{x})$ and potential energy $\mathcal{P}(x)$ of the system [26]. Since the planar snake robot is not subject to any potential field, i.e. $-\nabla\mathcal{P}(x) = 0$, we may write the Lagrangian equal to the kinetic energy, which is the sum of the translational and the rotational kinetic energy of the robot:

$$\mathcal{L}(q_a, \dot{x}) = \mathcal{K}(q_a, \dot{x}) = \frac{1}{2}m\sum_{i=1}^{N}\left(\dot{p}_{x,i}^2 + \dot{p}_{y,i}^2\right) + \frac{1}{2}J\sum_{i=1}^{N}\dot{\theta}_i^2 \tag{9}$$

where m and J denote the uniformly distributed mass and moment of inertia of the links, respectively. Using the Lagrangian function (9), we write the Euler-Lagrange equations of motion of the controlled system as

$$\frac{d}{dt}\left[\frac{\partial\mathcal{L}(q_a, \dot{x})}{\partial\dot{x}_i}\right] - \frac{\partial\mathcal{L}(q_a, \dot{x})}{\partial x_i} = \left(B(x)\tau - \tau_f\right)_i \tag{10}$$

where $i \in \{1, \ldots, N+2\}$, $B(x) = \left[e_j\right] \in \mathbb{R}^{(N+2)\times(N-1)}$ is the full column rank actuator configuration matrix, where e_j denotes the jth standard basis vector in $\mathbb{R}^{N+2}$. Moreover, $B(x)\tau \in \mathbb{R}^{N+2}$ with $\tau = [\tau_1, \ldots, \tau_{N-1}]^T \in \mathbb{R}^{N-1}$ stands for the generalized forces resulting from the control inputs. Furthermore, $\tau_f = \left[\tau_f^1, \ldots, \tau_f^{N+2}\right]^T \in \mathbb{R}^{N+2}$ denotes viscous and Coulomb friction forces acting on $(N+2)$ DOF of the system. The controlled Euler-Lagrange equations (10) can also be written in the form of a second-order differential equation as

$$M(q_a)\ddot{x} + C(x, \dot{x})\dot{x} = B(x)\tau - \tau_f \tag{11}$$

where $M(q_a) \in \mathbb{R}^{(N+2)\times(N+2)}$ is the positive definite symmetric inertia matrix, $C(x, \dot{x})\dot{x} \in \mathbb{R}^{N+2}$ denotes the generalized Coriolis and centripetal forces, and the right-hand side terms denote the external forces acting on the system. The fact that the inertia matrix is only a function of the directly actuated shape variables q_a is a direct consequence of the invariance of the Lagrangian function (9) w.r.t. the position variables. Moreover, since rank $[B(x)] <$ dim(x), the system is underactuated. This underactuation represents the lack of direct control on the head angle and the position of the CM of the robot.

The dynamic model (11) perfectly agrees with the equations of motion which are derived based on the Newton-Euler formulation in previous works (see, e.g. [16]). In order to validate the model, in the last section of this paper, we present simulation results which are obtained using the dynamic model (11) together with experimental results for the locomotion of the robot which are obtained using a physical snake robot. The agreement between simulations and experiments shows that the dynamic model (11) accurately represents the motion of the robot.

The ground friction model

In this subsection, both viscous and Coulomb friction models are used for capturing the essential properties of the anisotropic ground friction forces. For modelling the friction, we first define the rotation matrix for mapping from the global frame to the local frame of link i (*cf.* Figure 1) as

$$R_i = \begin{bmatrix} \cos\theta_i & -\sin\theta_i \\ \sin\theta_i & \cos\theta_i \end{bmatrix} \tag{12}$$

Using (4)-(5) and (12), the velocities of the links in the local link frames can be written in terms of the velocities of the links in the global frame as

$$v^{\text{link},i} = \left[\, v_t^{\text{link},i} \;\; v_n^{\text{link},i} \,\right]^T = R_i^T \left[\, \dot{p}_{x,i} \;\; \dot{p}_{y,i} \,\right]^T \tag{13}$$

where $v_t^{\text{link},i}$ and $v_n^{\text{link},i}$ denote the linear velocity of the CM of the ith link in the tangential (along link x-axis) and normal (along link y-axis) direction of the link, respectively. The total friction force acting on link i is defined as the sum of the viscous and Coulomb friction forces, which are denoted by f_{v_i} and f_{c_i}, respectively, as

$$f^{\text{link},i} = f_{c_i} + f_{v_i} \tag{14}$$

Assuming equal friction coefficients for all the links, we write the model of the friction for each individual link i as

$$f_{c_i} = mg \left[\, \mu_t \text{sgn}\left(v_t^{\text{link},i}\right) \;\; \mu_n \text{sgn}\left(v_n^{\text{link},i}\right) \,\right]^T \in \mathbb{R}^2 \tag{15}$$

$$f_{v_i} = \left[\, c_t v_t^{\text{link},i} \;\; c_n v_n^{\text{link},i} \,\right]^T \in \mathbb{R}^2 \tag{16}$$

where $i \in \{1, \ldots, N\}$, m denotes the mass of a link, g denotes the acceleration due to gravity, and μ_t and μ_n denote Coulomb friction coefficients in the tangential and normal direction of the link, respectively. Furthermore, c_t and c_n denote viscous friction coefficients in the tangential and normal direction of the link, respectively. Thus, we map the friction force acting on the ith link to the global x-y frame as

$$f_{\text{global}}^{\text{link},i} = R_i f^{\text{link},i} \tag{17}$$

Finally, we can write τ_f in (11) as

$$\tau_f = \sum_{i=1}^{N} \mathcal{J}_i^T(x) f_{\text{global}}^{\text{link},i} \tag{18}$$

where

$$\mathcal{J}_i^T(x) = \left[\frac{\partial \dot{p}_{x,i}}{\partial \dot{x}_j}, \frac{\partial \dot{p}_{y,i}}{\partial \dot{x}_j} \right] \in \mathbb{R}^{(N+2)\times 2}, \qquad j \in \{1, \ldots, N+2\} \tag{19}$$

denotes the transpose of the Jacobian matrix of the CM of the ith link.

Remark 1. As argued in [16], the motion of a snake robot with anisotropic viscous ground friction is qualitatively (but not quantitatively) similar as with anisotropic Coulomb friction. However, a viscous friction model is less complex w.r.t. control design and analysis. Accordingly, we employ a viscous friction model for the control design in this paper.

Partial feedback linearization of the dynamic model

A common method for control of mechanical systems is full-state feedback linearization. This approach is not applicable for snake robots due to the underactuation. However, it is still possible to linearize the dynamics of the actuated DOF of the robot, which is called collocated partial feedback linearization, and can simplify the analysis as well as the control design. A similar approach is considered in [16], but for the sake of completeness, we present the approach here. To this end, we separate the dynamic equations of the robot given by (11) into two subsets by taking $x = [q_a, q_u]^T \in \mathbb{R}^{N+2}$, with $q_a \in \mathbb{R}^{N-1}$ and $q_u \in \mathbb{R}^3$ which were defined in the subsection describing the geometry of the problem:

$$m_{11}(q_a)\ddot{q}_a + m_{12}(q_a)\ddot{q}_u + h_1(x,\dot{x}) = \psi \in \mathbb{R}^{N-1} \tag{20}$$

$$m_{21}(q_a)\ddot{q}_a + m_{22}(q_a)\ddot{q}_u + h_2(x,\dot{x}) = 0_{3\times 1} \in \mathbb{R}^3 \tag{21}$$

where $m_{11} \in \mathbb{R}^{(N-1)\times(N-1)}$, $m_{12} \in \mathbb{R}^{(N-1)\times 3}$, $m_{21} \in \mathbb{R}^{3\times(N-1)}$, and $m_{22} \in \mathbb{R}^{3\times 3}$ denote the corresponding submatrices of the inertia matrix, and $0_{3\times 1} = [0,0,0]^T \in \mathbb{R}^3$. Furthermore, $h_1(x,\dot{x}) \in \mathbb{R}^{N-1}$ and $h_2(x,\dot{x}) \in \mathbb{R}^3$ include all the contributions of the Coriolis, centripetal, and friction forces. Moreover, $\psi \in \mathbb{R}^{N-1}$ denotes the non-zero part of the vector of control forces, i.e. $B(x)\tau = [\psi, 0_{3\times 1}]^T \in \mathbb{R}^{N+2}$. From (21), we have

$$\ddot{q}_u = -m_{22}^{-1}(h_2 + m_{21}\ddot{q}_a) \in \mathbb{R}^3 \tag{22}$$

Substituting (22) into (20) yields

$$\left(m_{11} - m_{12}m_{22}^{-1}m_{21}\right)\ddot{q}_a - \left(m_{12}m_{22}^{-1}\right)h_2 + h_1 = \psi \tag{23}$$

For linearizing the dynamics of the directly actuated DOF, we apply the global transformation of the vector of control inputs as

$$\psi = \left(m_{11} - m_{12}m_{22}^{-1}m_{21}\right)\vartheta - \left(m_{12}m_{22}^{-1}\right)h_2 + h_1 \tag{24}$$

where $\vartheta = [\vartheta_1, \vartheta_2, \ldots, \vartheta_{N-1}]^T \in \mathbb{R}^{N-1}$ is the vector of new control inputs. Consequently, the dynamic model (20)-(21) can be written in the following partially feedback linearized form

$$\ddot{q}_a = \vartheta \in \mathbb{R}^{N-1} \tag{25}$$

$$\ddot{q}_u = \mathcal{D}(x,\dot{x}) + \mathcal{C}(q_a)\vartheta \in \mathbb{R}^3 \tag{26}$$

with

$$\mathcal{D}(x,\dot{x}) = -m_{22}^{-1}(q_a)\,h_2(x,\dot{x}) = \left[f_{\theta_N}, f_x, f_y\right]^T \in \mathbb{R}^3 \quad (27)$$

$$\begin{aligned}\mathcal{C}(q_a) &= -m_{22}^{-1}(q_a)\,m_{21}(q_a)\\ &= \left[\beta_i(q_a), 0, 0\right]^T \in \mathbb{R}^{3\times(N-1)} \end{aligned} \quad (28)$$

where $\beta_i(q_a) : \mathcal{Q} \rightarrow \mathbb{R}$ is a smooth scalar-valued function. It can be numerically shown that the value of β_i is negative at any configuration $q_a \in \mathcal{Q}$. Furthermore, f_{θ_N}, f_x, and f_y denote the friction forces acting on θ_N, p_x, and p_y, respectively (f_{θ_N} also contains Coriolis forces besides the friction forces). For the aim of analysis and model-based control design, we write (25)-(26) in a more detailed form:

$$\ddot{q}_a = \vartheta \in \mathbb{R}^{N-1} \quad (29)$$

$$\ddot{\theta}_N = f_{\theta_N}(x,\dot{x}) + \beta_i(q_a)\vartheta^i \in \mathbb{R} \quad (30)$$

$$\ddot{p}_x = f_x(x,\dot{x}) \in \mathbb{R} \quad (31)$$

$$\ddot{p}_y = f_y(x,\dot{x}) \in \mathbb{R} \quad (32)$$

where the summation convention is applied in (30), and henceforth, to all the equations which contain repeated upper-lower indices (i.e. whenever an expression contains a repeated index, one as a subscript and the other as a superscript, summation is implied over this index [26]). The dynamical system (29)-(32) is in the form of a control-affine system with drift. In particular, the term

$$\mathcal{A}(x,\dot{x}) = \left[\dot{q}_a, \dot{q}_u, 0_{(N-1)\times 1}, \mathcal{D}(x,\dot{x})\right]^T \in \mathbb{R}^{2N+4} \quad (33)$$

is called the drift vector field, which specifies the dynamics of the robot when the control input is zero. Furthermore, the columns of the matrix

$$\mathcal{B}(q_a) = \begin{bmatrix} 0_{(N+2)\times(N-1)} \\ I_{N-1} \\ \left[\beta_1(q_a), \ldots, \beta_{N-1}(q_a)\right] \\ 0_{2\times(N-1)} \end{bmatrix} \in \mathbb{R}^{(2N+4)\times(N-1)} \quad (34)$$

are called the control vector fields, which enable us to control the internal configuration and consequently the orientation and the position of the robot in the plane.

Remark 2. The last two rows of the control vectors in (34) are composed of zero elements. This implies that the control forces have no direct effect on the dynamics of the position of the CM of the robot, i.e. (31)-(32). Furthermore, the dynamics of the position of the CM are coupled with the dynamics of the directly actuated shape variables q_a, i.e. (29), only through the friction forces. Accordingly, in the absence of the friction forces, the linear momentum of the robot is a conserved quantity, and the position of the CM of the robot is not controllable.

Control design objectives and the track-follow problem formulation

In this section, we state our control design objectives which will be followed throughout the remaining sections of the paper. In particular, we stress that for a complex mobile multi-link robotic structure such as a snake robot, formulating a pure path following, trajectory tracking, or maneuvering problem is unusual (for definitions of these problem formulations, see [27]). This is due to the fact that for a part of the state variables of the system (particularly the shape variables and the head angle), it is most natural to formulate the control problem as a trajectory tracking problem, while for the other state variables (particularly the position of the CM), we may formulate the problem as a path following or a maneuvering one.

To formulate a combinational *track-follow* problem for the snake robot, which we define as a trajectory tracking formulation for a subset of the state variables, together with a path following formulation for the remaining subset, we introduce the error variable for the ith joint of the robot as

$$y_i = q_i - \Phi_i \quad (35)$$

where $i \in \{1, \ldots, N-1\}$, and $\Phi_i \in \mathbb{R}$ denotes a function that defines the reference trajectory for the ith joint which will be chosen through the control design in the next section. The head angle error is defined as

$$y_N = \theta_N - \Phi_N \quad (36)$$

where $\Phi_N \in \mathbb{R}$ denotes the reference head angle for the robot.

We divide the control objectives into three main parts. In the first part, the goal is to make the shape variables of the robot track given bounded smooth time-varying references, i.e. asymptotic trajectory tracking problem, such that

$$\lim_{t\rightarrow\infty} \|y_i(t)\| = 0 \quad (37)$$

for all $i \in \{1, \ldots, N-1\}$. Furthermore, we seek to control the head angle of the robot. The second part of the control objective is thus to make the head angle of the robot track a desired head angle such that

$$\lim_{t\rightarrow\infty} \|y_N(t)\| = 0 \quad (38)$$

Moreover, we define a desired straight path that we want the CM of the snake robot to follow. This is defined as a smooth one-dimensional manifold $\mathcal{P} \subset \mathbb{R}^2$, with coordinates in the x-y plane given by the pair (p_{xd}, p_{yd}), which are parameterized by a scalar time-dependent variable $\Theta(t)$ as

$$\mathcal{P} = \left\{ \left(p_{xd}(\Theta), p_{yd}(\Theta)\right) \in \mathbb{R}^2 : \Theta \geq 0 \right\} \tag{39}$$

We define the vector of the path following error variables for the position of the CM of the robot as $\tilde{p} = \left[p_x(t) - p_{xd}(\Theta), p_y(t) - p_{yd}(\Theta)\right]^T \in \mathbb{R}^2$. Subsequently, the third part of the control objectives is defined as practical convergence (see, e.g. [4]) of the position of the CM of the robot to the desired path such that

$$\lim_{t\to\infty} \sup \|\tilde{p}(t)\| \leq \varepsilon \tag{40}$$

where $\varepsilon \in \mathbb{R}_{>0}$ is an arbitrary positive scalar. Moreover, we require that $\dot{\Theta}(t) \geq 0$ and $\lim_{t\to\infty} \Theta(t) = \infty$ (forward motion along the path), and boundedness of the states of the controlled system.

Path following control with virtual holonomic constraints

The idea of virtual holonomic constraints is particularly a useful concept for control of oscillations (see, e.g. [20-24]). We will in this section show how this approach can be used to solve the path following control problem of snake robots. In particular, we will show how, by designing the joint reference trajectories in (35) using virtual holonomic constraints and by combining this with virtual holonomic constraints motivated by line-of-sight (LOS) guidance for the head angle in (36), we are able to solve the path following control problem, i.e. achieving (40). Our main motivation for using this approach is the fact that while performing the gait pattern lateral undulation which consists of fixed periodic body motions, all the solutions of the snake robot dynamics have inherent oscillatory behaviour. Moreover, we will show how this behaviour can be analytically and *constructively* controlled based on virtual holonomic constraints. In particular, we use the word 'constructive' in the sense that through the feedback action, we shape the dynamics of the system such that it possesses the desired structural properties, i.e. positive invariance and exponential stability of an appropriately defined constraint manifold. To this end, we define a constraint manifold for the system, and we design the control input of (29) to exponentially stabilize the constraint manifold. The geometry of this manifold is defined based on specified geometric relations among the generalized coordinates of the system which are called virtual holonomic constraints. In particular, we call them virtual constraints because they do not arise from a physical connection between two variables but rather from the actions of a feedback controller [20].

Trajectory planning by virtual holonomic constraints

Virtual holonomic constraints are specified through C^1 coordinate-dependent functions $\Phi_i : \mathcal{Q} \to \mathbb{R}$ which are called the constraint functions, in the relations of the form $\Phi_i(x) = 0$, which can be enforced through the feedback action. In particular, for the snake robot, we define a vector-valued function

$$\Phi = [\Phi_1, \ldots, \Phi_N]^T \in \mathbb{R}^N \tag{41}$$

in which every element defines one constraint function for the corresponding angular coordinate of the system.

At this point, we augment the state vector of the system with three new states that in the following will be used in the control design. The introduction of these new variables to the state vector of the system, which will be used as constraint variables, is inspired by the notion of dynamic virtual holonomic constraints [21], i.e. virtual holonomic constraints which depend on the solutions of a dynamic compensator. The idea is to make the virtual holonomic constraints to depend on the variations of a dynamic parameter, which is used for controlling the system on the constraint manifold. The purpose of these additional states is explained below.

1. We introduce two new states $\left[\phi_o, \dot{\phi}_o\right]^T \in \mathbb{R}^2$ where the second-order time derivative of ϕ_o will be used as an additional control input that drives the snake robot towards the desired path by modifying the orientation of the robot in accordance with a path following guidance law.
2. In the previous section, we defined the control objective for the joints and the head angle of the robot as a trajectory tracking problem. However, it is known that holonomic constraints are coordinate-dependent equality constraints of the form $\Phi_i(x) = 0$, where Φ_i is a time-independent function [25]. Thus, we remove this explicit time dependency from the reference joint trajectories by augmenting the state vector of the system with a new variable η, with $\dot{\eta} = 2\pi/T$ and $\eta(0) = 0$, where T denotes the period of the cyclic motion of the shape variables of the robot.
   Subsequently, we denote the augmented coordinate vector of the system by

$$\hat{x} = \left[q_1, \ldots, q_{N-1}, \theta_N, p_x, p_y, \phi_o, \eta\right]^T \in \mathbb{R}^{N+4} \tag{42}$$

and the corresponding augmented state space by $T\hat{\mathcal{Q}}$.

Virtual holonomic constraints for the joint angles

A fundamental work in the area of snake robots was presented by Hirose [5]. In this work, Hirose considers empirical studies of biological snakes to derive a mathematical approximation of the most common gait pattern among biological snakes, known as lateral undulation. In particular, the shape of a snake conducting lateral undulation can be described by a planar curve (the serpenoid curve) with coordinates in the x-y plane along the curve at arc length s given by

$$x(s) = \int_0^s \cos\left(a\cos(bz) + cz\right) dz \quad (43)$$

$$y(s) = \int_0^s \sin(a\cos(bz) + cz) dz \quad (44)$$

where a, b, and c are positive scalars. Locomotion of a snake-like structure in accordance with the serpenoid curve, i.e. lateral undulation, is achieved if the joints of the robot move according to the reference joint trajectories in the form of a sinusoidal function with specified amplitude, frequency, and phase shift. In particular, using the foregoing defined new states, we define a constraint function for the ith joint of the snake robot by

$$\Phi_i = \alpha \sin\left(\eta + (i-1)\delta\right) + \phi_o \quad (45)$$

where $i \in \{1,\ldots,N-1\}$, α denotes the amplitude of the sinusoidal joint motion, and δ is a phase shift that is used to keep the joints *out of phase*. Moreover, ϕ_o is an offset value that is identical for all of the joints. It was illustrated in [16] how the offset value ϕ_o affects the orientation of the snake robot in the plane. Building further on this insight, we consider the second-order time derivative of ϕ_o in the form of a dynamic compensator, which will be used to control the orientation of the robot. In particular, through this control term, we modify the orientation of the robot in accordance with a reference orientation. This will be done by adding an offset angle to the reference trajectory of each joint. We will show that this will steer the position of the CM of the robot towards the desired path. The constraint function (45) is dynamic, since it depends on the solution of a dynamic compensator.

Virtual holonomic constraint for the head link angle

In this subsection, we define a constraint function for the head angle of the robot. In particular, we use a line-of-sight (LOS) guidance law as the reference angle for the head link. LOS guidance is a much-used method in marine control systems (see, e.g. [27]). In general, guidance-based control strategies are based on defining a reference heading angle for the vehicle through a guidance law and designing a controller to track this angle [27]. Motivated by marine control literature, in [17] based on a simplified model of the snake robot, using cascade systems theory, it was proved that if the heading angle of the snake robot was controlled to the LOS angle, then also the position of the CM of the robot would converge to the desired path. We will show that a similar guidance-based control strategy can successfully steer the robot towards the desired path. However, we perform the model-based control design based on a more accurate model of the snake robot which does not contain the simplifying assumptions of [17] which are valid for small joint angles.

To define the guidance law, without loss of generality, we assign the global coordinate system such that the global x-axis is aligned with the desired path. Consequently, the position of the CM of the robot along the y-axis, denoted by p_y, defines the shortest distance between the robot and the desired path, often referred to as the *cross-track error*. In order to solve the path following problem, we use the LOS guidance law as a virtual holonomic constraint, which defines the desired head angle as a function of the cross-track error as

$$\Phi_N = -\tan^{-1}\left(\frac{p_y}{\Delta}\right) \quad (46)$$

where $\Delta > 0$ is a design parameter known as the look-ahead distance. The idea is that steering the head angle of the snake robot such that it is headed towards a point located at a distance Δ ahead of the robot along the desired path will make the snake robot move towards the path and follow it.

Defining a constraint manifold

We collect all the foregoing defined constraint functions in the following vector-valued function

$$\Phi = \Big[\alpha\sin(\eta) + \phi_o, \ldots, \alpha\sin\left(\eta + (N-1)\delta\right) + \phi_o, \tan^{-1}\left(\frac{p_y}{\Delta}\right)\Big]^T \in \mathbb{R}^N \quad (47)$$

For trajectory planning using virtual holonomic constraints, we define the constraint manifold associated with the constraint functions (47) as

$$\Gamma = \left\{ \left(\hat{x}, \dot{\hat{x}}\right) \in T\hat{\mathcal{Q}} : q_i = \Phi_i\left(\eta, \phi_o\right), \theta_N = \Phi_N\left(p_y\right), \dot{q}_i = \dot{\eta}\frac{\partial \Phi_i}{\partial \eta} + \dot{\phi}_o\frac{\partial \Phi_i}{\partial \phi_o}, \dot{\theta}_N = \dot{p}_y\frac{\partial \Phi_N}{\partial p_y} \right\} \quad (48)$$

where $i \in \{1,\ldots,N-1\}$. The constraint manifold (48) is a six-dimensional submanifold of $\hat{\mathcal{Q}}$, since we have three different constraint variables, i.e. η, ϕ_o, p_y. The goal of the control input is to enforce the virtual holonomic constraints (47), by making Γ exponentially stable for the closed-loop system and thereby achieving the control objectives (37)-(38). To this end, we define the elements of a controlled output vector $y \in \mathbb{R}^N$ for the system (29)-(32) as the difference between the angular coordinates and their corresponding constraint functions as

$$y = \left[q_1 - \Phi_1(\eta, \phi_o), \ldots, q_{N-1} - \Phi_{N-1}(\eta, \phi_o), \theta_N - \Phi_N(p_y)\right]^T \in \mathbb{R}^N \tag{49}$$

We will achieve our control design objectives which we defined in the previous section, by designing the control inputs ϑ and $\ddot{\phi}_o$ such that $(y_i, \dot{y}_i) \to (0,0)$ for all $i \in \{1, \ldots, N\}$. To this end, we first need to ensure that the given relations in (47) are stabilizable, i.e. a suitable choice of feedback can make the constraint manifold exponentially stable for the closed-loop system. For simplicity of notation, we denote the following differentials:

$$d\Phi_i = \dot{\eta}\frac{\partial \Phi_i}{\partial \eta} + \dot{\phi}_o\frac{\partial \Phi_i}{\partial \phi_o} \tag{50}$$

$$d\Phi_N = \dot{p}_y\frac{\partial \Phi_N}{\partial p_y} \tag{51}$$

$$d^2\Phi_i = \ddot{\eta}\frac{\partial \Phi_i}{\partial \eta} + \dot{\eta}^2\frac{\partial^2 \Phi_i}{\partial \eta^2} + \ddot{\phi}_o\frac{\partial \Phi_i}{\partial \phi_o} + \dot{\phi}_o^2\frac{\partial^2 \Phi_i}{\partial \phi_o^2} \tag{52}$$

$$d^2\Phi_N = \ddot{p}_y\frac{\partial \Phi_N}{\partial p_y} + \dot{p}_y^2\frac{\partial^2 \Phi_N}{\partial p_y^2} \tag{53}$$

where $i \in \{1, \ldots, N-1\}$. The Lie derivative of (49) along the solutions of (29)-(32) is of the form

$$\dot{y} = \left[\dot{q}_1 - d\Phi_1, \ldots, \dot{q}_{N-1} - d\Phi_{N-1}, \dot{\theta}_N - d\Phi_N\right]^T \in \mathbb{R}^N \tag{54}$$

which lacks the control inputs. The Lie derivative of (54) along the solutions of (29)-(32) is of the form

$$\ddot{y} = \left[\vartheta_1 - d^2\Phi_1, \ldots, \vartheta_{N-1} - d^2\Phi_{N-1}, f_{\theta_N} + \beta_i\vartheta^i - d^2\Phi_N\right]^T \in \mathbb{R}^N \tag{55}$$

which contains the control inputs. Consequently, the controlled output vector (49) yields a well-defined vector relative degree $\{2, 2, \ldots, 2\}$ everywhere on the constraint manifold Γ. The virtual holonomic constraints satisfying this vector relative degree condition are called regular, and regular constraints are always feasible [22], i.e. there exists a smooth feedback such that Γ is positively invariant for the closed-loop system. Furthermore, regular constraints in parametric form (47) are always stabilizable [21].

The well-defined vector relative degree $\{2, 2, \ldots, 2\}$ on Γ implies that the system (29)-(32) with the controlled output function (47) is input-output feedback linearizable. Consequently, we can stabilize Γ with an input-output feedback linearizing controller.

Output regulation via input-output linearization

In this subsection, we will derive a control law for (29) such that the constraint manifold (48) with the constraint functions defined in (47) is globally exponentially stable for the closed-loop system. In particular, we use input-output linearization to stabilize the constraint manifold (48).

To stabilize $\Phi_i(\eta, \phi_o)$ for the ith joint, i.e. to make $(y_i, \dot{y}_i) \to (0,0)$ for all $i \in \{1, \ldots, N-1\}$, we define an exponentially stabilizing joint control law. The second-order time derivative of the ith joint tracking error, i.e. the ith element of (55), is of the form

$$\ddot{y}_i = \vartheta_i - d^2\Phi_i \tag{56}$$

We define the control input for the ith joint in (29) as

$$\vartheta_i = d^2\Phi_i - k_p y_i - k_d \dot{y}_i \tag{57}$$

where $k_p > 0$ and $k_d > 0$ are the joint controller gains. These gains are chosen similar for all the joints since the links have similar inertial parameters. Inserting (57) into (56) yields

$$\ddot{y}_i + k_d\dot{y}_i + k_p y_i = 0 \tag{58}$$

The tracking error dynamics of the ith joint angle (58) clearly has a globally exponentially stable equilibrium at the origin $(y_i, \dot{y}_i) = (0,0)$, which implies that every ith control input (57) exponentially stabilizes the constraint manifold for the ith joint, and the control objective (37) is achieved.

In the following, we aim to stabilize the constraint manifold for the head angle, i.e. to make $(y_N, \dot{y}_N) \to (0,0)$. The head angle error corresponds to the Nth element of the controlled output vector (49), and its second-order time derivative (i.e. the head angle error dynamics) is given by

$$\ddot{y}_N = f_{\theta_N} + \beta_i\vartheta^i - d^2\Phi_N \tag{59}$$

Inserting ϑ_i from (57) into (59) gives

$$\ddot{y}_N = f_{\theta_N} + \sum_{i=1}^{N-1}\beta_i\left(d^2\Phi_i - k_p y_i - k_d\dot{y}_i\right) - d^2\Phi_N \tag{60}$$

which is equivalent to

$$\ddot{y}_N = f_{\theta_N} + \sum_{i=1}^{N-1} \beta_i \left(\ddot{\eta}\frac{\partial \Phi_i}{\partial \eta} + \dot{\eta}^2 \frac{\partial^2 \Phi_i}{\partial \eta^2} + \ddot{\phi}_o \frac{\partial \Phi_i}{\partial \phi_o} + \dot{\phi}_o^2 \frac{\partial^2 \phi_i}{\partial \phi_o^2} - k_p y_i - k_d \dot{y}_i \right) - d^2 \Phi_N \quad (61)$$

For simplicity of notation, we denote the constraint function for the ith joint angle of the robot by $\Phi_i = S_i + \phi_o$, where $S_i = \alpha \sin(\eta + (i-1)\delta)$. Subsequently, based on the specified constraint functions in (47), i.e. since $\ddot{\eta} = 0$ and $\frac{\partial^2 \phi_i}{\partial \phi_o^2} = 0$, we may write (61) as

$$\ddot{y}_N = f_{\theta_N} + \sum_{i=1}^{N-1} \beta_i \left(-\dot{\eta}^2 S_i + \ddot{\phi}_o - k_p y_i - k_d \dot{y}_i \right) - d^2 \Phi_N \quad (62)$$

In order to stabilize the constraint function $\Phi_N(p_y)$ for the head angle, we define the second-order time derivative of the augmented coordinate ϕ_o in the form of a dynamic compensator as

$$\ddot{\phi}_o = \left(\sum_{i=1}^{N-1} \beta_i \right)^{-1} \left(\sum_{i=1}^{N-1} \beta_i \left(\dot{\eta}^2 S_i + k_p y_i + k_d \dot{y}_i \right) + d^2 \Phi_N - f_{\theta_N} - k_{p,\theta_N} y_N - k_{d,\theta_N} \dot{y}_N \right) \quad (63)$$

where $k_{p,\theta_N} > 0$ and $k_{d,\theta_N} > 0$ are the head angle controller gains. Notice that since β_i is negative-valued in any configuration, (63) is globally well-defined. Through numerical simulations, we show that the states of the dynamic compensator (63), i.e. $(\phi_o, \dot{\phi}_o)$, remain bounded. By inserting (63) into (62), the error dynamics of the head angle takes the form

$$\ddot{y}_N + k_{d,\theta_N} \dot{y}_N + k_{p,\theta_N} y_N = 0 \quad (64)$$

which clearly has a globally exponentially stable equilibrium at the origin $(y_N, \dot{y}_N) = (0, 0)$. Consequently, we have that $(y_N, \dot{y}_N) \to (0, 0)$ from any initial condition, and the control objective (38) will be achieved.

Finally, we conjecture that while the output trajectories are evolving on the constraint manifold (48), the internal dynamics (31)-(32), which has the form

$$\ddot{p}_x = f_x \left(\Phi, p_x, p_y, \Phi', \dot{p}_x, \dot{p}_y \right) \quad (65)$$

$$\ddot{p}_y = f_y \left(\Phi, p_x, p_y, \Phi', \dot{p}_x, \dot{p}_y \right) \quad (66)$$

converge to and follow the desired planar path. Analytically investigating the convergence of the snake robot position to the desired path is a topic of future work. As a preliminary support of this conjecture, we provide simulation and experimental results which show that the snake robot successfully converges to and follows the desired path.

Simulation results

In this section, we present simulation results which illustrate the performance of the proposed path following controller. We considered a snake robot with $N = 11$ links, $m = 1$ kg, $l = 0.07$ m, and $J = 0.0016$ kg m^2 (Figure 2). The friction coefficients were $c_n = 10$ and $c_t = 1$. The parameters of the joint constraint functions (45) were $\alpha = \pi/6$ rad, $\eta = 70\pi t/180$ rad, and $\delta = 36\pi/180$ rad. The controller gains in (57) and (63) were tuned as $k_p = 10$, $k_d = 5$, $k_{p,\theta_N} = 20$, $k_{d,\theta_N} = 1$, and $\Delta = 1.4$ m. In order to calculate $\dot{\Phi}_N$ and $\ddot{\Phi}_N$, we employed the approach taken in [27] by passing Φ_N through a low-pass filter of the form

$$\dot{\Omega} = \begin{bmatrix} 0 & 1 \\ -\omega_n^2 & -2\psi_f \omega_n \end{bmatrix} \Omega + \begin{bmatrix} 0 \\ \omega_n^2 \end{bmatrix} \Phi_N \quad (67)$$

with natural frequency $\omega_n = \pi/2$ rad, damping ratio $\psi_f = 1$, and initial condition $\Omega(0) = [0, 0]^T$. As seen from the simulation results which are presented in Figures 3, 4, 5, 6, 7, the snake robot successfully converges to and follows the desired path. In particular, Figure 3 shows that the solutions of the dynamic compensator (63) remain bounded. Figure 4 shows that the joint angles track the reference angles provided by the constraint functions (45), while the tracking errors converge exponentially to zero. Figure 5 shows that the head angle tracks the reference head angle provided by the constraint function (46), while the tracking error converges to zero exponentially fast. Finally, Figure 6 shows that the CM of the robot converges to and follows the desired straight path. Moreover, in order to show the performance of the proposed tracking control law (57) in the presence of angular position measurement noise, we subjected every ith joint angle q_i to an additive noise by using Matlab function *randn*() which generates normally distributed pseudorandom numbers

Figure 2 An illustration of the experimental setup. Snake robot Wheeko was used for the experiments.

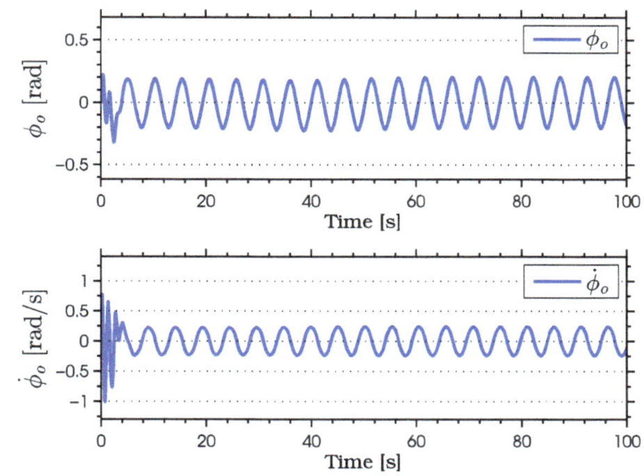

Figure 3 The states of the dynamic compensator in simulations. The states of the dynamic compensator remain bounded.

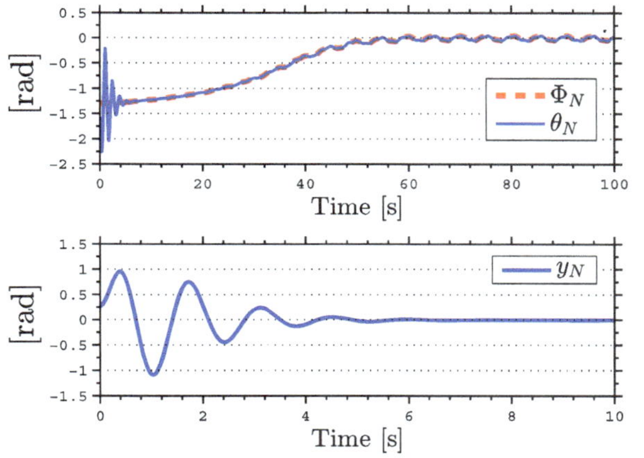

Figure 5 Head angle tracking error in simulations. The head angle tracking error converges exponentially to zero.

that can be considered as measurement noise for the joint angles. In particular, we added $randn(1)$ to each joint angle q_i in each integration step. The result of the simulation is presented in Figure 7, which shows that the joint tracking errors converge to a very small neighbourhood of zero in the presence of measurement noise.

Experimental results

In this section, we present results from an experimental investigation of the real-time performance of the proposed control strategy using a mechanical snake robot.

Experimental setup

The experiment was carried out using the snake robot Wheeko [16]. The robot, which is shown in Figure 2, has 10 identical joint modules, i.e. $N = 11$ links. Each joint module is equipped with a set of passive wheels which give the robot anisotropic ground friction properties during motion on flat surfaces. The wheels are able to slip sideways and thus do not introduce nonholonomic velocity constraints in the system. Each joint is driven by a Hitec servo motor (HS-5955TG; Hitec RCD USA, Inc., Poway, CA, USA), and the joint angles are measured using magnetic rotary encoders. The motion of the snake robot was measured using a camera-based motion capture system from OptiTrack of type Flex 13 (NaturalPoint, Inc., Corvallis, OR, USA). The system consists of 16 cameras which are sampled at 120 frames per second and which allow reflective markers to be tracked on a submillimetre level. During the experiment, reflective markers were mounted on the head link of the snake robot in order to measure

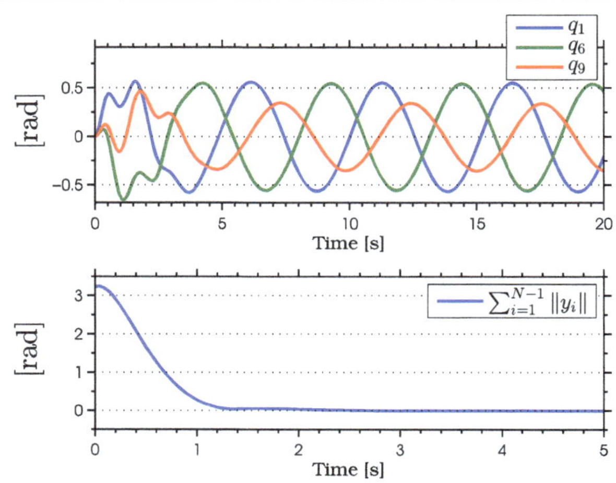

Figure 4 Joint angles tracking the reference joint angles in simulations. The joints of the robot track the sinusoidal motions (above). The joint tracking errors converge exponentially to zero (below).

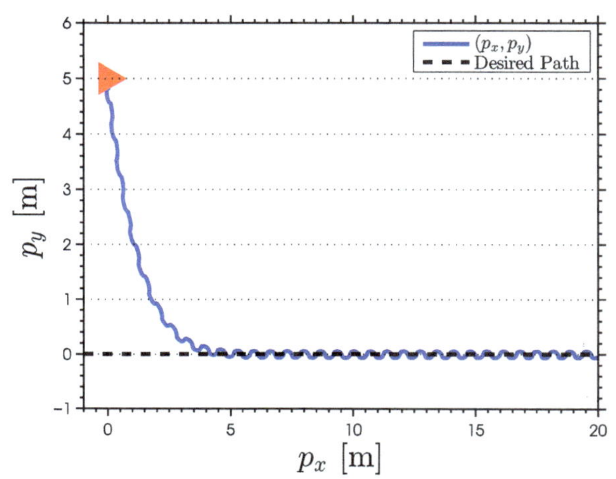

Figure 6 The motion of the center of mass in the *x*-*y* plane in simulations. The position of the CM of the robot (blue) converges to and follows the desired straight line path (the *x*-axis).

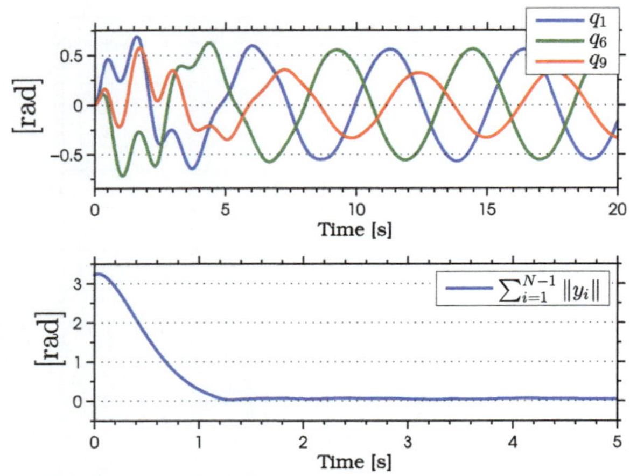

Figure 7 Joint angles tracking the reference joint angles in simulations in the presence of measurement noise. The joints of the robot track the sinusoidal motions in the presence of measurement noise (above). The joint tracking errors converge exponentially to a neighbourhood of zero in the presence of measurement noise (below).

the position $(p_{x,N}, p_{y,N})$ and orientation (θ_N) of the head. These measurements were combined with the measured joint angles $(q_1, \ldots, q_{N-1})$ of the snake robot in order to measure the absolute link angles (1) and the position of the CM (p_x, p_y) of the robot. In order to obtain the derivatives of the reference head angle (46), we used the same technique as in the simulations, i.e. passing Φ_N through a low-pass filter of the form (67). The parameters of the low-pass filter were set to $\omega_n = \pi/2$ and $\psi_f = 1$.

In the following, we elaborate on a few adjustments that were made in the implemented path following controller in order to comply with the particular properties and capabilities of the physical snake robot employed in the experiment. We conjecture that these adjustments only marginally affected the overall motion of the robot. The successful path following behaviour of the robot demonstrated below supports this claim. Since the experimental setup only provided measurements of the joint angles and the position and orientation of the head link, we chose to implement the joint controller in (57) as

$$\vartheta_i = -k_p y_i \tag{68}$$

where $i \in \{1, \ldots, 10\}$. We conjecture that eliminating the joint angular velocity terms from (57) did not significantly change the dynamic behaviour of the system since the joint motion was relatively slow during the experiment. The main consequence of excluding the velocity terms from (57) is that we potentially introduce a steady-state error in the tracking of the joint angles. Consequently, since with the joint control law (68) the derivative terms in (63) are identically zero, they need not to be linearized in the head angle dynamics by the dynamic compensator.

As the result, we implemented the dynamic compensator of the form

$$\ddot{\phi}_o = \left(\sum_{i=1}^{N-1} \beta_i\right)^{-1} \left(-f_{\theta_N} + d^2\Phi_N - k_{p,\theta_N} y_N - k_{d,\theta_N} \dot{y}_N\right) - k_p \phi_o - k_d \dot{\phi}_o \tag{69}$$

where the controller gains were $k_{p,\theta_N} = 20$, $k_{d,\theta_N} = 1$, $k_p = 10$, and $k_d = 5$. We saturated the joint angle offset ϕ_o according to $\phi_o \in [-\pi/6, \pi/6]$, in order to keep the joint reference angles within reasonable bounds w.r.t the maximum allowable joint angles of the physical snake robot. Moreover, from Figure 2, it can be seen that the head link of the physical snake robot does not touch the ground since the ground contact points occur at the location of the joints. As a results, we implemented (69) with $f_{\theta_N} \equiv 0$. The solutions of the dynamic compensator (69) were obtained by numerical integration in *LabVIEW* which was used as the development environment.

We chose the look-ahead distance of the path following controller as $\Delta = 1.4$ m. The initial values for the configuration variables of the snake robot were $q_i = 0$ rad, $\theta_N = -\pi/2$ rad, $p_x = 0.3$ m, and $p_y = 1.7$ m, i.e. the snake robot was initially headed towards the desired path (the x-axis), and the initial distance from the CM to the desired path was 1.7 m. Furthermore, the parameters of the constraint functions for the joint angles, i.e. (45), were $\alpha = \pi/6$, $\eta = 70\pi t/180$, and $\delta = 36\pi/180$, while the ground friction coefficients were $c_t = 1$ and $c_n = 10$ (i.e identical to the simulation parameters).

Results and discussion

The results of the experiments are illustrated in Figures 8, 9, 10, 11, 12, 13. In particular, Figure 8 shows that the

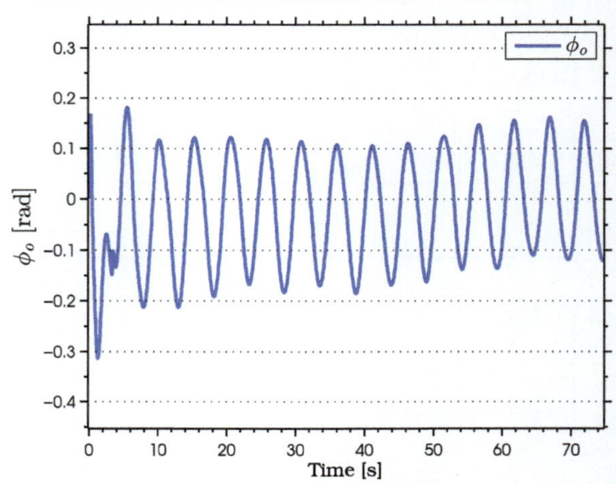

Figure 8 The solution of the dynamic compensator in experiments. The solution of the dynamic compensator remains bounded during the experiments.

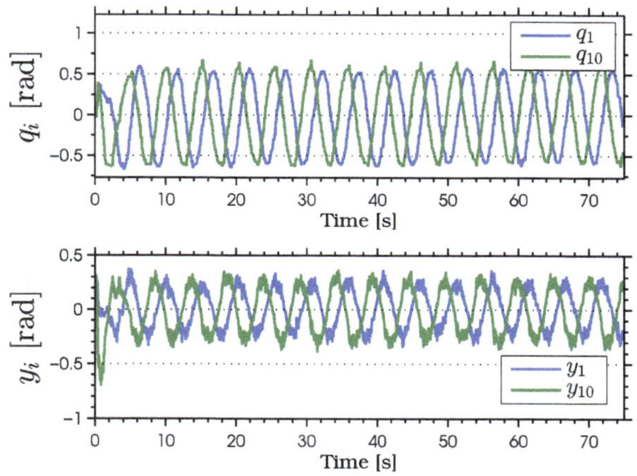

Figure 9 Joint angles tracking the reference joint angles during the experiments. The joints of the robot track the sinusoidal motions (above). The tracking errors converge to a neighbourhood of zero during the experiment (below).

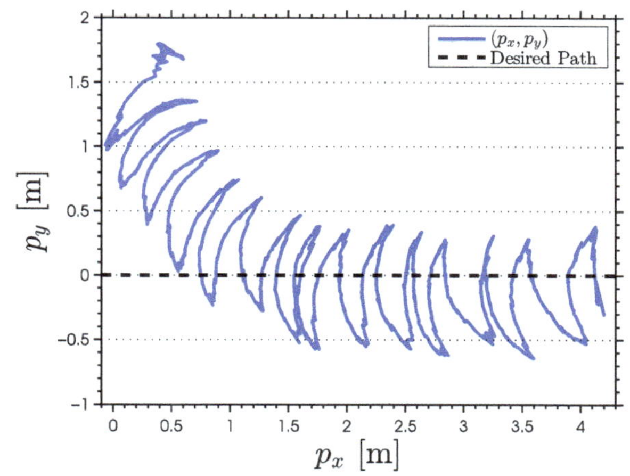

Figure 11 The motion of the center of mass in the *x*-*y* plane during the experiments. The position of the CM of the robot (blue) converges to and follows the desired straight path (the *x*-axis).

solution of the dynamic compensator remained bounded. Figure 9 shows that the joints of the robot tracked the sinusoidal reference angles provided by the constraint functions (45) and that the tracking error converged to a neighbourhood of the origin. As discussed above, this is probably due to the modification of the joint controller (68) due to the lack of velocity measurements in the lab. Figure 10 shows that the head angle of the robot tracked the reference head angle defined by the constraint function (46) and that the tracking error converged to a neighbourhood of the origin. Figure 11 shows the motion of the CM of the robot in the *x*-*y* plane, which converged to and followed the desired path. Figure 12 compares the motion of the CM during the simulation and the experiment, which were performed using the same controller parameters in order to obtain comparable results. In particular, from Figure 12, it can be seen that the physical snake and the simulated snake follow almost the same path. However, due to precise measurement and a more accurate joint control law for the simulated snake, the path following error converges to a smaller neighbourhood of the origin. Figure 13 shows screenshots from a video recording of the experiment.

Conclusions

This paper has considered path following control of planar snake robots by using virtual holonomic constraints. The equations of motion of the snake robot were derived using

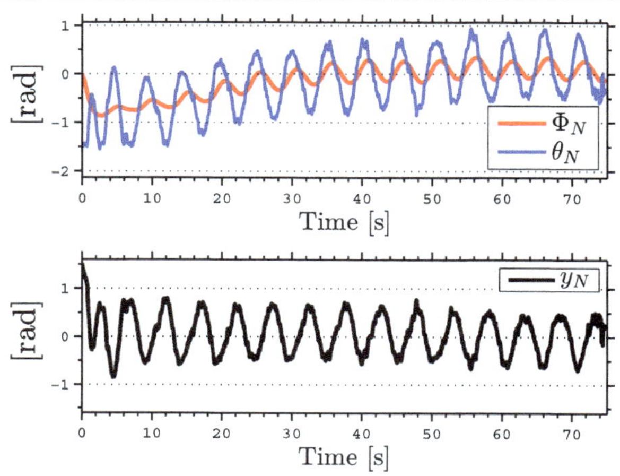

Figure 10 Head angle tracking error during the experiments. The head angle tracks the reference head angle (above). The head angle tracking error converges to a neighbourhood of zero (below).

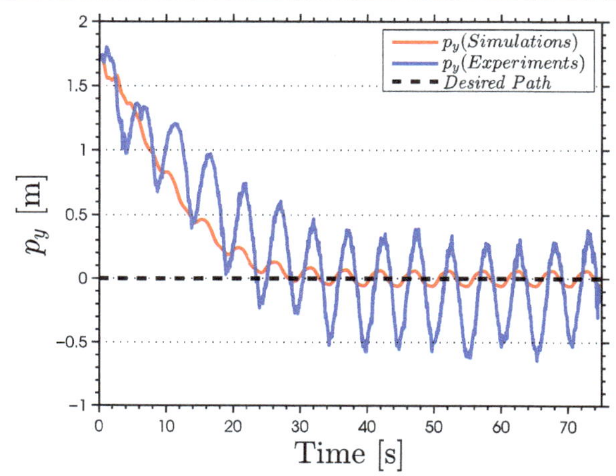

Figure 12 Comparison between experiments and simulations. Comparison of the convergence of the cross-track error during simulations (red) and experiments (blue).

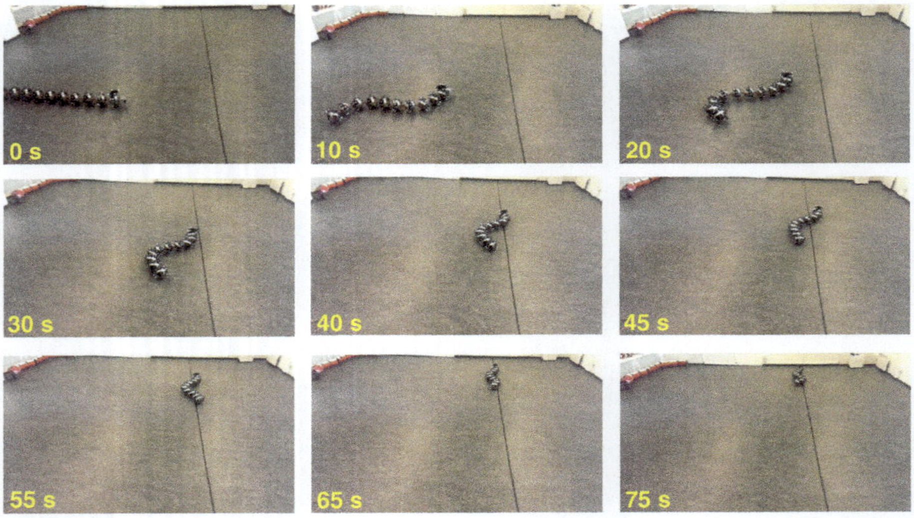

Figure 13 An image of the motion of the robot during the experiments. The robot converges to and follows the desired path.

a Lagrangian framework. We then introduced virtual holonomic constraints that defined the geometry of a constraint manifold for the robot. We showed that the constraint manifold can be made positively invariant by a suitable choice of feedback, and we designed an input-output feedback linearizing control law to exponentially stabilize the constraint manifold for the system. We presented simulation and experimental results which validated the theoretical design. In particular, the robot successfully converged to and followed a desired straight path.

As a topic of future work, we aim to prove the practical stability of the desired path with the proposed control approach. Furthermore, a formal proof for boundedness of the solutions of the dynamic compensator remains as a topic of future work. Moreover, application of the proposed control strategy for more complex paths such as curved paths, using different path following guidance laws, remains as a topic of future work.

Competing interests
The authors declare that they have no competing interests.

Authors' contributions
All authors contributed equally and significantly in the writing of this paper. All authors read and approved the final manuscript.

Acknowledgements
This research was partly supported by the Research Council of Norway through the Centres of Excellence funding scheme, project no. 223254 AMOS and project no. 205622.

Author details
[1]Department of Engineering Cybernetics, Norwegian University of Science and Technology (NTNU), Trondheim 7491, Norway. [2]Department of Applied Cybernetics, SINTEF ICT, Trondheim 7465, Norway.

References
1. Liljebäck P, Stavdahl Ø, Beitnes A (2006) SnakeFighter - development of a water hydraulic fire fighting snake robot. In: Proc. IEEE international conference on control, automation, robotics, and vision ICARCV, Singapore
2. Wang Z, Appleton E (2003) The concept and research of a pipe crawling rescue robot. Adv Robot 17.4339–358
3. Fjerdingen SA, Liljebäck P, Transeth AA (2009) A snake-like robot for internal inspection of complex pipe structures (PIKo). In: Proc. IEEE/RSJ international conference on intelligent robots and systems, St. Louis, MO, USA
4. Dacic DB, Nesic D, Teel AR, Wang W (2011) Path following for nonlinear systems with unstable zero dynamics: an averaging solution. IEEE Trans Automatic Control 56880–886
5. Hirose S (1993) Biologically inspired robots: snake-like locomotors and manipulators. Oxford University Press, Oxford, England
6. Matsuno F, Sato H (2005) Trajectory tracking control of snake robots based on dynamic model. In: Proc. IEEE international conference on robotics and automation, pp 3029–3034. 18-22 April 2005
7. Date H, Hoshi Y, Sampei M (2000) Locomotion control of a snake-like robot based on dynamic manipulability. In: Proc. IEEE/RSJ international conference on intelligent robots and systems, Takamatsu, Japan
8. Tanaka M, Matsuno F (2008) Control of 3-dimensional snake robots by using redundancy. In: Proc. IEEE international conference on robotics and automation, Pasadena, CA, pp 1156–1161
9. Ma S, Ohmameuda Y, Inoue K, Li B (2003) Control of a 3-dimensional snake-like robot. In: Proc. IEEE international conference on robotics and automation, vol. 2, Taipei, Taiwan, pp 2067–2072
10. Tanaka M, Matsuno F (2009) A study on sinus-lifting motion of a snake robot with switching constraints. In: Proc. IEEE international conference on robotics and automation, pp 2270–2275. 12-17 May 2009
11. Prautsch P, Mita T, Iwasaki T (2000) Analysis and control of a gait of snake robot. Trans IEE J Ind Appl Soc 120-D372–381
12. McIsaac K, Ostrowski J (2003) Motion planning for anguilliform locomotion. IEEE Trans Robot Automation 19637–652
13. Hicks G, Ito K (2005) A method for determination of optimal gaits with application to a snake-like serial-link structure. IEEE Trans Automatic Control 501291–1306
14. Ma S, Ohmameuda Y, Inoue K (2004) Dynamic analysis of 3-dimensional snake robots. In: Proc. IEEE/RSJ international conference on intelligent robots and systems, pp 767–772. 28 Sept.-2 Oct. 2004
15. Ma S (2001) Analysis of creeping locomotion of a snake-like robot. Adv Robot 15(2)205–224

16. Liljebäck P, Pettersen KY, Stavdahl Ø, Gravdahl JT (2013) Snake robots - modelling, mechatronics, and control. Advances in industrial control. Springer
17. Liljebäck P, Haugstuen IU, Pettersen KY (2012) Path following control of planar snake robots using a cascaded approach. IEEE Trans Control Syst Technol 20:111–126
18. Rezapour E, Pettersen KY, Liljebäck P, Gravdahl JT (2013) Path following control of planar snake robots using virtual holonomic constraints. Paper presented at the IEEE international conference on robotics and biomimetics, Shenzhen, China
19. Liljebäck P, Pettersen KY, Stavdahl Ø, Gravdahl JT (2013) Controllability and stability analysis of planar snake robot locomotion. IEEE Trans Automatic Control 56(6):1365–1380
20. Westervelt ER, Grizzle JW, Chevallereau C, Choi JH, Morris B (2007) Feedback control of dynamic bipedal robot locomotion. CRC press, Boca Raton
21. Maggiore M, Consolini L (2013) Virtual holonomic constraints for Euler-Lagrange systems. IEEE Trans on Automatic Control 58(4):1001–1008
22. Consolini L, Maggiore M (2010) Control of a bicycle using virtual holonomic constraints. In: Proc. 49th IEEE conference on decision and control, Atlanta, Georgia, USA, December 15-17, 2010
23. Shiriaev A, Perram JW, Canudas-de-Wit C (2005) Constructive tool for orbital stabilization of underactuated nonlinear systems: virtual constraints approach. IEEE Trans Automatic Control 50(8):1164–1176
24. Freidovich L, Robertsson A, Shiriaev A, Johansson R (2008) Periodic motions of the Pendubot via virtual holonomic constraints: theory and experiments. Automatica 44(3):785–791
25. Spong MW, Hutchinson S, Vidyasagar M (2006) Robot modeling and control. John Wiley and Sons, New York
26. Bullo F, Lewis A (2005) Geometric control of mechanical systems. Springer
27. Fossen TI (2002) Marine control systems: guidance, navigation and control of ships, rigs and underwater vehicles. Trondheim, Norway, Marine Cybernetics

State of the art: micro-nanorobotic manipulation in single cell analysis

Yajing Shen[1*] and Toshio Fukuda[2]

Abstract

Single cell analysis is an essential research approach to reveal the secret of life. At such a small scale, it puts forward higher demand on the accuracy of the cell manipulation system. This paper reviews the state of the art of micro-nanorobotic manipulation in single cell analysis. First, the key applications of the micro-nanorobotic manipulation system in single cell analysis are introduced, including the single cell injection, positioning, characterization, assembly, and the development of biomedical device. Then, the current key techniques, challenges, and future trends in micro-nanorobot are discussed from the aspects of actuating, sensing, controlling, system integration, and commercialization. To meet the requirement of cell biology, the next generation micro-nanorobot should have small, automatic, high integration and applicable features in multiscale.

Keywords: Single cell analysis; Micro-nanorobot; Micro-nanomanipulation; Biomedical

Review

Introduction

The single cell is the smallest unit of living things and is often called the building block of life [1]. This small organizational system maintains a highly complex and hierarchical architecture of interconnected molecular networks and represents life and its related consciousness [2]. The understanding of such a complex system will help us to elucidate the functioning of complete organs and will pave new ways to cure diseases or to develop personalized medicine.

One contribution of single cell analysis relies on the existence of cellular heterogeneity within the same population of cells. Bulk-scale measurements, which rely on data averaged across thousands or millions of cells, report only average values for the population. Thus, the characteristics of a group can obscure the differences between the individuals in it. Because studies made at the single cell level are not subject to the averaging effect characteristic of bulk-phase population-scale methods, they offer a level of discrete observation that is unavailable with traditional biological methods [3]. Secondly, single cell analysis provides an approach to identify the differences between individual cells. By characterizing the single cells at different conditions, such as health and unhealthy, we can take the single cell as an ideal model to detect disease, test medicines, and develop drugs. Furthermore, analyzing the content within single cells would benefit the deeper understanding of cellular functionalities and behaviors. For example, in the field of immunology, single cell analysis can help us to monitor gene expression in complex tissues, to define different cell subsets, to get insights into both their functional capacity and their specificity, and to track antigen-specific cells [2]. Therefore, nowadays, techniques for single cell analysis have attracted increasing interests in cell biology and the life science.

Single cell analysis is a broad subject that covers many research areas, such as single cell genomics, immunology, diagnostics, and so on. Many review articles have been reported to discuss the problems and trends of single cell research from the biological view [4-8]. This article will not address the above biological problems deeply but will mainly focus on the engineering problems behind those biological issues. To perform single cell analysis, the analysis platform should have both accurate observation and manipulation abilities. For observation, various effective microscopes have been developed, such as optical microscopy (OM), atomic force microscopy (AFM), environmental scanning electron

* Correspondence: yajishen@cityu.edu.hk
[1]Department of Mechanical and Biomedical Engineering, City University of Hong Kong, Tat Chee Avenue, Kowloon, Hong Kong 99907, China
Full list of author information is available at the end of the article

microscope (ESEM), scanning ion conductance microscopy (SICM), and so on. Generally, each observation system has its own advantages and issues for the single cell analysis; for instance, OM is easy to use but cannot provide high image resolution, AFM can provide high image resolution but limited to the nonreal-time observation, and ESEM allows real-time high resolution imaging but cannot for *in vivo* cell research. The review about the observation system for single cell analysis at nanoscale can be found from the article report by Zheng [9]. These different types of observation systems make high demand on the manipulation system correspondingly. Here, micro-nanorobot technique is widely employed owing to its accuracy, controllability, flexibility, and other unique features. Many articles have been reported to discuss the application of the micro-nanorobot in the biomedical field. For example, Dario reviews robotics and microsystems in biomedical applications [10,11], Park reports on the biomedical robot for intravascular therapy [12], Nelson reviews microrobotics for minimally invasive medicine [13], Lenaghan reviews the art and challenge of nanorobot for cancer therapy [14], and so on. For the specific topic of single cell analysis, the micro-nanorobot plays more and more important roles in recent years but few review papers have been reported.

Micro-nanorobotic technique allows the precise manipulation at a small scale. It provides essential engineering support for the biological research at the single cell level, including the cell injection, positioning, characterization, surgery, and so on, which gives us a novel approach to elucidate the fundamental and practical principles within the cell. In this article, we will review the state of the art of micro-nanorobotic manipulation for single cell analysis, including its key applications, main techniques, the current challenges, and the trends in the future.

Key applications

With the rapid growth of nanotechnology in the last few decades, the techniques for precise positioning and manipulation also have a fast development. Micro-nanorobotic manipulation is an advance method by integrating the traditional robotic technique and micro-nanotechnology together. Generally, the micro-nanorobotic manipulation system allows precise positioning ranged from several nanometers to micrometers and provides flexible movement ability with multiple degrees of freedom (DOFs) in three-dimensional space. In addition, it has high controllability that can automatically perform some tasks based on computer programming. Moreover, the micro-nanorobotic manipulation can integrate with various commercial microscopes, such as OM, SEM, and AFM, to solve different biological problems effectively. Owing to these advantages, micro-nanorobotic manipulation has been widely used in the biological fields. In this section, we will focus on some key applications of the micro-nanorobotic manipulation system in single cell analysis.

Cell injection

Technologies of delivering exogenous materials into cells play a central role in experimental cell biology, such as cell cloning, DNA therapy, and so on. As one type of nonviral vector-mediated method, micropipette injection has been widely accepted owing to its low toxicity and high delivery efficiency. Microinjection is normally performed with a micromanipulator, which refers to the process of using a glass micropipette to insert into or suck out substances from a single living cell [15-17]. In this process, two key procedures are how to move the pipette to the cell precisely and how to penetrate the cell membrane safely, which directly determine the injection success and efficiency. At such a small scale, it becomes very inefficient and ineffective to perform the injection task by hand directly. Therefore, precise methods based on the micro-nanorobotic manipulation system are widely used. Two main types of injection systems are illustrated in Figure 1.

The glass pipette has been used for drug delivery since several decades ago. Normally, the single cell is handled by a pipette straw and the microinjector is assembled on the micromanipulator. Thus, the position of the micropipette can be controlled via the manipulator precisely, which usually has a positioning resolution higher than 2 μm and more than three DOFs [20]. In the traditional methods, the micromanipulator is only taken as a precision tool, which lacks intelligence. To improve the accuracy and to reduce the time cost in the injection process, the automatic injection system is developed based on robotic control technique in recent years. For example, Sun takes the images of the objects, including the cell and glass pipette, as the feedback signals to monitor the injection process dynamically. As a result, an automatic injection process can be realized, which greatly improves the injection success rate and reduces the time cost [18]. In addition, a force sensor is used to measure the penetration force, which is taken as the feedback to ensure the injection safety and improve the injection efficiency. For instance, Xie proposes a two-loop force control framework to regulate the penetration force in the fish embryos injection [21]. Yanagida proposes a piezo-micromanipulator to prevent the large deformation of the cell during the injection [22]. It utilizes the rapid deformation of a piezoelectric element to realize smooth insertion in intracytoplasmic sperm injection, which reduces the penetration force and improves the survival rate. Tan reports a mechanical model to elucidate the cell deformation in the penetration process,

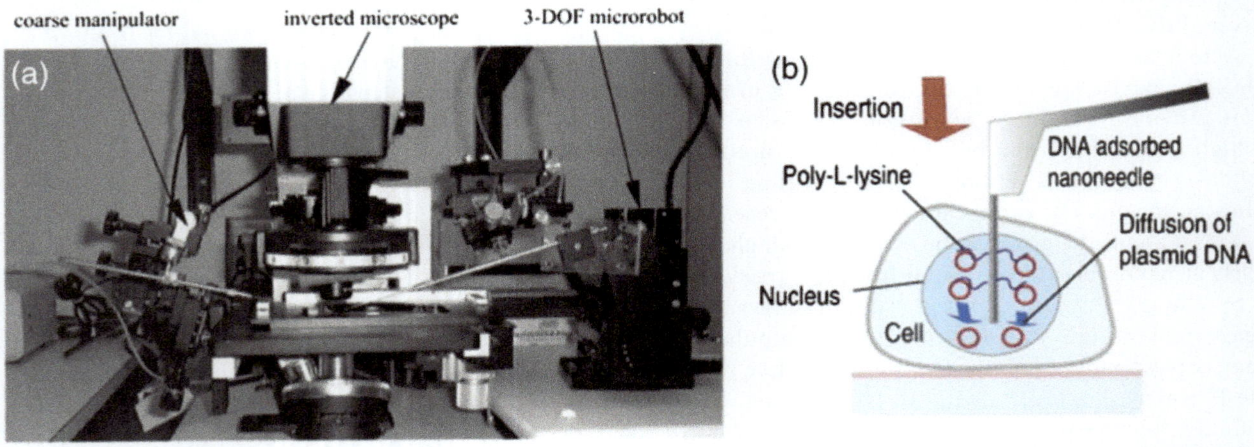

Figure 1 Micro-nanorobotic manipulation system for single cell injection. (a) Autonomous embryo injection system by glass micropipette [18]. **(b)** Schematics of an AFM cantilever-based nanoinjection [19].

which paves a way to optimize the injection process theoretically [23]. The usage of the robotic technique, including image processing and force feedback, could improve the traditional injection system greatly. However, there are still some intrinsic problems. One serious problem is that the injection is performed under OM, which has an imaging resolution limitation (approximately 200 nm); therefore, it makes challenges to the image processing. The limited visual range of the microscopy is another challenge to the micro-nanorobotic system. It is clear that a wider visual range will help us to find the object easier, but the low magnification image will result in the loss of image details. Even though the automatic single cell injection has been realized, this technique is usually performed under an invariable magnification. Integrating the automagnification technique with the micro-nanorobotic manipulation technique will improve the productivity, which should be considered in the future.

In addition to the glass pipette, nanoinjection methods based on an AFM cantilever nanoprobe are also proposed recently [19,24]. In these techniques, a smaller injector with nanometer size is used to penetrate the cell and deliver the materials into the cell. The manipulation strategy of the nanoneedle is similar to that of the glass pipette for single cell injection. However, there are few automation techniques reported for the nanoinjection process. We think one main reason is that the nanoinjection technique is still at the stage of laboratory research; therefore, it does not have an urgent demand on the automatic system. However, in the long term, the improvement of the nanoinjection system is important for practical applications.

Consequently, micro-nanoinjection is a simple process in which a needle penetrates the cell membrane and/or the nuclear envelope. The micro-nanorobotic manipulation system provides an effective platform for the injection process. Comparing with manual injection, a robotic manipulation system has the advantages of high quality, productivity, and repeatability. In the future, a commercialized automatic-injection equipment that allows a user to operate it easily without specialized training is expected to be developed.

Cell positioning

Cell positioning is an essentially important technique, since it is the basis of cell array and single cell analysis in biomedical research. One of the main challenges for cell positioning is how to transfer a single cell to a certain place precisely and effectively. Nowadays, to address the above issues in cell positioning, several micro-nanorobotic manipulation systems have been proposed. Figure 2 illustrates some micro-nanorobotic manipulation systems for single cell positioning.

One direct method for cell positioning is based on microrobot and microtools. In this design, the microtools, such as the micropipette and microgripper, are assembled on the microrobotic manipulation system. The microtool has the capability to handle and release the cell and the microrobot can control the position of the microtool precisely. Therefore, this integrated system can pick up the single cell and place it on the desired position [25,30,31]. To improve the accuracy and reproducibility, force feedback and version feedback are also employed in the cell positioning process. For example, Kim reports a nano-Newton-force-controlled micromanipulator for the soft object manipulation, which can detect the contact force fast and handle the object more safely [32]. Fatikow proposes an automatic control frame for the micro-object manipulation and positioning based on vision feedback, which greatly improves the working efficiency [33]. The microtool manipulation approach

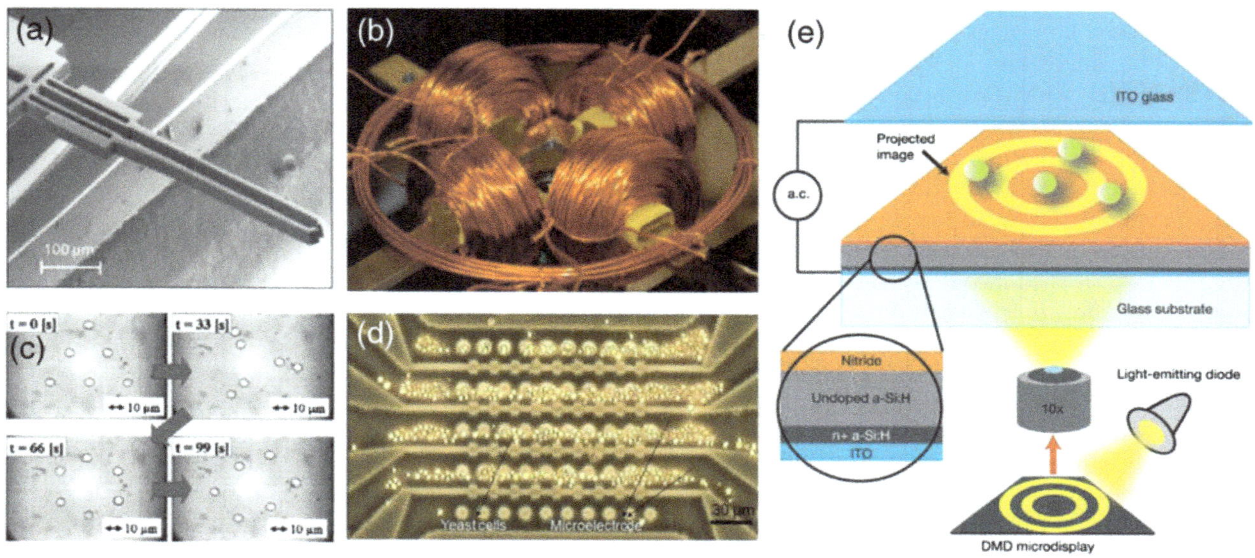

Figure 2 Micro-nanorobotic manipulation system for single cell positioning. (a) Scanning electron micrographs of a microgripper [25]. **(b)** Five electromagnetic coils are mounted on an optical microscope to actuate the magnetic robot in the center [26]. **(c)** Microbeads dancing controlled by the optical tweezers [27]. **(d)** Cell positioning by dielectrophoresis (DEP) chip [28]. **(e)** Schematic drawing of the optoelectronic tweezers (OETs) [29].

enables the pick-transport-place of a specific single cell precisely and effectively. However, the issue of this method lies in that the microtool has to contact with the cell during the manipulation process. Therefore, it is difficult, if not impossible, to implement this platform in a closed environment, as in the microfluidic channel.

A magnetic robot is actuated by the magnetic field remotely, by which it avoids the usages of electric wires [26]. Therefore, it is able to manipulate and position cells in both open and closed environment. In this kind of manipulation system, the cell movement is usually achieved by the pushing action of the robot. Therefore, this method has a higher requirement on the control system. We have to control the location of the robot and the pushing direction dynamically, which is not convenient for flexible manipulation, especially for the mass cell manipulation.

Optical tweezers are instruments that use a highly focused laser beam to hold and manipulate a small object. In the late 1980s, Arthur and Joseph demonstrated the first application of optical tweezers to the biological sciences, using it to trap an individual tobacco mosaic virus and *Escherichia coli* bacterium [34]. Today, optical tweezers have been a comment scientific instrument in trapping cells to study a variety of biological problems [35-37]. Fukuda develops a microrobotic manipulation system based on the controlling of the light path of the optical tweezers. It is able to manipulate small objects in 3D space flexibility, such as microbead and single cell [27]. Sun studies the dynamics behavior of the cell in the optical tweezers manipulation and then designs a closed-loop control algorithm to transport a single cell or multi-cells to a certain place [38]. Because optical tweezers can manipulate the object precisely without physical contact, it is very powerful for the cell positioning inside the microchip. However, the main drawback of the optical tweezers is that the generated force by laser is usually limited to nano-Newton, which is not large enough to drive the non-suspending cells sometimes.

Dielectrophoresis (DEP) is the phenomenon in which a force is exerted on a dielectric particle when it is subjected to a nonuniform electric field [39-41]. DEP has the advantages of noncontact manipulation, good controllability, and high-integration ability within the microchip. Therefore, it has been widely used for cell trapping and sorting, especially for the cell analysis in the microfluidic chip [28,42]. One problem of DEP lies in that the strength of the force strongly depends on the medium and particles' electrical properties, on the particles' shape and size, as well as on the frequency of the electric field. Another problem is that it is difficult to make the size of the electrode to nanoscale, which affects the manipulation accuracy.

To overcome the drawback of DEP, especially to avoid the electrode fabrication, a new concept called optoelectronic tweezers (OETs) is proposed [29]. OETs use projected optical images to grab and corral tiny particles with sizes ranging from hundreds of micrometers to tens of nanometers. In this method, light first creates some 'virtual electrodes' on the substrate. Then, the image in conjunction with an externally applied electrical bias creates the localized DEP traps in the illuminated areas. It combines the advantages of optical tweezers and electrode-based DEP, which is expected to have a bright future.

As a summary, cell positioning is a very important process in biomedical research. Nowadays, many different types of manipulation strategies have been proposed to deal with this problem. Each method has its own features that can be applied for some certain cases efficiently. Understanding their advantages and drawbacks is very helpful for us to choose the proper approach and develop new methods for the single cell analysis. Considering that the on-chip analysis is more and more important nowadays, a micro-nanorobotic manipulation system that allows cell positioning in a closed environment will be a trend in the future.

Cell characterization

Single cell characterization provides a quantitative method to study the cell's behavior and to elucidate the cell's activate. For example, the mechanical properties of the single cell can reflect the cell's healthy condition, which benefits disease diagnosis and drug testing [43]. The electrical property of the single cell can reflect the open or close status of the ion channels of the cell membrane, which benefits the understanding of cell transportation [44]. In single cell characterization, micro-nanorobotic manipulation plays an important role owing to its flexible manipulation capability. Many types of micro-nanorobots and micro-nanotools are used to handle, manipulate, and characterize the single cell to obtain various types of properties of the single cell, including strength, stiffness, resistance, and so on, as illustrated in Figure 3.

The cell's activities, such as mitosis, morphogenesis, cell orientation, cell motility, and survival, depend on the attachment to the extracellular matrix or neighboring cells. Characterization of adhesion strength at the single cell level provides an effective approach to elucidate these activates. Micropipette aspiration is one traditional technique for cell-cell adhesion study [48]. In this method, two aspiration pipettes are assembled on two micromanipulators. Then, the pipettes are used to handle and separate two single cells driven by the micromanipulator. In this process, the cell-cell adhesion strength can be calculated from the suction pressure applied through the pipette. The AFM system is another powerful system for the single-cell-adhesion characterization, which enables to measure the cell-substrate or the cell-cell adhesion force. For example, one common method is to put the cell on the AFM sample stage, which is taken as the micromanipulator (x-y-z). Then, the cell is moved against the AFM cantilever driven by the stage, and the adhesion strength can be measured from the deflection of the AFM cantilever [49]. Puech reports a method to measure the cell-cell adhesion strength [50,51]. Here, one cell is immobilized on a modified AFM cantilever tip by chemical glue. Then, the adhesion force between this cell and the other cell on AFM sample stage is obtained by separating these two cells. The AFM system also allows the measurement of the adhesive molecules on the cell membrane [52,53]. Comparing with the aspiration pipette, the AFM system has the advantages of higher positioning accuracy and higher force resolution. However, one drawback is that the manipulation depends on the movement of the AFM sample stage, which is not as flexible as end-effector manipulators. Another issue is that the AFM system cannot scan the sample to get a high resolution image during the characterization process. Environmental SEM (ESEM) is one special type of SEM that can maintain a wet environment within the sample chamber, which provides a novel platform for single cell analysis. Shen reports a single-cell-adhesion study system based on the nanorobotic manipulation system inside ESEM, which can measure both the cell-cell adhesion and cell-substrate adhesion strength [45,54-56]. This manipulation system

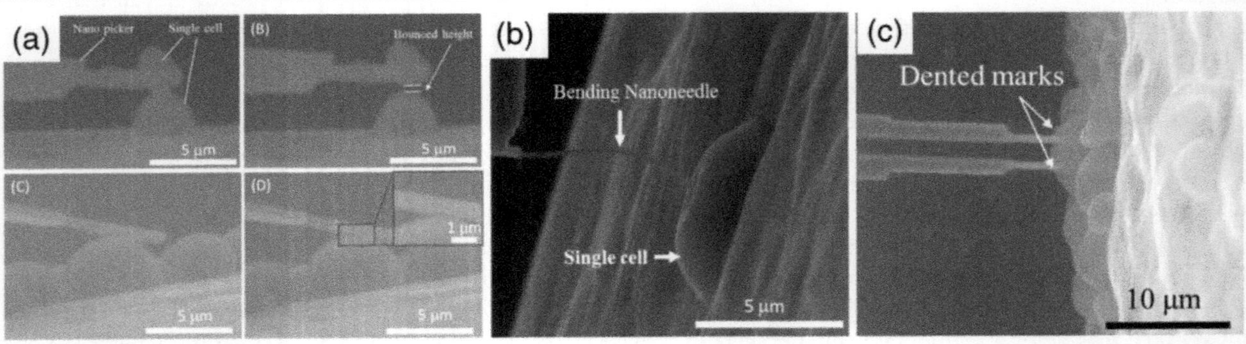

Figure 3 Examples of the single cell characterization by micro-nanorobot. (a) Cell-adhesion-force characterization [45]. **(b)** Cell mechanical property characterization [46]. **(c)** Cell electricity characterization [47].

has seven DOFs in total and has the positioning resolution approximately 30 nm. More important, this system allows the real-time imaging at nanometer resolution during the manipulation and characterization. These features could provide more information in the characterization process, which could benefit the dynamic single cell study in the future.

Single-cell-mechanical-property characterization, such as stiffness and elasticity, is another example that benefits from micro-nanorobotic manipulation system. Cell stiffness value closely relates to the structure of the cell's membrane and the cytoskeleton within the cell. Therefore, the stiffness of a single cell reflects the cell function in many aspects [57,58]. For example, the stiffness of the mammalian cell shows obvious difference after medical treatment or after death, which could benefit disease diagnosis, drug development, and disease therapy [59-61]. Recently, various methods have been reported to characterize the single cell's stiffness based on micro-nanorobotic manipulation technique, including micro-electromechanical systems (MEMS) tool, optical tweezers, magnetic beads [62,63], AFM system, ESEM nanotool, and so on. Nelson assembles a MEMS probe on a microrobotic manipulator to measure the stiffness of the pollen tube, where the MEMS probe is taken as the force sensor and the microrobotic manipulator is used to control the position of the probe [64]. Dao studies the mechanics of the human red blood cell by optical tweezers [65]. Optical tweezers enable high positioning accuracy and high force resolution for the stiffness characterization. However, the relative low force ability (approximately piconewton), the light-induced damage, and the observation accuracy (approximately 200 nm) are thought as the challenges for optical tweezers. Magnetic manipulation takes usage of the magnetic field to drive a small bead to deform the cell, which avoids potential light-induced damage as in laser trapping. Bausch measured the local viscoelasticity of the surface of adhering fibroblasts by controlling the position of the magnetic beads via magnetic fields [63]. The modeling and controlling of the nonhomogeneous magnetic field are considered as the challenge to get accuracy data for the magnetic manipulation method. The high force sensitivity of the AFM system makes it particularly appropriate for studying cell stiffness [66-68]. Here, the AFM cantilever is used as a nanoindentor to deform the cell. As a result, the stiffness can be calculated based on the cell deformation and the applied force [69,70]. Despite the AFM technique having been well-developed and having shown to be noninvasive to the cell if applied correctly, one existing issue for the AFM cantilever is that the deflection of the cantilever beam may cause slipping on the cell surface during the measurement process. Shen reports a nanoneedle to overcome this drawback [71]. They used the nanorobotic manipulator to control the movement of the nanoneedle with nanoscale resolution inside an ESEM chamber, which allows the characterization of the local region of the cell surface. Consequently, the application of micro-nanorobotic manipulation has enhanced the single cell stiffness characterization greatly. However, the aim of the research is to exactly elucidate how the stiffness value relates to the active cells. From the mechanical information, we expect to know how healthy the cell is rather than if the cell is alive or not. This requirement is bringing new challenges to the current micro-nanorobotic manipulation system, i.e., much faster, more accuracy, and more intelligent.

Biological cells, which consist of a conductive salt-water interior coved by a relatively insulated outer sheath, continuously exchange ions with surroundings. Therefore, the electrical property of a single cell can reflect the physiological state of the cell. Nowadays, the characterization of a cell's electroconductivity has attracted increasing interest in the cell viability identification, ion channel monitoring, drug testing, and other biological fields. Ahmad designs a dual-probe and integrates it with a nanorobotic manipulation system [47]. This system is used to measure the electrical resistance of the single cell, and the results clearly show the variability between living and dead cells. Zhang studies the electrical response of the human embryonic kidney cell corresponding to external mechanic stimulation based on a robot-assisted AFM manipulation system and a patch-clamp system, which can monitor the ion transfer through the ion channels in the cell membrane [72]. This system has been expected to test the effectiveness of drugs in the future. Although micro-nanorobotic manipulation allows the characterization of the single cell electrical conductivity, there are still some challenges. For example, monitor of ion transfer in one single ion channel is still difficult, since it requires higher imaging, positioning, and sensing ability to the robotic manipulation system. Another challenge is the high throughput, i.e., how to find the regions and characterize the electrical property of the single cell faster, which has a higher requirement on the robot intelligence and automatics.

As a summary, micro-nanorobotic manipulation is the essential technique for accuracy positioning and manipulation. It has been widely used in single cell characterization, including the cell's mechanical property, electrical property, its interaction with the environment, and other features. In the future, the micro-nanorobotic manipulation system should be more accurate and intelligent to provide more information for single cell analysis.

3D cell assembly

As the basic unit of life, the single cell is the elementary block to construct tissues and organs, which show complex structures in three-dimensional space. It is clear that cell culture in 3D space is much more close to the practical comparing with the traditional 2D plate culture. Many researches have also proved that the cell growth in 3D space has a different behavior with that in the 2D plate, such as the cancer cell [73]. Moreover, the 3D cell culture is also an essential technique to develop the artificial organs. For example, an artificial ear has been created successfully by culturing the cells on a 3D scaffold [74]. Therefore, recently, increasing interests have been focused on how to build a 3D structure from a single cell. This research topic, called as '3D cell culture', is quit broad that covers many research areas, including but not limited to cell culture, growth, physiology, and so on (Figure 4a) [75]. In this review, we will not discuss the biological problems but will discuss this topic from the engineering point of view, i.e., '3D cell assembly'. It is subject to '3D cell culture', but specifically focuses on the 3D manipulation and assembly methods, in which micro-nanorobotic manipulation involves deeply.

One widely used method for 3D cell assembly is to fabricate a biocompatible 3D 'scaffold' to support the growth of the cell [78]. Thus, the key issue becomes how to fabricate the 3D scaffold with an accurate structure. One main scaffold fabrication technique is based on 3D printing. Here, the scaffold material is deposited layer by layer controlled by a flexible manipulation system. For example, Yan manufactures 3D architectures through a controllable manipulation system with three DOFs (x-y-z) and successfully uses it to study the growth of hepatocytes. Klein fabricates one type of two-component polymer scaffolds through laser, directly writing based on nanomanipulation [76]. This scaffold is successfully applied to study the growth of a single fibroblast cell in 3D space. In addition to 3D printing, another scaffold fabrication technique is also proposed based on physical or chemical reaction. For example, some materials are sensitive to light; that means the biocompatible liquid will be solidified after exposure to light. Based on this feature, the 3D scaffold can be fabricated by controlling the projection area of the light. For example, Liu reports a method by light-induced gel to generate a 3D scaffold, where the shape and height of the gel can be controlled precisely by controlling the light illumination area and the exposure time [79].

Another assembly method is to build the small blocks with cells into a complex 3D structure, as building a house from blocks. In this kind of approach, the biological cells are first capsulated in series gel blocks by biochemical process. Then, a microrobot is used to manipulate these blocks to build the 3D structure. In this method, it is able to change the shape of the structure easily. For example, Tasoglu reports an approach that uses a magnetic crawling microrobot to assemble cell-encapsulating gel blocks into a 3D structure, by which various structures can be built from the same series of blocks (Figure 4b) [77].

The 3D cell assembly is a relative new research area that is rising with the fast growth of the tissue engineering and biomedical engineering research area. Many assembly techniques are benefited from the micro-nanorobotic manipulation system more or less. However, the final goal of 3D cell assembly is to construct the functional artificial structures, which are able to replace

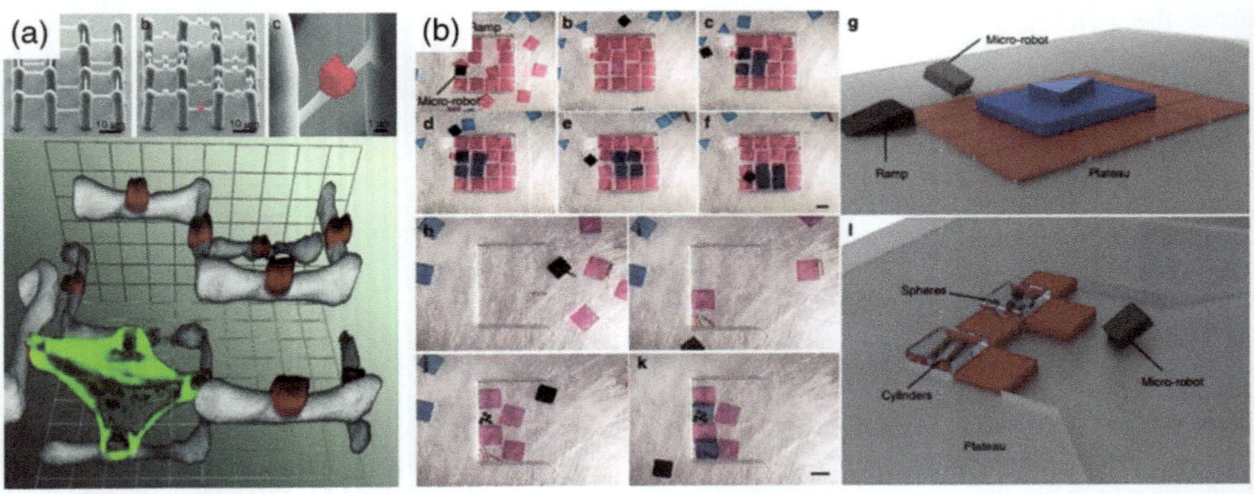

Figure 4 Examples of the 3D cell assembly based on micro-nanorobotic manipulation. (a) Scaffold for single cell 3D growth study by 3D printing [76]. **(b)** 3D Cell assembly by microrobot [77].

the natural tissue and organ. Therefore, it is important to find the way to assemble different types of cells together, which puts forward higher requirement on the intelligence and flexibility of the micro-nanorobot.

Biomedical device

Micro-nanorobotics is not only an essential supporting technique for single cell research but also could be a miniature device for biomedical application directly. For instance, the endoscopic capsule robot is one of the most popular miniature devices based on robot technique. It integrates the actuating, sensing, and controlling system in one capsule within tens of a millimeter [80]. This kind of robot can not only take images inside our body but can also address the diagnostic or treatment issue automatically. However, these techniques are still difficult to apply at the single cell level owing to the small size of the cell. Zhang reports an artificial bacterial-inspired robot. This robot is fabricated from thin films of semiconductors and its movement is controlled by a magnetic field [81]. The length of this small robot is only several tens of a micrometer, which is within the same scale of a single cell. It has the ability to move outside or inside the human body, which can be used for an invasive medical operation in the future. However, these techniques are not yet mature nowadays, because the robot has to face the actuating and sensing challenges within a small space. The development of small biomedical robot that can work as an independent device will be an essential step for the practical applications for single cell analysis. In the future, we should think of a way to make the robot much smaller, have higher integration, and more functionality.

Current techniques, challenges, and future trends

Actuating

A large type of the micro-nanorobotic actuating depends on the mechatronic principle, such as hydraulic motor, electric motor, electrostatic actuator, piezo actuator, and so on. Hydraulic and electric motors have the ability to generate a large force output and provide long movement distance. However, they cannot provide accurate positioning resolution at nanometer scale. Therefore, they are usually taken as the stage to drive the micro-nanomanipulator mode coarsely. Electrostatic actuation is one widely used technique in the development of MEMS device, which allows both the precise actuating and sensing. For example, many micro-nanomanipulators, like the microgripper, are designed based on electrostatic mechanism [82,83]. However, the short moving distance is one problem for electrostatic actuation, which is usually less than several millimeters. Therefore, these kinds of manipulators are usually taken as the end effector for the micro-nanorobot. Piezo actuation is an effective approach that converts electrical power to mechanical movement (Figure 5). It can provide a large movement distance with several centimeters and a positioning resolution as high as several nanometers. Nowadays, the piezo-based manipulator has been well developed and available in market. The mechatronic-principle-based manipulator has a relatively mature physical model; therefore, it is easy to control and implete. One main issue for this type of actuation is the existence of wires, which make it not applicable for work in a closed environment, such as inside a microchip or inside our body.

Wireless actuation-based manipulation avoids the usage of electrical wire to connect the power supplier. Therefore, it provides the possibility to work in a closed environment. The commonly used wireless actuation techniques include optical tweezers, DEP, magnetic driving, and so on. As discussed in the 'Cell positioning' section, these actuators are driving by physical fields, like light field or electromagnetic field. Thus, they can manipulate the object in a narrow space with high positioning accuracy, which are very important for the *in situ* analysis in the microchip. For the wireless actuation, the controllability and repeatability are big challenges, since it is difficult to build an exact physical model to control the statues of the small object in the fields. Another issue is that the force generated by these methods is usually smaller than the mechatronic actuation due to the physical principle.

In addition to the above two types of actuating methods, the biochemical actuator also receives more and more attentions nowadays. In these techniques, creatures or part of creatures are taken as the actuator directly, such as sperm, bacterial, flagella, protein, DNA, and so on, which have the moveable ability naturally. If a proper stimulation is given, the movement of these biorobots can be controlled effectively. For example, Nogawa

Figure 5 A typical micro-nanorobotic manipulation system based on piezo actuation.

uses the ion concentration to control the movement direction of bacterial, which can be used as a robot to deliver drugs [84]. Martel uses magnetic field to control the movement of the flagellated magnetotactic bacteria, which can perform the medical operation in the human microvasculature [85]. Scientists also design many prototypes of the nanorobot based on the unique features of protein or DNA, which can work well in a natural biochemical environment and show its significant potential in the drug encapsulation, drug delivery, and disease therapy [86-89]. These biochemical nanorobots have many advantages, such as small size, simple structure, self-assembly, biocompatible, and so on. The problem is that these types of robot are usually sensitive to the environment, such as the temperature, pH value, chemical solution concentration, and so on. Nevertheless, natural creatures provide us a good example to learn from, which usually have high efficiency and can work better in a biological environment. We believe the biochemical-inspired micro-nanorobot should be a new trend in the future, especially in the biomedical field.

As a summary, actuation is one of the most important techniques for micro-nanorobotic manipulation, which involves power supply, energy transmission, driven approach, and so on. The current actuating techniques have provided many solutions for the single cell analysis. However, there are still some challenges that need to be addressed, including the optimization of size, working time, output capability, efficiency, and so on. The biological sample is the nanorobot in nature, which has a high working efficiency and is exactly compatible for bioapplication. 'learn from nature' will drive new inspiration in the micro-nanorobot development, which is expected to benefit the biomedical field in the future.

Sensing

Unlike the traditional robot sensing, only a few sensing technologies can be used at the micro-nanoscale owing to the specific working environment. Because single cell analysis is usually performed under microscopy, the micro-nanorobotic manipulation has to be workable under microscopy. In this specific case, vision and force are the most effective sensing methods.

In vision sensing, the sample images are captured dynamically from the microscope, firstly. Then, the objects are identified and their positions are calculated based on image processing. After that, the position information is taken as the feedback signal to control the movement of the micro-nanomanipulator. This kind of sensing and control strategy has been successfully used for automatic single cell injection to improve the injection efficiency and reduce the injection time (Figure 6a) [90]. One challenge for the vision feedback lies on the quality of the microscope image. The microscope images usually can only provide a few information about the environment due to the small observation area. Moreover, depth information is hard to obtain from the microscope image directly. Therefore, the object's identification and positioning becomes much more difficult than that at the macroscale.

Force sensing is usually used with vision sensing together to improve the performance of the micro-nanorobot. For example, in single cell injection, the force feedback has been applied successfully to improve the injection process (Figure 6b) [21,91]. The challenge for force sensing lies on the lack of force/torque information from different directions. Generally, six DOFs, three force and three torque, are required to describe the status of one point clearly. However, it is quite difficult for the force-sensing system at micro-nanoscale, which usually can only provide one to three DOFs.

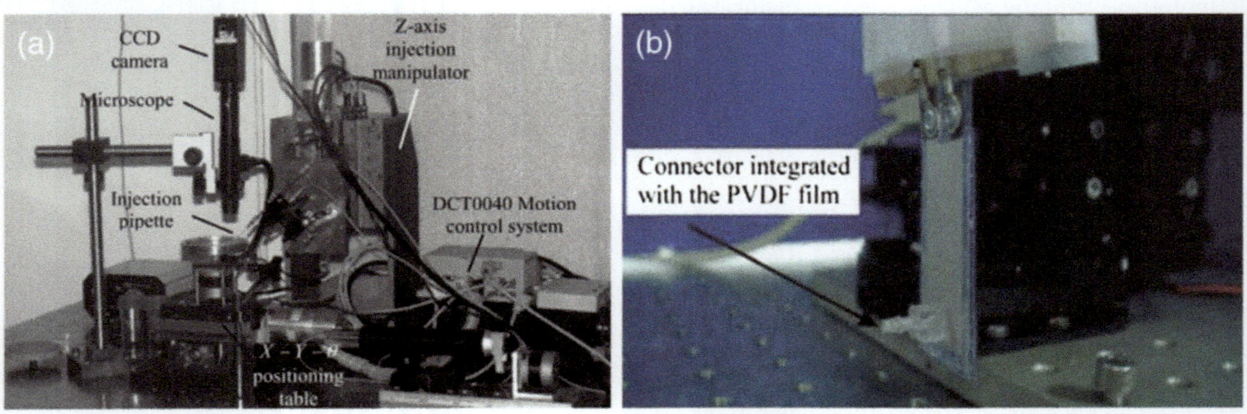

Figure 6 Examples of the sensing system for a micro-nanorobot. (a) A vision feedback system based on the CCD camera for single cell injection [90]. **(b)** A force feedback system based on a PVDF (Polyvinylidene fluoride) piezoelectric polymer film for single cell injection [21].

Therefore, the development of a MEMS force sensor that can work well under microscopy is urgently demanded for the micro-nanorobot sensing.

Control

At micro-nanoscale, the surface force rather than the inertia force becomes dominate, fluids appear as viscous as molasses, and Brownian motion makes everything incessantly shake [92]. These unique physics make the behavior of the object become sophisticated under manipulation. For example, in the single cell gripping process, the cell may adhere to the gripper surface due to the electrostatic force rather than the clamp force, and therefore, the cell cannot be released after the gripper is opened. Thus, the modeling of the control object becomes a big challenge, which is usually nonlinear, dynamic, randomness, and sometimes even unpredictable. In addition, only a few feedback-sensing techniques, visual and force, can work at the micro-nanoscale. Therefore, controlling is a big challenge for the micro-nanorobot owing to the modeling and sensing problems.

Nowadays, the micro-nanorobot can perform some simple tasks automatically, for example, the single cell injection. Here, the mechanical behavior of the cell can be modeled, so the traditional visual feedback, force feedback, with PID controller, and impedance control strategy can be used (Figure 7). However, in most of the cases, the single cell analysis requires complex manipulation at a much smaller scale. It is hard to build the physical model to describe the controlled objected exactly. Thus, most of the control process has to be performed based on the master-slave manipulation, in which the decisions are made by a human rather than the robot itself.

Consequently, the main difficulty for the micro-nanorobotic control is not the control algorithm but lies in the exact modeling of the controlled object. We must consider the unnormal physical phenomena at a small scale. In addition, the lack of powerful feedback signals is also considered as a drawback for the micro-nanorobot system. In future work, molecular dynamics modeling or quantum electrodynamical models should be considered to better describe the property and behavior of the object at small scale. In addition, the small and intelligent sensor for micro-nanorobot should be another essential way to solve current control problem.

System integration

The micro-nanorobotic manipulator has the ability to move in micro-nanometer scale accurately, but it does not mean the size of the robot system itself is within micro-nanometer size. Actually, to the contrary, a large space is always required to support the robot to work well, since all the driving, sensing, and controlling modules need space to put equipment. However, on the other hand, we expect the robot can be as small as possible to improve its movability, especially its working ability in a closed environment. Therefore, system integration is a big issue in micro-nanorobot design.

Various integration methods have been reported to address different kinds of problems in single cell analysis. For the research in an open environment, such as single cell injection and mechanical characterization, it is easy to find extra space for the auxiliary equipment. The micro-nanomanipulators can be fixed on the microscopy stage and connected with the auxiliary equipment in other places by electrical wires. Contrarily, for the single cell analysis in a closed environment, it is difficult to connect wires directly. In this case, wireless actuation methods are usually used, such as optical tweezers and DEP. However, the wireless connection method brings difficulties to force sensing, because there is no force sensor and the force has to be estimated from the deformation of the object or the applied voltage indirectly. The lack of force information in this integration system increases the control difficulty to the micro-nanorobot. Arai reports a magnetically driven microtool (MMT) chip by introducing the MEMS technique into the robot [94]. They fabricated the magnetic-sensitive microtool based on MEMS technique. Then, they take it as the

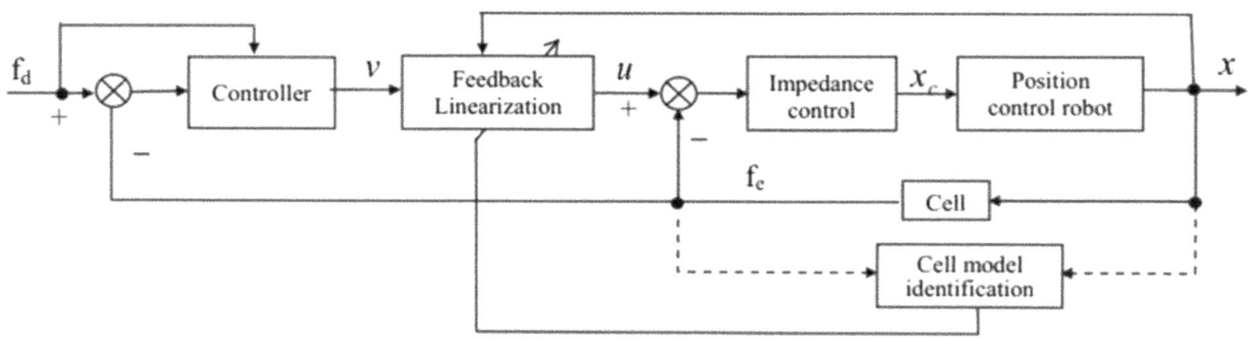

Figure 7 An example of a force feedback control diagram for single cell injection [93].

end effector of the robot and assemble it inside the microchip. This kind of robot can work inside the microchip since it can be controlled by the magnetic field remotely. Furthermore, it has the sensing ability owing to its deformable structure. The integration of the MEMS technique and robot technique will promote the micro-nanorobot research. We think this kind of method will be a trend for the single cell analysis in the future.

The micro-nanosystems mentioned above can be roughly divided to two parts, i.e., the functional end effector and other auxiliary parts. They communicate with each other either by wire or wireless field. How to integrate them together into a whole system is a big challenge. If we take a look at the internal of creatures, we can find many sophisticated robots in nature. Therefore, we think in addition to the physical way, the biological and chemical principles may also be helpful to design a novel micro-nanorobot. For example, the bacteria themselves have the actuating and sensing abilities and even have their own intelligence. Nowadays, artificial bacteria have been designed based on bacteria's actuating mechanism, which can be injected into the body and swim driven by magnetic field. DNA is another excellent robot mode for us. Scientists have developed the walking robot, DNA computer, and DNA cages by programming, which are the first crucial steps towards the small micro-nanorobotic systems.

Consequently, the next generation of the micro-nanorobot will require on-board sensors, functional tools, computing components, and so on. Combining the robot manipulation technique and MEMS fabrication technique and referring the biological and chemical technique could be the way to create a much smaller and highly integrated robot in the future.

Commercialization

One significant goal of the micro-nanorobotic research is to solve problems at the single cell scale and finally benefit society. Therefore, how to make the micro-nanorobot system more acceptable should be considered, especially for the acceptance by the biological scientist and the biological company. The automatic single cell injection system is only a handful of successful commercial systems that use micro-nanorobots. However, currently, most of the commercialized micro-nanorobots are only at the stage of precision manipulators. They are able provide accurate positioning capabilities but cannot solve a specific biological problem directly, which are still far from commercialization. Nowadays, most of the application research are still at the laboratory stage, which mainly focuses on the reliability but without considering the cost and productivity. From the commercial point of view, the system should be reliable, low cost, and high production. Thus, in the future research, these aspects should be considered more.

Future trends

Nowadays, the micro-nanorobot has achieved some level of success in single cell analysis. It provides much higher manipulation accuracy and stability than manual operation. Thus, the implement of many complex tasks becomes possible, and a lot of exciting new findings have been brought. However, we should realize that the micro-nanorobotic manipulation system still has shortages in many aspects, including low efficient actuation, lacking sensing approaches, high control difficulty, low integration, and it is far from commercialization. A combination of the traditional robotic technique and the micro-nanofabrication technique should be a way to solve the current problems. Next, although there are many micro-nanorobot prototypes, the mathematical model and the control strategy are not mature at the current stage, since it is highly interdisciplinary. Therefore, the cooperation and effort from different research fields is definitely required, including robotic, physics, chemical, biological, and so on. Next, at the small scale, sometimes the manipulation method based on biological and chemical principles is more efficient than the traditional mechanical and electrical method. Therefore, referring biological and chemical principles to improve the current robotic technique should be another trend in the future. Last but not least, extending the application of the micro-nanorobot from laboratory to practical applications should be taken as a significant issue all of the time.

Conclusions

As the basic unit of life, the single cell receives increasing interest nowadays. The micro-nanorobotic manipulation technique provides an effective approach for single cell analysis. It has been widely used for single cell injection, positioning, characterization, 3D assembly, and other related researches, by which it improves the manipulation accuracy and repeatability and increases the working efficiency greatly. We should realize that despite the fact that the micro-nanorobot has achieved many successes, there are still many challenges that need to be addressed in the single cell analysis field, including the actuating, sensing, controlling, integrating, and commercialization. The future trend of the micro-nanorobot should be much smaller, more efficient, more intelligent, widely applicable, and practical and acceptable by the market.

Competing interests

The authors declare that they have no competing interests.

Authors' contributions

YS and TF determined the architectural and contents of the manuscript. YS carried out the data collection and drafted the manuscript. Both authors read and approved the final manuscript.

Acknowledgement
This publication is supported by the National Natural Science Foundation of China (61403323), the Early Career Scheme from UGC of Hong Kong (21201314), and the Applied Research Grant of City University of Hong Kong (6000475, 9610300, 7200356).

Author details
[1]Department of Mechanical and Biomedical Engineering, City University of Hong Kong, Tat Chee Avenue, Kowloon, Hong Kong 99907, China. [2]School of Mechatronic Engineering, Beijing Institute of Technology, 5 South Zhongguancun Street, Haidian District, Beijing 100081, China.

References

1. Alberts B, Johnson A, Lewis J, Raff M, Roberts K, Walter P (2002) Molecular Biology of the Cell. Garland Science, New York
2. Anselmetti D (2009) Single Cell Analysis: Technologies and Applications. Vch Pub, German
3. Brehm-Stecher BF, Johnson EA (2004) Single-cell microbiology: tools, technologies, and applications. Microbiol Mol Biol Rev 68:538
4. Schubert C (2011) The deepest differences. Nature 480:133–137
5. Macaulay IC, Voet T (2014) Single cell genomics: advances and future perspectives. PLoS Genet 10:e1004126
6. Polonsky M, Zaretsky I, Friedman N (2013) Dynamic single-cell measurements of gene expression in primary lymphocytes: challenges, tools and prospects. Brief Funct Genomics 12:99–108
7. Ning L, Liu G, Li G, Hou Y, Tong Y, He J (2014) Current challenges in the bioinformatics of single cell genomics. Frontiers in Oncology 4:7
8. Wang D, Bodovitz S (2010) Single cell analysis: the new frontier in 'omics'. Trends Biotechnol 28:281–290
9. Zheng XT, Li CM (2012) Single cell analysis at the nanoscale. Chem Soc Rev 41:2061–2071
10. Dario P, Guglielmelli E, Allotta B (1994) Robotics in medicine. In: Intelligent Robots and Systems' 94.'Advanced Robotic Systems and the Real World', IROS'94. Proceedings of the IEEE/RSJ/GI International Conference on, Munich, Germany, pp 739–752
11. Dario P, Carrozza MC, Benvenuto A, Menciassi A (2000) Micro-systems in biomedical applications. J Micromech Microeng 10:235
12. Park S, Park J-O (2008) Frontier research program on biomedical microrobot for intravascular therapy. In: Biomedical Robotics and Biomechatronics, 2008. BioRob 2008, 2nd edn. 2nd IEEE RAS & EMBS International Conference on, Scottsdale, USA, pp 360–365
13. Nelson BJ, Kaliakatsos IK, Abbott JJ (2010) Microrobots for minimally invasive medicine. Annu Rev Biomed Eng 12:55–85
14. Lenaghan SC, Wang Y, Xi N, Fukuda T, Tarn T, Hamel WR, Zhang M (2013) Grand challenges in bioengineered nanorobotics for cancer therapy. IEEE Transactions on Bio-medical Engineering 60:667–673
15. Pepperkok R, Schneider C, Philipson L, Ansorge W (1988) Single cell assay with an automated capillary microinjection system. Exp Cell Res 178:369–376
16. Matsuoka H, Komazaki T, Mukai Y, Shibusawa M, Akane H, Chaki A, Uetake N, Saito M (2005) High throughput easy microinjection with a single-cell manipulation supporting robot. J Biotechnol 116:185–194
17. Zhang Y, Yu LC (2008) Single-cell microinjection technology in cell biology. Bioessays 30:606–610
18. Sun Y, Nelson BJ (2002) Biological cell injection using an autonomous microrobotic system. Int J Robotics Res 21:861–868
19. Lamontagne CA, Cuerrier CM, Grandbois M (2008) AFM as a tool to probe and manipulate cellular processes. Pflügers Archiv European J Physiology 456:61–70
20. Lacal JC, Perona R, Feramisco J (1999) Microinjection: XD-US vol. 1. Springer, Spain
21. Xie Y, Sun D, Liu C, Tse HY, Cheng SH (2010) A force control approach to a robot-assisted cell microinjection system. Int J Robotics Res 29:1222–1232
22. Yanagida K, Katayose H, Yazawa H, Kimura Y, Konnai K, Sato A (1999) The usefulness of a piezo-micromanipulator in intracytoplasmic sperm injection in humans. Hum Reprod 14:448–453
23. Tan Y, Sun D, Huang W, Cheng SH (2008) Mechanical modeling of biological cells in microinjection. NanoBioscience IEEE Transactions on 7:257–266
24. Han S-W, Nakamura C, Kotobuki N, Obataya I, Ohgushi H, Nagamune T, Miyake J (2008) High-efficiency DNA injection into a single human mesenchymal stem cell using a nanoneedle and atomic force microscopy. Nanomedicine 4:215–225
25. Chronis N, Lee LP (2005) Electrothermally activated SU-8 microgripper for single cell manipulation in solution. J Microelectromech Syst 14:857–863
26. Sakar MS, Steager EB, Cowley A, Kumar V, Pappas GJ (2011) Wireless manipulation of single cells using magnetic microtransporters. In: Robotics and Automation (ICRA). IEEE International Conference on, Shanghai, China, pp 2668–2673
27. Arai F, Yoshikawa K, Sakami T, Fukuda T (2004) Synchronized laser micromanipulation of multiple targets along each trajectory by single laser. Appl Phys Lett 85:4301–4303
28. Yue T, Nakajima M, Tajima H, Fukuda T (2013) Fabrication of microstructures embedding controllable particles inside dielectrophoretic microfluidic devices. Int J Adv Robot Syst 10:132
29. Chiou PY, Ohta AT, Wu MC (2005) Massively parallel manipulation of single cells and microparticles using optical images. Nature 436:370–372
30. Beyeler F, Neild A, Oberti S, Bell DJ, Sun Y, Dual J, Nelson B (2007) Monolithically fabricated microgripper with integrated force sensor for manipulating microobjects and biological cells aligned in an ultrasonic field. J Microelectromech Syst 16:7–15
31. Chan H-Y, Li WJ (2003) A thermally actuated polymer micro robotic gripper for manipulation of biological cells. In: Robotics and Automation. Proceedings. ICRA'03. IEEE International Conference on, Taipei, Taiwan, pp 288–293
32. Kim K, Liu X, Zhang Y, Sun Y (2008) Nanonewton force-controlled manipulation of biological cells using a monolithic MEMS microgripper with two-axis force feedback. J Micromech Microeng 18:055013
33. Fatikow S, Rembold U (1996) An automated microrobot-based desktop station for micro assembly and handling of micro-objects. In: EFTA '96. Proceedings, 1996 IEEE Conference on. vol.2., pp 586–592
34. Ashkin A, Dziedzic J (1987) Optical trapping and manipulation of viruses and bacteria. Science 235:1517–1520
35. Block SM, Goldstein LSB, Schnapp BJ (1990) Bead Movement by Single Kinesin Molecules Studied With Optical Tweezers. Nature 348:348–352
36. Wang MD, Yin H, Landick R, Gelles J, Block SM (1997) Stretching DNA with optical tweezers. Biophys J 72:1335–1346
37. Block SM, Blair DF, Berg HC (1989) Compliance of Bacterial Flagella Measured With Optical Tweezers. Nature 338:514–518
38. Hu S, Sun D (2011) Automatic transportation of biological cells with a robot-tweezer manipulation system. Int J Robotics Res 30:0278364911413479
39. Castellanos A, Ramos A, Gonzalez A, Green NG, Morgan H (2003) Electrohydrodynamics and dielectrophoresis in microsystems: scaling laws. J Phys D Appl Phys 36:2584
40. Jones TB (2005) Electromechanics of Particles. Cambridge Univ Pr, UK
41. Chang HC, Yeo L (2008) Electrokinetically-Driven Microfluidics and Nanofluidics. Cambridge University Press, USA
42. Hunt T, Westervelt R (2006) Dielectrophoresis tweezers for single cell manipulation. Biomed Microdevices 8:227–230
43. Lee GY, Lim CT (2007) Biomechanics approaches to studying human diseases. Trends Biotechnol 25:111–118
44. Hamill O, Marty A, Neher E, Sakmann B, Sigworth F (1981) Improved patch-clamp techniques for high-resolution current recording from cells and cell-free membrane patches. Pflugers Arch 391:85–100
45. Shen Y, Nakajima M, Kojima S, Homma M, Fukuda T (2011) Study of the time effect on the strength of cell–cell adhesion force by a novel nano-picker. Biochem Biophys Res Commun 409:160–165
46. Ahmad MR, Nakajima M, Kojima S, Homma M, Fukuda T (2011) Buckling nanoneedle for characterizing single cells mechanics inside environmental SEM. Nanotechnology IEEE Transactions on 10:226–236
47. Ahmad MR, Nakajima M, Kojima M, Kojima S, Homma M, Fukuda T (2012) Instantaneous and quantitative single cells viability determination using dual nanoprobe inside ESEM. Nanotechnology IEEE Transactions on 11:298–306
48. Chu YS, Dufour S, Thiery JP, Perez E, Pincet F (2005) Johnson-Kendall-Roberts theory applied to living cells. Phys Rev Lett 94:28102
49. Yamamoto A, Mishima S, Maruyama N, Sumita M (1998) A new technique for direct measurement of the shear force necessary to detach a cell from a material. Biomaterials 19:871–879
50. Puech P-H, Poole K, Knebel D, Muller DJ (2006) A new technical approach to quantify cell–cell adhesion forces by AFM. Ultramicroscopy 106:637–644

51. Krieg M, Arboleda-Estudillo Y, Puech P-H, Käfer J, Graner F, Müller DJ, Heisenberg C-P (2008) Tensile forces govern germ-layer organization in zebrafish. Nature Publishing Group 10:429–436
52. Marshall BT, Long M, Piper JW, Yago T, McEver RP, Zhu C (2003) Direct observation of catch bonds involving cell-adhesion molecules. Nature 423:190–193
53. Benoit M, Gabriel D, Gerisch G, Gaub HE (2000) Discrete interactions in cell adhesion measured by single-molecule force spectroscopy. Nat Cell Biol 2:313–317
54. Shen Y, Nakajima M, Ridzuan Ahmad M, Kojima S, Homma M, Fukuda T (2011) Effect of ambient humidity on the strength of the adhesion force of single yeast cell inside environmental-SEM. Ultramicroscopy 111:1176–1183
55. Shen Y, Ahmad MR, Nakajima M, Kojima S, Homma M, Fukuda T (2011) Evaluation of the single yeast cell's adhesion to ITO substrates with various surface energies via ESEM nanorobotic manipulation system. NanoBioscience IEEE Transactions on 10:217–224
56. Shen Y, Nakajima M, Kojima S, Homma M, Kojima M, Fukuda T (2011) Single cell adhesion force measurement for cell viability identification using an AFM cantilever-based micro putter. Meas Sci Technol 22:115802
57. Ning W, James PB, Donald EI (1993) Mechanotransduction across the cell surface and through the cytoskeleton. Science 260:21
58. Wang K, Sun D (2012) Influence of semiflexible structural features of actin cytoskeleton on cell stiffness based on actin microstructural modeling. J Biomech 45:1900–1908
59. Li M, Liu L, Xi N, Wang Y, Dong Z, Tabata O, Xiao X, Zhang W (2011) Imaging and measuring the rituximab-induced changes of mechanical properties in B-lymphoma cells using atomic force microscopy. Biochem Biophys Res Commun 404:689–694
60. Lam WA, Rosenbluth MJ, Fletcher DA (2007) Chemotherapy exposure increases leukemia cell stiffness. Blood 109:3505–3508
61. Sarah EC, Yu-Sheng J, Qing-Yi L, JianYu R, James KG (2011) Green tea extract selectively targets nanomechanics of live metastatic cancer cells. Nanotechnology 22:215101
62. Wu Y, Sun D, Huang W (2011) Mechanical force characterization in manipulating live cells with optical tweezers. J Biomech 44:741–746
63. Bausch AR, Ziemann F, Boulbitch AA, Jacobson K, Sackmann E (1998) Local measurements of viscoelastic parameters of adherent cell surfaces by magnetic bead microrheometry. Biophys J 75:2038–2049
64. Vogler H, Draeger C, Weber A, Felekis D, Eichenberger C, Routier-Kierzkowska AL, Boisson-Dernier A, Ringli A, Nelson BJ, Smith RS, Grossniklaus U (2013) The pollen tube: a soft shell with a hard core. Plant J 73:617–627
65. Dao M, Lim C, Suresh S (2003) Mechanics of the human red blood cell deformed by optical tweezers. J Mechanics and Physics of Solids 51:2259–2280
66. Haga H, Sasaki S, Kawabata K, Ito E, Ushiki T, Sambongi T (2000) Elasticity mapping of living fibroblasts by AFM and immunofluorescence observation of the cytoskeleton. Ultramicroscopy 82:253–258
67. Cross SE, Jin YS, Rao J, Gimzewski JK (2007) Nanomechanical analysis of cells from cancer patients. Nat Nanotechnol 2:780–783
68. Callies C, Schön P, Liashkovich I, Stock C, Kusche-Vihrog K, Fels J, Sträter JF, Oberleithner H (2009) Simultaneous mechanical stiffness and electrical potential measurements of living vascular endothelial cells using combined atomic force and epifluorescence microscopy. Nanotechnology 20:175104
69. Oberleithner H (2005) Aldosterone makes human endothelium stiff and vulnerable. Kidney Int 67:1680–1682
70. Sarah EC, Yu-Sheng J, Julianne T, Roger W, JianYu R, James KG (2008) AFM-based analysis of human metastatic cancer cells. Nanotechnology 19:384003
71. Shen Y, Nakajima M, Yang Z, Tajima H, Najdovski Z, Homma M, Fukuda T (2013) Single cell stiffness measurement at various humidity conditions by nanomanipulation of a nano-needle. Nanotechnology 24:145703
72. Zhang C, Li P, Liu L, Wang Y, Gao Z, Li G (2014) Development of mechanostimulated patch-clamp system for cellular physiological study. Mechatronics IEEE/ASME Transactions on 19:1138–1147
73. Pampaloni F, Reynaud EG, Stelzer EH (2007) The third dimension bridges the gap between cell culture and live tissue. Nat Rev Mol Cell Biol 8:839–845
74. Cao Y, Vacanti JP, Paige KT, Upton J, Vacanti CA (1997) Transplantation of chondrocytes utilizing a polymer-cell construct to produce tissue-engineered cartilage in the shape of a human ear. Plast Reconstr Surg 100:297–302
75. Haycock JW (2011) 3D cell culture: a review of current approaches and techniques. In: 3D Cell Culture. Springer, USA, pp 1–15
76. Klein F, Richter B, Striebel T, Franz CM, Freymann GV, Wegener M, Bastmeyer M (2011) Two-component polymer scaffolds for controlled three-dimensional cell culture. Adv Mater 23:1341–1345
77. Tasoglu S, Diller E, Guven S, Sitti M, Demirci U (2014) Untethered micro-robotic coding of three-dimensional material composition. Nat Commun 5:3124
78. Carletti E, Motta A, Migliaresi C (2011) Scaffolds for tissue engineering and 3D cell culture. In: 3D Cell Culture. Springer, USA, pp 17–39
79. Liu N, Liang W, Liu L, Wang Y, Mai JD, Lee G-B Li WJ (2014) Extracellular-controlled breast cancer cell formation and growth using non-UV patterned hydrogels via optically-induced electrokinetics. Lab Chip 14:1367–1376
80. Moglia A, Menciassi A, Schurr MO, Dario P (2007) Wireless capsule endoscopy: from diagnostic devices to multipurpose robotic systems. Biomed Microdevices 9:235–243
81. Peyer KE, Zhang L, Nelson BJ (2013) Bio-inspired magnetic swimming microrobots for biomedical applications. Nanoscale 5:1259–1272
82. Kometani R, Morita T, Watanabe K, Hoshino T, Kondo K, Kanda K, Haruyama Y, Kaito T, Fujita J, Ishida M, Ochiai Y, Matsui S (2004) Nanomanipulator and actuator fabrication on glass capillary by focused-ion-beam-chemical vapor deposition. J Vac Sci Technol B 22:257–263
83. Aoyama H, Fuchiwaki O (2001) Flexible micro-processing by multiple microrobots in SEM. In: Robotics and Automation. Proceedings 2001 ICRA. IEEE International Conference on, Seoul, Korea, pp 3429–3434
84. Nogawa K, Kojima M, Nakajima M, Kojima S, Homma M, Fukuda T (2009) Rotational speed control of Na-driven flagellar motor by dual pipettes. NanoBioscience IEEE Transactions on 8:341–348
85. Martel S, Mohammadi M, Felfoul O, Lu Z, Pouponneau P (2009) Flagellated magnetotactic bacteria as controlled MRI-trackable propulsion and steering systems for medical nanorobots operating in the human microvasculature. Int J Robotics research 28:571–582
86. Ding B, Seeman NC (2006) Operation of a DNA robot arm inserted into a 2D DNA crystalline substrate. Science 314:1583–1585
87. Zheng J, Birktoft JJ, Chen Y, Wang T, Sha R, Constantinou PE, Ginell SL, Mao C, Seeman NC (2009) From molecular to macroscopic via the rational design of a self-assembled 3D DNA crystal. Nature 461:74–77
88. Andersen ES, Dong M, Nielsen MM, Jahn K, Subramani R, Mamdouh W, Golas MM, Sander B, Stark H, Oliveira CLP, Pedersen JS, Birkedal V, Besenbacher F, Gothelf KV, Kjems J (2009) Self-assembly of a nanoscale DNA box with a controllable lid. Nature 459:73–76
89. Zhang Z, Olsen EM, Kryger M, Voigt NV, Tørring T, Gültekin E, Nielsen M, Zadegan RM, Andersen ES, Nielsen MM, Kjems J, Birkedal V, Gothelf KV (2011) A DNA tile actuator with eleven discrete states. Angew Chem Int Ed 50:3983–3987
90. Haibo H, Dong S, Mills JK, Li WJ, Shuk Han C (2009) Visual-based impedance control of out-of-plane cell injection systems. Automation Science and Engineering IEEE Transactions on 6:565–571
91. Pillarisetti A, Anjum W, Desai JP, Friedman G, Brooks AD (2005) Force feedback interface for cell injection. In: Eurohaptics Conference, 2005 and Symposium on Haptic Interfaces for Virtual Environment and Teleoperator Systems. World Haptics 2005. First Joint, Pisa, Italy, pp 391–400
92. Mallouk TE, Sen A (2009) Powering nanorobots. Sci Am 300:72–77
93. Xie Y, Sun D, Liu C, Cheng SH, Liu YH (2009) A force control based cell injection approach in a bio-robotics system. In: Robotics and Automation. ICRA'09. IEEE International Conference on, Kobe, Japan, pp 3443–3448
94. Hagiwara M, Kawahara T, Yamanishi Y, Masuda T, Feng L, Arai F (2011) On-chip magnetically actuated robot with ultrasonic vibration for single cell manipulations. Lab Chip 11:2049–2054

18

Enhanced assessment of limb neuro-mechanics via a haptic display

Alejandro Melendez-Calderon[1,6*†], Davide Piovesan[2,6†], James L Patton[1,3,4,5,6] and Ferdinando A Mussa-Ivaldi[1,4,6,7]

Abstract

Current state-of-the-art quantitative assessments of abnormal neuro-mechanics (e.g., spasticity, rigidity, dystonia) require sophisticated measurement systems that, together with the lengthiness of the data acquisition, make these approaches impractical for the clinical setting. Our long-term goal is to create a tool that enhances the capability of therapists to diagnose abnormal neuro-mechanics without the need of lengthy procedures and sophisticated equipment. Towards this direction, we present a haptic display of abnormal limb neuro-mechanics. This haptic display can be used as a training device for clinicians to understand the causality of the haptic sensation associated with neurological impairments and abnormal biomechanics during manipulation of a patients' limb. Other uses include the testing of operational research algorithms to maximize the exchange of haptic information, reproducing conditions of human-human interaction, and a test bed for developing novel assessment techniques and targeted interventions. Experimental results support the hypothesis that the best assessment of abnormalities can be obtained via specific manipulation strategies.

Keywords: Biomechanics; Haptic display; Hypertonia

Background

A common consequence of many neurological disorders like cerebral palsy, Parkinson's disease, or stroke (to name a few) is the appearance of atypical neuro-mechanical conditions such as an increase in muscle tone (hypertonia) due to active contraction of the muscle, hyperexcitability of motoneurons causing excessive co-activation, muscle contractures, etc. It is common in today's clinical practice to manipulate the patient's limbs to diagnose these abnormal neuro-mechanical changes. Through such physical interaction, clinicians extract as much information as possible to infer the patients' condition, such as in the assessment of hypertonic conditions [1-3] (e.g., spasticity, rigidity, dystonia, among others). This approach is subjective [4], which may be a limitation when inferring the nature and gravity of the impairment since the effect generated by pluri-articular muscles to multi-joint mechanics can be misjudged. A misjudgment on the initial or intermediate assessments can lead to selecting a non-optimal intervention, which can have negative consequences, such as increasing intervention time, rising in treatment cost, or even worsening the condition. Hence, the capability to discriminate between different types and levels of impairments is important for the clinical intervention to be chosen and planned accordingly.

Traditional engineering approaches to assess differences in limb mechanics such as non-parametric [5,6] and parametric [7,8] system identification techniques provide a less subjective, quantitative measure of biomechanical variables. However, the necessity of sophisticated measurement systems, such as stiff robotic devices, makes most of these classical engineering methods impractical for the clinical setting due to a high cost/benefit ratio. In addition, the majority of studies applying such purely robotic-driven approaches have been limited to single joints (e.g., knee, ankle, elbow) as more complex joints (e.g., shoulder) pose significant challenges [9]. Yet, abnormal neuro-mechanical conditions such as hypertonia can encompass alteration of inter-muscular (heteronymous) reflexes [10] or abnormal multi-joint couplings due to increased rigidity [11,12]. Consequently, some of the

* Correspondence: alejandro.melendez@northwestern.edu
†Equal contributors
[1]Department of Physical Medicine and Rehabilitation, Northwestern University, 60611 Chicago, IL, USA
[6]Sensory Motor Performance Program, Rehabilitation Institute of Chicago, 60611 Chicago, IL, USA
Full list of author information is available at the end of the article

most important clinical problems have been ignored in these engineering approaches.

Our approach

A recent review on spasticity assessment [13] concluded that there is a need for simple instruments that can provide reliable measures in a practical setting. Therefore, our goal is to extend the interactive nature of traditional assessment techniques by providing clinicians with a method that enables them to appreciate and quantify hypertonic phenomena during manipulation of the affected limb in multiple degrees of freedom (DOF). Towards this end, we propose a haptic display, in which a model of an impaired limb is haptically rendered on a robotic device in which clinicians can physically interact with. This haptic display allows us to evaluate the threshold of human perceptions to different pathological neuro-mechanics at different levels of severity, which in turn can translate into different strategies that enhance the human perception of these impairments. This will likely provide therapists with better training and improved classification, measurement, and treatment of these disorders.

Haptic rendering of physical impairments of stroke survivors have been reported previously (e.g., [14,15]). Such approaches, however, have been limited to only one DOF and have focus on mimicking current clinical assessment techniques (e.g., Ashworth test). It is essential to extend the manual assessment of hypertonia to more than one DOF so as to capture the effect of pluri-articular muscles and multi-joint coupling. Differently from the previous work, our approach allows us to characterize how clinicians perceive different types of hypertonia in a multi-DOF environment. The interaction in our haptic display occurs through a single haptic port under the rationale that assessment techniques derived from the use of this tool should be directly applicable to both on-site and remote scenarios (Figure 1). Extending traditional manipulation methods in tele-interaction scenarios requires the use of an exoskeleton on the patient side and a robotic mock-up of the patient's limb on the clinician side so that the different limb segments can be manipulated independently. We suggest the implementation of simpler configurations where the interaction occurs through a single port from both sides, allowing the use of commercially available robots. Our hypothesis is that the individualization of abnormalities through a single interaction port in patient-clinician physical interaction can be done more objectively via specific manipulation strategies (*probing motions*) derived from understanding the forces at the point of interaction produced by the different types of impairment. The force-motion generated by the patient-clinician interaction along these motions could uniquely characterize the nature and severity of the impairment. We expect that our haptic display will help us better understand how to improve human perception during physical manipulation.

Methods

Haptic display of limb neuro-mechanics

The haptic display comprises a multi-joint robotic manipulator that is programmed to output mechanical forces that match the forces a clinician would feel when moving a limb (e.g., arm, leg, hand) of a patient (*model*). To interact with the model, subjects seat in front of the robot and hold the robot's end effector. The end effector of the robot is virtually connected to the model via a two-way connection that can be rendered, for instance, through a shared virtual environment which includes a simulated mechanical element [16,17]. The limb model

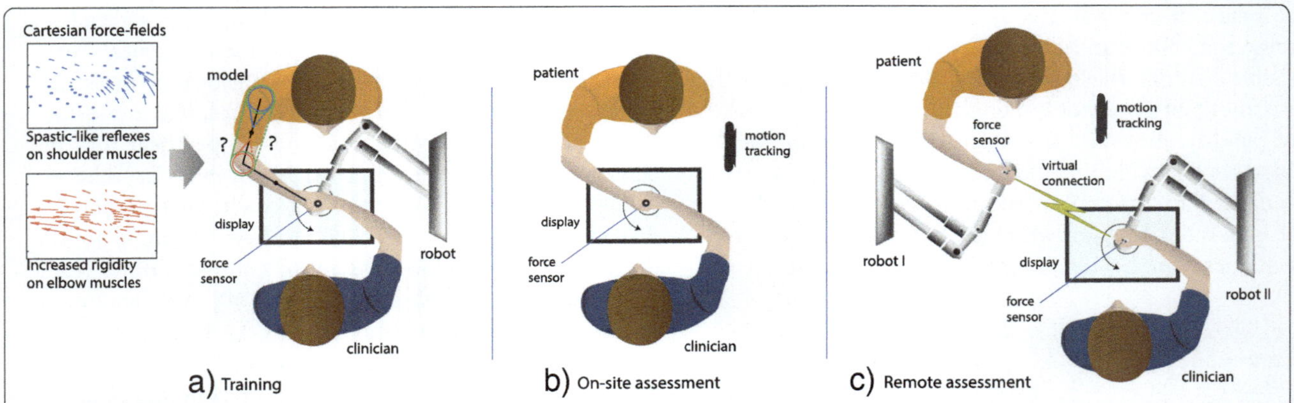

Figure 1 Different uses of the proposed method include assessment training and on-site and remote assessment scenarios. (a) *Training* - a clinician interacting with a simulated patient rendered on a robotic device (haptic display). Different abnormal biomechanics can be rendered by the haptic display so the therapist can be trained in assessing these conditions. Manipulation techniques for assessment derived from the use of the haptic display can be extended to both on-site and remote scenarios. **(b)** *On-site assessment* - a clinician interacting with a patient using a system to assist in the assessment and determination of the abnormal limb biomechanics of a patient. **(c)** *Remote assessment* - a clinician interacting with a patient using a non-co-located setup of the system described in (b).

is rendered as if the patient were seated opposite the participants or at any other desired configuration. Both the subjects' hand and the robotic arm can be covered by an opaque horizontal screen, on which an image of the rendered virtual limb is projected. A superimposed image representing the workspace can also be projected. This configuration can also be reproduced during tele-assessment, where instead of a simulated arm, the users interact remotely with another person.

Neuro-mechanical model of a human limb

The force produced by muscles can be expressed in terms of passive (i.e., intrinsic physical properties) and active (i.e., reflexes and voluntary activation) components. The combination of forces generated by different muscle groups gives rise to characteristic force patterns at a limb's endpoint. The set of forces associated with a set of limb states (e.g., position, velocity) in the Cartesian space is a *force field* [18,19]. In our haptic display, the force of each muscle group is modeled as a linear combination of non-linear passive and active components. The passive component is position dependent and it is a function of the muscle length, whereas the active component depends on reflexive activation and is both position and velocity dependent. The resulting force field is the linear superposition of the two force components. This simple, yet representative, approach allows us to simulate different normal and hypertonic-like forces at the point of interaction by modifying few weighting coefficients. The different abnormal force fields generated by a hypertonic limb can be rendered on a robotic device (*haptic display*) and so the haptic perception to the nature and severity of hypertonic phenomena can be investigated systematically (e.g., [20,21]).

As a proof of concept, we emulated a simple representation of the human arm with six different muscle groups that are embedded in a planar, two-joint mechanism. The model considers rigid body dynamics and muscle intrinsic and active properties. The model was implemented in Simulink and was rendered on a two revolute joint planar robot running in real-time using xPC Target at a rate of 1 kHz. The dynamics of the virtual arm moving in a horizontal plane while interacting with the subject were modeled as

$$H[q]\ddot{q} + C[q,\dot{q}]\dot{q} = J_q{}^T[q]\cdot F_{\text{external}} - J_\lambda{}^T[q]\cdot\left(\Phi\left[\lambda, u\left[\dot{\lambda}\right]\right] + \Psi[\lambda]\right) \tag{1}$$

where $H[q]$ represents the arm inertia matrix of a double pendulum system; q denotes the vector of shoulder and elbow joint angles [rad]; $C[q,\dot{q}]\dot{q}$ is the term corresponding to Coriolis and centripetal forces; $J_q[q]$ is the Jacobian matrix transforming an endpoint force, F_{external}, into joint torque; λ and $\dot{\lambda}$ correspond to the muscle stretch and muscle stretch velocity, respectively; Φ is a vector of muscle forces produced by motor signals u; and ψ is a vector of muscle forces that are intrinsic to the biomechanical properties of the muscle. Simulation-specific parameters for the limb centers of mass and inertia were estimated using anthropometrical tables from Winter [22]. The Jacobian matrix $J_\lambda[q]$ transforms the muscle force into joint torque and contains the muscle moment arms ρ at any particular position, q:

$$J_\lambda[q] = \begin{pmatrix} -\rho_{\text{sf}} & \rho_{\text{se}} & 0 & 0 & -\rho_{\text{bf}_1} & \rho_{\text{be}_1} \\ 0 & 0 & -\rho_{\text{ef}} & \rho_{\text{ee}} & -\rho_{\text{bf}_2} & \rho_{\text{be}_2} \end{pmatrix}^T \tag{2}$$

The sub-indexes correspond to sf, shoulder adductors (deltoid anterior, coracobrachialis, pectoralis major clav.); se, shoulder abductors (deltoid posterior); ef, elbow flexors (biceps long, brachialis, brachioradialis); ee, elbow extensors (triceps lateral, anconeus); bf, bi-articular flexors (biceps short); and be, bi-articular extensors (triceps long) muscle groups.

The force of each muscle group is modeled as a linear combination of active (i.e., produced by a motor command u) and passive (i.e., produced by intrinsic rigidity of the muscles and connective tissue) components [23,24] associated with intrinsic rigidity of the muscles and connective tissue. The force produced by each muscle in response to the motor signal u is

$$\begin{aligned} \Phi\left[\lambda, u\left[\dot{\lambda}\right]\right] &= \left(\varphi_{\text{sf}} \quad \varphi_{\text{se}} \quad \varphi_{\text{ef}} \quad \varphi_{\text{ee}} \quad \varphi_{\text{bf}} \quad \varphi_{\text{be}}\right)^T \\ \varphi_i &= \max\left(\begin{matrix} 0, \\ \alpha_i u\left[\dot{\lambda}_i\right]\cdot\Gamma\cdot e^{-\Gamma^2} \end{matrix}\right) \\ \Gamma &= \left(\frac{\lambda_i - \lambda_{\min,i}}{\lambda_{\max,i} - \lambda_{\min,i}} - \frac{\lambda_{\text{rest},i}}{\lambda_{\max,i}}\left(1 - u\left[\dot{\lambda}_i\right]\right)\right) \end{aligned} \tag{3}$$

The term u corresponds to an active motor signal (e.g., reflexes, voluntary commands); the term α corresponds to the stiffness of the muscle when the system is linearized at its equilibrium position when the activation is maximum (i.e., $u = 1$).

Typical assessment of hypertonia is performed by asking the patient to not voluntarily intervene. Under this assumption, the term $u\left[\dot{\lambda}\right]$ corresponds to an active motor signal that depends on the muscle stretch velocity $\dot{\lambda}$ (e.g., tonic reflex), which we defined as

$$u\left[\dot{\lambda}_i\right] = \beta_i \min\left(1, \max\left(0, \frac{\dot{\lambda}_i}{\dot{\lambda}_{\max,i}}\right)\right) \quad \Big| \quad \beta_i \in [0,1] \tag{4}$$

where β corresponds to a 'stretch gain'. The variables

$\lambda_{\text{rest},i}$, $\lambda_{\max,i}$, and $\lambda_{\max,i}$ correspond to the length of the $i$thmuscle group at rest, its maximum length and maximum rate of length change, respectively. The maximum rate of length change was calculated as the one obtained by moving the endpoint of the virtual arm along a circle of 35 cm in diameter at a frequency of 2 Hz. The maneuver was assumed to be a good compromise between the amplitude and frequency of a very fast movement in endpoint space.

The force produced by the intrinsic rigidity of the muscles and connective tissue is a function of the muscles' lengths:

$$\Psi[\lambda] = (\psi_{\text{sf}} \quad \psi_{\text{se}} \quad \psi_{\text{ef}} \quad \psi_{\text{ee}} \quad \psi_{\text{bf}} \quad \psi_{\text{be}})^T$$
$$\psi_i = \max\begin{pmatrix} 0, \\ K_{m,i}(\lambda_i - \lambda_{\text{rest},i})e^{-(\lambda_i-\lambda_{\text{rest},i})^2} \end{pmatrix} \quad (5)$$

The term $\Psi[\lambda]$ can be multiplied by a generalized logistic function to avoid sharp discontinuities when adding both passive and active components in Equation 1. The term λ represents the muscle length, and K_m corresponds to the muscle stiffness linearized at the muscle's rest position.

Haptic rendering of abnormal limb neuro-mechanics

Haptic rendering of the correct ranges of forces that impaired individuals can exhibit is essential so that valid psychometric studies on haptic perception can be done in a haptic display. These ranges of forces depend on both the intrinsic physical properties (rigidity) of the limb and the influence of muscle reflexes on the limb mechanics. However, the separation of these two terms is an open, difficult, and controversial issue [6,25-28]. However, assuming that both intrinsic and reflexive stiffness are linearly combined, we could get a good approximation of the order of magnitude of these two terms by means of measuring *postural* and *stiffness* maps. A *postural map* represents the stationary force field (i.e., with zero velocity) produced by the intrinsic properties of the limb measured at specific locations [11,29,30]. It is representative of the intrinsic properties of the arm. The integral of this force field with respect to the position workspace is the potential energy associated with the passive mechanics of the limb. A *stiffness map* represents the Hessian of the potential of active muscle force at a specified velocity obtained by perturbing the limb at different positions and frequencies. Given the rapid change in velocity during the acquisition of these maps, the reflexive stiffness is expected to rise and be somehow distinguishable from the intrinsic rigidity, as demonstrated in [11]. By integrating each map twice with respect to the position, a potential describing the active component of the muscles at the specified velocity is obtained.

Given a database of different force fields generated by different impaired individuals in the form of *postural* and *stiffness* maps, one can use optimization algorithms (e.g., simulated annealing) to tune the sensitive parameters of the model, such as the stiffness of the connective tissue and the reflex gain for each muscle, to generate a force field that matches the measured ones (see Figure 2). Therefore, one could create and reproduce any arbitrary mixture of representative reflex and intrinsic stiffness by means of our haptic display.

Haptic rendering of hypertonic-like forces

Among the numerous factors that characterize hypertonia, we are interested in training the ability to discriminate forces produced by increased muscle rigidity and the non-linear phenomena associated with velocity-dependent stiffness. Another way of simulating hypertonic-like forces is to assume that these could be achieved by increasing both α and K_m. in Equations 3 and 5. Measurements of joint rigidity during passive movements are available in the literature for both stroke survivors and unimpaired individuals [8,31]. Based on such data, the average joint passive linear stiffness of unimpaired individuals can be taken as the *lower boundary* of joint rigidity. The *upper boundary* of joint rigidity corresponds to the passive linear stiffness recorded on stroke survivors with the Modified Ashworth Score equal to 4 [11,32]. To render intermediate levels of severity, one could consider the just noticeable difference (JND) or the Weber fraction, which is an important index representing the sensitivity of the subject to stiffness stimuli. Different stiffness JND have been obtained empirically, depending on the experimental protocol used. For palpation with a fixed displacement, the value of stiffness JND is around 8% [33]. For free exploration, the JND is much higher and can be up to 67% [34]. Since the rigidity discrimination in the clinical setting is performed with a free movement, a stiffness JND = 60% could be chosen to segment the whole range of rigidity between the two aforementioned boundary conditions.

Psychophysical experiments

We used our haptic display to investigate whether specific exploration patterns or probing motions can be used to discriminate the salient features of a given set of force fields associated with different hypertonic-like conditions. In particular, we present a study that evaluates the capabilities of (non-therapists) individuals in discriminating between different hypertonic virtual arms with diverse muscle groups affected. Three different muscle groups were tested - *i)* hypertonic shoulder muscles, *ii)* hypertonic elbow muscles, and *iii)* hypertonic bi-articular muscles - at four levels of severity - *i)* very mild, *ii)* mild, *iii)* moderate, and *iv)* severe, following a Weber Fraction of 0.6.

Twelve subjects (non-therapists), naïve to the experimental conditions, participated in the study. Subjects

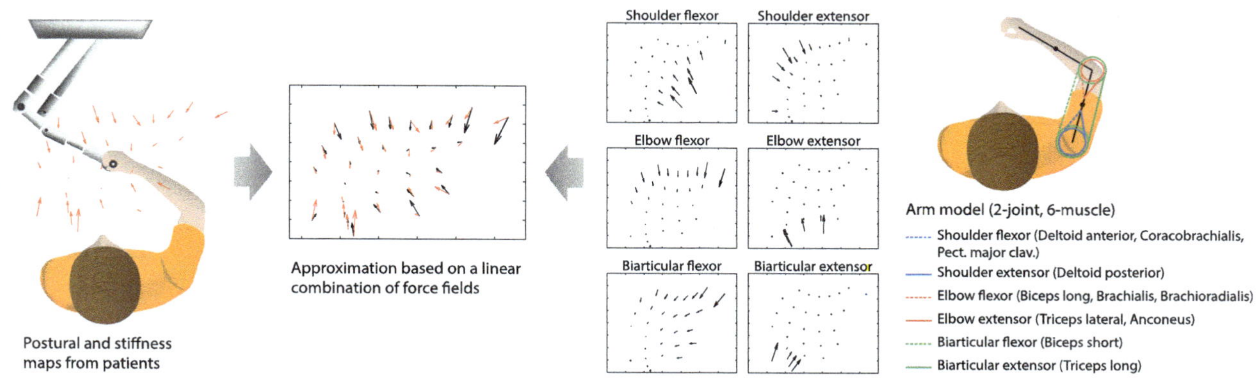

Figure 2 Modeling of limb neuro-mechanics. A good approximation of the order of magnitude of the intrinsic properties of the muscles can be obtained by measuring *postural* maps. *Far left.* Postural map (force field) obtained from a stroke patient. *Far right.* Decomposition of postural map into force fields generated by each muscle group. *Middle.* Comparison between real and approximated force fields.

gave informed consent prior to participation. Experiments were approved by the Northwestern University's Institutional Review Board. Participants were divided in two groups: i) non-instructed and ii) instructed. *Non-instructed* subjects were required to discriminate between the different conditions using their own-developed manipulation technique during a familiarization phase. *Instructed* subjects were required to manipulate the haptic display with specific probing motions that highlight the salient features of each abnormal condition (pulling and semi-elliptical motions - see 'Selection of specific probing motions').

Experimental protocol

The experiment was divided in two consecutive phases: *a)* familiarization and *b)* assessment. During *familiarization*, participants interacted with four different virtual arms: *i)* unimpaired arm, *ii)* maximally impaired shoulder, *iii)* maximally impaired elbow, and *iv)* maximally impaired bi-articular muscles. Subjects interacted sequentially with each condition in blocks of 30 s for eight trials (8 trials × 4 conditions =32 trials). During this phase, a legend appeared on the top right corner of the screen indicating the condition that the user was experiencing. In order to familiarize with the different haptic sensations produced by the hypertonic muscle groups, *non-instructed* subjects were free to interact with those four conditions as they wished, while *instructed* subjects were specifically asked to move the arm with a pulling and semi-elliptical probing motions. During *assessment*, all subjects were randomly presented with different abnormal virtual arms of varied hypertonic severity. Their task was to identify which group of muscles was impaired. All subjects were presented each condition nine times in trials of 15 s (3 hypertonic conditions × 4 levels of severity × 9 trials = 108 trials). The only information available during this phase was the configuration of the arm and the rendered force at the end effector of the haptic display. After physical manipulation of the virtual arm, subjects were required to select one of the three options presented on the screen - *shoulder, elbow*, or *both*. Participants could take as much time as they wanted to give their assessments and start a new trial. The whole experiment lasted for about 45 min. Simulation-specific parameters were the same as the ones described in detail in [20].

Selection of specific probing motions

As a first approximation, we planned the probing motion for this experiment considering only the effect of the position-dependent muscle force Ψ. Given the three force fields associated with *severe* hypertonic *shoulder, elbow*, and *bi-articular* muscle groups, our first goal was to find the minimum number of probing motions that, executed within each force field, would produce a different force profile. Such force profiles along the trajectories ought to be easily identifiable from one another. One possible strategy is to include the locations of the workspace in which the minimum and maximum force magnitude occurs for each field in each probing trajectory. We called this strategy the *min-max paradigm*. The set of probing trajectories should be designed so that, along the same probing motion, the minimum and maximum force magnitude occurs at different locations for each force field. Figure 3(a) represents the passive force Ψ for severe hypertonic muscle groups, generated by a circular probing motion. It can be seen that the minimum and maximum force occurs in different locations along the circles for hypertonic *shoulder* and *elbow* force fields. Yet, there is not much difference in the location between hypertonic *shoulder* and *bi-articular*. Hence, an extra probing motion would allow disambiguation between conditions. It can be demonstrated that given n types of force fields, when using the *min-max paradigm*, it is

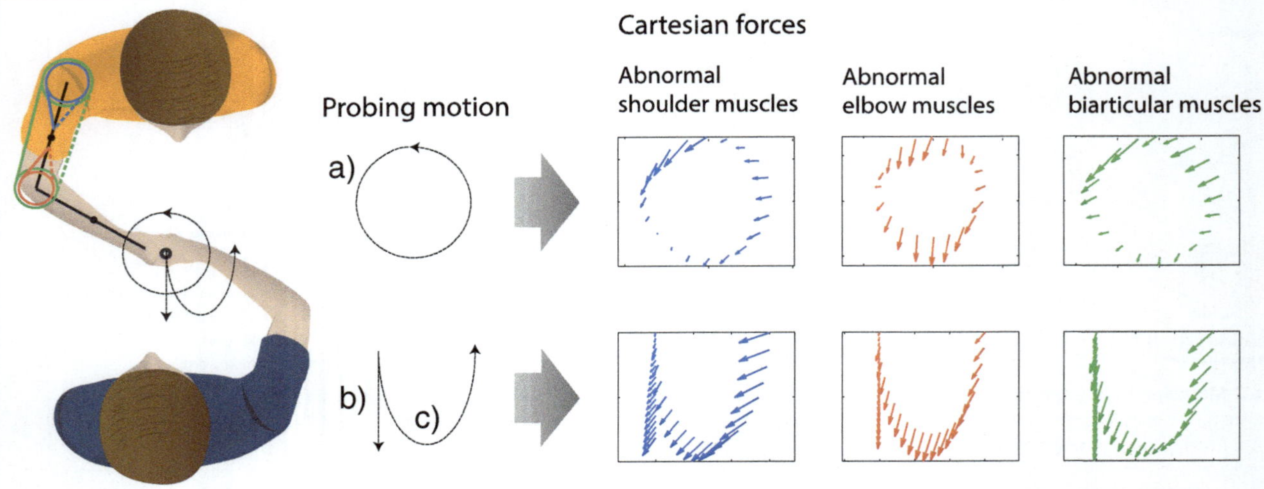

Figure 3 Forces experienced by the person manipulating an impaired limb vary depending on the manipulation strategy. Different abnormal conditions can be undistinguishable if forces produced by the movement are not sufficiently different. **(a)** *Circular* probing motion (note that there is not much difference between the forces produced by abnormal *shoulder* and *bi-articular* muscles; **(b)** *pulling* and **(c)** *semi-elliptical* probing motions (note that the forces produced by the different abnormal conditions can be discriminated by these two motions).

always possible to identify all impairments by exploring $n-1$ trajectories.

While circles could completely characterize the location and intensity of the maximum force associated with a specific impairment, the path to be covered is quite long. This is a limitation since the effect of the reflexes is attenuated after prolonged repetitive movements. To avoid this, we tried to reduce the length of the movements by incorporating simple strokes. According to the *min-max paradigm*, the arm's equilibrium position (i.e., the center of each circle) ought to be included in the probing motion as it is the point of absolute minimum force for all conditions. On the other hand, it should be noted that the magnitude of the force is not the only variable that can be discriminated. The orientation of the force is also important as it allows for an unequivocal haptic cue. Pulling the endpoint of impaired arm from its equilibrium position along the mid-sagittal plane in a fast and straight motion (Figure 3(b)), we can find that the orientation of the force profile generated by the hypertonic *shoulder* force field is completely different from those produced in the other two fields. Hence, we chose this movement for discriminating the shoulder force field from all the other. It is important to notice that the orientation and magnitude of the force would change depending on the impairment. We refer to these probing strategies as the *orientation paradigm*. By pulling the endpoint of impaired arm along a semi-elliptical trajectory as shown in Figure 3(c), the orientation of the corresponding force profile is quite different if the movement is executed when interacting with hypertonic *elbow* or *bi-articular* conditions. This is the second trajectory that we chose to complete the set of probing motion. Since the number of impairments was restricted to three in the current experiment, we hypothesize that subjects would be able to discriminate between the conditions by using only two probing motions.

Results and discussion

Psychophysical experiments

Figure 4a shows stereotypical probing motions executed by both groups, and Figure 4b shows a comparison of the reliability between the two manipulation techniques. It can be seen that the reliability of the users' discrimination was better when they followed a specific probing motion as compared to random exploration. These experimental results support the hypothesis that better assessment of abnormalities can be obtained via specific probing motions that highlight haptic features of the different abnormalities, thus enhancing human perception.

Other experiments

In addition to this experiment, we are currently using this haptic display to *i)* assess the capabilities of naïve subjects and expert therapists to correctly identify the nature and severity (see [20,21]) of hypertonic impairments and to *ii)* assess how the perception of hypertonia is affected by the impedance introduced by the virtual connection between a patient and a clinician interacting remotely [35]. The haptic display allows to tackle several practical issues. For example, it can be used to investigate the effects of transmission delays on haptic perception of abnormal biomechanics or to derive specific probing motions that would enhance recognition of such disorders.

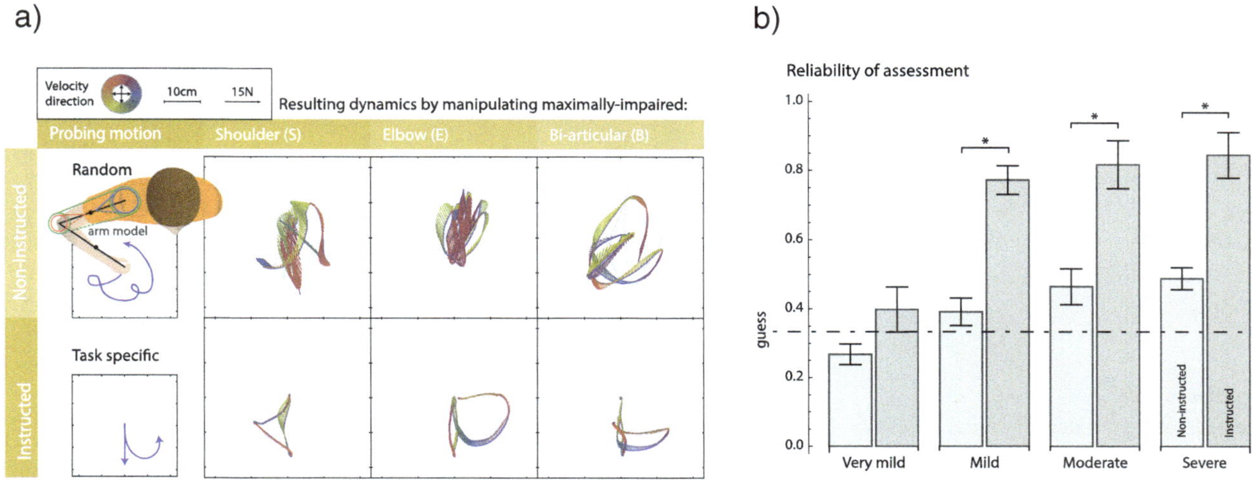

Figure 4 Better assessment of abnormal neuro-mechanics can be obtained via designed probing motions. (a) Stereotypical probing motions executed by experimental subjects. Non-instructed subjects manipulated the hypertonic arm model mainly with random motions. Instructed subjects were encouraged to use the two task-specific probing motions (pulling and semi-elliptical motions - see *'Selection of specific probing motions'*). **(b)** Mean and standard error (n =6 per group) of the assessment reliability as calculated in [20]; specific probing motions significantly improved the assessment. Note that a random guess approach has a reliability of 1/3.

Discussion

This paper presents a haptic display for assessing neuromechanical pathologies such as hypertonia. An innovative step in our haptic display is to give clinicians the capacity to translate clinical parameters into a haptic sensation. This could also fulfill the need for a training system that can help clinicians to better understand the causality of the haptic sensation experienced during physical manipulation of a patient's limb afflicted by a particular impairment, for instance, to find specific probing motion(s) that are better at discriminating between different pathologies. While in current clinical practice the assessment of different aspects of hypertonia are expressed by the subjective judgment of the therapist [4], to our knowledge, there is no scientific literature that has systematically investigated a therapist's perceptual identification of the nature and severity of multi-joint hypertonic phenomena. Some studies have tried to categorize the sensitivity of therapists to non-biological stiffness [36,37]; however, these studies have been based on identifying the sensitivity of tactile sensations produced by surfaces. Our haptic display allows us to systematically study these conditions.

A guided approach to manipulation

The capability to discriminate between two haptic stimuli is not just a function of touch sensitivity. While direct touch conveys cutaneous information about local surface features at the points of contact, probing movements (such as in hypertonia assessments) convey information mostly from proprioceptors [38]. The manner in which a user chooses to actively probe can influence the information gained and quality of the assessment. Visually impaired individuals use different exploration strategies which allow them to gain richer information content, and when sighted subjects are required to use similar strategies, their performance is radically improved [39]. Our haptic display is therefore a valuable tool to explore the haptic sensation of different neuro-mechanics and to determine the most informative probing motion(s) to be used in order to discriminate them

It should be noted, however, that the number of probing motions to distinguish different abnormalities could be quite large and the search could be long and tedious. This poses a problem, as humans can handle only limited amounts of information at once and their memory is often not reliable. Our current research focuses on the development of an operational research algorithm that searches for the 'most informative probing motion', i.e., the one that produces the most characteristic haptic sensation when interacting with different stereotypical force fields generated by patients.

Autonomous guided vehicles operating in dynamic and unknown environments are often faced with the problem of deciding where to go to get the most relevant information [40]. The dynamic environment is assumed to be unknown, but the automaton has a sensor to measure the environment. The environment can be described using a potential function such as an elastic energy, where maxima of the potential are associated with repulsive forces which represent obstacles to avoid. Minima in the potential correspond to attractive forces representing waypoints.

This framework closely resembles the problem a clinician faces when testing a patient. The total potential, which gradient is composed of the characteristic forces of the patient's hypertonic arm, is unknown to the clinician (i.e. they need to be assessed). Conversely, the clinician has the capability to 'perceive' the force generated by hypertonic arms, as the autonomous guided vehicles can sense the environment. We wish for the clinician to learn the structure of the patients' total potential by identifying the areas within the subjects' reachable space that are dynamic and the rate of change for these areas. Given the structure of the total potential, we conceive a path for the clinicians to allow them sensing such dynamic areas.

Future directions

Non-robotic assessment techniques

A derivation from the knowledge gained from the studies using our haptic display is an assessment approach in which clinicians and computational algorithms participate in the diagnosis process. The idea is to have clinicians physically interacting with the patient while kinematic and dynamic information is fed into a computational algorithm. The algorithm would suggest optimal manipulation strategies to the clinician for the identification of the impairment. These optimal strategies would consider the non-linear effects of human perception and stereotypical forces produced by the different types of impairment.

An expert clinician's experience should provide an educated guess of what the impairment is. This can be immediately translated into a total potential that can be searched by the informative path planning algorithm, which will suggest to the clinician the best trajectory to follow for the assessment.

Our goal is to close the loop between human experts and a database of stereotypical force fields generated by different types of impairment via a mathematical algorithm, essentially augmenting human perception and interpretation capabilities with computer algorithms. The therapist would be able to select the initial choices of most probable impairments, and the algorithm will suggest a minimum amount of probing for identification. This way, the experience of the therapists will be embedded in a supervised algorithm, where both human and classification algorithms interact. This approach would allow us to identify exploration strategies (easily recognizable in intensity, location, and time course) that maximize exchange of information and can uniquely discriminate among different types of hypertonia. In addition, this would provide clinicians with the capability of testing a patient easily in a routine clinical setting with the ability to support the perceptual findings by quantitative methods. This approach is innovative as it shifts the analysis of impaired behavior from the mere perception of force to the characterization of contact forces, motions, and haptic perception produced in the patient-therapist interaction.

Robotic-aided targeted intervention

Given a set of possible interventions to treat specific impairments, clinicians could analyze their impact directly on the haptic display. For example, two common interventions for the treatment of severe hypertonia in children cerebral palsy are surgery or injections of botulin toxin. The invasiveness of these procedures makes them a one-time intervention. Therefore, careful planning is required. With our haptic display, clinicians could test the biomechanical effects of the different kinds of intervention without even touching the patient. This can be of great help on these cases, helping clinicians to take a more reliable decision and with less margin of error.

Conclusions

We propose a haptic display that can serve as an aiding tool to *i)* synthesize the impairment based on direct measurements of limb mechanics; *ii)* train clinicians in diagnosing different types of abnormal biomechanics, affecting different muscle groups; *iii)* analyze the impact of targeted intervention on specific aspects of limb biomechanics; *iv)* establish and customize a series of most informative probing motions that would maximize the information regarding the impairment so as to assess abnormal biomechanics in a useful, valid, and reliable manner without the need of sophisticated equipment on-site; or *v)* allow the assessment of abnormal biomechanics via remote physical interaction. We expect that this will transform the diagnostic practice by the combination of manual assessment techniques with the computational and mechanical simulation of specific abnormal biomechanics.

Abbreviations

be: bi-articular extensors (triceps long) muscle groups; bf: bi-articular flexors (biceps short); DOF: degrees of freedom; ee: elbow extensors (triceps lateral, anconeus); ef: elbow flexors (biceps long, brachialis, brachioradialis); JND: just noticeable difference; se: shoulder abductors (deltoid posterior); sf: shoulder adductors (deltoid anterior, coracobrachialis, pectoralis major clav).

Competing interests

Part of the work in this manuscript has been used to redact the following provisional patent application: U.S. Patent App. No. 61/991,370, HAPTIC DEVICE AND METHODS FOR ABNORMAL LIMB BIOMECHANICS (2014) to AM-C, DP and FAM-I. JLP has no competing interests. AM-C is a current employee at Hocoma AG, a medical technology company. However, the work presented in this paper has been performed before any relationship between AM-C and Hocoma AG.

Authors' contributions

AM-C and DP contributed equally to this work. AM-C, DP, and FAM-I conceived the idea. AM-C and DP participated in the rendering of the haptic display. DP and AM-C performed the psychophysical experiments. AM-C and DP analyzed the data. AM-C, DP, JLP, and FAM-I contributed the methods and tools. AM-C and DP wrote the paper. All authors read and approved the final manuscript.

Authors' information

AM-C received a B.Sc. degree in Mechatronics and a Pg.Cert in Control Engineering from Tecnológico de Estudios Superiores de Monterrey-Campus Estado de México (ITESM-CEM) in 2006; an M.Sc. degree in Intelligent Autonomous Systems from Aalborg University in 2007, working on the control of powered transfemoral prostheses; and a PhD degree in Bioengineering from Imperial College of Science, Technology and Medicine in 2011, working at the interface of rehabilitation robotics, human-human interaction, and human motor control and learning. He worked as a Postdoctoral Research Fellow at the Rehabilitation Institute of Chicago and Northwestern University from 2012 to 2013. He is currently a Project Manager at Hocoma AG and Adjunct Assistant Professor at the Department of Physical Medicine & Rehabilitation at Northwestern University. His research has demonstrated how visual cues can compensate for the lack of proprioceptive error and basic mechanisms of human-human physical interaction. His main research interests include human-robot interfaces and motor control/learning.

DP received his M.S.M.E. in 2003 and D.Eng. in Mechanical Measurement in 2007 at the University of Padova, Italy. His dissertation presented a set of experimental and analytical validation techniques for human upper limb models. From 2004 to 2008, he was a visiting scholar and postdoctoral fellow at the Ashton Graybiel Spatial Orientation Lab at Brandeis University, under the supervision of Professors Paul DiZio and James R. Lackner. There, he worked on the mechanics of movement adaptation in non-inertial environments as part of a NASA extramural funding program. He joined Northwestern University in 2008, working as a Postdoctoral Fellow at the Rehabilitation Institute of Chicago under the supervision of Professor Ferdinando (Sandro) Mussa-Ivaldi. DP is currently an Assistant Professor in the Mechanical Engineering Department at Gannon University and a Director of the Biomedical Engineering Program. His main interest is to gain insights on the role of biomechanics in the neural control of movements, with applications to rehabilitation engineering.

JLP received a BS degree in Mechanical Engineering and Engineering Science from the University of Michigan, an MS degree in Theoretical Mechanics from Michigan State University, and a PhD degree in Biomedical Engineering from Northwestern University. Currently, he is working as an Associate Professor of Bioengineering at The University of Illinois at Chicago and as an Associate Director of the Center for Rehabilitation Robotics at the Rehabilitation Institute of Chicago (RIC). He also holds affiliate positions in Physical Medicine and Rehabilitation and Mechanical and Biomedical Engineering at Northwestern University. He has worked in manufacturing for Ford and as a cyclotron operator in nuclear medicine before turning his attention to biomechanics and control of human movement. His general interests include robotic teaching, dynamic balance control, haptics, modeling of the human-machine interface, and robot-facilitated recovery from a brain injury. His research has demonstrated how control is made robust by avoiding undesirable situations. His more recent research has demonstrated how one can exploit the nervous system's natural capacity to adapt in order to teach movements and help people to recover from brain injury. He is on the International Advisory Committee for the IEEE-EMBS and also chairs the technical committee on biomedical robotics.

FAM-I received a Laurea degree in Physics from the University of Torino, Turin, Italy, in 1978 and a PhD degree in Biomedical Engineering from the Politecnico of Milano, Milan, Italy, in 1987. He is a Professor of Physiology, Physical Medicine and Rehabilitation, and Biomedical Engineering at Northwestern University, IL, USA. He is currently a Senior Research Scientist with the Rehabilitation Institute of Chicago, IL, USA, where he founded and directs the Robotics Laboratory. His areas of interest and expertise include robotics, neurobiology of the sensory-motor system, and computational neuroscience. FAM-I serves on the Editorial Boards of the Journal of Neural Engineering and the Journal of Motor Behavior and is a Member of the Society for Neuroscience and of the Society for the Neural Control of Movement. His achievements include the first measurement of human arm multi-joint impedance; the development of a technique for investigating the mechanisms of motor learning through the application of deterministic force fields; the discovery of a family of integrable generalized inverses for redundant kinematic chains; the discovery of functional modules within the spinal cord that generate a discrete family of force fields; the development of a theoretical framework for the representation, generation, and learning of limb movements; and the development of the first neurorobotic system in which the brainstem of a lamprey controls the behavior of a mobile robot through a closed-loop interaction. He has 120 full-length publications and 100 abstracts.

Acknowledgements
AM-C and DP were with the Sensory Motor Performance Program, Rehabilitation Institute of Chicago at the time this research was performed. This work was partially supported by NINDS grants 2R01NS035673-10A2 and R01-NS053606-04 and the Coolidge fellowship. The authors thank Felix Huang for the fruitful discussions.

Author details
[1]Department of Physical Medicine and Rehabilitation, Northwestern University, 60611 Chicago, IL, USA. [2]Biomedical Program, Department of Mechanical Engineering, Gannon University, 16541 Erie, PA, USA. [3]Department of Bioengineering, University of Illinois at Chicago, 60607 Chicago, IL, USA. [4]Department of Biomedical Engineering, Northwestern University, 60208 Evanston, IL, USA. [5]Department of Mechanical Engineering, Northwestern University, 60208 Evanston, IL, USA. [6]Sensory Motor Performance Program, Rehabilitation Institute of Chicago, 60611 Chicago, IL, USA. [7]Department of Physiology, Northwestern University, 60611 Chicago, IL, USA.

References
1. Haugh AB, Pandyan AD, Johnson GR (2006) A systematic review of the Tardieu Scale for the measurement of spasticity. Disabil Rehabil 28(15):899–907, doi:10.1080/09638280500404305
2. Bohannon RW, Smith MB (1987) Interrater reliability of a modified Ashworth scale of muscle spasticity. Phys Ther 67(2):206–207
3. Jethwa A, Mink J, Macarthur C, Knights S, Fehlings T, Fehlings D (2010) Development of the Hypertonia Assessment Tool (HAT): a discriminative tool for hypertonia in children. Dev Med Child Neurol 52(5):e83–e87, doi:10.1111/j.1469-8749.2009.03483.x
4. Brashear A, Zafonte R, Corcoran M, Galvez-Jimenez N, Gracies JM, Gordon MF, McAfee A, Ruffing K, Thompson B, Williams M, Lee CH, Turkel C (2002) Inter- and intrarater reliability of the Ashworth Scale and the Disability Assessment Scale in patients with upper-limb poststroke spasticity. Arch Phys Med Rehabil 83(10):1349–1354, doi:S0003999302000497
5. de Vlugt E, de Groot J, Schenkeveld K, Arendzen JH, van der Helm F, Meskers C (2010) The relation between neuromechanical parameters and Ashworth score in stroke patients. J NeuroEng Rehab 7(1):35
6. Mirbagheri MM, Alibiglou L, Thajchayapong M, Rymer WZ (2008) Muscle and reflex changes with varying joint angle in hemiparetic stroke. J NeuroEng Rehab 5:6–6
7. Piovesan D, Morasso P, Giannoni P, Casadio M (2012) Arm Stiffness during Assisted Movement after Stroke: the Influence of Visual Feedback and Training. In: IEEE Transactions on Neural Systems and Rehabilitation Engineering. doi:10.1109/TNSRE.2012.2226915
8. Li-Qun Z, Hyung-Soon P, Yupeng R (2009) Shoulder, elbow and wrist stiffness in passive movement and their independent control in voluntary movement post stroke. In: IEEE international conference on rehabilitation robotics (ICORR). Kyoto, pp 805–811, doi:10.1109/icorr.2009.5209489
9. Wood D, Burridge J, Wijck FV, McFadden C, Hitchcock R, Pandyan A, Haugh A, Salazar-Torres J, Swain I (2005) Biomechanical approaches applied to the lower and upper limb for the measurement of spasticity: a systematic review of the literature. Disabil Rehabil 27(1–2):19–33, doi:10.1080/09638280400014683
10. Dyer J-O, Maupas E, de Andrade MS, Bourbonnais D, Fleury J, Forget R (2009) Transmission in heteronymous spinal pathways is modified after stroke and related to motor incoordination. PLoS One 4(1):e4123, doi:10.1371/journal.pone.0004123
11. Piovesan D, Casadio M, Mussa-Ivaldi FA, Morasso P (2012) Comparing two computational mechanisms for explaining functional recovery in robot-therapy of stroke survivors. In: 4th IEEE RAS & EMBS international conference on biomedical robotics and biomechatronics (BioRob). Rome, pp 1488–1493, doi:10.1109/BioRob.2012.6290914
12. Mayer NH, Esquenazi A, Childers MK (1997) Common patterns of clinical motor dysfunction. Muscle Nerve Suppl 6:S21–S35
13. Biering-Sorensen F, Nielsen JB, Klinge K (2006) Spasticity-assessment: a review. Spinal Cord 44(12):708–722
14. Grow DI, Mengnan W, Locastro MJ, Arora SK, Bastian AJ, Okamura AM (2008) Haptic Simulation of Elbow Joint Spasticity. In: Symposium on Haptic Interfaces for Virtual Environment and Teleoperator Systems (Haptics 2008). Reno, pp 475–476, doi:10.1109/haptics.2008.4479997
15. Hyung-Soon P, Jonghyun K, Damiano DL (2012) Development of a Haptic Elbow Spasticity Simulator (HESS) for improving accuracy and reliability of clinical assessment of spasticity. IEEE Trans Neural Syst Rehabil Eng 20(3):361–370, doi:10.1109/tnsre.2012.2195330
16. Kim J-P, Ryu J (2010) Robustly stable haptic interaction control using an energy-bounding algorithm. Int J Rob Res 29(6):666–679, doi:10.1177/0278364909338770

17. Carignan CR, Olsson PA (2004) Cooperative Control of Virtual Objects over the Internet using Force-Reflecting Master Arms. In: 2004 IEEE International Conference on Robotics and Automation, 2004. Proceedings. ICRA'04, Barcelona, vol 1222, pp 1221–1226. doi:10.1109/robot.2004.1307991
18. Mussa-Ivaldi FA (1992) From basis functions to basis fields: vector field approximation from sparse data. Biol Cybern 67(6):479–489
19. Mussa-Ivaldi FA, Giszter SF (1992) Vector field approximation: a computational paradigm for motor control and learning. Biol Cybern 67(6):491–500
20. Melendez-Calderon A, Piovesan D, Mussa-Ivaldi FA (2013) Therapist discrimination of impaired muscle groups in simulated multi-joint hypertonia. In: 13th IEEE international conference on rehabilitation robotics (ICORR). Seattle, doi:10.1109/ICORR.2013.6650425
21. Piovesan D, Melendez-Calderon A, Mussa-Ivaldi FA (2013) Haptic Discrimination of Multijoint Dystonia and Spasticity in the Assessment of Hypertonia. In: 13th IEEE International Conference on Rehabilitation Robotics (ICORR). Seattle, doi:10.1109/ICORR.2013.6650449
22. Winter DA (2004) Biomechanics and motor control of human movement, 3rd edn. John Wiley & Sons, New York
23. Hatze H (1977) A myocybernetic control model of skeletal muscle. Biol Cybernetics 25(2):103–119, doi:10.1007/bf00337268
24. Hatze H (1978) A general myocybernetic control model of skeletal muscle. Biol Cybernetics 28(3):143–157, doi:10.1007/bf00337136
25. Galiana L, Fung J, Kearney R (2005) Identification of intrinsic and reflex ankle stiffness components in stroke patients. Exp Brain Res 165(4):422–434
26. Granata KP, Wilson SE, Massimini AK, Gabriel R (2004) Active stiffness of the ankle in response to inertial and elastic loads. J Electromyogr Kinesiol 14(5):599–609, doi:10.1016/j.jelekin.2004.03.005 S1050641104000458
27. Mirbagheri MM, Barbeau H, Kearney RE (2000) Intrinsic and reflex contributions to human ankle stiffness: variation with activation level and position. Exp Brain Res 135(4):423–436
28. Pisano F, Miscio G, Del Conte C, Pianca D, Candeloro E, Colombo R (2000) Quantitative measures of spasticity in post-stroke patients. Clin Neurophysiol 111(6):1015–1022, doi:S1388-2457(00)00289-3
29. Piovesan D, Casadio M, Mussa-Ivaldi FA, Morasso PG (2011) Multijoint arm stiffness during movements following stroke: implications for robot therapy. In: 2011 IEEE international conference on rehabilitation robotics (ICORR). Zurich, doi:10.1109/ICORR.2011.5975372
30. Kanade P, Scheidt RA (2009) Spatial mapping of posture-dependent endpoint forces in the hypertonic arm post-stroke. Doctoral dissertation, Marquette University, USA
31. Watts RL, Wiegner AW, Young RR (1986) Elastic properties of muscles measured at the elbow in man: II. Patients with parkinsonian rigidity. J Neurol Neurosurg Psychiatry 49(10):1177–1181
32. Given JD, Dewald JP, Rymer WZ (1995) Joint dependent passive stiffness in paretic and contralateral limbs of spastic patients with hemiparetic stroke. J Neurol Neurosurg Psychiatry 59(3):271–279, doi:10.1136/jnnp.59.3.271
33. Tan HZ, Durlach NI, Beauregard GL, Srinivasan MA (1995) Manual discrimination of compliance using active pinch grasp: the roles of force and work cues. Percept Psychophys 57(4):495–510
34. Youngung S, McMains S (2004) Evaluation of drawing on 3D surfaces with haptics. IEEE Comput Graph Appl 24(6):40–50, doi:10.1109/mcg.2004.49
35. Piovesan D, Melendez-Calderon A, Mussa-Ivaldi FA (2013) Haptic perception of multi-joint hypertonia during simulated patient-therapist physical tele-interaction. Conf Proc IEEE Eng Med Biol Soc 2013:4143–4147, doi:10.1109/EMBC.2013.6610457
36. Nicholson L, Adams R, Maher C (1997) Reliability of a discrimination measure for judgements of non-biological stiffness. Man Ther 2(3):150–156, doi:10.1054/math.1997.0295
37. Crutchfield CA (1976) The perception of tactile stimuli by physical therapy students before and after haptic training. Doctoral dissertation, West Virginia University, USA
38. Gibson JJ (1966) The senses considered as perceptual systems. Houghton Mifflin, Boston
39. Davidson PW, Whitson TT (1974) Haptic equivalence matching of curvature by blind and sighted humans. J Exp Psychol 102(4):687–690, doi:10.1037/h0036245
40. Soltero DE, Schwager M, Rus D (2012) Generating Informative Paths for Persistent Sensing in Unknown Environments. In: 2012 IEEE/RSJ International Conference on Intelligent Robots and Systems (IROS). Vilamoura, doi:10.1109/IROS.2012.6385730

Magnetic microrobot and its application in a microfluidic system

Jingyi Wang[1,2], Niandong Jiao[1*], Steve Tung[1,3] and Lianqing Liu[1*]

Abstract

This paper researches the design and control method of a microrobot in a microfluidic system by electromagnetic field. The microrobot can move along the microchannel to a required position, and by changing the magnetic torque, the microrobot can also rotate in the microfluidic chip. As an application of the microrobot, it is used as a mobile micromixer to mix two solutions in the microfluidic chip, and the experimental results verify its effectiveness.

Keywords: Microrobot; Magnetic drive; Microfluidic; Micromixer

Background

Microrobot is a hot research field in robotics. Using microrobots as a tool to surgery *in vivo* has the advantages of small trauma, safety, reliability, and significant reduction of medical costs and rehabilitation duration, so entering the liquid environment of the human body and making *in vivo* detection and treatment [1-3] are an important developing direction of microrobots. Recently, many researches on microrobots have carried out a lot of various actuation methods for microrobots, such as electromagnetic actuators [4,5], chemical bubble actuators [6,7], swimming tail actuators [8-10], bacterial actuators [11-13], and so on. Especially, the electromagnetic driving method has many advantages. First, electromagnetic field can be used to drive the microrobot with high controllability through changing the currents of the electromagnetic coils. Second, magnetic field will not cause harm or side effects to humans.

Among the researches of magnetic controlled microrobots, Bradley J. Nelson's group, inspired by the natural design of bacterial flagella, proposed an artificial bacterial flagella microrobot [14] that could swim in a controllable fashion using weak applied magnetic fields. The flagella microrobot consisted of a helical tail and a small soft magnetic material as a head on one end. The flagella microrobot's helical propulsion could be precisely controlled by electromagnetic coils. They also proposed porous microniches using 3D laser lithography as a transporting microrobot for targeted cell delivery [15]. Metin Sitti proposed a microrobot driven by magnetic torque. It could be controlled by two pairs of Helmholtz coils and a pair of clamping coils with a stick-slip motion [16]. The horizontal movement was controlled by the Helmholtz coils, and the clamping coils made the driving signals for the stick-slip motion. Sukho Park proposed a kind of microrobot controlled by magnetic force. The driving system consisted of two pairs of Helmholtz coils and two pairs of Maxwell coils [17]. The magnetic force was provided by the Maxwell coils, and the Helmholtz coils were used to generate the torque for the rotation of the microrobot. However, in manipulating microrobots to explore the human body, there are still a lot of works that need to be done.

In order to control the microrobot to fulfill a specific function in the internal liquid environment of the human body, it is necessary to firstly control the microrobot's movement in the microfluidic environment simulating the vessels. Since a microfluidic system has an effective functional zone of microscale as well as precise control of flow rate, it can well simulate the vessel liquid environment [18].

This paper researches the design and fabrication of a microrobot controlled by outside electromagnetic field and proposes a control method for the motion of the microrobot in the microfluidic system. This microrobot can move along the microchannel to a required position. By changing the magnetic torque, the microrobot can

* Correspondence: ndjiao@sia.cn; lqliu@sia.cn
[1]State Key Laboratory of Robotics, Shenyang Institute of Automation, Chinese Academy of Sciences, No.114 Nanta Street, Shenyang 110016, China
Full list of author information is available at the end of the article

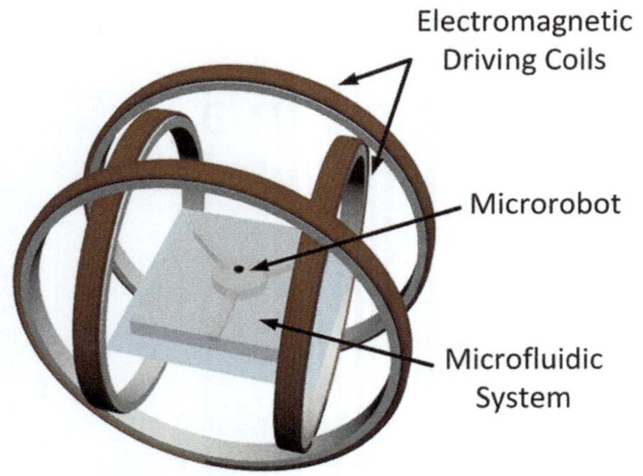

Figure 1 Magnetic driving system for the microrobot in a microfluidic system.

also rotate in the microfluidic chip. As an application of the microrobot, it can be used as a mobile micromixer in the microfluidic chip.

Methods

Theoretical model

In order to study the movement of the microrobot in a microfluidic system, the force acting on the microrobot and the action of the magnetic field are analyzed. The microrobot is designed to be a hemispherical structure and located in the magnetic driving system, which has four independent controlling electromagnetic coils. The system structure is shown in Figure 1. When the microrobot is suspended in the liquid, gravity is equal to buoyancy. So force analysis is mainly in the horizontal direction. When the microrobot is immersed in the liquid, magnetic force is the main driving force that works on the robot. And the viscous force has a relationship with the velocity of the microrobot:

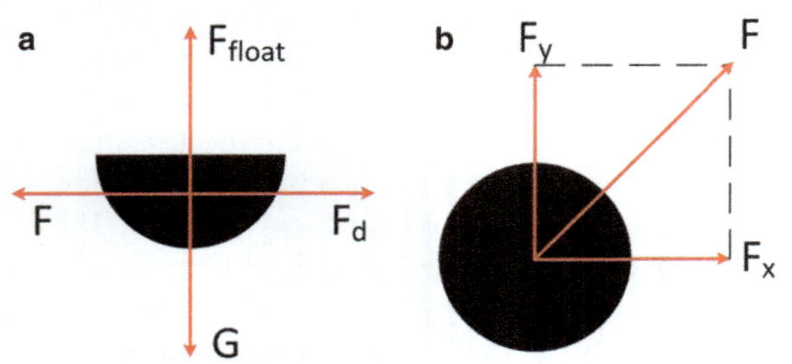

Figure 2 Force analysis of the microrobot suspends in liquid. **(a)** Side view of the microrobot. The forces acting on the microrobot in the horizontal direction are magnetic force and viscous force. In the vertical direction, buoyancy and gravity are equal when the microrobot is suspended in the channel. **(b)** Top view of the microrobot. Magnetic force can be decomposed into two components in the *x*, *y* direction.

$$F + F_d + m\frac{dv}{dt} = 0 \tag{1}$$

where v is the velocity of the microrobot. Force analysis of the microrobot immersed in liquid is shown in Figure 2.

In the horizontal direction, the resistance force works against the motion of the microrobot. In order to manipulate the microrobot, the input magnetic force must overcome the resistance:

$$F \geq F_d \tag{2}$$

where F is the magnetic force exerted on the microrobot and F_d is the resistance of the fluid. The magnetic force is given by:

$$F = V(M \cdot \nabla)B \tag{3}$$

where V is the volume of the magnetic microrobot, M is the uniform magnetization of the magnetic microrobot, and B is the magnetic flux. ∇ is the gradient operator:

$$\nabla = \begin{bmatrix} \frac{\partial}{\partial x} \\ \frac{\partial}{\partial y} \end{bmatrix} \tag{4}$$

$$M = \begin{bmatrix} M_x \\ M_y \end{bmatrix}, B = \begin{bmatrix} B_x \\ B_y \end{bmatrix} \tag{5}$$

Only considering in the xy plane, magnetic force can be expressed as the following vector form:

$$F = \begin{bmatrix} F_x \\ F_y \end{bmatrix} = V \begin{bmatrix} \frac{\partial B_x}{\partial_x} & \frac{\partial B_y}{\partial_x} \\ \frac{\partial B_x}{\partial_y} & \frac{\partial B_y}{\partial_y} \end{bmatrix} \begin{bmatrix} M_x \\ M_y \end{bmatrix} = V \begin{bmatrix} M_x & \frac{\partial B_x}{\partial_x} \\ M_y & \frac{\partial B_y}{\partial_y} \end{bmatrix} \tag{6}$$

The counter-diagonal of ∇B can be considered to be 0, and M_x and M_y are constants in the magnetic field simultaneously.

Therefore, the main factors of the direction and the force are $\frac{\partial B_x}{\partial x}$ and $\frac{\partial B_y}{\partial y}$. To generate the required gradient magnetic field, magnetic coils are needed. The magnetic field generated by the current flow through a circular electromagnetic coil can be calculated by the Biot-Savart law as follows:

$$B = \frac{\mu_0 I}{4\pi} \int_L \frac{dl \times r}{|r|^3} \tag{7}$$

where I is the current through the coil, L is the integral path, r is the full displacement vector from the wire

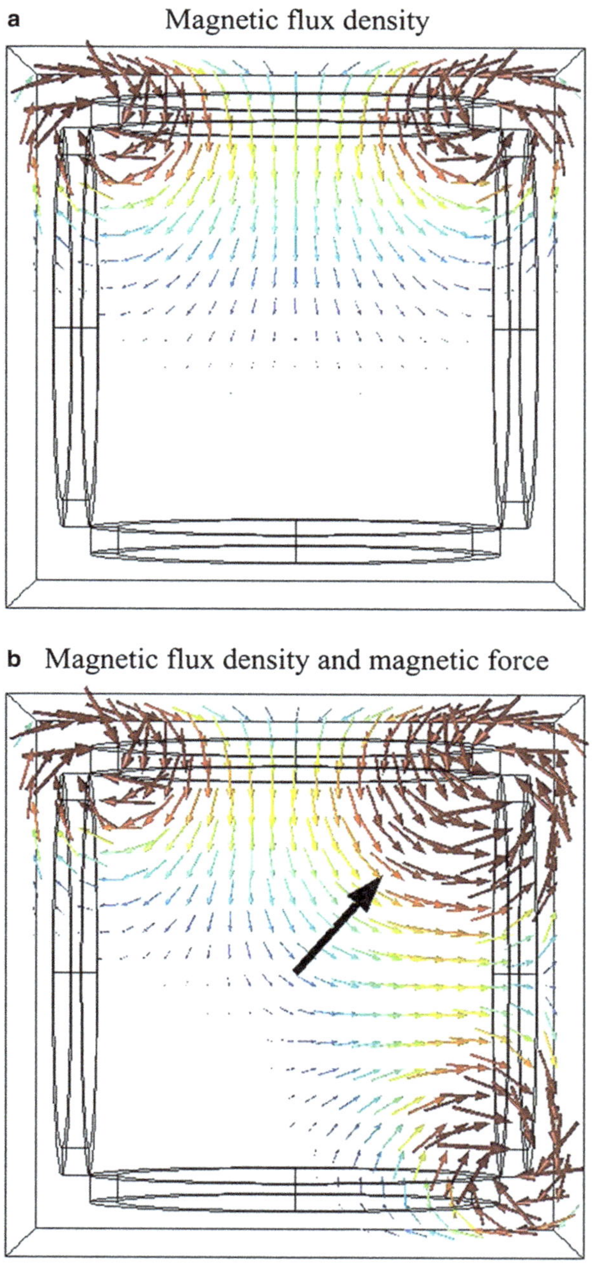

Figure 3 Simulation of the magnetic field generated in the *xy* plane. (a) Magnetic field generated by the top electromagnetic coil. **(b)** Magnetic field generated by the top and right electromagnetic coils. The black arrow indicates the resultant magnetic force generated by the two coils.

element to the point where the field needs to be calculated, dl is a vector of the differential element of the current through the wire, and μ_0 is the magnetic permeability ($\mu_0 = 4\pi \times 10^{-7}$ H/m).

To manipulate the microrobot accurately, the magnetic field produced by the electromagnetic coils must be controlled nicely. The microrobot is placed in the controlled area that is surrounded by four electromagnetic coils, and each electromagnetic coil can be controlled accurately according to the current outputting by the PC. Four coils can be controlled separately or together. COMSOL multiphysics is used to simulate the magnetic field and the force generated by the electromagnetic coils.

Figure 3 shows the simulation of the magnetic field generated by the electromagnetic coils in the *xy* plane. It is the top view of the four coils. Figure 3a shows the

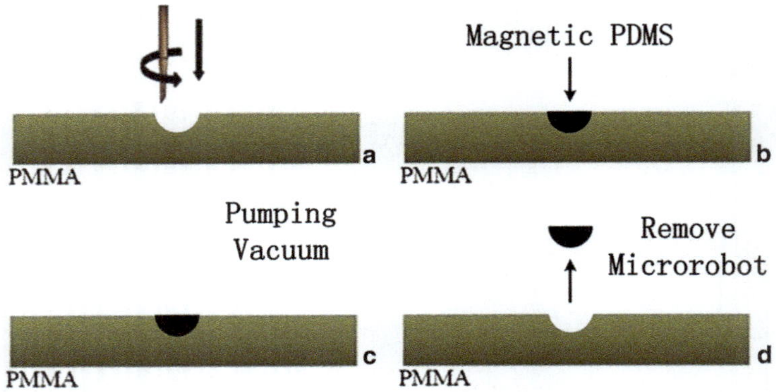

Figure 4 Process of microrobot fabrication. (a) Use an engraving machine curving the microrobot's mold. **(b)** Put the magnetic PDMS into the mold. **(c)** Degas and cure. **(d)** Remove the microrobot from the mold.

magnetic field generated by the top electromagnetic coil; the length and the direction of the arrow indicate the strength and the direction of the magnetic field, respectively. Figure 3b shows the magnetic field generated by the top and right electromagnetic coils; the black arrow indicates the resultant magnetic force generated by the top and right electromagnetic coils. By controlling the magnetic field, the strength and the direction of the resultant magnetic force can be regulated; therefore, the microrobot in the field can move freely in the plane.

Microrobot and system configuration

Fabrication of the microrobot

Polydimethylsiloxane (PDMS) has great plasticity and can be shaped to any structures depending on the needed size, shape, and so on. The microrobot is made of magnetic PDMS. To make magnetic PDMS, magnetic microparticles are added into PDMS. The microrobot is a hemispherical structure with a radius of 750 μm. Figure 4 shows the main production process of the microrobot:

1) Carve the microrobot's mold out of the acrylic plate using an engraving machine (Benchtop Engravers EGX-600/400, Roland DGA Corporation, Irvine, CA, USA) which is equipped with a milling cutter with a diameter of 200 μm.
2) Mix the polydimethylsiloxane and its curing agent at a mass ratio of 10:1 and then stir for 30 min.
3) Put NdFeB magnetic microparticles into mixed PDMS and then put them into the microrobot's mold.
4) Degas for 30 min in a vacuum box to remove bubbles.
5) Put the microrobot into an oven to heat for 3 h at 60°C.
6) Remove the microrobot from the mold with tweezers after magnetic PDMS is cured.

Setup of the magnetic driving system

The magnetic driving system consists of a PC, vision acquisition camera, data acquisition card (NI PCI-6229 DAQ Card, National Instruments, Austin, TX, USA), DC power supply, array of air-core electromagnetic coils, micron three-dimensional positioning stage, and power amplifier. The electromagnetic coils are supported on aluminous frames which will not affect the magnetic field by their nonmagnetic conductive property. The electromagnetic coils, the micron three-dimensional positioning stage, and the power amplifier are assembled on an aluminous platform with heat sinks. The schematic diagram of the platform is shown in Figure 5.

The detailed specification of the electromagnetic coils is described in Table 1. The picture of the whole magnetic driving system is shown in Figure 6.

The location information of the microrobot is acquired by the vision acquisition camera and sent to the PC. The

Figure 5 Schematic diagram of the magnetic driving platform. (1) An array of air-core electromagnetic coils. (2) A micron three-dimensional positioning stage. (3) Controlling area. (4) A power amplifier.

Table 1 Parameters of coils

Coils	Coil turns	Insider radius/outsider radius (mm)	Diameter of copper wires (mm)	Resistance (Ω)
x-axis	250	43.5/56.5	0.51	5
y-axis	250	62/75	0.51	6.5

control algorithm programmed by LabVIEW processes the location information and generates an output signal and then transmits it to the data acquisition card. The output signal is amplified through the power amplifier to generate the required magnetic field. When using the joystick to point a desired direction, the electromagnetic coils generate the needed magnetic force to control the microrobot. The microrobot in the microfluidic system is placed on the micron three-dimensional positioning stage in the center of the electromagnetic coils. The hardware connection of the magnetic driving system is shown in Figure 7.

Results and discussion

Motion control of the microrobot

To study the motion of the microrobot in the microfluidic channel, a microfluidic chip with cross-shaped channels is designed and fabricated. Potassium chloride solution is added into the microfluidic channels, so the microrobot can be suspended in the channel well. The solution in the microfluidic chip is in static state. The microfluidic chip is placed on the micron three-dimensional positioning stage and the positioning stage is adjusted so that the microfluidic chip is located in the magnetic driving system's controlling area. The width and depth of the microfluidic chip are both 3 mm. Figure 8 shows the motion of the microrobot in the microfluidic chip controlled by a joystick. The microrobot is located at the far left of the microfluidic chip. When using the joystick to control the microrobot to move to the center, the electromagnetic coil on the right generates the magnetic force. Then the microrobot is controlled to move to the top.

The microrobot starts to move in the channel due to the magnetic force. In the initial case, the magnetic force is greater than the liquid viscous force, with the increase of microrobot speed; fluid viscous force is gradually increased until the magnetic force is equal to the liquid viscous force. The microrobot reaches equilibrium.

The velocity of the microrobot is determined by the magnetic force, and the magnetic force can be changed by changing the currents in the electromagnetic coils. Figure 9 shows the velocities under different current values. With the increase of the current, the velocity of

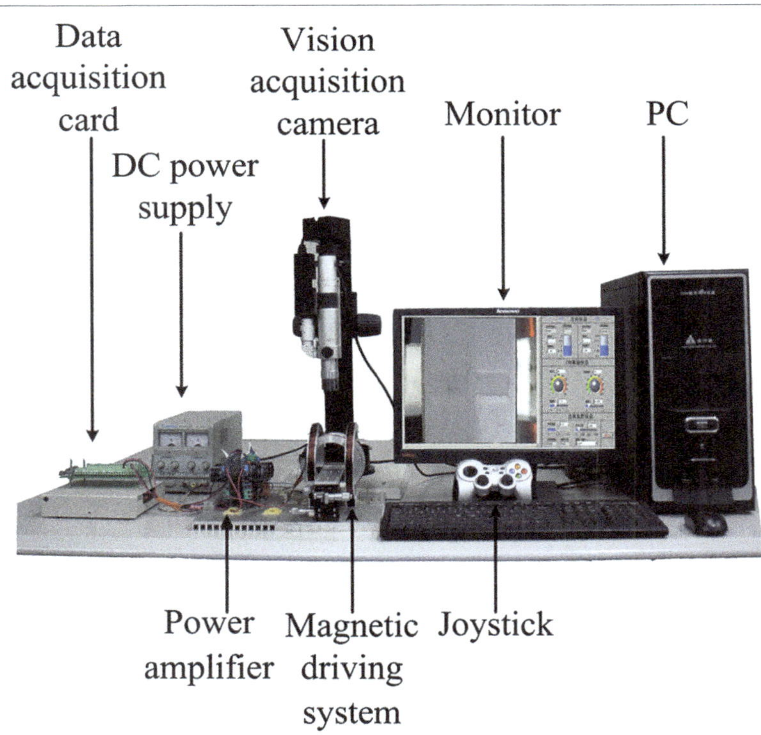

Figure 6 The whole magnetic driving system.

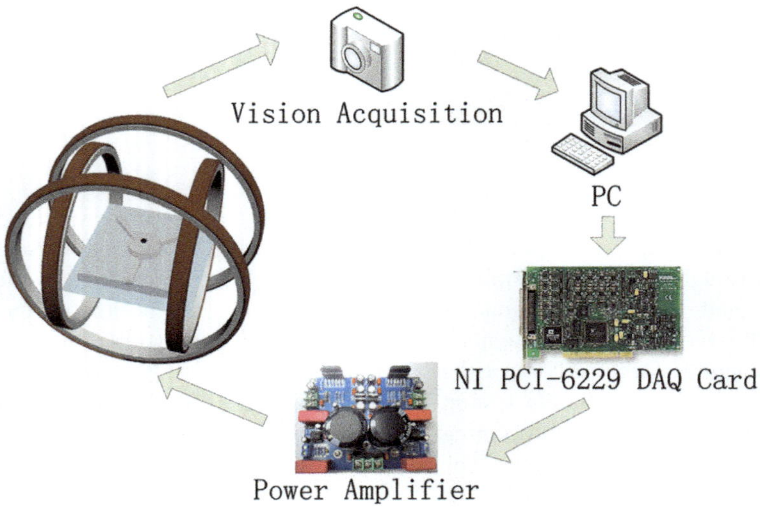

Figure 7 Hardware connection of the magnetic driving system.

the microrobot increases gradually, so the microrobot can be controlled precisely by changing the current.

The microrobot works as a mobile micromixer

The microrobot can move to any position in the microfluidic chip and can be controlled to produce rotary motion, so the microrobot can be used as a mobile micromixer in microfluidic chips.

In the existing micromixing research in microfluidics, two main kinds of micromixing technologies have been developed to improve mixing efficiency: passive and active micromixing. Passive micromixing generally has microchannels with complex design, special geometry [19], or additional complex obstacles [20] to increase contact with the liquid. Active micromixing mainly relies on the physical field that is exerted outside the microfluidic system. In accordance with the physical field exerted outside, the active micromixer is mainly divided into ultrasonic mixing [21], magnetic mixing [22,23], and so on. Active micromixing generally spends less time and has more simple structures. In this paper, a new mobile active micromixer has been proposed. A magnetic microrobot injected by a pipette into the microfluidic chip through the inlet can be controlled to move along the channel in a microfluidic chip, and when the desired location is reached, it can rotate to complete the mixing function as a mobile active micromixer. To prove that the microrobot can be used as a mobile micromixer, we fabricated a microfluidic chip and made a comparison by mixing two solutions with and without the magnetic mobile active micromixer. The microfluidic chip is a PDMS-glass chip that is used for mixing in the lab. The depth and radius of the mixing chamber are 1.2 and 3 mm,

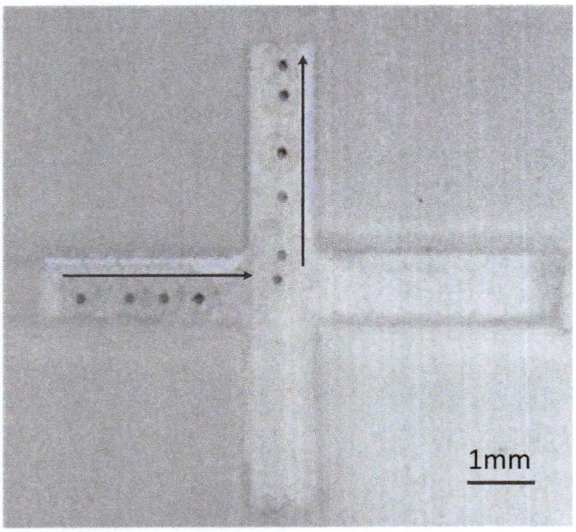

Figure 8 Motion of the microrobot in a fully static cross-shaped microfluidic channel. The width and depth of the microfluidic chip are both 3 mm. The microrobot is a hemispherical magnetic PDMS with a radius of 750 μm.

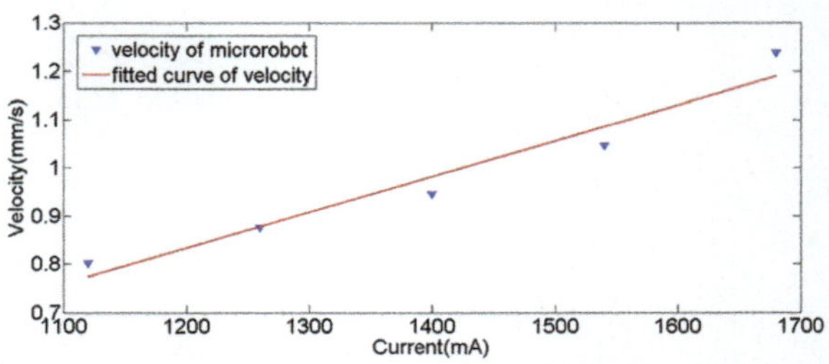

Figure 9 Velocity of the microrobot, along with the current in the electromagnetic coils. The red line indicates that the velocity increases with the current.

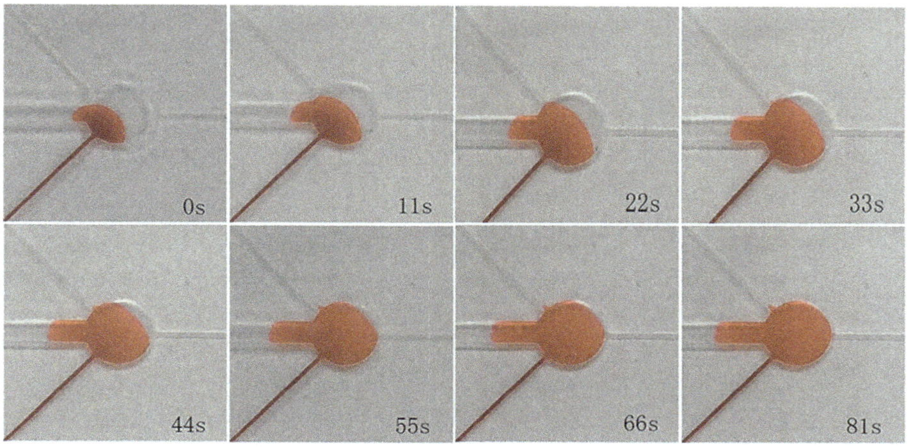

Figure 10 **Natural mixing of potassium chloride solution and red pigment.**

respectively. Colorless potassium chloride solution and red pigment were added into the mixing chamber for natural mixing, and the process was recorded by the vision acquisition camera. Figure 10 shows the process of natural mixing. Natural mixing mainly relies on the diffusion of liquid molecules. It takes 81 s until the diffusion is completed.

In the same condition with natural mixing, the microrobot moved through a microchannel (left side of the mixing chamber in Figure 11) into the mixing chamber of the microfluidic chip and then started to rotate clockwise with a frequency of 2 Hz to accelerate the mixing speed. Figure 11 shows the process of mixing with the microrobot. It takes only 32 s to complete the mixing.

The rotation made by the microrobot can generate significant tensile deformation and folding effect on the fluid streamline, so it increases the contact area between different liquids effectively. The rotational frequency of the microrobot can be increased by increasing the frequency of the sinusoidal signals provided in electromagnetic coils. To generate a rotating magnetic field, the sinusoidal signal is provided in each of the four coils, whose phase difference is 90°. When the rotational frequency increases, the microrobot can stir the liquid more quickly. Therefore, the microrobot can enhance the mixing efficiency with a larger rotational frequency. The microrobot can also move in the mixing chamber, and the motion can accelerate the mixing speed and enhance the mixing efficiency.

Conclusions

This paper researches the design and control method of a microrobot in microfluidics. A theoretical analysis is performed for the microrobot and magnetic driving system. The microrobot is made of magnetic PDMS. With four electromagnetic coils and other attachments, the magnetic driving system is constructed. By controlling the coils, the microrobot can move and rotate in a microfluidic chip. The

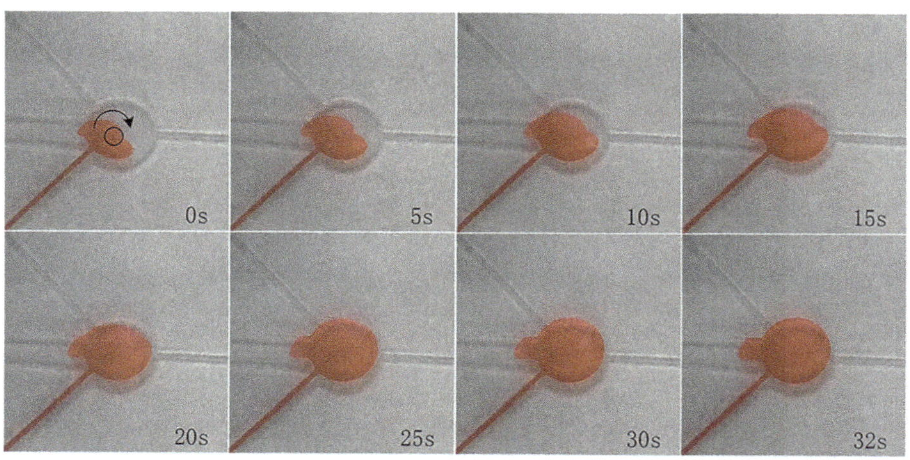

Figure 11 **Process of the microrobot's mixing in the mixing chamber.** The microrobot rotated clockwise with a frequency of 2 Hz. The black circle indicates the position of the microrobot, and the arrow shows the direction of the rotation.

microrobot can be used as a mobile micromixer, and a mixing experiment is accomplished inside a microfluidic mixing chip, which verifies the microrobot's effectiveness. Future studies will focus on the motion control of the microrobot in 3D and its interaction with live cells in a vessel-like microfluidic system.

Competing interests
The authors declare that they have no competing interests.

Authors' contributions
JW and NJ participated in the setup of the magnetic driving system and drafted the manuscript. JW carried out the analysis of the system and performed the experiments. ST provided advices and made contributions to the revision of the draft. LL proposed the basic idea and supervised the project. All authors read and approved the final manuscript.

Acknowledgements
This paper is supported by the National Natural Science Foundation of China (Grant Nos. 61106109 and 61304251) and the CAS FEA International Partnership Program for Creative Research Teams.

Author details
[1]State Key Laboratory of Robotics, Shenyang Institute of Automation, Chinese Academy of Sciences, No.114 Nanta Street, Shenyang 110016, China. [2]University of Chinese Academy of Sciences, No.19A Yuquan Road, Beijing 100049, China. [3]Department of Mechanical Engineering, University of Arkansas, W Dickson Street, Fayetteville, AR 72701, USA.

References

1. Dario P, Hannaford B, Menciassi A (2003) Smart surgical tools and augmenting devices. Robot Autom IEEE Trans 19(5):782–792
2. Ponsky JL (2006) Endoluminal surgery: past, present and future. Surg Endosc Intervent Tech 20(2):S500–S502
3. Cuschieri A, Melzer A (1997) The impact of technologies on minimally invasive therapy. Surg Endosc 11(2):91–92
4. Xuanying L, Zhenbo L, Ling M, Dawei Z, Xiaoning T, Jiapin C (2013) Manufacturing and Wireless Driving of a Permanent Magnetic Micro-robot. Robot 35(5):513–520
5. Vollmers K, Frutiger DR, Kratochvil BE, Nelson BJ (2008) Wireless resonant magnetic microactuator for untethered mobile microrobots. Appl Phys Lett 92(14):144103
6. Sul OJ, Falvo MR, Taylor RM II, Washburn S, Superfine R (2006) Thermally actuated untethered impact-driven locomotive microdevices. Appl Phys Lett 89(20):203512
7. Búzás A, Kelemen L, Mathesz A, Oroszi L, Vizsnyiczai G, Vicsek T, Ormos P (2012) Light sailboats: laser driven autonomous microrobots. Appl Phys Lett 101(4):041111
8. Solovev AA, Mei Y, Bermúdez Ureña E, Huang G, Schmidt OG (2009) Catalytic microtubular jet engines self-propelled by accumulated gas bubbles. Small 5(14):1688–1692
9. Hwang G, Braive R, Couraud L, Cavanna A, Abdelkarim O, Robert-Philip I, Régnier S (2011) Electro-osmotic propulsion of helical nanobelt swimmers. Int J Robot Res 30(7):806–819
10. Edd J, Payen S, Rubinsky B, Stoller ML, Sitti M (2003) Biomimetic propulsion for a swimming surgical micro-robot. In: Proceedings of the IEEE/RSJ international conference on intelligent robots and systems, Las Vegas, 2003 (IROS 2003), vol 3., pp 2583–2588
11. Yamazaki A, Sendoh M, Ishiyama K, Ichi Arai K, Kato R, Nakano M, Fukunaga H (2004) Wireless micro swimming machine with magnetic thin film. J Magn Magn Mater 272:E1741–E1742
12. Dreyfus R, Baudry J, Roper ML, Fermigier M, Stone HA, Bibette J (2005) Microscopic artificial swimmers. Nature 437(7060):862–865
13. Darnton N, Turner L, Breuer K, Berg HC (2004) Moving fluid with bacterial carpets. Biophys J 86(3):1863–1870
14. Yesin KB, Vollmers K, Nelson BJ (2006) Modeling and control of untethered biomicrorobots in a fluidic environment using electromagnetic fields. Int J Robot Res 25(5–6):527–536
15. Kim S, Qiu F, Kim S, Ghanbari A, Moon C, Zhang L, Choi H (2013) Fabrication and characterization of magnetic microrobots for three-dimensional cell culture and targeted transportation. Adv Mater 25(41):5863–5868
16. Pawashe C, Floyd S, Sitti M (2009) Modeling and experimental characterization of an untethered magnetic micro-robot. Int J Robot Res 28(8):1077–1094
17. Choi J, Choi H, Jeong S, Park B, Ko S, Park J, Park S (2013) Position-based compensation of electromagnetic fields interference for electromagnetic locomotive microrobot. Proceedings of the Institution of Mechanical Engineers, Part C: Journal of Mechanical Engineering Science 227(9):1915–1926
18. Yeom E, Kang YJ, Lee SJ (2014) Changes in velocity profile according to blood viscosity in a microchannel. Biomicrofluidics 8(3):034110
19. Tran-Minh N, Frank K, Tao D (2014) A simple and low cost micromixer for laminar blood mixing: design, optimization, and analysis. Biomed Informat Technol 404:91–104
20. Murakami Y (2014) Mems mixer as an example of a novel construction method of microfluidics by discrete microparts. Sensors and Actuators B: Chemical 194:528–533
21. Yaralioglu GG, Wygant IO, Marentis TC, Khuri-Yakub BT (2004) Ultrasonic mixing in microfluidic channels using integrated transducers. Anal Chem 76(13):3694–3698
22. Lu LH, Ryu KS, Liu C (2002) A magnetic microstirrer and array for microfluidic mixing. Microelectromechanical Syst J 11(5):462–469
23. Wang Y, Zhe J, Chung BT, Dutta P (2008) A rapid magnetic particle driven micromixer. Microfluid Nanofluid 4(5):375–389

The FES-assisted control for a lower limb rehabilitation robot: simulation and experiment

Yixiong Chen, Jin Hu, Long Peng and Zeng-guang Hou*

Abstract

To design a control strategy for iLeg, an exoskeleton robot developed for lower limb rehabilitation aiming at investigating the feasibility of integrating functional electrical stimulation (FES) with robot-based rehabilitation training, an FES-assisted training strategy combined with impedance control, has been proposed in this paper. Through impedance control, an active compliance of the robot is established, and the patient's voluntary effort to accomplish the training task is inspired. During the training process, the patient's related muscles are applied with FES which provides an extra assistance to the patient. The intensity of the FES is properly chosen in order to induce a desired active torque which is proportional to the voluntary effort extracted from the electromyography signals of the related muscles using back propagation neural networks. This kind of enhancement serves as a positive feedback which reminds the patient of the correct attempt to fulfill the desired motion. FES control is conducted by a combination of neural network-based feedforward controller and a PD feedback controller. Simulation conducted using Matlab and the experiment with a spinal cord injury subject and a healthy subject have shown satisfactory results which verify the feasibility of this control strategy.

Keywords: Rehabilitation robot; FES; Impedance control

Introduction

Background and literature review

Spinal cord injury (SCI) and stroke are the two main types of central nervous system injury which usually results in motor disorder such as paralysis and hemiparesis, cognitive dysfunction, as well as psychological problems. SCI is mainly caused by various kinds of accidents, while stroke is caused by an interruption of blood flow to the brain or an obstruction within a blood vessel in supplying blood to the brain [1,2]. Rehabilitation is one of the prime treatments for SCI and stroke patients, aiming to provoke the motor plasticity and to actuate these patients to relearn the use of their limbs. According to the rehabilitation theory, the resulting sensory feedback of the patient is associated with the cortical changes that can bring recovery of functional movement during the rehabilitation process so that the functional deficits and the risk of consequential injuries can be minimized [3-6].

Traditional rehabilitation is labor-intensive and requires several therapists to manually help the patient to perform training, especially for lower limb rehabilitation. Therefore, it imposes an enormous economic burden to any country's health care system [5]. Meanwhile, there are a number of factors which have been found to contribute to faster motor recovery, all of which have not been taken full advantage of due to the lack of rehabilitation services [7]. Modern robotics technology has offered us a more convenient and efficient opportunity to carry out rehabilitation training, known as rehabilitation robot. Rehabilitation robot is a mechatronics system which could be used to replace the physical training effort of the therapists, allowing more intensive and repetitive motions [5]. Since research has identified the treatment intensity as a critical element for successful therapeutic outcomes, the robotic-assisted rehabilitation is likely to increase the training effect [8,9]. What is more, various kinds of sensors installed on the robot feed back the signals of interest,

*Correspondence: hou@compsys.ia.ac.cn
State Key Laboratory of Management and Control for Complex Systems, 95 Zhongguancun East Road, Beijing 100190, China

such as the force signal, electromyography (EMG) signal, and heart rate. Thereby, the states of the patient during training can be monitored. By measuring the force or EMG signal, it is possible for the robot to perceive the voluntary effort of the patient and to adjust the training pattern accordingly in order to actuate the participation of the patient [10,11].

Several training strategies for rehabilitation robot have been presented in recent years. These control strategies can be categorized into two types. The first one is the position-based passive training which is widely used for its simplicity. However, robot based on this training strategy only moves the patient along predefined, fixed trajectories, and the guidance provided by the robot is so strong that it suppresses the voluntary motor control of the patient [12]. Therefore, this strategy is less likely to motivate the patient. The second one is the impedance-based active training. The rehabilitation robot designed for this type executes a task by amplifying human force or supplying the human operator with an assisting force [13,14]. Under this type of control strategy, an active compliance environment between the patient and the robot is established, while the assistant level can be adjusted by the impedance parameters. When the movement of the patient deviates from the predefined trajectory, a restoring force is generated using an appropriately designed impedance and then applied to the patient [6]. Unlike position-based training, impedance-based training is position error-tolerated, because research has emphasized that the kinematic errors generated during movement are a fundamental neural signal that drives motor adaptation [15].

On of the most commonly used technology for rehabilitation is the functional electrical stimulation (FES), which uses short electrical pulses to generate FES-induced contraction of the paralyzed muscles, and the level of the contraction can be controlled by modulating the intensity of FES [16], and the therapeutic effect of FES in rehabilitation is known to be increased when associated with a person's voluntary effort [17]. For this reason, some researchers has integrated FES into robotic-based rehabilitation to improve the training efficiency [3,18,19], in which an extra actuator besides the motor of the robot is available (considering the muscle under FES as a 'soft motor'). Therefore, the robot-generated assistance can be partially compensated by the FES-induced contraction of the muscle resulting in an increase of the muscle activity. If the level of FES assistance is regulated by the voluntary effort of the patient, a higher degree of functional recovery can be expected [3,20].

iLeg is a horizontal exoskeleton robot designed for lower limb rehabilitation. To develop a control strategy for iLeg, an FES-assisted training strategy combined with impedance control is introduced in this paper, aiming to achieve two goals. The first one is to create a compliance environment to inspire the voluntary effort of the patient, and the second one is to provide assistance through FES rather than purely through robot. Active compliance is guaranteed by impedance control in order to establish a soft interaction between the patient and the robot. The velocity of the movement depends on the measured voluntary effort. Basically, if no voluntary effort is measured, the rehabilitation robot only provides minimal assistance which is just sufficient to counterbalance the weight of the limb. The FES assistance, which is proportional to the patient's voluntary effort, is applied to the patient. This kind of enhancement serves as a positive feedback which reminds the patient of the correct attempt to fulfill the desired motion. To ensure that the active torque induced by FES follows the desired torque calculated from the voluntary effort, a combination of an inverse model-based feedforward and a PD feedback controller is implemented to compose the FES controller.

The rest of the paper is organized as follows. In the 'Methods' section, the rehabilitation robot is described in detail; the dynamic model of the leg-robot hybrid system is introduced and analyzed, while the model of the FES-stimulated muscle is also given. Then the control strategy of both impedance control and FES control are proposed, and the methods used for estimating muscle torques by system identification process and the method used for extracting voluntary effort by EMG signals are also described. The simulation conducted using Matlab and experimental results are depicted in the 'Results' section. Finally, conclusions are drawn in the last section.

Research design and methods

Methods

Model description

iLeg lower limb rehabilitation robot

iLeg is a horizontal exoskeleton lower limb rehabilitation robot designed by our institute. Considering that the movement of the human lower limbs in the sagittal plane is the basic movement form, and many other lower limbs rehabilitation robots such as Locomat and MotionMaker follow this principle, iLeg is also designed with three degrees of freedom (DOF) that can realize the movement of the hip joint, the knee joint, and the ankle joint in the sagittal plane for each side. The prototype of the iLeg is showed in Figure 1.

As it can be seen in Figure 2, the electrical control system of iLeg consists of the following main modules: a host PC, a motion control subsystem, an analog signal acquisition subsystem, a safe-security subsystem, an EMG acquisition subsystem, and a functional electrical stimulator. The motion control subsystem contains motion control card, motor drivers, and DC motors for each joint of both

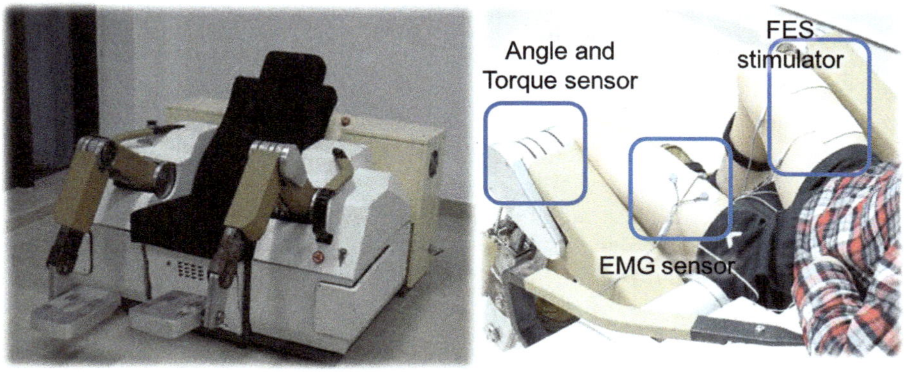

Figure 1 The prototype of iLeg lower limb rehabilitation robot.

sides. The hip joint is driven by chain after a reduction ratio of 1:120. The knee joint is driven through harmonic reducer (1:188) and a timing belt transmission. The ankle joint is driven directly through timing belt transmission. Besides, both of the thigh length and the shank length are adjustable through DC motors to ensure perfect alignment of each joint. The acquisition of the torques and the interaction force between the human leg and iLeg are carried out through analog signal acquisition subsystem. This subsystem contains torque sensors equipped at each joint and force sensor installed at the pedal. The safe-security subsystem is responsible for monitoring whether each joint are working within their safety ranges. If a joint is exceeding the safety range, the corresponding limit switch will be triggered, and the brake of this joint will be turned on to prevent secondary damage to the patient. To extract the voluntary intention of the patient, an eight-channel self-made EMG acquisition subsystem is integrated. Since EMG signal contains most of its power in the frequency range of 5 to 500 Hz, the sampling

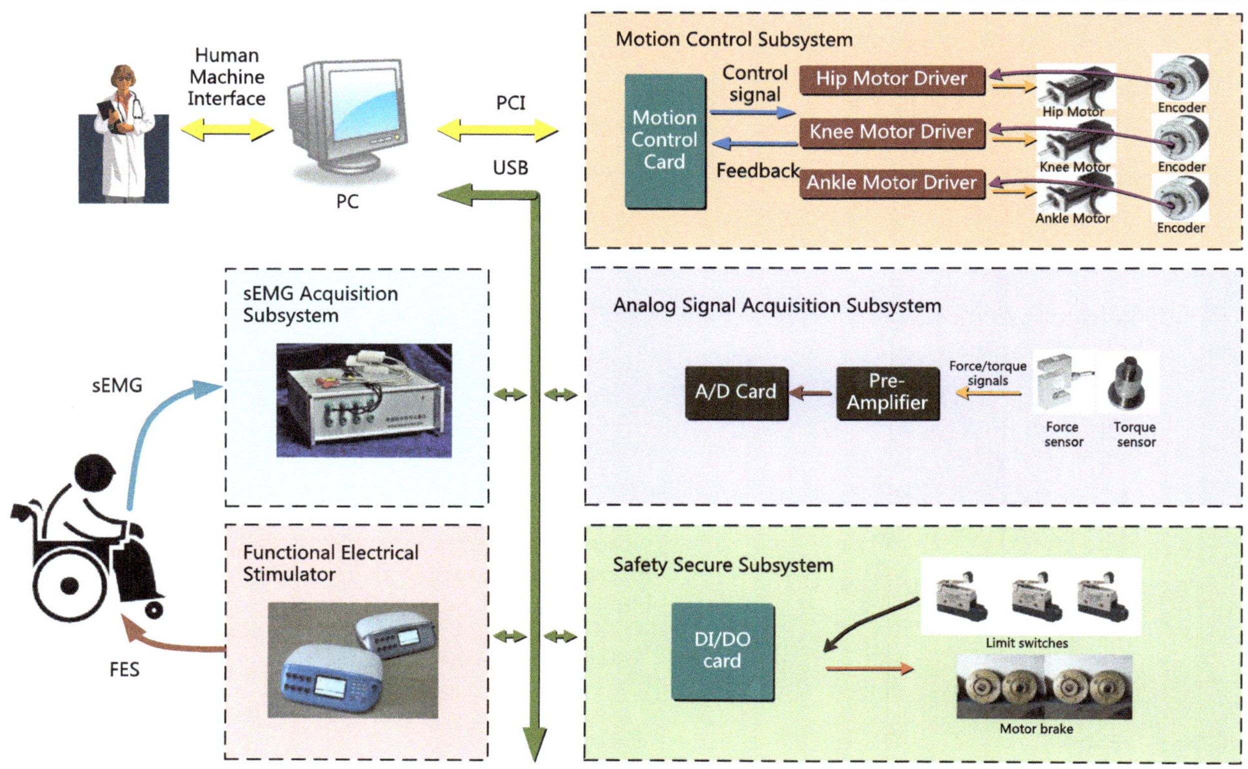

Figure 2 The electrical control system of iLeg.

rate of each channel is set to be 2 kHz. To enhance the muscle contraction during rehabilitation training, a functional electrical stimulator is also employed in iLeg. The self-made functional electrical stimulator has eight channels and is able to apply dual-phase current pulse to the patient's muscle. Each channel of the stimulator can be controlled independently with pulse frequency of 0 to 100 Hz, pulse width of 0 to 1,000 μs and pulse amplitude of 0 to 100 mA. All these subsystems are connected to the host PC whose main tasks are system management and algorithm processing via PCI bus and universal serial bus (USB).

Kinematic and dynamic of human leg and iLeg

The simplified model of human leg and iLeg is shown in Figure 3. Since the ankle joint is fix at 90° during the training process and the joints of human leg and iLeg are well aligned, the leg-robot system can be considered as two parallel two-link plants. The patient's foot is fastened to the pedal, and the physical interaction occurs at the endpoint. Gluteus maximus and quadriceps femoris are the two muscle groups stimulated by FES to generate active torques on the hip joint and the knee joint. In Figure 3, q_i is the angle of joint i; m_i^r, m_i^h, and l_i represent the mass of the robot, the mass of the leg, and the length of link i, respectively. F_{r2h} and F_{h2r} denote the interaction forces between the robot and the human leg. The forward kinematics of the system can be calculated by the following:

$$X = \begin{Bmatrix} x \\ y \end{Bmatrix} = \begin{Bmatrix} l_1 \cos q_1 + l_2 \cos(q_1 + q_2) \\ l_1 \sin q_1 + l_2 \sin(q_1 + q_2) \end{Bmatrix} \quad (1)$$

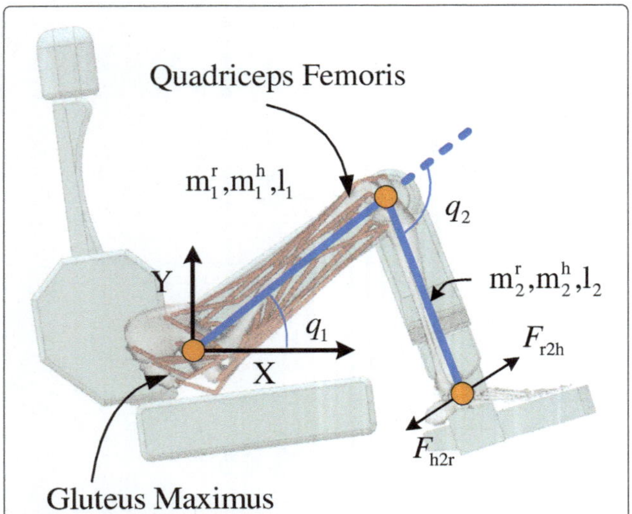

Figure 3 Simplified model of iLeg and the lower limb.

Accordingly, the speed of the end-point $\dot{X}$ in the Cartesian coordinate is determined by the differential kinematics equation:

$$\dot{X} = J\dot{q} \quad (2)$$

where $\dot{q}$ is the angular velocity in the joint space, and J is the Jacobian matrix represented as follows:

$$J = \begin{bmatrix} -l_1 \sin q_1 - l_2 \sin(q_1 + q_2) & -l_2 \sin(q_1 + q_2) \\ l_1 \cos q_1 + l_2 \cos(q_1 + q_2) & l_2 \cos(q_1 + q_2) \end{bmatrix} \quad (3)$$

Inverse kinematics of the simplified model can be derived from the forward kinematics equation and expressed as follows:

$$q = \begin{Bmatrix} q_1 \\ q_2 \end{Bmatrix} = \begin{Bmatrix} \tan^{-1}\frac{y}{x} + \cos^{-1}\left(\frac{y^2+x^2+l_1^2-l_2^2}{2l_1\sqrt{y^2+x^2}}\right) \\ -\cos^{-1}\left(\frac{y^2+x^2-l_1^2-l_2^2}{2l_1l_2}\right) \end{Bmatrix} \quad (4)$$

In generalized form, the dynamic model of the robot system is expressed by the following:

$$M_r(q)\ddot{q} + C_r(q,\dot{q})\dot{q} + G_r(q) + \tau_{rv}(\dot{q}) + \tau_{rf}(\dot{q}) = \tau_r - \tau_{h2r} \quad (5)$$

where $\tau_{rv}(\dot{q}) = k_{vis}\dot{q}$ and $\tau_{rf}(\dot{q}) = k_{fri}\mathrm{sgn}(\dot{q})$ represent the viscosity and friction of the robot; τ_{h2r} represents the external torque exert by human; $M(q)$, $C(q,\dot{q})$, and $G(q)$ are the inertia, Coriolis/centripetal, and gravity matrixes given as

$$M_{1,1}^r = \left(\frac{1}{3}m_1^r + m_2^r\right)l_1^2 + \frac{1}{3}m_2^r l_2^2 + m_2^r l_1 l_2 \cos q_2,$$
$$M_{1,2}^r = M_{2,1}^r = \frac{1}{3}m_2^r l_2^2 + \frac{1}{2}m_2^r l_1 l_2 \cos q_2,\ M_{2,2}^r = \frac{1}{3}m_2^r l_2^2,$$
$$C_{1,1}^r = -\frac{1}{2}m_2^r l_1 l_2 \dot{q}_2 \sin q_2,\ C_{2,1}^r = \frac{1}{2}m_2^r l_1 l_2 \dot{q}_1 \sin q_2,$$
$$C_{1,2}^r = -\frac{1}{2}m_2^r l_1 l_2 (\dot{q}_1 + \dot{q}_2) \sin q_2,\ C_{2,2}^r = 0,$$
$$G_{1,1}^r = \left(\frac{1}{2}m_1^r + m_2^r\right) g l_1 \cos q_1 + \frac{1}{2}m_2^r g l_2 \cos(q_1 + q_2),$$
$$G_{2,1}^r = \frac{1}{2}m_2^r g l_2 \cos(q_1 + q_2) \quad (6)$$

where m_i^r, l_i represent the mass and the length of link i. Similar to the robot dynamic, the dynamic model of human leg can be described as

$$M_h(q)\ddot{q} + C_h(q,\dot{q})\dot{q} + G_h(q) + \tau_{hv}(\dot{q}) + \tau_{he}(q) = \tau_{r2h} + \tau_m \quad (7)$$

where $M_h(q)$, $C_h(q,\dot{q})$, and $G_h(q)$ are the inertia matrix, the Coriolis/centripetal matrix, and the gravity vector of human leg, respectively. Their expressions are similar to those in the robot dynamic, but the only difference is that the masses of each robot link in (6) is replaced by the masses of human' thigh and shank. $\tau_{hv}(\dot{q})$ and $\tau_{he}(q)$ describe the passive viscous and passive elastic characters of the lower limb, which are described in detail in [21]. Unlike robot dynamics, the human leg can be considered as driven by redundant actuators (muscle contraction and robot assistance). The muscle contraction contributes to τ_m in Equation 7, while τ_{r2h} represents the assistance from the robot, which can be measured by the force sensor attached on the end effector using the following equation:

$$\tau_{r2h} = J^T F_{r2h} \quad (8)$$

where F_{r2h} is the force imposed to the leg due to interaction between human and robot, and J is the Jacobian matrix. By adding Equations 7 and 5, we can obtain the dynamic model of the leg-robot hybrid system which is represented as

$$M(q)\ddot{q} + C(q, \dot{q})\dot{q} + G(q) + \tau_{pas}(q, \dot{q}) = \tau_r + \tau_m \quad (9)$$

where $\tau_{pas} = \tau_{hv}(\dot{q}) + \tau_{he}(q) + \tau_{rv}(\dot{q}) + \tau_{rf}(\dot{q})$, and $M_h(q)$, $C_h(q, \dot{q})$ and $G_h(q)$ are the inertia matrix, the Coriolis/centripetal matrix, and the gravity vector of the hybrid system, respectively. According to Equation 9, there are two types of inputs which contribute to the input torque of the system: the one from DC motor and the one from the muscle contraction.

FES-stimulated skeletal muscle

During the training stage, two human muscles (gluteus maximus and quadriceps femoris) will be stimulated to produce partial torques around the hip joint and the knee joint. Therefore, a biological model is needed to depict the muscle response to FES, i.e., how the FES delivered to the relative muscle is converted into an active moment at the joint. Figure 4 shows a biological model of FES-stimulated muscle developed by Doctor Riener which is used in this study [22,23]. The active torque produced by the FES-induced muscle contraction is described as two parts, namely, activation dynamics and contraction dynamics, while the passive torque of the muscle is described as passive elastic and passive viscous properties [22]. Note that the original model has two inputs for activation dynamics which are stimulation pulse width and frequency; however, for the sake of simplicity, we fix the stimulation frequency for each muscle constantly at 50 Hz while regulating the intensity of FES by controlling the pulse width of the stimulation using FES controller.

The active torque of FES-induced muscle contraction is the product of contraction dynamics and activation dynamics given as

$$\tau_{fes} = a_{act} f_{fv} f_{fl} F_{max} \mathrm{ma} \quad (10)$$

In the contraction dynamics, four parts are involved, i.e., moment arm, ma, maximum isometric force F_{max}, force-length relation f_{fl}, and force-velocity relation f_{fv}. The product of these four parts denotes the maximum torque of the stimulated muscle at certain joint angle and angular velocity. The force-length relation is calculated using

$$f_{fl} = \exp\left\{-\left[(\bar{l} - 1)/\varepsilon\right]^2\right\} \quad (11)$$

where $\bar{l}$ is the muscle length normalized with respect to the optimal muscle length $_{opt}$, and ε is a shape factor. The force-velocity relation is determined by

$$f_{fv} = 0.54 \arctan(5.69\bar{v} + 0.51) + 0.745 \quad (12)$$

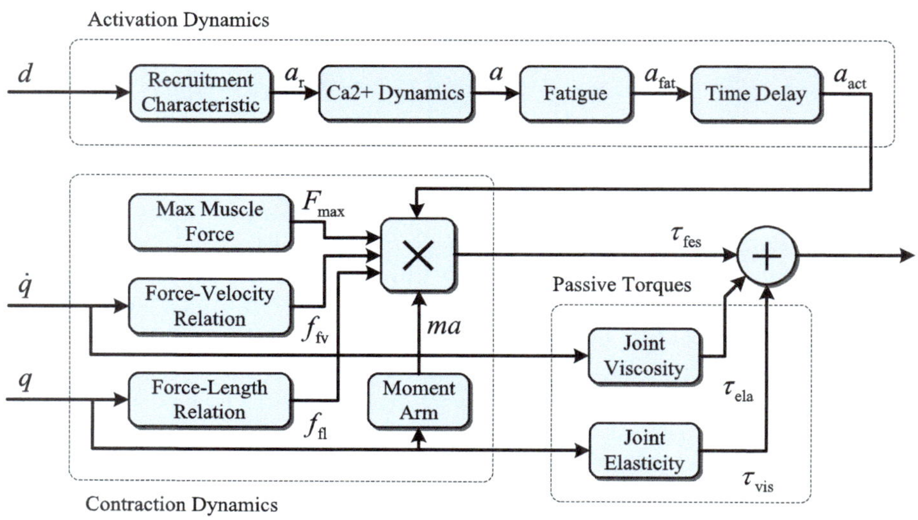

Figure 4 Activation dynamic and contraction dynamic of the muscle.

where $\bar{v}$ is the muscle velocity normalized with respect to the maximum contraction velocity v_m of the muscle. The muscle length and velocity can be calculated using joint angle, angular velocity, and moment arm of the muscle described in detail in [22]. The activation dynamics determines how the motion unit is recruited by FES in the final form of a_{act}. Basically, the activation dynamics is composed of four segments. They are the recruitment characteristic, the Calcium dynamic, the fatigue process, and a delay of about 25 ms. The recruitment characteristic shows the relationship between the FES pulse width and the percentage of the motor unit activated by FES, and it is given as

$$a_r = a_f \{c_1 \{(d - d_{thr}) \arctan [k_{thr} (d - d_{thr})] - (d - d_{sat}) \arctan [k_{sat} (d - d_{sat})]\} + c_2\} \quad (13)$$

where d is the pulse width of FES, d_{thr} and d_{sat} denote pulse width values corresponding to threshold and saturation. The shape of the recruitment curve is describe using c_1, c_2, k_{thr}, and k_{sat}. a_f is introduced to represent the frequency characteristic which is a function of the stimulation frequency in [22]. Since the stimulation frequency is fixed at 50 Hz, here a_f is a constant. Calcium dynamics presents the mechanism that calcium ion released from sarcoplasmic reticulum and is expressed using the following two-order linear relation:

$$T_{Ca}^2 \ddot{a} + 2T_{Ca}\dot{a} + a = a_r \quad (14)$$

where T_{Ca} is the time constant, and a denotes non-fatigue muscle activation. Fitness function is introduced to depict the effect of muscle fatigue and recovery phenomenon using the following first-order relation:

$$\frac{d\text{fit}}{dt} = \frac{0.55a\,(\text{fit}_{min} - \text{fit})}{T_{fat}} + \frac{(1 - \text{fit})(1 - 0.55a)}{T_{rec}} \quad (15)$$

where fit is the fitness of the muscle, while fit_{min} is the minimum fitness. Time constants for fatigue and recovery are given as T_{fat} and T_{rec}. Final activation a_{act} is the product of fitness fit and non-fatigue activation a with a constant time delay T_{del}. After the maximum torque of the stimulated muscle at certain joint angle and angular velocity are being scaled by the activation a_{act}, the active torque of the stimulated muscle can be obtained.

Control strategy

The control strategy contains two main parts, namely, the impedance control and FES control. The goal of the first one is to achieve certain desired active compliance on the robot, while the goal of the second one is to generate desired FES-induced torques around the joint by regulating the FES intensity. The architecture of the control strategy is shown in Figure 5.

Impedance control

The aim of an impedance controller is to establish a mass-damper-spring relationship between the position and the force so that an active compliance between the robot and the leg can be achieved [24-26]. The desired impedance relationship can be expressed as follows:

$$M_d (\ddot{x} - \ddot{x}_d) + B_d (\dot{x} - \dot{x}_d) + K_d (x - x_d) = F_{ext} \quad (16)$$

where M_d, B_d, K_d are positive definite matrices representing the desired inertia, damping, and stiffness of the system, respectively, and $\ddot{x}_d$, $\dot{x}_d$, x_d are the references for the acceleration, the velocity, and the position of the end-effector in the Cartesian space, while $\ddot{x}$, $\dot{x}$ and x stand for the actual acceleration, velocity and position, respectively. F_{ext} is the external force imposed on the end effector. In the case of our rehabilitation robot, F_{ext} means F_m in Figure 5, which is the force produced by muscle contraction, either by voluntary-controlled contraction or by FES-induced contraction. Usually, the desired impedance is chosen to be a second-order linear equation, as in mass-spring-damper system; however, in majority of the kinds of rehabilitation exercises, the speed of the robot is relatively low, so the acceleration effect can be neglected [27,28], resulting Equation 16 to be degenerated to the following equation:

$$B_d (\dot{x} - \dot{x}_d) + K_d (x - x_d) = F_m \quad (17)$$

If the manipulator follows a speed reference given by

$$\dot{x}_{ref} = \dot{x}_d + B_d^{-1} [F_m - K_d (x - x_d)] \quad (18)$$

the rehabilitation robot will behave as described by Equation 17. The speed tracking task is achieved by a PI speed controller in the joint space given as

$$\tau_{PI} = K_P (\dot{q}_{ref} - \dot{q}) + K_I \int (\dot{q}_{ref} - \dot{q})\, dt \quad (19)$$

where, K_P and K_I are the proportional and integral gains of the speed controller. $\dot{q}_{ref}$ is the reference of the angular velocity given by $\dot{q}_{ref} = J^{-1}\dot{x}_{ref}$, where J represents the Jacobian matrix.

The impedance control strategy is conducted in a leg press exercise, in which a predefined trajectory of the end-point (the ankle joint of the patient) is predefined. The patient is encouraged to complete a leg press motion and is also expected to maintain his ankle on the predefined trajectory during his effort. $\dot{q}_d$ is set to be zero in this control, so basically the rehabilitation robot behaves as a gravity canceler. If there is no muscular force measured, the robot does not assist the patient to move towards the target point. x_d in Equation 18 is the desired position for the endpoint on the predefined trajectory at a specific time point, and x_d is moving towards the target point at

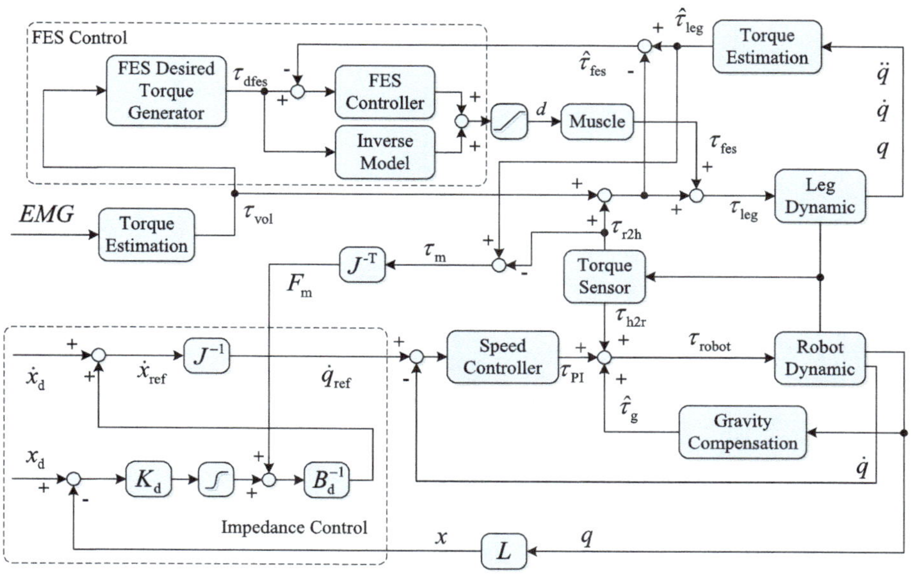

Figure 5 Control architecture of FES-assisted training strategy.

a constant speed, meaning a 'virtual moving wall' which is pushing the patient towards the target [12]. A waiting window for delayed assistance is also employed, embodied by adding a modification factor to the stiffness component of the impedance control described in Equation 17. The factor mod is given by a sigmoid function expressed as follows:

$$\text{mod} = \frac{1}{1 + e^{-a \cdot \text{dis} + b}} \tag{20}$$

where dis represents the distance between the actual and desired position at a specific time point; a, b define the shape of the sigmoid function. The task is error-tolerated, but if the endpoint deviates too far from the desired position, the patient will feel harder to maintain the incorrect effort, so he has to make a proper adjustment in order to accomplish the task. From the perspective of the patient, this impedance control provides a virtual channel with the predefined trajectory in its center, in which the motion is partially constrained in the desired direction.

FES control

Since the patient has insufficient voluntary contraction of his muscle, FES is implemented to enhance the contraction, aiming to assist the patient to accomplish the leg press task. To be more specific, the desired FES-induced force is proportional to the decomposed voluntary force whose direction is alongside with the direction of the predefined trajectory. Because this part of the voluntary force is considered as the 'correct' effort for the task, the decomposed voluntary force which is vertical to the predefined path is considered as the 'incorrect' effort and will not be enhanced. The voluntary force F_{vol} can be transferred from the voluntary contraction torque τ_{vol}, which is estimated by the EMG signals of the related muscle groups. The calculation of the desired FES-induced torque $\tau_{\text{dfes}}$is given as

$$\tau_{\text{dfes}} = J^{\mathrm{T}} \begin{bmatrix} \cos\theta & \sin\theta \\ -\sin\theta & \cos\theta \end{bmatrix} J^{-\mathrm{T}} \tau_{\text{vol}} \tag{21}$$

where J is the Jacobian matrix, and θ represents the angle of F_{vol} to the predefined trajectory.

As shown in Figure 5, the FES torque control is achieved with the combination of a feedforward controller and a feedback controller. An inverse model of the stimulated muscle and a PD controller are employed and serve as the feedforward controller and the feedback controller, respectively. The advantage of using the combination of these two controllers compared with only using one of them is significant. The delay and highly nonlinear characters of muscle response to FES are easy to cause oscillation when the FES is regulated merely by the feedback controller. On the other hand, the feedforward controller is incapable of dealing with disturbance and eliminating error [23,29].

A three-layer neural network is introduced to capture the nonlinear relation of the inverse model. For each stimulated muscle (gluteus maximus and quadriceps femoris), a three-layer back propagation (BP) neural network is constructed whose structure is shown in Figure 6. Since the inverse model of muscle requires joint angle, joint angular velocity, and desired torque to be the input, the

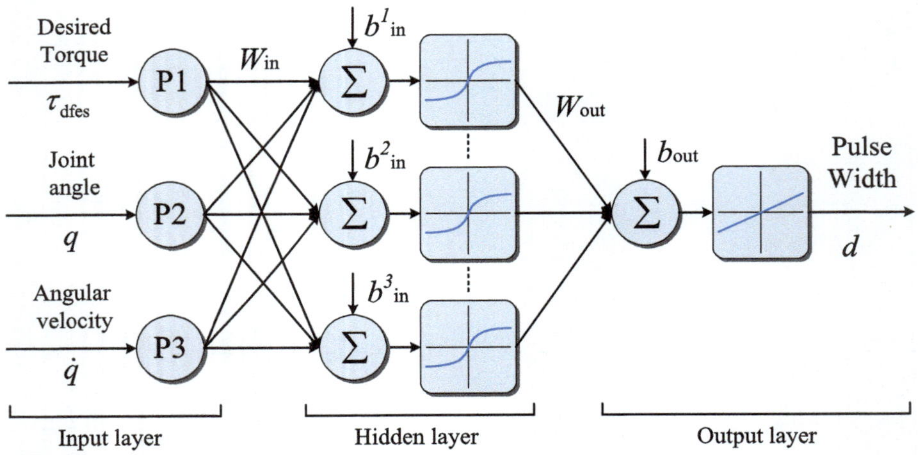

Figure 6 Neural network structure of the inverse model for the stimulated muscle.

first layer of neural network consists of three input neurons. The neuron in the output layer calculates the pulse width of the FES with the linear function. The weight matrixes between the input layer, the hidden layer, and the output layer are adjusted using back propagation algorithm.

The PD controller which serves as the feedback controller is defined as follows:

$$d = K_P e + K_D \dot{e}, \quad e = \tau_{dfes} - \tau_{fes} \tag{22}$$

where d is the pulse width of the FES, and K_P, K_D are proportional and derivative gains of the feedback controller, respectively. e represents the error between desired torque calculated in (12) and the actual torque induced by FES.

FES-induced torque estimation

In the case of applying FES to a moving limb, it is hard to measure the actual active torque τ_{fes} induced by FES directly. So an estimation of τ_{fes} should be made. The total input torque of the leg-robot hybrid system which is the right side of Equation 9 can be rewritten as

$$\tau_{total} = \tau_r + \tau_m = \tau_r + \tau_{fes} + \tau_{vol} \tag{23}$$

That is to say, the torque caused by muscle contraction τ_m is considered as a combination of two parts: the one from voluntary-controlled contraction τ_{vol} and the one from FES-induced contraction τ_{fes}. Therefore, to acquire the value of τ_{fes}, the other three items have to be obtained. τ_r can be measured via torque sensors, and τ_{vol} can be predicted by the EMG signal. As for τ_{total}, an system identification process is adopted so that the estimation of τ_{fes} can be calculated from

$$\hat{\tau_{fes}} = \hat{\tau_{total}} - \tau_r - \tau_{vol} \tag{24}$$

Hybrid-activated torque

The first step to obtain the FES-induced torque is to calculate the muscle torque τ_m which is produced by muscle contraction either it is FES-induced or voluntary-controlled, so τ_m is considered as hybrid-activated torque. A system identification process is introduced to make an estimation of this hybrid-activated torque. We rewrite Equation 9 as

$$D(q, \dot{q}, \ddot{q})\, \chi = \tau_r + \tau_m \tag{25}$$

where $D(q, \dot{q}, \ddot{q})$ is the 2×8 regression matrix associated with the 8×1 dynamic parameters χ. For the identification process, no force is applied on the robot [30], thus, $\tau_m = 0$ and

$$D(q, \dot{q}, \ddot{q})\, \chi = \tau_r \tag{26}$$

$D(q, \dot{q}, \ddot{q})$ is given as

$$\begin{aligned}
&D_{11} = \ddot{q}_1, D_{12} = g\cos(q_1), D_{13} = \ddot{q}_2,\\
&D_{14} = l_1(2\ddot{q}_1 + \ddot{q}_2)\cos(q_2) - l_1(\dot{q}_2^2 + 2\dot{q}_1\dot{q}_2)\sin(q_2)\\
&\qquad + g\cos(q_1 + q_2),\\
&D_{15} = \mathrm{sgn}(\dot{q}_1), D_{16} = 0, D_{17} = \dot{q}_1, D_{18} = 0,\\
&D_{21} = 0, D_{22} = 0, D_{23} = \ddot{q}_1 + \ddot{q}_2,\\
&D_{24} = l_1\ddot{q}_1\cos(q_2) + l_1\dot{q}_1^2\sin(q_2) + g\cos(q_1 + q_2),\\
&D_{25} = 0, D_{26} = \mathrm{sgn}(\dot{q}_2), D_{27} = 0, D_{28} = \dot{q}_2
\end{aligned} \tag{27}$$

and χ is defined as

$$\begin{aligned}
&\chi_1 = I_1 + I_2 + l_1^2 m_2, \chi_2 = l_1(m_2 + m_1/2), \chi_3 = I_2,\\
&\chi_4 = m_2 l_2/2, \chi_5 = k_{sf1}, \chi_6 = k_{sf2}, \chi_7 = k_{df1}, \chi_8 = k_{df2}
\end{aligned} \tag{28}$$

where I_i is the inertia moment of the ith link; k_{sfi} and k_{dfi} are the Coulomb and viscous friction of the ith joint.

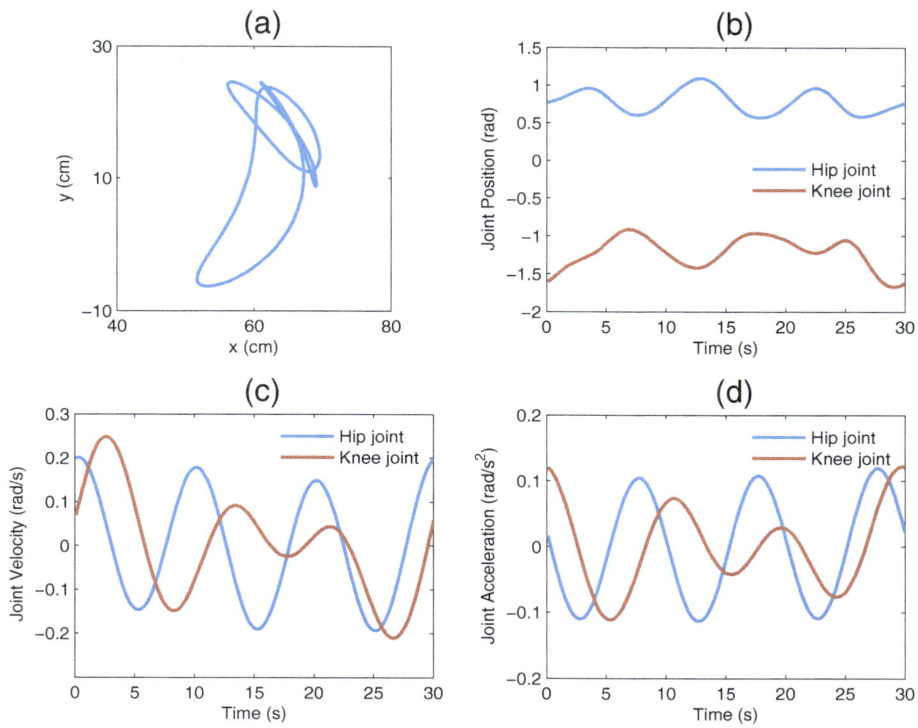

Figure 7 Optimized exciting trajectory for the identification of the dynamic parameters. (a) Trajectory of the endpoint in the Cartesian space. **(b, c, d)** The joint angle, the joint velocity, and the joint acceleration of each joint during the identification process.

For identification, the motion states ($q, \dot{q}, \ddot{q}$ of each joint) and the torques measured by sensors should be obtained during the robot running on an exciting trajectory. Samplings are carried out at the time of $t_1, t_2 \ldots t_n$ so that an overdetermined equation is constructed as

$$W\chi = \Gamma \tag{29}$$

where W is the observation matrix defined as

$$W = \begin{bmatrix} D\left(q\left(t_1\right), \dot{q}\left(t_1\right), \ddot{q}\left(t_1\right)\right) \\ D\left(q\left(t_2\right), \dot{q}\left(t_2\right), \ddot{q}\left(t_2\right)\right) \\ \vdots \\ D\left(q\left(t_n\right), \dot{q}\left(t_n\right), \ddot{q}\left(t_n\right)\right) \end{bmatrix}$$

and

$$\Gamma = \left[\tau_r\left(t_1\right)^T \tau_r\left(t_2\right)^T \ldots \tau_r\left(t_n\right)^T\right]^T$$

Thus, the dynamic parameters χ can be estimated using least square estimation method given by

$$\chi = \left(W^T W\right)^{-1} W^T \Gamma \tag{30}$$

To improve the accuracy of the estimation, the exciting trajectory should be optimized for the sufficient excitation of the hybrid system. In this study, finite Fourier series (FFS) is introduced to parameterized the exciting trajectory [31]. For each joint, the position at time t is a sum of sine waves expressed as

$$q_i(t) = \sum_{l=1}^{N} \left(\frac{a_{li}}{\omega_f l} \sin\left(\omega_f l t\right) - \frac{b_{li}}{\omega_f l} \cos\left(\omega_f l t\right) \right) + q_{i0} \tag{31}$$

where ω_f defines the period of the exciting trajectory; N is the number of FFS; q_{i0} is the initial position of the

Table 1 Identification results

	Parameter	Value	Unit		Parameter	Value	Unit
χ_1	$l_1 + l_2 + l_1^2 m_2$	57.3328	kg m^2	χ_5	k_{sf1}	−1.1060	Nm
χ_2	$l_1 (m_2 + m_1/2)$	2.6204	kg m	χ_6	k_{sf2}	0.1335	Nm
χ_3	l_2	6.3912	kg m^2	χ_7	k_{df1}	−4.0587	Nm s/rad
χ_4	$m_2 l_2/2$	5.2661	kg m	χ_8	k_{df2}	2.5287	Nm s/rad

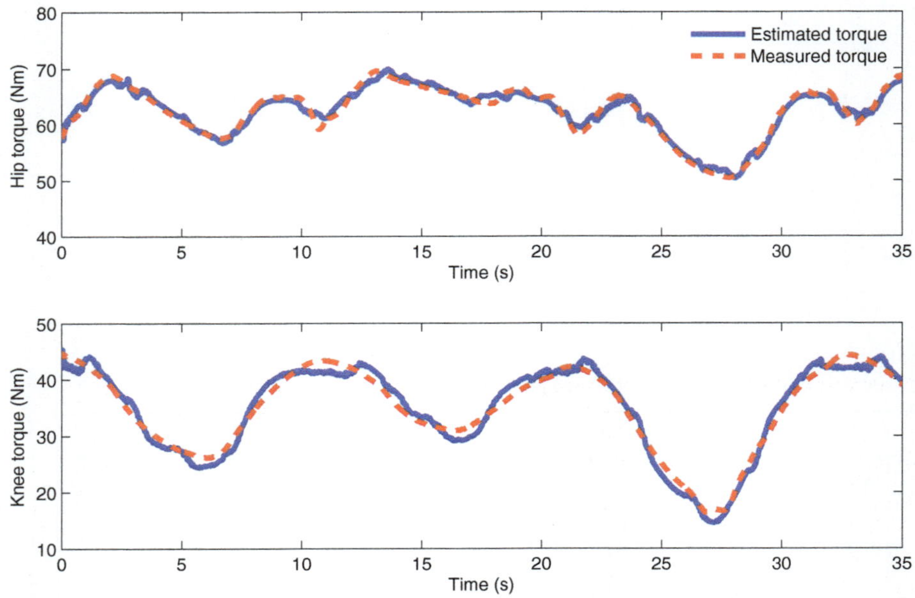

Figure 8 Measured torques and estimated torques of each joint during the validation process. Top, hip torque. Bottom, knee torque.

*i*th joint. The optimization of the exciting trajectory is carried out by choosing the proper a_{li}, b_{li} for each joint so that the condition number of the observation matrix can be minimized. The condition number of the observation matrix represents the upper bound for input/output error. It directly affects the convergence rate and noise immunity of the identification experiment [32]. This optimization problem can be solved using genetic algorithm (GA). During the GA optimization, variables a_{li}, b_{li} are the decision variables which meet the constrain that the trajectory defined by them should not exceed the workspace of the robot, and the condition number of the observation matrix is chosen as the fitness function. The optimized exciting trajectory is shown in Figure 7.

The samples used for the identification are obtained by performing the exciting trajectory with the patient involved in the pre-training stage. The subject of this study is an SCI patient who has been injured in T11 for 16 months. The exciting trajectory is performed for five times. According to the method described above, the estimation of the dynamic parameters of the leg-robot hybrid system is achieved, which are given in Table 1.

An unoptimized FFS trajectory is utilized to validate the identification results. The comparison between the measured torques and the estimated torques is shown in Figure 8. Root mean square (RMS) error of the estimation torques for both exiting trajectory and validation trajectory is given in Table 2. Both Figure 8 and Table 2 have shown satisfactory results which prove that the proposed method is capable of estimating the muscle torques of the patient.

Table 2 The RMS error of the estimation torque (Nm)

	Hip	Knee
Exiting trajectory	0.4057	0.3168
Validation trajectory	0.8716	0.5483

Voluntary torque

The second step to obtain the FES-induced torque is to calculate the voluntary-controlled muscle torque τ_{vol}. EMG signal is a weak electrical potential generated by the muscle cells when these cells are activated [33], so it has the potential to reflect the level of voluntary intention and can be used to estimate the joint angle or torque [34-36]. In this study, EMG signal is also introduced to achieve joint torque estimation. As depicted in the 'FES-stimulated skeletal muscle' subsection, the active torque is the product of the contraction dynamics and the activation of the muscle. If the activation is induced by FES, the activation can be calculated using Equations 13, 14, and 15; if the activation is voluntary-controlled, the EMG signal can be utilized as the index of the level of activation. The nonlinear relationship of the contraction dynamics of the muscle is approximated by BP neural networks. For the hip joint and the knee joint, there are six muscle groups involved. They are gluteus maximus (GM), iliopsoas (LP), biceps femoris short head (BFs), rectus femoris (RF), rectus lateralis (RL), and vastus medialis (VM).

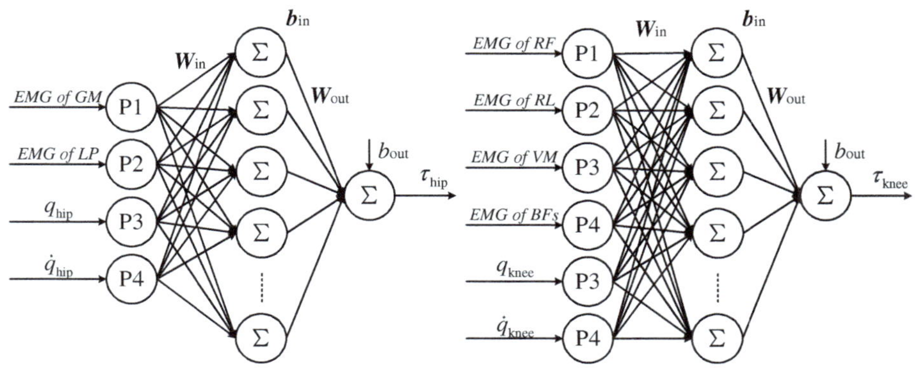

Figure 9 Neural network structures of the hip joint and the knee joint for voluntary torque estimation.

We construct two independent neural networks for the torque estimation of the hip joint and the knee joint. The structures of these two networks are shown in Figure 9. Each BP neural network has three layers, and the weight matrixes between the input layer, the hidden layer, and the output layer are adjusted using back propagation algorithm. Note that these EMG signals which served as the inputs of the neural networks are root mean square (RMS) EMG signals defined as

$$\mathrm{RMS}(t) = \sqrt{\frac{1}{N}\sum_{i=t-N+1}^{t} y^2(i)} \tag{32}$$

where RMS(t) is the RMS EMG signal at time t; $y(i)$ is the ith sample of the raw EMG signal; N is the length of sliding window which contains samples of 200 ms before time t.

The torques of the hip joint and the knee joint as well as the EMG signals are acquired during the robot performing the predefined trajectory, and at the same time, the subject makes random efforts. Ten groups of samples are collected, and leave-one-out cross validation process is carried out: nine groups of the samples are used for training and the rest for validation, then the data set for training and validation are changed; this procedure is executed ten times, which means that every group will be served as validation group for one time. The overall estimation error is the average error of each times.

Figure 10 shows the EMG signals and torques of the hip joint and the knee joint acquired during the subject making random efforts. The curves of the measured torques and the estimated torques on the hip joint and the knee joint are shown in Figure 11. The average RMS error

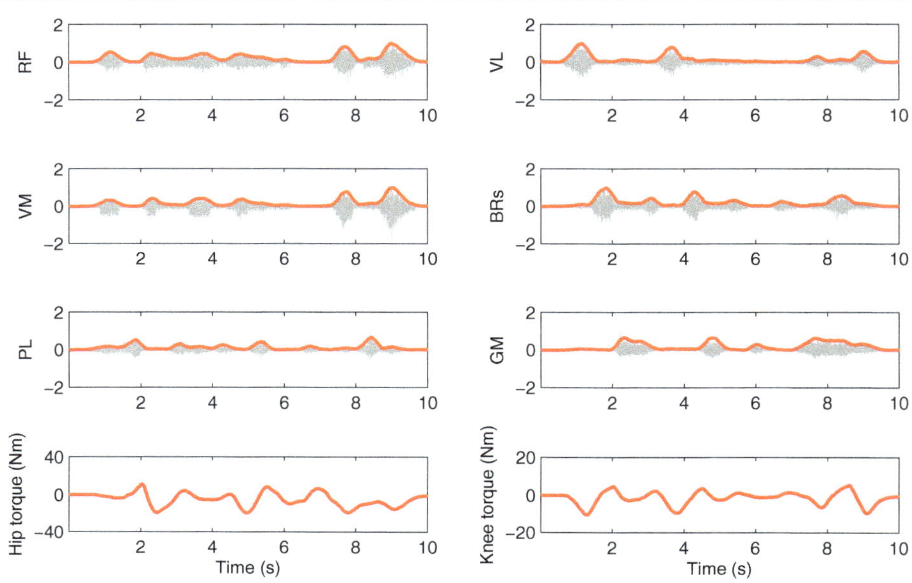

Figure 10 EMG signals and joint torques acquired during the subject making random efforts. Left, hip torque. Right, knee torque.

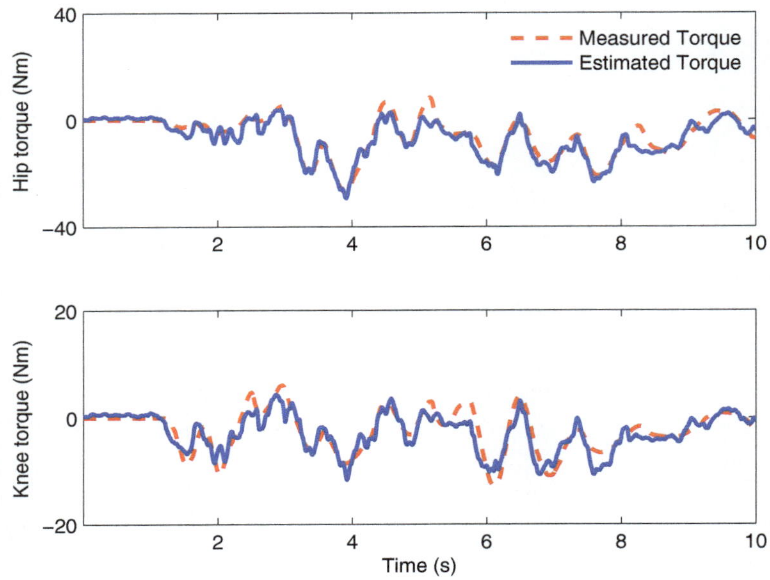

Figure 11 The curves of the measured and estimated torques on the hip knee joints. Top, hip torque. Bottom, knee torque.

is 0.8862 Nm for the hip joint and 1.3424 Nm for the knee joint. These errors are in an acceptable range, so the voluntary-controlled torque τ_{vol} can be estimated using EMG signals and the proposed BP neural networks.

Results

Simulation results

The above control strategy is verified using Simulink toolbox of Matlab. The link lengths and the masses are set to be $l_1 = 0.5$ m, $l_2 = 0.45$ m, $m_1^h = 10$ kg, $m_2^h = 4.5$ kg, $m_1^r = 13$ kg, and $m_2^r = 12$ kg. The parameters to construct the simulation models of gluteus maximus and quadriceps femoris are chosen according to [22,37]. For the leg press exercise, the patient is requested to accomplish the motion following a predefined trajectory substantially, and the predefined trajectory is a linear path with the start point set as (0.5, 0.1), and target point set as (0.8, 0) in the Cartesian space coordinates. The patient is encouraged to accomplish the task within about 10 s.

The PI parameters for the speed controller described in Equation 19 are set as $K_P = \text{diag}(500, 500)$ and $K_I = \text{diag}(100, 100)$. The stiffness and damping parameters in Equation 18 are set as $K_d = \text{diag}(60, 60)$ and $B_d = \text{diag}(30, 30)$. $\dot{x}_d$ is set as zero, and x_d is defined as

$$x_d(t) = [0.5 + 0.03t\cos\theta \quad 0.1 - 0.03t\sin\theta]^T \tag{33}$$

where t represents time, and θ is the angle of the desired path to the horizontal.

To simulate the voluntary effort of the patient, random voluntary torques around the hip joint and the knee joint are implemented in the simulation, which is shown in Figure 12. In real situation, voluntary torque has to be estimated by means of the EMG signals, however, for simplicity reason, this step has been skipped. Note that the voluntary effort of the patient starts at $t = 1$ s, and withdrawals at $t = 6$ s.

Figure 13a shows the actual trajectory of the endpoint during the leg press exercise. The color of each point on the trajectory represents the speed of the endpoint at the very position. Hotter color indicates a relatively higher speed, and cooler color indicates a relatively lower speed. Virtual channel is also visualized in Figure 13a, with the predefined path at the center of the channel. Figure 13b is the zoom-in view of the trajectory for the first 2 s. The muscular force which is the resultant force of voluntary force and FES-induced force is presented as a black arrow, pointing to the direction of the force, with the length denoting the strength. As the endpoint deviates from the predefined path, the component of force perpendicular to the path increases. This means the patient has to apply more force to counterbalance the 'virtual force' which drags the endpoint back to the predefined path according to the impedance mechanism, therefore reminding the patient of the correct direction of the excise. From the patient's perspective, it is much easier to press his leg towards the target point than towards the incorrect direction. As voluntary effort withdraws at $t = 6$ s, the endpoint stops at (0.74, 0.02). Since the position error is modified by the mod factor in Equation 20 whose parameters are set as $a = -200$, $b = 6$, the robot does not

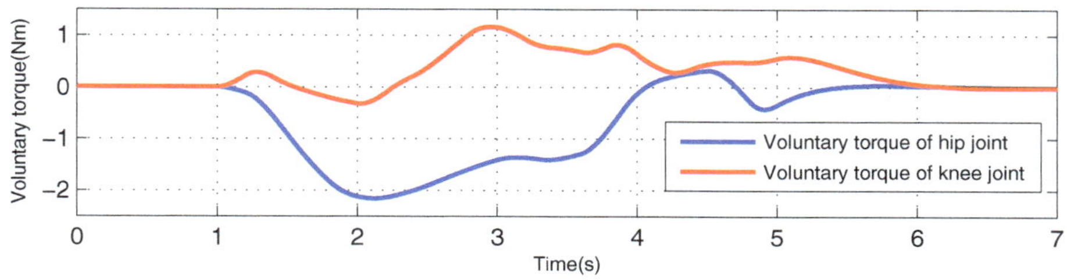

Figure 12 Voluntary torques generated by patient.

assist immediately. However, once the distance between the desired position and the actual position increases to a certain degree, the robot assists the patient to complete the task. The angular speeds of the hip joint and the knee joint are shown in Figure 14. When muscular force is applied on the robot, the robot begins to move (after $t = 1$ s); when it is withdrawn, the speed of the robot drops to zero (after $t = 6$ s). Followed by a waiting of about 4 s, expecting the patient to make attempt again, however, in this simulation, if the patient does not take acting, the robot begins to assist and finally 'drag' the leg to reach the target point. The above results prove the first goal which is to create a compliance environment to inspire the voluntary effort of the patient is achieved.

The FES control is conducted first by determining the desired torque induced by FES. The idea is to amplify the voluntary effort towards the target point. Using Equation 21, the force component of the voluntary effort parallel to the predefined trajectory is calculated, amplified, and then translated to the desired active torque for FES control in joint space. The blue arrows in Figure 13b represent the desired active force expected being induced by FES. All of them point to the same direction, the direction of the target point. This indicates that the FES assistance is not only proportional to the voluntary effort but also provided in the correct direction. The blue dash line in Figure 13a shows the trajectory of the endpoint when the same voluntary effort is made, but this time, no FES assistance is applied. The comparison of these two trajectories verifies the contribution of FES-induced torque which assists the patient during the leg press task by enhance the 'correct part' of the voluntary effort.

The FES control contains two parts: feedforward control and feedback control. For the training of the neural network-based inverse muscle models, samples are collected at pre-training stage, in which the muscles are applied with FES of random intensity while the robot is running in the predefined trajectory. The pulse width d of the FES uniformly distributed on the interval [0 500] μs. Off-line training process is conducted with 5,014 sampling, and both neural networks for gluteus maximus and quadriceps femoris converge after 250 iterations with the mean square errors below 5.47 μs for gluteus maximus

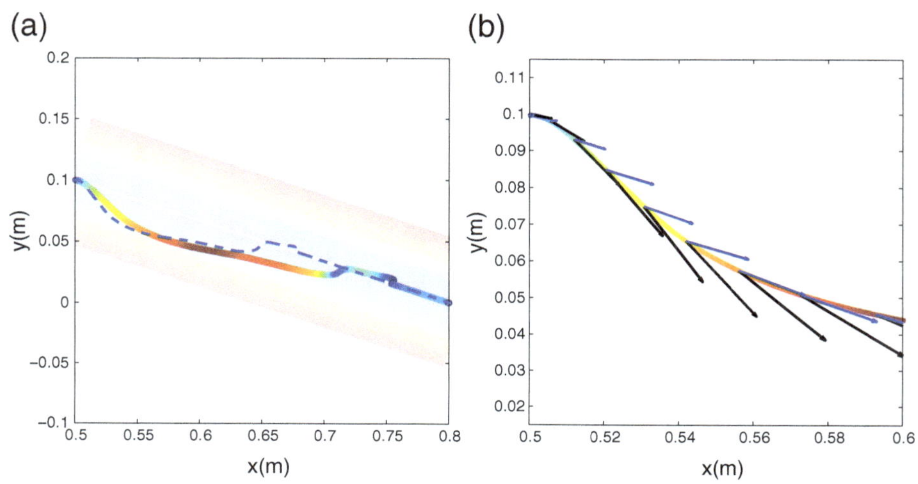

Figure 13 Trajectory of the endpoint in Cartesian space. (a) Entire trajectory of the endpoint with the color indicating the speed and **(b)** zoom-in view of the trajectory for the first 2 s.

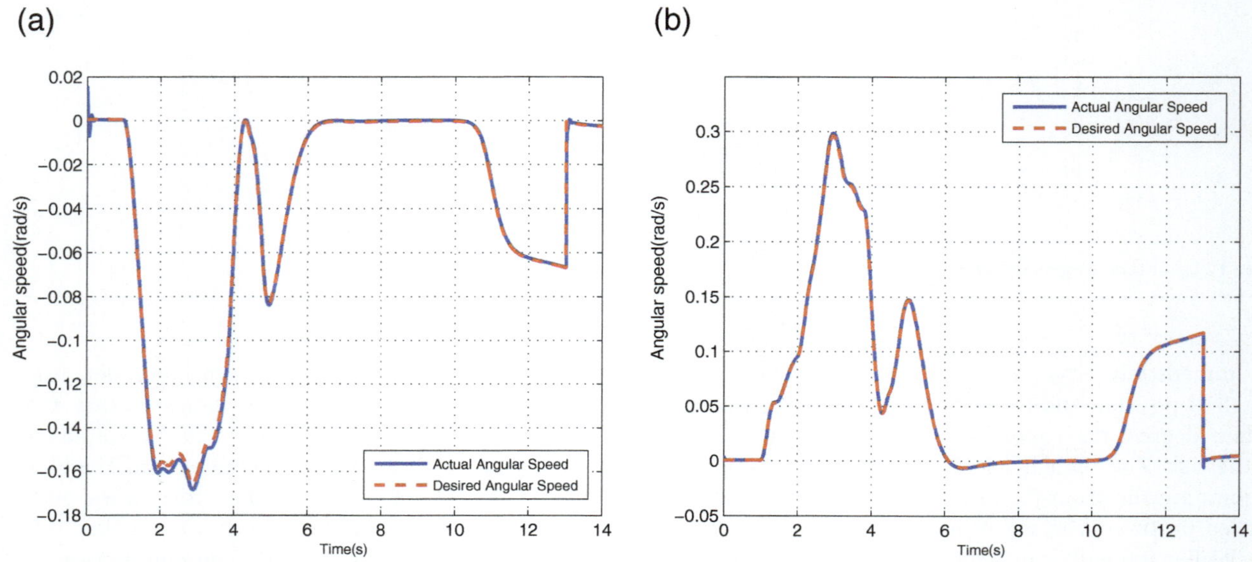

Figure 14 Speed tracking performance of the robot. (a) Curve of the angular velocity of the hip joint. **(b)** Curve of the angular velocity of the knee joint.

and 3.98 μs for the quadriceps femoris. The effectiveness of the inverse model is shown in Figure 15. Random desired active torque (0 to 3 Nm for quadriceps femoris and −3 to 0 Nm for gluteus maximus) is input into the inverse model when the robot is running on the predefined trajectory for validation, and the outputs are used to set the intensity of the FES which is applied on the muscles. The approximation of the profiles for desired torque and active torque in both Figure 15a,b shows that the inverse model is capable of outputting the proper pulse width for FES to generate the desired torque. As described in the former FES control section, $\hat{\tau}_{fes}$ is implemented as the estimation of actual active torque, and the error between τ_{dfes} and $\hat{\tau}_{fes}$ serves as the input of feedback controller described in Equation 22 with the PD parameter set as $K_P = \text{diag}(3,000; 3,000), K_D = \text{diag}(400, 400)$.

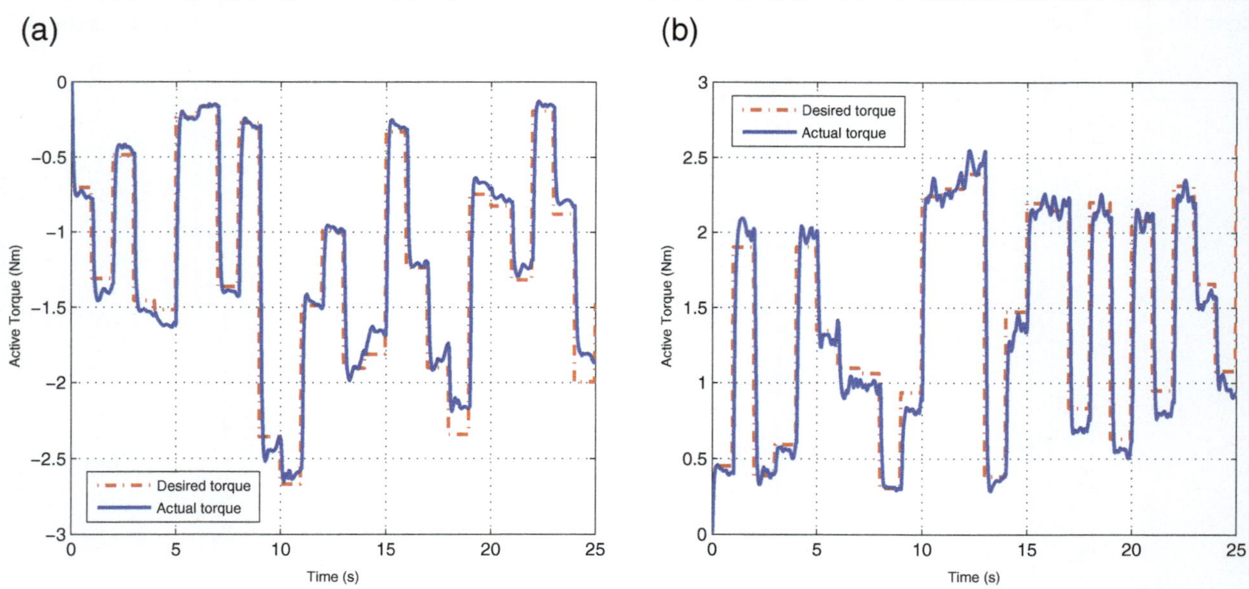

Figure 15 Results of the inverse muscle model for generating desired active torque. Validation results of the inverse models for the gluteus maximus **(a)** and quadriceps femoris **(b)**.

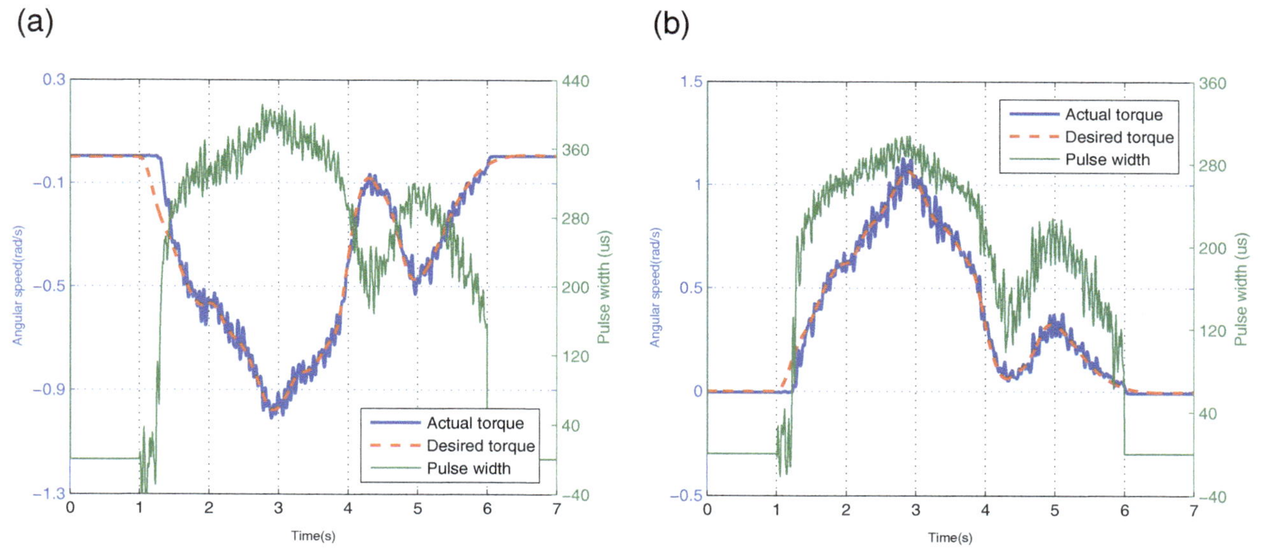

Figure 16 FES-induced torque tracking performance. Tracking performance of the gluteus maximus **(a)** and quadriceps femoris **(b)**.

Figure 16 shows the desired torque τ_{dfes} and the estimated active torque $\hat{\tau}_{fes}$ produced by the muscle contraction response to FES. The green line is the pulse width calculated by the FES controller, and FES whose intensity depends on this pulse width is applied on the muscle to generate the desired active torque. The profiles of the red line and the blue line indicate that the error between the desired torque and the actual active torque is in an acceptable range which proves that the FES controller is capable of inducing desired active torque so that a precise assistance through FES can be delivered to the patient.

Experiment results

The proposed control strategy is also verified by the experiment conducted on iLeg lower limb rehabilitation robot. The method described in 'FES-induced torque estimation' subsection is utilized to obtain voluntary torque τ_{vol} and FES-induced torque τ_{fes}. An SCI subject (male, age 65) and a healthy subject (male, age 28) are invited to participate in the experiment. The SCI patient has been injured in T11 for 16 months, and he remains with partial muscle force. The speed control of the robot is carried out by the PID controller of the motor drivers. The stiffness

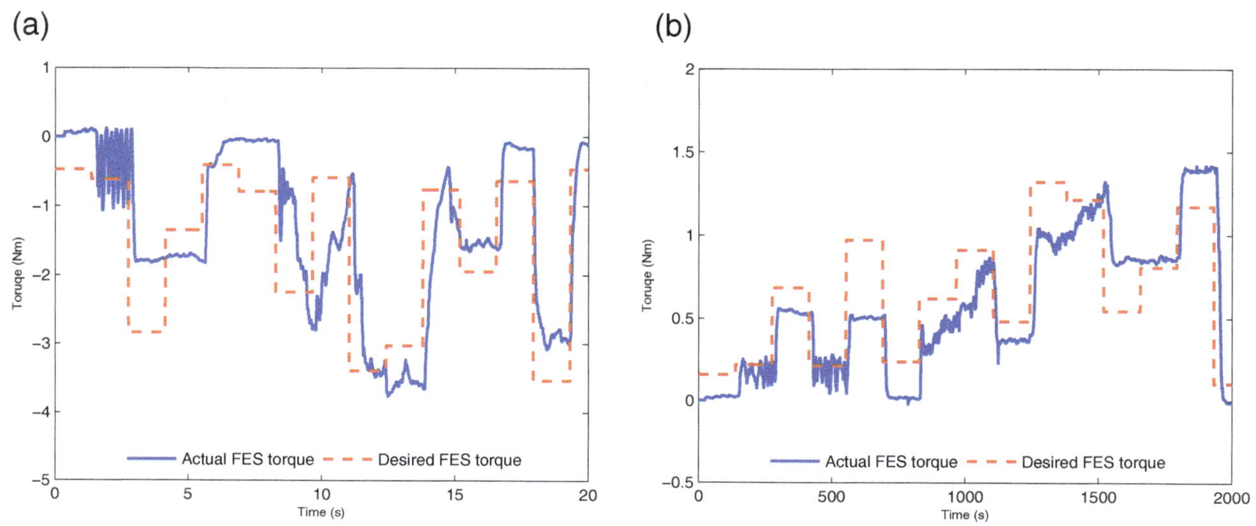

Figure 17 Results of the inverse muscle model for generating desired active torque (SCI subject). Validation results of the inverse models for the gluteus maximus **(a)** and quadriceps femoris **(b)**.

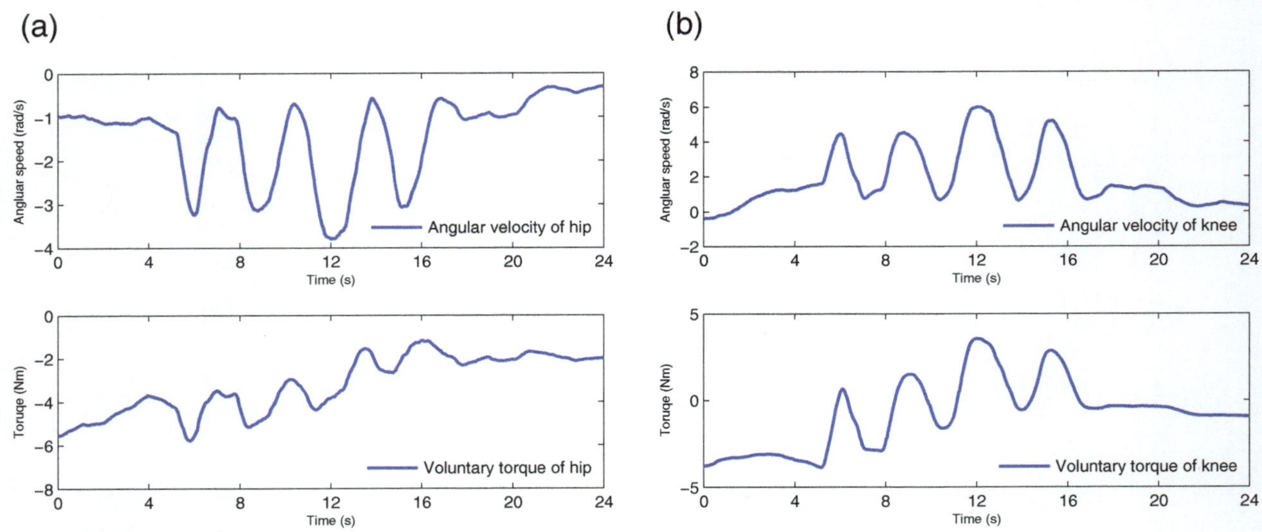

Figure 18 Curves of the voluntary torques and angular velocities of each joint (SCI subject). (a) Hip. **(b)** Knee.

and damping parameters in Equation 18 are set as $K_{\mathrm{d}} = \mathrm{diag}(2,500; 2,500)$ and $B_{\mathrm{d}} = \mathrm{diag}(2,000; 2,000)$. $\dot{x}_{\mathrm{d}}$ is set as zero. Considering that the SCI subject's motion ranges of his lower limb are limited, we decrease the length of the predefined trajectory, and set the start point to be (0.5, 0.1) and the target point to be (0.7, 0).

At the pre-training stage, the tolerance of the subject applied with FES is tested. For each stimulated muscle, the pulse width of FES is firstly fixed at 700 μs, and the amplitude of FES is increased from 15 mA until obvious muscle contraction being observed. Then the amplitude of the FES is fixed during training, and the intensity of FES is regulated by adjusting the pulse width through FES controller. To construct the BP neural networks which serve as the FES feedfoward controller, the muscles are applied with FES of random intensity while the robot running in the predefined trajectory. The pulse width d of the FES uniformly distributed on the interval [150 500] μs.

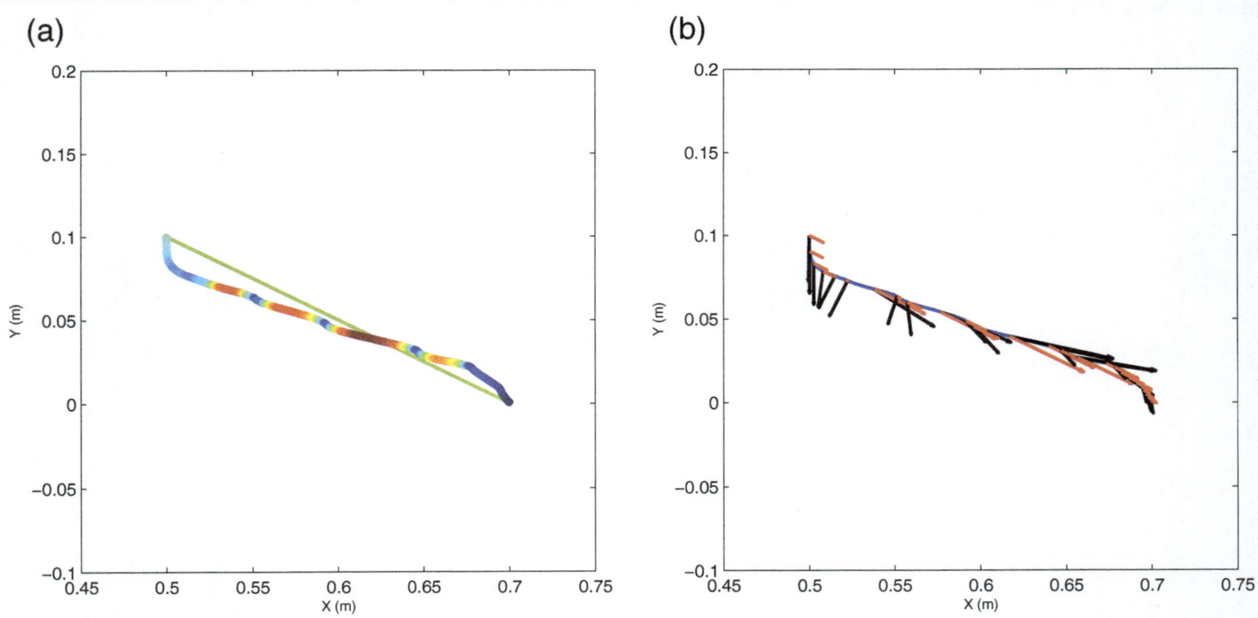

Figure 19 Trajectory of the endpoint in Cartesian space (SCI subject). (a) Entire trajectory of the endpoint with the color indicating the speed. **(b)** Voluntary forces of the subject and the FES-assisted forces using black arrows and red arrows, respectively.

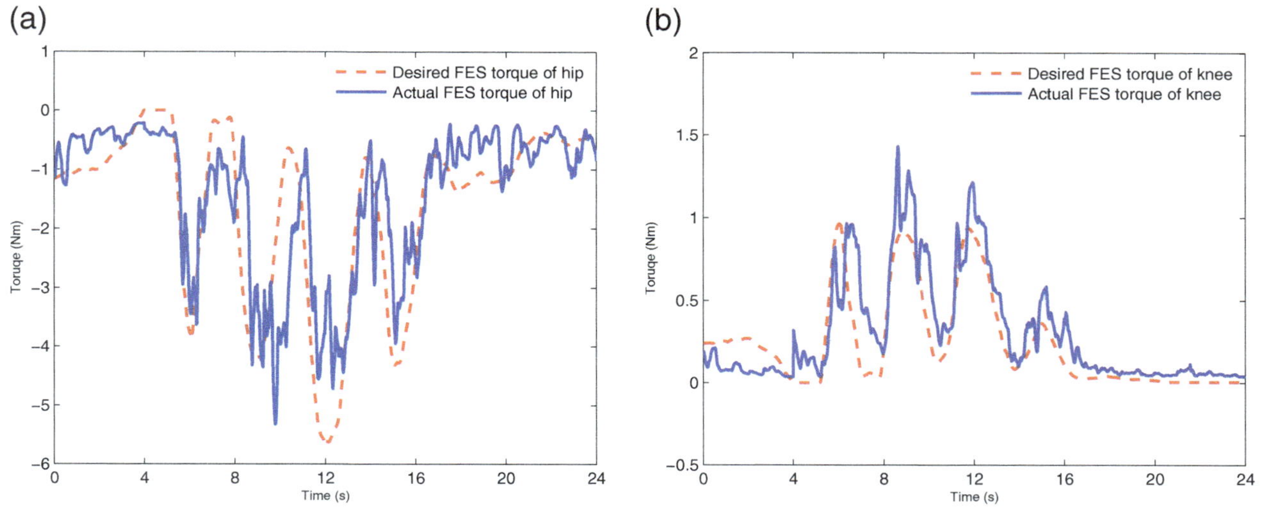

Figure 20 FES-induced torque tracking performance (SCI subject). Tracking performance of the gluteus maximus **(a)** and quadriceps femoris **(b)**.

The effectiveness of the inverse model is shown in Figure 17. Random desired active torques (0 to 1.5 Nm for quadriceps femoris and −3.5 to 0 Nm for gluteus maximus) are input into the inverse model when the robot is running on the predefined trajectory, and the outputs are used to set the intensities of the FES applied on the muscles. Compared to the simulation results, the muscle delay is relatively serious and the results are not so satisfactory which reflects that the FES-induced muscle contraction is unstable. However, the error is acceptable and can be eliminated by the PD feedback controller.

Figure 18 shows the curves of voluntary torques estimated by EMG signals and the angular velocities of each joint. As the voluntary torques vary, the angular velocities of each joint change accordingly, which follows the principle of impedance control. The trajectory of the end-point is shown in Figure 19. There are four obvious efforts represented by four red sections in Figure 19a and four peaks in the curve of angular velocity in Figure 18. As the same as in Figure 13b, the FES-assisted forces and voluntary forces are drawn as several arrows in Figure 19b from which the same conclusion can be drawn: the FES assistance is not only proportional to the voluntary effort but is also provided in the correct direction. The tracking performances of the FES controller which is the combination of a feedforward controller and a PD feedback controller

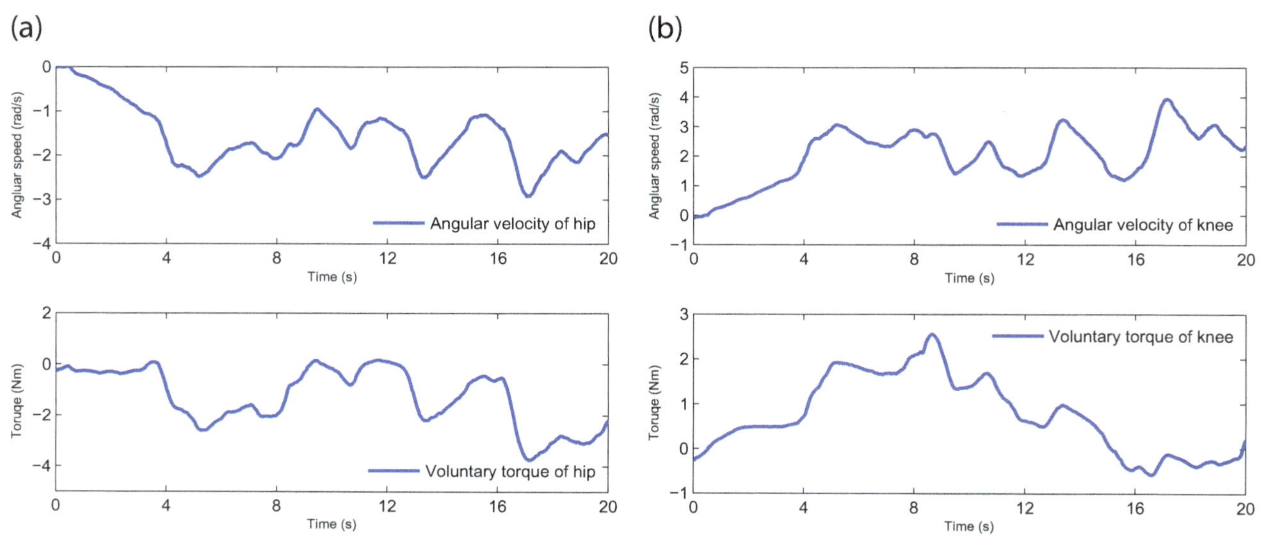

Figure 21 Curves of the voluntary torques and angular velocities of each joint (healthy subject). (a) Hip. **(b)** Knee.

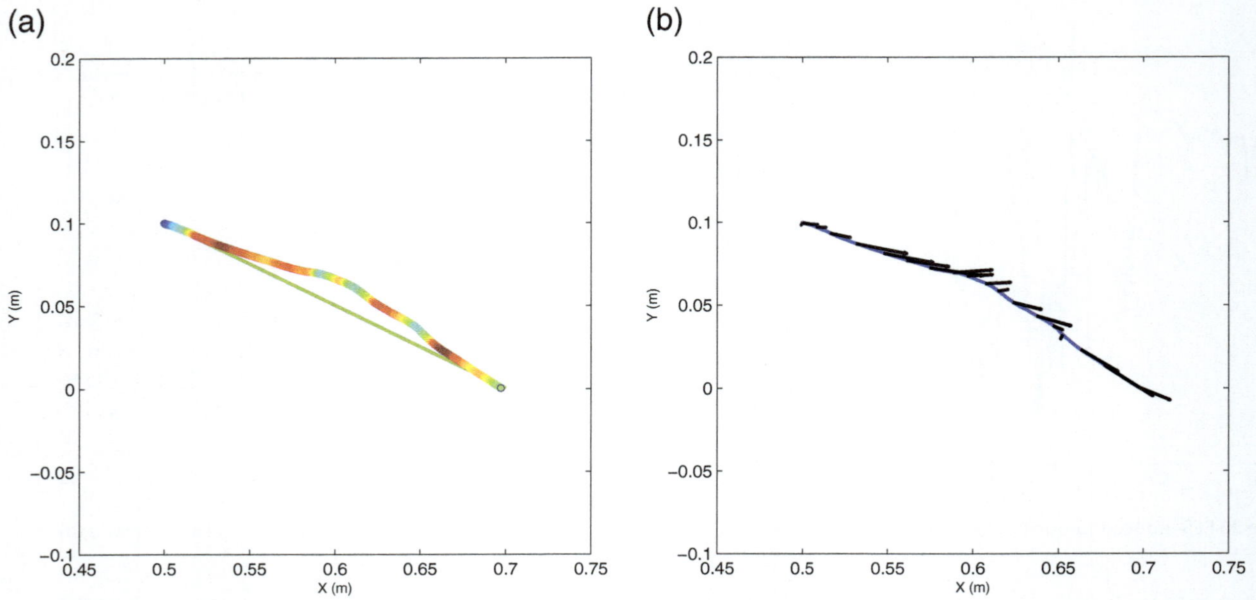

Figure 22 Trajectory of the endpoint in Cartesian space (healthy subject). (a) Entire trajectory of the endpoint with the color indicating the speed. **(b)** Voluntary forces of the subject using red arrows.

are represented in Figure 20. Even though the tracking is not as precise as in the simulation, these results indicate the FES controller is capable of generating desired FES-induced torque to apply assistance to the patient.

One healthy subject is also involved in this experiment. However, during the training process, no FES assist is applied on the healthy subject, only the effect of the impedance control is validated. Figures 21 and 22 are the curves of the voluntary torques and the angular velocities of each joint, the trajectory of the endpoint, respectively. Besides the similar conclusion that we can obtain from Figures 18 and 19, we can also notice that the voluntary

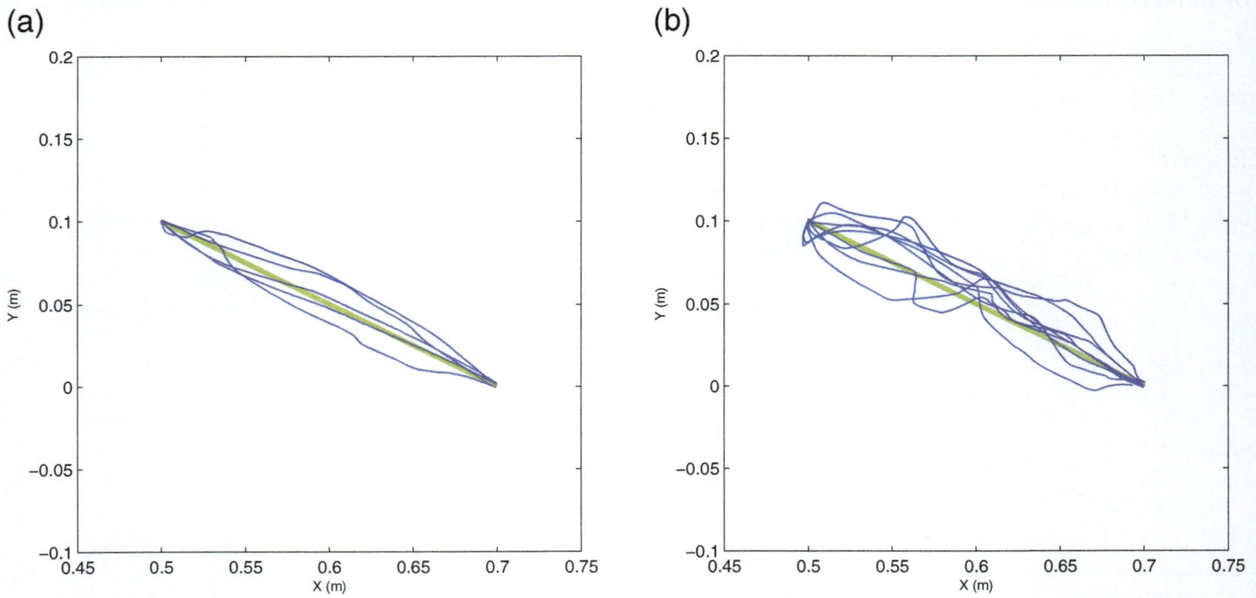

Figure 23 Comparison of the trajectories of the endpoint. The end-point trajectories of the healthy subject **(a)** and the SCI subject **(b)** in several leg press trials.

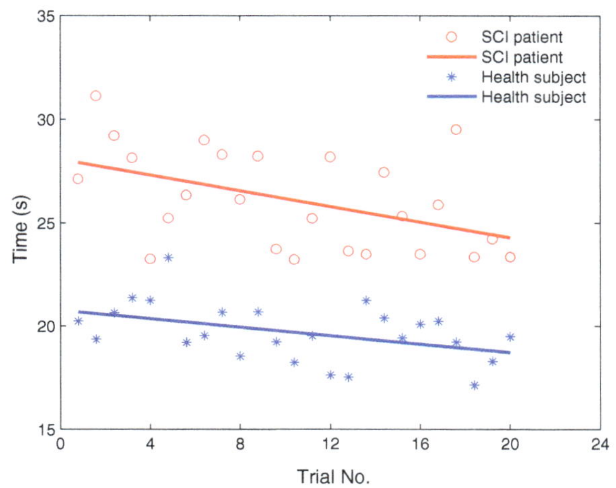

Figure 24 The time consumption for 20 leg press trials.

efforts are smoother compared with the effort made by the SCI subject, resulting a smooth trajectory of the endpoint.

The end-point trajectories of several trials for both the SCI subject and the healthy subject are drawn in Figure 23. The end-point trajectories of the healthy subject are much smoother than the end-point trajectories of the SCI subject, reflecting that the SCI subject's controlling ability of his lower limb is insufficient.

Figure 24 represents the time consumption for 20 leg press trials. As we can see from Figure 24, due to lack of muscle strength, the average time consumption of the SCI subject is more than that of the healthy subject. We also notice that as the trial number increases, the time consumption decreases. This phenomenon indicates that through repetitive training, both subjects gradually adjust the patten of making effort so that the performance of the leg press exercise is improved.

Conclusions

In this paper, an FES-assisted training strategy combined with impedance control for the lower limb rehabilitation robot is presented and the controller used to accomplish this task is developed, aiming to achieve two goals. The first one is to create a compliance environment to inspire the voluntary effort of the patient, and the second one is to provide assistance through FES rather than purely through robot. Impedance control is ideal for establishing active compliance of the robot, and through the mechanism of impedance control, the patient has to make effort to accomplish the task rather than passively waiting the assistance from the robot so that it increases the participation of the patient's voluntary effort. Two muscle groups are stimulated with FES to provide extra assistance to the patient during the training process. The intensity of the FES is properly chosen so as to induce the desired active torque which is proportional to the voluntary effort of the patient. The control of active torque of FES-stimulated muscle is achieved using the combination of a neural network-based feedforward controller and a PD feedback controller which have been proven to be competent for this task. The dynamic parameters of the leg-robot hybrid system are estimated using system identification method; the voluntary torques are also estimated according to the EMG signals acquired from the related muscle groups. These two methods are utilized to obtain the actual FES-induced torque so that the FES control can be accomplished. Both simulation conducted in Matlab and experiments conducted in iLeg reveal satisfactory results, from which the feasibility and effectiveness of the proposed control strategy are verified. In the future, more SCI and stroke subjects will be recruited to carry out a 3 to 6-month training session. During this follow up study, we will evaluate whether the proposed control strategy is capable of improving the rehabilitation effects.

Abbreviations

BFl: biceps femoris long head; BFs: biceps femoris short head; DOF: degrees of freedom; EMG: electromyography; FES: functional electrical stimulation; GM: gluteus maximus; RF: rectus femoris; RL: rectus lateralis; RMS: root mean square; VM: vastus medialis.

Competing interests

The authors declare that they have no competing interests.

Authors' contributions

YC conceived the concept of the study, participated in acquisition and analysis of data, and drafted the manuscript. HJ participated in the development of the robot platform. LP participated in acquisition of data and made contributions to the revision of the draft. ZH made contributions to the revision of the draft. All authors read and approved the final manuscript.

Acknowledgements

This work was supported in part by the National Natural Science Foundation of China (Grants 61225017, 61175076) and the International S&T Cooperation Project of China (Grant 2011DFG13390).

References

1. Marino R, Barros T, Biering-Sorensen F, Burns S, Donovan W (2003) International standards for neurological classification of spinal cord injury. J Spinal Cord Med 26:50–56
2. Sims NR, Muyderman H (2010) Mitochondria, oxidative metabolism and cell death in stroke. Biochim Biophys Acta - Molecular Basis of Disease 1802(1):80–91
3. Freeman C, Tong D, Meadmore K, Hughes AM, Rogers E, Burrideg J (2012) FES based rehabilitation of the upper limb using input output linearization and ILC. In: Proceedings of the 2012 American Control Conference, pp 4825–4830. http://ieeexplore.ieee.org/Xplore/home.jsp
4. Lo HS, Xie SQ (2012) Exoskeleton robots for upper-limb rehabilitation: state of the art and future prospects. Med Eng Phys 34:261–268
5. Diaz I, Gil JJ, Sanchez E (2011) Lower-limb robotic rehabilitation: literature review and challenges. J Robot 2011:1–10
6. Crespo LM, Reinkensmeyer DJ (2009) Review of control strategies for robotic movement training after neurologic injury. J Neuroeng Rehabil 6:20
7. Jung S, Hsia T (2000) Robust neural force control scheme under uncertainties in robot dynamics and unknown environment. IEEE Trans Ind Electron 47(2):403–412

8. Barreca S, Wolf S, Fasoli S, Bohannon R (2003) Treatment interventions for the paretic upper limb of stroke survivors: a critical review. Neurorehabil Neural Repair 17(4):220–226
9. Kwakkel G, Kollen BJ, Wagenaar RC (1999) Therapy impact on functional recovery in stroke rehabilitation: a critical review of the literature. Physiotherapy 85(7):377–391
10. Fiering-Sorensen F, Andersen OT, Hansen SD (1997) Functional neuromuscular stimulation controlled by surface electromyographic signals produced by volitional activation of the same muscle: adaptive removal of the muscle response from the recorded EMG-signal. IEEE Trans Rehabil Eng 5(2):195–206
11. Giuffrida JP, Crago PE (2001) Reciprocal EMG control of elbow extension by FES. IEEE Trans Neural Syst Rehabil Eng 9(4):338–345
12. Duschau-Wicke A, von Zitzewitz J, Caprez A, Lunenburger L, Riener R (2010) Path control: a method for patient-cooperative robot-aided gait rehabilitation. IEEE Trans Neural Syst Rehabil Eng 18:38–48
13. Kosuge K, Fujisawa Y, Fukuda T (1993) Mechanical system control with man-machine-environment interactions. In: Proceedings of the 1993 IEEE international conference on robotics and automation, pp 239–244. http://ieeexplore.ieee.org/Xplore/home.jsp
14. Tsumugiwa T, Yokogawa R, Hara K (2002) Variable impedance control based on estimation of human arm stiffness for human-robot cooperative calligraphic task. In: Proceedings of the IEEE international conference on robotics and automation, pp 644–650. http://ieeexplore.ieee.org/Xplore/home.jsp
15. Emken JL, Reinkensemeyer DJ (2005) Robot-enhanced motor leaning: accelerating internal model formation during locomotion by transient dynamic amplification. IEEE Trans Neural Syst Rehabil Eng 13:33–39
16. Lynch CL, Popovic MR (2012) A comparison of closed-loop control algorithms for regulating electrically stimulated knee movements in individual with spinal cord injury. IEEE Trans Neural Syst Rehabil Eng 20(4):539–548
17. Kroon JR, IJzerman MJ, Chae J, Lankhorst GJ, Zilvold G (2005) Relation between stimulation characteristics and clinical outcome in studies using electrical stimulation to improve motor control of the upper extremity in stroke. J Rehabil Med 37:65–74
18. Metrailler P, Brodard R, Clavel R, Frischknecht R (2006) Closed loop electrical muscle stimulation in spinal cord injured rehabilitation. Mediterr J Phys Rehabil Med 42(4):72–83
19. Freeman CT, Hughes AM, Burridge JH, Chappell PH, Lewin PL, Rogers E (2009) A robotic workstation for stroke rehabilitation of the upper extremity using FES. Med Eng Phys 31:364–373
20. Rushton DN (2003) Functional electrical stimulation and rehabilitation - an hypothesis. Med Eng Phys 25:75–78
21. Riener R, Edrich T (1999) Identification of passive elastic joint moments in the lower extremities. J Biomech 32:539–544
22. Riener R, Fuhr T (1998) Patient-driven control of FES-supported standing up: a simulation study. IEEE Trans Rehabil Eng 6(2):113–124
23. Ferrarin M, Palazzo F, Riener R, Quintern J (2001) Model-based control of FES-induced single joint movements. IEEE Trans Neural Syst Rehabil Eng 9(3):245–257
24. Almeida F, Lopes A, Abreu P (1999) Force-impedance control: a new control strategy of robotic manipulators. In: Recent advances in mechatronics. Springer, Singapore, pp 126–137
25. Anam K, Al-Jumaily AA (2012) Active exoskeleton control systems: State of the art. Procedia Eng 41:988–994
26. Aguirre-Ollinger G, Colgate JE, Peshkin MA, Goswami A (2007) Active-impedance control of a lower-limb assistive exoskeleton. In: Proceedings of the IEEE 10th international conference on rehabilitation robotics, pp 188–195. http://ieeexplore.ieee.org/Xplore/home.jsp
27. Yang Y, Wang L, Tong J, Zhang L (2006) Arm rehabilitation robot impedance control and experimentation. In: Proceedings of the IEEE international conference on robotics and biomimetics, pp 914–918. http://ieeexplore.ieee.org/Xplore/home.jsp
28. Zhang L, Sun H, Li C (2010) Experiment study of impedance control on horizontal lower limbs rehabilitation robot. In: Proceedings of the 2010 IEEE international conference on information and automation, pp 1421–1425. http://ieeexplore.ieee.org/Xplore/home.jsp
29. Lynch CL, Popovic MR (2008) Functional electrical stimulation. Contr Syst Mag 28(2):40–50
30. Denève A, Moughamir S, Moughamir A, Zaytoon J (2008) Control system design of a 3-DOF upper limbs rehabilitation robot. Comput Methods Programs Biomed 89(2):202–214
31. Swevers J, Ganseman C, Schutter JD, Brussel HV (1996) Experimental robot identification using optimised periodic trajectories. Mech Syst Signal Process 10(5):561–577
32. Wu W, Zhu S, Wang X, Liu H (2012) Closed-loop dynamic parameter identification of robot manipulators using modified fourier series. Int J Adv Robotic Syst 9:561–577
33. Kamen G, Gabriel DA (2010) Essentials of electromyography. Human Kinetics, Champaign
34. Sartori M, Reggiani M, Pagello E, Lloyd DG (2012) Modeling the human knee for assistive technologies. IEEE Trans Biomed Eng 59(9):2642–2649
35. Clancy EA, Liu L, Liu P, Moyer DVZ (2012) Identification of constant-posture EMG–torque relationship about the elbow using nonlinear dynamic models. IEEE Trans Biomed Eng 59(1):205–212
36. Clancy EA, Bouchard S, Rancourt D (2001) Estimation and application of emg amplitude during dynamic contractions. IEEE Eng Med Biol 20(6):47–54
37. Delp SL (1990) Surgery simulation: a computer graphics system to analyze and design musculoskeletal reconstructions of the lower limb. Ph.D. Dissertation

Permissions

All chapters in this book were first published in RB, by Springer; hereby published with permission under the Creative Commons Attribution License or equivalent. Every chapter published in this book has been scrutinized by our experts. Their significance has been extensively debated. The topics covered herein carry significant findings which will fuel the growth of the discipline. They may even be implemented as practical applications or may be referred to as a beginning point for another development.

The contributors of this book come from diverse backgrounds, making this book a truly international effort. This book will bring forth new frontiers with its revolutionizing research information and detailed analysis of the nascent developments around the world.

We would like to thank all the contributing authors for lending their expertise to make the book truly unique. They have played a crucial role in the development of this book. Without their invaluable contributions this book wouldn't have been possible. They have made vital efforts to compile up to date information on the varied aspects of this subject to make this book a valuable addition to the collection of many professionals and students.

This book was conceptualized with the vision of imparting up-to-date information and advanced data in this field. To ensure the same, a matchless editorial board was set up. Every individual on the board went through rigorous rounds of assessment to prove their worth. After which they invested a large part of their time researching and compiling the most relevant data for our readers.

The editorial board has been involved in producing this book since its inception. They have spent rigorous hours researching and exploring the diverse topics which have resulted in the successful publishing of this book. They have passed on their knowledge of decades through this book. To expedite this challenging task, the publisher supported the team at every step. A small team of assistant editors was also appointed to further simplify the editing procedure and attain best results for the readers.

Apart from the editorial board, the designing team has also invested a significant amount of their time in understanding the subject and creating the most relevant covers. They scrutinized every image to scout for the most suitable representation of the subject and create an appropriate cover for the book.

The publishing team has been an ardent support to the editorial, designing and production team. Their endless efforts to recruit the best for this project, has resulted in the accomplishment of this book. They are a veteran in the field of academics and their pool of knowledge is as vast as their experience in printing. Their expertise and guidance has proved useful at every step. Their uncompromising quality standards have made this book an exceptional effort. Their encouragement from time to time has been an inspiration for everyone.

The publisher and the editorial board hope that this book will prove to be a valuable piece of knowledge for researchers, students, practitioners and scholars across the globe.

List of Contributors

Jianyang Liu
Department of Applied Physics, Sichuan University, Chengdu, China

Youfu Li
Department of Applied Physics, Sichuan University, Chengdu, China

Hiu Man Yip
Department of Mechanical and Automation Engineering, The Chinese University of Hong Kong, Shatin, N.T., Hong Kong

Peng Li
Department of Mechanical and Automation Engineering, The Chinese University of Hong Kong, Shatin, N.T., Hong Kong

David Navarro-Alarcon
Department of Mechanical and Automation Engineering, The Chinese University of Hong Kong, Shatin, N.T., Hong Kong

Yun-hui Liu
Department of Mechanical and Automation Engineering, The Chinese University of Hong Kong, Shatin, N.T., Hong Kong

Veronika Magdanz
Institute for Integrative Nanosciences, Leibniz Institute for Solid State and Materials Research Dresden, 01069 Dresden, Germany

Maria Guix
Institute for Integrative Nanosciences, Leibniz Institute for Solid State and Materials Research Dresden, 01069 Dresden, Germany

Oliver G Schmidt
Institute for Integrative Nanosciences, Leibniz Institute for Solid State and Materials Research Dresden, 01069 Dresden, Germany
Material Systems for Nanoelectronics, Chemnitz University of Technology, 09126 Chemnitz, Germany
Center for Advancing Electronics Dresden, Dresden University of Technology, 01187 Dresden, Germany

Ronghao Zheng
Department of Mechanical and Biomedical Engineering, City University of Hong Kong, 83 Tat Chee Avenue, Hong Kong 999077, China

Dong Sun
Department of Mechanical and Biomedical Engineering, City University of Hong Kong, 83 Tat Chee Avenue, Hong Kong 999077, China

Evandro M Ficanha
Department of Mechanical Engineering-Engineering Mechanics, Michigan Technological University, Houghton, Michigan 49931, USA

Mohammad Rastgaar
Department of Mechanical Engineering-Engineering Mechanics, Michigan Technological University, Houghton, Michigan 49931, USA

Kenton R Kaufman
Motion Analysis Laboratory, Department of Orthopedic Surgery, Mayo Clinic and Mayo Foundation, 201 W. Center St., Rochester, MN 55902, USA

Abhra Roy Chowdhury
Department of Electrical and Computer Engineering, National University of Singapore, Singapore 117576, Singapore

Vinoth Kumar
Department of Electrical and Computer Engineering, National University of Singapore, Singapore 117576, Singapore

Bhuneshwar Prasad
Department of Electrical and Computer Engineering, National University of Singapore, Singapore 117576, Singapore

Rajesh Kumar
Electrical Engineering Department, Malviya National Institute of Technology, Jaipur, India

Sanjib Kumar Panda
Department of Electrical and Computer Engineering, National University of Singapore, Singapore 117576, Singapore

Wenqi Hu
Department of Electrical Engineering, University of Hawaii at Manoa, 2540 Dole Street, Holmes Hall 483, Honolulu, HI 96822, USA

Qihui Fan
Department of Mechanical Engineering, University of Hawaii at Manoa, 2540 Dole Street, Holmes Hall 300, Honolulu, HI 96822, USA

Aaron T Ohta
Department of Electrical Engineering, University of Hawaii at Manoa, 2540 Dole Street, Holmes Hall 483, Honolulu, HI 96822, USA

Milad S Malekzadeh
Department of Advanced Robotics, Istituto Italiano di Tecnologia (IIT), Via Morego 30, 16163 Genoa, Italy

Sylvain Calinon
Department of Advanced Robotics, Istituto Italiano di Tecnologia (IIT), Via Morego 30, 16163 Genoa, Italy
Idiap Research Institute, Rue Marconi 19, CH-1920 Martigny, Switzerland

Danilo Bruno
Department of Advanced Robotics, Istituto Italiano di Tecnologia (IIT), Via Morego 30, 16163 Genoa, Italy

Darwin G Caldwell
Department of Advanced Robotics, Istituto Italiano di Tecnologia (IIT), Via Morego 30, 16163 Genoa, Italy

Yudong Luo
Department of Electrical and Biomedical Engineering, University of Nevada at Reno, 1664 N Virginia Street, Reno, NV 89557, USA

SankuNiu
Department of Electrical and Biomedical Engineering, University of Nevada at Reno, 1664 N Virginia Street, Reno, NV 89557, USA
Department of Mechatronic Engineering, Beijing Institute of Technology, 5 South Zhongguancun Street, Haidian, Beijing 100081, China

Jose Cordero
Department of Electrical and Biomedical Engineering, University of Nevada at Reno, 1664 N Virginia Street, Reno, NV 89557, USA

HongbinDeng
Department of Electrical and Biomedical Engineering, University of Nevada at Reno, 1664 N Virginia Street, Reno, NV 89557, USA
Department of Mechatronic Engineering, Beijing Institute of Technology, 5 South Zhongguancun Street, Haidian, Beijing 100081, China

Yantao Shen
Department of Electrical and Biomedical Engineering, University of Nevada at Reno, 1664 N Virginia Street, Reno, NV 89557, USA

Tao Guan
College of Electronic Science and Engineering, National University of Defense Technology, Kaifu, Changsha 410073, Hunan, China

Dongxiang Zhou
College of Electronic Science and Engineering, National University of Defense Technology, Kaifu, Changsha 410073, Hunan, China

Chao Xu
College of Electronic Science and Engineering, National University of Defense Technology, Kaifu, Changsha 410073, Hunan, China

Yunhui Liu
Department of Mechanical and Automation Engineering, The Chinese University of Hong Kong, Shatin, Hong Kong, China

Yi Shen, Xuyan Hou
School of Mechatronics Engineering, Harbin Institute of Technology, West Dazhi Street, 15001 Harbin, China

Yiwei Qin
School of Mechatronics Engineering, Harbin Institute of Technology, West Dazhi Street, 15001 Harbin, China

Shengyuan Jiang
School of Mechatronics Engineering, Harbin Institute of Technology, West Dazhi Street, 15001 Harbin, China

Zongquan Deng
School of Mechatronics Engineering, Harbin Institute of Technology, West Dazhi Street, 15001 Harbin, China

Weiyang Lin
Shenzhen Graduate School, Harbin Institute of Technology, Shenzhen University Town, Xili, 518055 Shenzhen, China
School of Astronautics, Harbin Institute of Technology, Xidazhi Street, 150001 Harbin, China
Mechanical Automation of Engineering, the Chinese University of Hong Kong, NT Road, 999077 Shatin, Hong Kong

Peng Xu
Shenzhen Graduate School, Harbin Institute of Technology, Shenzhen University Town, Xili, 518055 Shenzhen, China
Department of Industrial and Systems Engineering, the Hong Kong Polytechnic University, Yuk Choi Road, 999077 Kowloon, Hong Kong

Bing Li
Shenzhen Graduate School, Harbin Institute of Technology, Shenzhen University Town, Xili, 518055 Shenzhen, China

Xiaojun Yang
Shenzhen Graduate School, Harbin Institute of Technology, Shenzhen University Town, Xili, 518055 Shenzhen, China

Masataka Yoshioka
Department of Systems Life Engineering, Maebashi Institute of Technology, Kamisadori 460-1, 371-0816 Maebashi, Gunma, Japan

Chi Zhu
Department of Systems Life Engineering, Maebashi Institute of Technology, Kamisadori 460-1, 371-0816 Maebashi, Gunma, Japan

Kazuyuki Imamura
Department of Systems Life Engineering, Maebashi Institute of Technology, Kamisadori 460-1, 371-0816 Maebashi, Gunma, Japan

FengWang
Department of Systems Life Engineering, Maebashi Institute of Technology, Kamisadori 460-1, 371-0816 Maebashi, Gunma, Japan

Haoyong Yu
Department of Bioengineering, Faculty of Engineering, National University of Singapore, 9, Engineering Drive 1, 117576 Singapore City, Singapore

Feng Duan
Department of Automation, College of Information Technical Science, Nankai University, No. 94 Weijin Road, 300071 Tianjin, People's Republic of China

Yuling Yan
Department of Bioengineering, School of Engineering, Santa Clara University, 500, El Camino Real, Santa Clara, CA 95053, USA

Runhuai Yang
Department of Mechanical and Biomedical Engineering, City University of Hong Kong, Tat Chee Ave, Kowloon, Hong Kong

Chi Ho Tam
Department of Mechanical and Biomedical Engineering, City University of Hong Kong, Tat Chee Ave, Kowloon, Hong Kong

Ka Lun Cheung
Department of Mechanical and Biomedical Engineering, City University of Hong Kong, Tat Chee Ave, Kowloon, Hong Kong

Kai Chun Wong
Department of Mechanical and Biomedical Engineering, City University of Hong Kong, Tat Chee Ave, Kowloon, Hong Kong

Ning Xi
Michigan State University, Trowbridge Rd, East Lansing, MI 48824, USA

Jie Yang
University of Science and Technology of China, Jinzhai RdShushan, Hefei, China

King Wai Chiu Lai
Department of Mechanical and Biomedical Engineering, City University of Hong Kong, Tat Chee Ave, Kowloon, Hong Kong

Arnab Sinha
Singapore University of Technology and Design, 20 Dover Drive, Singapore, Singapore

Ning Tan
Singapore University of Technology and Design, 20 Dover Drive, Singapore, Singapore

Rajesh Elara Mohan
Singapore University of Technology and Design, 20 Dover Drive, Singapore, Singapore

Ehsan Rezapour
Department of Engineering Cybernetics, Norwegian University of Science and Technology (NTNU), Trondheim 7491, Norway

Kristin Y Pettersen
Department of Engineering Cybernetics, Norwegian University of Science and Technology (NTNU), Trondheim 7491, Norway

Pål Liljebäck
Department of Engineering Cybernetics, Norwegian University of Science and Technology (NTNU), Trondheim 7491, Norway
Department of Applied Cybernetics, SINTEF ICT, Trondheim 7465, Norway

Jan T Gravdahl
Department of Engineering Cybernetics, Norwegian University of Science and Technology (NTNU), Trondheim 7491, Norway

Eleni Kelasidi
Department of Engineering Cybernetics, Norwegian University of Science and Technology (NTNU), Trondheim 7491, Norway

Yajing Shen
Department of Mechanical and Biomedical Engineering, City University of Hong Kong, Tat Chee Avenue, Kowloon, Hong Kong 99907, China

Toshio Fukuda
School of Mechatronic Engineering, Beijing Institute of Technology, 5 South Zhongguancun Street, Haidian District, Beijing 100081, China

Alejandro Melendez-Calderon
Department of Physical Medicine and Rehabilitation, Northwestern University, 60611 Chicago, IL, USA
Sensory Motor Performance Program, Rehabilitation Institute of Chicago, 60611 Chicago, IL, USA

Davide Piovesan
Biomedical Program, Department of Mechanical Engineering, Gannon University, 16541 Erie, PA, USA
Sensory Motor Performance Program, Rehabilitation Institute of Chicago, 60611 Chicago, IL, USA

James L Patton
Department of Physical Medicine and Rehabilitation, Northwestern University, 60611 Chicago, IL, USA
Department of Bioengineering, University of Illinois at Chicago, 60607 Chicago, IL, USA
Department of Biomedical Engineering, Northwestern University, 60208 Evanston, IL, USA
Department of Mechanical Engineering, Northwestern University, 60208 Evanston, IL, USA
Sensory Motor Performance Program, Rehabilitation Institute of Chicago, 60611 Chicago, IL, USA

Ferdinando A Mussa-Ivaldi
Department of Physical Medicine and Rehabilitation, Northwestern University, 60611 Chicago, IL, USA
Department of Biomedical Engineering, Northwestern University, 60208 Evanston, IL, USA
Sensory Motor Performance Program, Rehabilitation Institute of Chicago, 60611 Chicago, IL, USA
Department of Physiology, Northwestern University, 60611 Chicago, IL, USA

Jingyi Wang
State Key Laboratory of Robotics, Shenyang Institute of Automation, Chinese Academy of Sciences, No.114 Nanta Street, Shenyang 110016, China
University of Chinese Academy of Sciences, No.19A Yuquan Road, Beijing 100049, China

Niandong Jiao
State Key Laboratory of Robotics, Shenyang Institute of Automation, Chinese Academy of Sciences, No.114 Nanta Street, Shenyang 110016, China

Steve Tung
State Key Laboratory of Robotics, Shenyang Institute of Automation, Chinese Academy of Sciences, No.114 Nanta Street, Shenyang 110016, China
Department of Mechanical Engineering, University of Arkansas, W Dickson Street, Fayetteville, AR 72701, USA

Lianqing Liu
State Key Laboratory of Robotics, Shenyang Institute of Automation, Chinese Academy of Sciences, No.114 Nanta Street, Shenyang 110016, China

Yixiong Chen
State Key Laboratory of Management and Control for Complex Systems, 95 Zhongguancun East Road, Beijing 100190, China

Jin Hu
State Key Laboratory of Management and Control for Complex Systems, 95 Zhongguancun East Road, Beijing 100190, China

Long Peng
State Key Laboratory of Management and Control for Complex Systems, 95 Zhongguancun East Road, Beijing 100190, China

Zeng-guang Hou
State Key Laboratory of Management and Control for Complex Systems, 95 Zhongguancun East Road, Beijing 100190, China